Akzo Che...

ISBN 978-0444415...

COMPREHENSIVE ANALYTICAL CHEMISTRY

ELSEVIER SCIENTIFIC PUBLISHING COMPANY
335 JAN VAN GALENSTRAAT
P.O. BOX 211, AMSTERDAM, THE NETHERLANDS

Distributors for the United States and Canada:

ELSEVIER/NORTH-HOLLAND INC.
52, VANDERBILT AVENUE
NEW YORK, N.Y. 10017

LIBRARY OF CONGRESS CARD NUMBER: 58-10158

ISBN 0-444-41523-8

WITH 204 ILLUSTRATIONS AND 56 TABLES

PRINTED IN THE NETHERLANDS

COMPREHENSIVE ANALYTICAL CHEMISTRY

ADVISORY BOARD

Contributors to Volume VIII

G.G. Guilbault, Chemistry Department, Louisiana State University in New
Orleans, New Orleans, U.S.A.

M.A. Leonard, Chemistry Department, Queen's University, Belfast, N. Ireland

W. Nebe, Haselstrauchweg 31, 69 Jena, D.D.R.

Wilson and Wilson's

COMPREHENSIVE ANALYTICAL CHEMISTRY

Edited by

G. SVEHLA, PH.D., D.SC., F.R.I.C.

Reader in Analytical Chemistry
The Queen's University of Belfast

VOLUME VIII

Enzyme Electrodes in Analytical Chemistry
Molecular Fluorescence Spectroscopy
Photometric Titrations
Analytical Applications of Interferometry

ELSEVIER SCIENTIFIC PUBLISHING COMPANY
AMSTERDAM   OXFORD   NEW YORK
1977

WILSON & WILSON'S

COMPREHENSIVE ANALYTICAL CHEMISTRY

VOLUMES IN THE SERIES

Vol. IA   Analytical Processes
                 Gas Analysis
                 Inorganic Qualitative Analysis
                 Organic Qualitative Analysis
                 Inorganic Gravimetric Analysis

Vol. IB   Inorganic Titrimetric Analysis
                 Organic Quantitative Analysis

Vol. IC   Analytical Chemistry of the Elements

Vol. IIA   Electrochemical Analysis
                 Electrodeposition
                 Potentiometric Titrations
                 Conductometric Titrations
                 High-frequency Titrations

Vol. IIB   Liquid Chromatography in Columns
                 Gas Chromatography
                 Ion Exchangers
                 Distillation

Vol. IIC   Paper and Thin-Layer Chromatography
                 Radiochemical Methods
                 Nuclear Magnetic Resonance and Electron Spin
                     Resonance Methods
                 X-Ray Spectrometry

Vol. IID   Coulometric Analysis

Vol. III   Elemental Analysis with Minute Samples
                 Standards and Standardization
                 Separations by Liquid Amalgams
                 Vacuum Fusion Analysis of Gases in Metals
                 Electroanalysis in Molten Salts

Vol. IV   Instrumentation for Spectroscopy
       Atomic Absorption and Fluorescence Spectroscopy
       Diffuse Reflectance Spectroscopy

Vol. V   Emission Spectroscopy
       Analytical Microwave Spectroscopy
       Analytical Applications of Electron Microscopy

Vol. VI   Analytical Infrared Spectroscopy

Vol. VII   Thermal Methods in Analytical Chemistry
       Substoichiometric Analytical Methods

Vol. VIII   Enzyme Electrodes in Analytical Chemistry
       Molecular Fluorescence Spectroscopy
       Photometric Titrations
       Analytical Applications of Interferometry

Preface

In *Comprehensive Analytical Chemistry* the aim is to provide a work which, in many instances, should be a self-sufficient reference work; but where this is not possible, it should at least be a starting point for any analytical investigation.

It is hoped to include the widest selection of analytical topics that is possible within the compass of the work, and to give material in sufficient detail to allow it to be utilised directly, not only by professional analytical chemists, but also by those workers whose use of analytical methods is incidental to their work rather than continual. Where it is not possible to give details of methods, full reference to the pertinent original literature is made.

Volume VIII contains four chapters. The first, on the application of enzyme electrodes, covers a growing subject, which will interest the biochemist first of all. Fluorescence spectroscopy and photometric titrations, described in Chapters 2 and 3, respectively, are well established methods which are used a great deal in chemical laboratories at the present time. The fourth chapter on interferometry describes a technique which offers quick and precise answers to certain analytical problems. The up-to-date discussions presented here will, I am sure, be read by a large number of analytical chemists. Our three contributors, coming from American and European academic and industrial institutions, are internationally known experts in their fields.

Dr. C.L. Graham of the University of Birmingham assisted in the production of the present volume; his contribution is acknowledged with many thanks.

September, 1975   G. Svehla

Contents

Preface ix

Chapter I. ENZYME ELECTRODES IN ANALYTICAL
CHEMISTRY, by G.G. Guilbault 1

1. Introduction 1
   (A) The electrode 1
   (B) The enzyme as reagent 2
   (C) Ion-selective electrodes 4
   (D) Assay of enzymes using ion-selective electrodes 11
2. Preparation and properties of enzyme electrodes 15
   (A) Immobilization methods 15
       (1) General, 15 — (2) Immobilization techniques, 16 — (3) Proper-
       ties of immobilized enzymes, 20
   (B) Construction of enzyme electrodes 21
       (1) Step 1, 21 — (2) Step 2, 22 — (3) Step 3, 22 — (4) Step 4, 24
   (C) Performance characteristics of electrodes 27
       (1) Stability, 27 — (2) Response time, 33 — (3) Factors that affect
       response time, 36 — (4) Other characteristics, 41 — (5) Effect of
       interferences, 44
3. Analytical applications of enzyme electrodes 49
   (A) Electrodes for assay of substrates using immobilized en-
       zymes 49
       (1) Glucose electrodes, 49 — (2) Urea electrodes, 51 — (3) Amino
       acid electrodes, 53 — (4) Alcohol electrodes, 54 — (5) Uric acid
       electrode, 55 — (6) Lactic acid electrode, 56 — (7) Amygdalin elec-
       trode, 56 — (8) Penicillin electrode, 57 — (9) Acetic, formic and
       succinic acid electrodes, 58 — (10) Phosphate ion electrode, 58 —
       (11) Nitrate ion electrode, 58
   (B) Substrate electrodes for the assay of enzymes 59
       (1) Urease electrode, 59 — (2) Cholinesterase electrode, 59

(C) Insolubilized co-enzymes in electrodes   60
(D) Comparison of electrodes   61
   (1) Enzyme electrodes for substrates, 61 — (2) Substrate electrodes
   for enzymes, 63
4. The future of enzyme electrodes   65
5. Acknowledgement   65
References   66

Chapter II. MOLECULAR FLUORESCENCE SPECTROSCO-
          PY, by G.G. Guilbault   71

1. Introduction to luminescence   71
(A) History of luminescence   71
(B) Theory of luminescence   72
(C) Types of fluorescence and emission processes   73
(D) Excitation spectrum   74
(E) Emission spectrum   75
(F) Fluorescence quantum efficiency and lifetime of the ex-
    cited state   78
(G) Relation between fluorescence intensity and concentra-
    tion   80
(H) Introduction to experimentation   82
(I) Use of luminescence   82
(J) Practical considerations   84
   (1) Advantages of fluorescence, 84 — (2) Limitations of fluores-
   cence, 85
(K) Types of luminescence   90
2. Instrumentation   91
(A) Component parts   91
   (1) Light sources, 92 — (2) Monochromators, 95 — (3) Cell compartment, 102 — (4) Cell configuration, 103 — (5) Slits, 103 — (6)
   Detectors, 105
(B) Basic instruments 107
   (1) Filter instruments, 107 — (2) Grating instruments, 111
3. Practical considerations 114
(A) Cuvettes 114
   (1) Emission from cuvettes, 114 — (2) Care of cells, 115
(B) Solvents 116
(C) Sources and detectors 118
(D) Temperature 119

(E) Filter vs. grating instruments 119
(F) Care of filters 120
(G) Selection of the excitation and emission wavelengths . . . 120
(H) Light scattering 122
(I) Photodecomposition 123
(J) Standardization 124
(K) Calibration of wavelength 125
(L) Correction of spectra 125
4. Structural and environmental effects 126
(A) Structural effects 126
(B) Environmental effects 131
(1) Solvent effects, 131 — (2) Effect of external heavy atoms and paramagnetic ions, 133 — (3) Effect of pH, 134 — (4) Hydrogen bonding, 136 — (5) Effect of temperature, 140 — (6) Effect of oxygen, 141 — (7) Fluorescence quenching by metal ions, 142 — (8) Effect of solute concentration on fluorescence, 143
5. Phosphorescence 144
(A) General 144
(B) Theoretical considerations 146
(C) Structural effects 148
(D) Instrumentation 149
(1) General, 149 — (2) The phosphoroscope, 150 — (3) The sample cell, 153 — (4) Solvents, 154 — (5) Commercial equipment, 155
(E) Analytical applications 156
(1) Practical aspects of measurement, 156 — (2) Analytical determinations, 163
6. Determination of inorganic ions 167
(A) General considerations 167
(B) Direct analysis of inorganic substances 169
(C) Fluorescent chelates and quenching reactions 169
7. Determination of organic compounds 178
8. Assay of enzymes 186
(A) Hydrolytic enzymes 188
(1) Cholinesterase, 188 — (2) β-D-galactosidase, 189 — (3) β-glucuronidase, 190 — (4) Lipase, 190 — (5) Phosphatase, 191
(B) Dehydrogenases and transaminases 193
(1) Methods based on the fluorescence of NADH or NADPH, 193 — (2) Resazurin method, 194
(C) Oxidative enzymes 195
(1) Peroxidase, 195 — (2) Monoamine oxidase, 196
(D) Enzyme cycling for pyridine nucleotides 196

References 198

Chapter III. PHOTOMETRIC TITRATIONS, by M.A. Leonard 207

1. Introduction 207
   (A) Definition 207
   (B) History 207
2. Theory 216
   (A) Physico-chemical background 216
       (1) Absorption of electromagnetic radiation by solutions, 216 —
       (2) Solution fluorescence, 218 — (3) Scattering of radiation by sus-
       pensions, 221
   (B) General theory of photometric titrations 224
3. Instrumentation 236
4. Acid—base photometric titrations 251
   (A) Self-indicating titrations 251
   (B) Indicated acid—base titrations 257
       (1) Type I plot, 258 — (2) Type II plot, 260 — (3) Type III plot,
       261
5. Photometric complexometric titrations 270
   (A) Slope and self-indicating EDTA titrations 274
   (B) Step (indicator) EDTA titrations 284
   (C) Complexometric titrations involving a photometric rea-
       gent as titrant (one phase) 302
   (D) Spectrophotometric extractive titrations using high ab-
       sorbance reagents as titrants 308
   (E) Photometric monodentate ligand complex-forming titra-
       tions 314
6. Precipitation titrations 314
   (A) General considerations 314
       (1) Heterometry, 320 — (2) Surface active agents, 320 — (3) Poly-
       mer titrations, 321 — (4) Indicated titrations, 321
   (B) The photometric titration of fluoride and sulphate 321
7. Redox titrations 331
   (A) General survey 331
   (B) Theory of photometric redox titrations 335
8. Spectrofluorimetric phototitrations 342
9. Organic functional group photometric titrations 356
10. Miscellaneous photometric titration methods 365
   (A) Spectropolarimetry 365

(B) Atomic absorption inhibition titration 369
(C) Chelate exchange titrimetry 372

(1) Titration of dithizone in benzene, 373 — (2) Titration of diethyldithiocarbamate in ethanol, 373 — (3) Titration of metal
dithizonates, 3733

(D) Determination of critical micelle concentration 375
(E) Phosphate determination 376
(F) The Karl Fischer determination 377
(G) Ion-pair partition titrations 377
References 380

Chapter IV. ANALYTICAL APPLICATIONS OF INTERFE-
ROMETRY, by W. Nebe 391

1. Historical background of analytical interferometry 391
2. Symbols 394
3. Theoretical background of interference measurements 397
(A) The wave nature of light 397
(B) Monochromaticity and coherence 400
(C) Interference 402
(D) Interference arrangements 403

(1) Wave-front division, 403 — (2) Amplitude division, 407 — (3)
Wave-front shearing, 411 — (4) Diffraction, 413

(E) The refractive index 418
(F) The measurement of refractive index by interference 423
4. Apparatus and accessories 424
(A) Laboratory interferometers for the measurement of
homogeneous substances 424

(1) The Rayleigh—Löwe interferometer, 424 — (2) The Jamin interferometer, 428 — (3) The Michelson—Kinder interferometer, 430

(B) Portable interferometers for the measurement of homoge-
neous substances 431

(1) The Rayleigh—Löwe autocollimating interferometer, 431 — (2)
The Jamin—Doi interferometer, 432

(C) Interferometers for the measurement of stratified sub-
stances 433
(D) Interferometers for the measurement of randomly heter-
ogeneous substances 437

(E) Accessories for interferometric measurements 441
    (1) Light sources, 441 — (2) Temperature control, 444 — (3) Gas
    mixer for calibration purposes, 444 — (4) Sorption media, 446
5. Methods for interferometric analysis of homogeneous substances 452
  (A) Properties of substances 453
  (B) Calibration procedures 455
  (C) Temperature and pressure 459
  (D) Determination of concentration 462
    (1) Binary gas mixtures, 462 — (2) Ternary gas mixtures, 469 —
    (3) Liquids, 475
6. Applications of interferometry for the analysis of homoge-
  neous substances 480
  (A) Technical gases and vapours 481
  (B) Explosive gases and vapours 482
  (C) Toxic gases and vapours 487
  (D) Biological and clinical gases 488
  (E) Special investigations on gases in research 492
  (F) Water, heavy water and aqueous solutions 494
  (G) Other liquids 496
  (H) Optical glasses and thin layers 498
7. Interferometric analysis of inhomogeneous substances 502
  (A) Methods of interferometric measurement on stratified
    substances 502
  (B) Electrophoresis 509
  (C) Diffusion 513
  (D) Sedimentation 521
  (E) Interferometric methods for investigating the random dis-
    tribution of refractive index in a sample 525
  (F) Some applications of interferometry in the investigation
    of inhomogeneous distribution of refractive index 531
References 536

Index 547

Chapter I

Enzyme electrodes in analytical chemistry

G.G. GUILBAULT

1. Introduction

(A) THE ELECTRODE

An enzyme electrode is the union of an enzyme, that biological catalyst which acts sensitively and specifically with almost all organic and inorganic compounds in nature, with an electronic sensor (an ion-selective electrode). The result is an electrode which is useful for the assay of organic and inorganic compounds, in a manner as simple as a pH measurement with a glass electrode.

In the past decade we have seen marketed over fifty new ion-selective electrodes, which work well for many types of cations and anions. However, with the exception of the Corning acetylcholine electrode [1], these electrodes have been only for inorganic species, not organic. Such much needed organic electrodes are possible using enzyme electrodes.

The principle of the enzyme electrode is simple, an enzyme is used which reacts with the compound to be assayed either specifically or highly selectively. This enzyme, in an immobilized or insolubilized form, is placed onto a conventional ion-selective electrode which measures either the decrease of one of the reactants (i.e. O_2 in the oxidation of uric acid or glucose) or a product (CO_2 from an amino acid). The substance to be assayed diffuses into the enzyme layer in the electrode producing or consuming an electroactive substance; this is sensed by the base electrode. The potential or current produced is a function of the concentration of the substance assayed. Thus a single self-contained electrode is placed into a solution, and the concentration is measured directly. Eliminated is the necessity for reagent preparation, construction of standard curves, etc. The only reagent required is a buffer. Furthermore, the same electrode can be used for many assays, decreasing the cost per test tremendously.

In this chapter the reader will be introduced to the concept of the enzyme electrode. He will be taught how to make and use enzyme electrodes and what can be expected from such electrodes; the types of electrode that have been described will be listed.

Since the enzyme electrode is a union of an ion-selective electrode and an enzyme, we shall now discuss each of these elements in some detail. From a good knowledge of each part, we should be able to recognize what can be expected from the total — the enzyme electrode.

(B) THE ENZYME AS REAGENT

An enzyme is a biological catalyst present in nature, that enables the many complex chemical reactions, upon which depends the very existence of life as we know it, to take place at ordinary temperatures.

Because enzymes work in complex living systems, one of their outstanding properties is specificity. An enzyme is capable of catalyzing a particular reaction of a particular substrate, even though other isomers of that substrate or similar substrates may be present. An example of the specificity of enzymes with respect to a particular substrate is found in luciferase, which catalyzes the oxidation of luciferin to oxyluciferin [2]. A rather complete study of many compounds similar in structure to luciferin showed that the catalytic oxidation resulting in the production of the green luminescence

$$\text{luciferin} + O_2 \xrightarrow[\text{Mg}^{2+}, \text{ ATP}]{\text{luciferase}} \text{oxyluciferin}$$
$$\text{(Green luminescence)}$$

occurs only with luciferin. Substitution of an amino group for a hydroxyl group or addition of another hydroxyl group to the luciferin molecule alters the enzymic action, and the green luminescence is not produced. Another example of the specificity of enzymes is glucose oxidase, which catalyzes the oxidation of β-D-glucose to gluconic acid. In a study of about 60 oxidizable sugars and their derivatives it was shown that only 2-deoxy-D-glucose is catalyzed at a

2

rate comparable to that of β-D-glucose. The anomer α-D-glucose is oxidized catalytically less than one percent as rapidly as the β-anomer [3]. Urease, which catalyzes the hydrolysis of urea, is even more specific.

Enzymes exhibit specificity with respect to one particular reaction with elimination of many side reactions. If one attempted to determine glucose by oxidation in an uncatalyzed way, for example, by heating a solution of glucose and an oxidizing agent like ceric perchlorate, other side reactions would occur to yield products in addition to gluconic acid. With glucose oxidase catalysis is so effective even at room temperature and a neutral pH that the rates of the other thermodynamically possible reactions are negligible.

This specificity of enzymes, and their ability to catalyze reactions of substrates at low concentrations, is of great use in chemical analysis. Enzyme-catalyzed reactions have been used for analytical purposes for a long time for the determination of substrates, activators and inhibitors [4]. Until recently, however, the disadvantages associated with the use of enzymes have seriously limited their usefulness. Frequently cited objections to the use of enzymes for analytical purposes have been their unavailability, instability, poor precision, and the time required to perform the analysis. While these objections were valid earlier, numerous enzymes are now available in purified form, with high specific activity, at reasonable prices.

There are several hundred purified enzymes available today from companies such as Sigma (St. Louis), Boehringer—Mannheim (Mannheim, Germany and New York City), Worthington (Freehold, N.J.), Calbiochem (Los Angeles), Nutritional Biochemicals (Cleveland), Miles (Elkhart, Indiana) and others. A listing of enzymes available through 1975 can be found in a book by Guilbault on Enzymic Methods of Analysis [4]. In addition to these, there are many other hundreds of enzymes known which can be isolated and purified for use in constructing enzyme electrodes.

The instability of the enzyme, and the fact that it is "consumed" in a reaction in the soluble form, is a problem. A continuous or semi-continuous routine analysis using enzymes would require large amounts of these materials, quantities greater than can be reasonably supplied, and quantities that would represent a prohibitive expenditure in many cases. If, however, the enzyme could be prepared in an immobilized (insolubilized) form without loss of activity, so that one sample could be used continuously for many days, a considerable advantage would be realized. The immobilized enzyme can be used analytically in much the same way that the soluble enzyme is used, that is, to determine the concentration of a substrate that is acted upon by the enzyme.

There are several ways that this immobilization of the enzyme can be effected, providing a product useful in an enzyme electrode. These will be discussed later. Furthermore, there are a number of immobilized enzymes sold commercially by companies such as Boerhringer (catalase, chymotrypsin, glucose oxidase, papain, ribonuclease, trypsin urease); Koch-Light (Buckinghamshire, England) or Aldrich (Milwaukee, Wisc.) (glucose oxidase, catalase, dextranase, β-glucosidase and uricase); Miles Labs (Elkhart, Ind.), and Corning (Corning, N.Y.) who offers to provide any enzyme insolubilized on glass bands.

In constructing an enzyme electrode one need only (1) pick an enzyme that reacts with the substance to be determined, (2) obtain that enzyme from commercial sources or isolate it oneself, (3) immobilize the enzyme by standard techniques, or, if possible, buy it already immobilized, and (4) place the immobilized enzyme around the appropriate electrode to monitor the reaction that occurs. (Note: this will probably be the limiting factor in the construction of an enzyme electrode since steps 1—3 are always possible.) These steps will be discussed more thoroughly later.

(C) ION-SELECTIVE ELECTRODES

The glass electrode, selective for hydrogen ion, a reference electrode and a pH meter combine to form an extremely useful analytical tool for assay of pH, one of the most commonly made measurements. The advantages of this measuring system, often taken for granted, are speed, sensitivity, cost, reliability and the sample is not destroyed or consumed in the process. The same advantages apply to other ion-selective electrodes which have become available for the past few years and to the enzyme electrode. Ion-selective electrodes, sensitive to particular cations and anions, can be purchased or constructed at moderate cost. The analytically useful range of these sensors is generally from 10^{-1} M to 10^{-5} M although there are several sensors which are useful at much lower concentrations. Since the response of ion-selective electrodes is logarithmic, the precision of measurements is constant over their dynamic range. Ion-selective electrodes have

4

been widely used in chemical, clinical and environmental analysis.

The hydrogen-selective or pH electrode is the best known ion-selective electrode. Its discovery is credited to Cremer [5] and Haber and Klemasiewicz [6] who found that certain glasses respond to hydrogen ion activity. The response of the glass electrode was commonly believed to be a result of migration of hydrogen ions through the thin glass membrane. The studies carried out by Karremas and Eisenman [7] and the work of Nicolsky and co-workers [8] provided the insight necessary for the development of new ion-selective electrodes. Ion-selective electrodes for many ions are now available. These are listed in Table 1. The electrodes are available from several manufacturers and as newer methods of preparation of ion-selective electrodes have been developed several kits for the preparation of different electrodes using a common body or housing have been introduced. With the exception of the single crystal F^- electrode and the glass electrodes, all electrodes can be easily "home-made" with a minimum of time and effort.

The field of ion-selective electrodes was reviewed by Pungor [9] and by Rechnitz [10]. More recent developments and applications of

TABLE 1

List of commercially available electrodes

Cations		Anions		Gases
Ammonium	G, S, L	Bromide	S	CO_2
Hydrogen	G	Chloride	S, L	NH_3
Cadmium	S	Cyanide	S	NO_x
Calcium	L	Fluoride	S	SO_2
Copper	S	Fluoroborate	L	H_2S
Divalent ions	L	Hydroxide	G	HCN
Lead	S	Iodide	S	HF
Monovalent	G	Nitrate	L	O_2
Potassium	G, S, L	Perchlorate	L	
Silver	G, S	Sulfide	S	
Sodium	G, L	Thiocyanate	S	
Rubidium	L			

G = Glass; S = Solid; L = Liquid membranes.
Gas electrodes all available from Orion (Cambridge, Mass.) and CO_2 from Radiometer (Copenhagen, Denmark) or Instrumentation Labs (Mass.). Other electrodes are available from many companies such as Orion, Corning (Corning, N.Y.), Beckman (Fullerton, Calif.), or Radiometer.

ion-selective electrodes may be found in the report of the Symposium held at the National Bureau of Standards, Gaithersburg, Maryland [11] or the article by Moody and Thomas [12].

An ion-selective electrode may be defined as a device that develops an electrical potential proportional to the logarithm of the activity of an ion in solution. The term "specific" is sometimes used to describe an electrode; this term indicates that the electrode responds to only one particular ion. Since no electrode is truly specific for one ion, the term ion-selective is recommended as more appropriate.

The basic equation for the response of an ion-selective electrode to an ion A, of activity a_A and charge z_A, is given by the expanded Nernst equation

$$E = \text{constant} + \frac{2.303\,RT}{zF}\,\log a_A + k_{A,B}(a_B)^{z_A/z_B} \tag{1}$$

in which E is the potential measured, R is the gas constant and is equal to 0.314 J deg^{-1} mol^{-1}, T is the absolute temperature in Kelvin, F is the Faraday constant equal to 96487 C equiv^{-1}, $k_{A,B}$ is the selectivity coefficient, B is any interfering ion of charge z_B and activity a_B.

The selectivity coefficient is a numerical description of the preferential response of an ion-selective electrode to the major ion, A, in the presence of the interfering ion B. The lower the numerical value of $k_{A,B}$ for a particular ion-selective electrode the greater the concentration of B that can be tolerated before causing errors in the measurement. Values of $k_{A,B}$ can be calculated from

$$\frac{E_2 - E_1}{2.303\,RT/zF} = \log k_{AB} + \frac{z_A}{z_B} - 1(\log a_A) \tag{2}$$

in which E_1 and E_2 are the potential measurements of separate solutions of the principle ion and interfering ion, respectively, at the same activity. A better calculation of k_{AB} may be made by taking measurements of the potential of an ion-selective electrode in solutions of constant interferent activity, a_B, and changing primary ion activity, a_A. Then, k_{AB} may be determined by

$$a_A = k_{A,B}(a_B)^{z_A/z_B} \tag{3}$$

The value of a_A is taken at the point where serious deviation from Nernstian response is noted [12].

6

At present, ion-selective electrodes can be divided into several categories according to the composition of their sensor membranes: (1) solid membranes, homogeneous (glass or crystal membrane) or heterogeneous; (2) liquid membranes (charged or uncharged), and (3) gas electrodes.

Glass electrodes are available which are sensitive to H^+ (pH electrodes), and to cations in the order $Ag^+ > H^+ > K^+ > NH_4^+ > Na^+ >>$ Li^+, Ca^{2+}, Mg^{2+} over a concentration range $10^{-1}-10^{-5}$ M. These are a type of solid homogeneous membrane electrode.

Solid-state electrodes are ion-selective electrodes in which the sensor is a thin layer of a single or mixed crystal or precipitate which is an ion conductor. Two classes of these electrodes are distinguished: homogeneous and heterogeneous. Homogeneous or crystal membrane electrodes refer to those electrodes in which the membrane is a pellet prepared from a precipitate, mixture of precipitates or a single crystal. In the heterogeneous electrodes a precipitate or mixture of precipitates is dispersed in an inert supporting matrix such as silicone rubber or polyvinylchloride (PVC).

Liquid membrane electrodes, which can be either liquid or "solid", are prepared by dissolving an organic exchanger in an appropriate solvent. The solution is held in an inert matrix. Such exchangers presently used in the preparation of these electrodes may be a ligand association complex such as those formed by the transition metals with derivatives of 1,10-phenanthroline, quarternary ammonium salts, organic phosphate complexes and antibiotics. In some cases the exchanger and solvent are entrapped in an inert polymer matrix such as PVC or polymethylmethacrylate, and can also be coated on a platinum wire or graphite rod ("solid" liquid membrane electrodes).

The final type of electrodes are those which employ a coating over the membrane of an ion-selective electrode. The coating may be a gas-permeable membrane in which case electrodes sensitive to CO_2 or NH_3 are the result. The gas diffuses through the membrane and alters the pH of an internal filling solution. The pH change is measured with a glass electrode and is proportional to the concentration of gas which enters the membrane. Such gas electrodes are available for a number of substances as shown in Table 1.

When using an ion-selective electrode to make potentiometric measurements of the activity of a given ion in solution, it is important to remember that the device is affected by the activity of the ion. Therefore, species which may complex the ion of interest and lower its activity must either be removed or masked. It is often necessary to use a buffer solution to control ionic strength, pH and to prevent changes in the activity of the ion being measured by oxidation—reduction or complexation.

There have been a large number of reports of applications and progress in the design and manufacture of ion-selective electrodes in the literature. An important development has been the use of PVC in the preparation of ion-selective electrodes. Electrodes manufactured with PVC are much lower in cost, have essentially the same response characteristics, except for better selectivities, and can usually be used for longer periods of time than previous electrode assemblies. Moody et al. constructed a calcium-selective electrode using a liquid ion exchanger incorporated in a PVC matrix [13]. The optimum concentration of calcium exchanger used in the preparation was described by Griffiths et al. [14]. The electrode constructed in this manner gave a near-Nernstian response (30 mV per pCa unit) over the range 2.6×10^{-2}—6.0×10^{-5} M in $CaCl_2$ solution. Davies et al. prepared nitrate ion-selective electrodes by incorporating commercially available liquid ion exchangers in PVC [15]. These electrodes overcome the problem of leakage that is associated with other liquid ion exchange assemblies. Furthermore, these "solid" liquid membrane electrodes have a greater selectivity for the ion of interest than the corresponding "liquid" electrode.

Pick et al. have used valinomycin in a variety of neutral carriers to prepare ion-selective electrodes for potassium [16]. The response time of this type of electrode is usually less than 3 s and the useful range of the electrodes is 10^{-1}—10^{-5} M. The electrode prepared from this material is exceptional in that there is very little or no drift in potential over a three-day period. As the working characteristics of electrodes are improved and new electrodes are introduced, many new applications of these devices can be expected. It is the author's opinion that all second generation "liquid" electrodes will be "solid" and the classical "liquid" electrode will disappear from the scientific scene. Any of these electrodes can be prepared by the reader very simply following the method of Moody and Thomas [13,15].

The family of electrodes prepared from silver sulfide alone or mixed with halogen salts of silver is well characterized [9,10]. The precipitates are used either in the form of a pellet or mixed in an inert supporting matrix such as silicone rubber. When silver sulfide is used alone, the electrode responds to sulfide and silver ions over the

8

concentration range 10^0—10^{-7} M.

When a particular silver halide is mixed with silver sulfide to form a sensor membrane, the membrane behaves as though it was composed of the halide salt alone. Electrodes for iodide, bromide and chloride have been prepared in this manner. The iodide electrode also responds to cyanide. An electrode for thiocyanate can be prepared by mixing silver thiocyanate with silver sulfide to form a sensor membrane. The selectivity coefficients of the halide and pseudo-halide electrodes can be estimated by

$$k_{A,B} = \frac{\text{Solubility product of AgA}}{\text{Solubility product of AgB}} \tag{4}$$

Another group of electrodes is prepared by mixing copper, cadmium or lead sulfide with silver sulfide. The electrodes constructed from these mixed sulfides are sensitive to Cu^{2+}, Cd^{2+} and Pb^{2+}, respectively, over the concentration range 10^0—10^{-7} M. Silver and mercury(II) are serious interferences and high levels of ferric ion also interfere, but this interference can be overcome by the addition of fluoride to complex the Fe(III).

Any of these solid electrodes can be prepared by pressing a thin wafer of freshly precipitated salt (i.e., Ag_2S for the sulfide electrode) using high pressure in an IR pellet press. The wafer is then sealed onto a glass tube of appropriate diameter using epoxy cement. The electrode is completed by adding 0.1 M electrolyte and an inner reference electrode (i.e., Ag/AgCl).

Cation-selective electrodes for H^+ and monovalent ions (i.e., Ag^+, Na^+, K^+, NH_4^+) are available from glass electrode companies such as Corning, Beckman or Orion. These electrodes are rugged, have a response time of a few seconds and a linear range of 10^0—10^{-5} M.

The text edited by Eisenmann [17] and the chapter of the N.B.S. publication written by Khuri [18] are recommended to those who are interested in more information on glass electrodes. For use with enzymes it is recommended that a flat combination (glass—calomel) electrode be used. Such electrodes are available from Radiometer (Copenhagen) and Ingold (Zurich, Switzerland).

Antibiotics and similar compounds have been used successfully to prepare cation-selective electrodes. The calcium ion-selective electrode developed by Ammann et al. is reported to have superior selectivity for calcium over sodium and magnesium [19]. Pioda et al. have studied the properties of the antibiotic nonactin for use as a sensor membrane [20]. An electrode prepared from this material in an inert matrix is sensitive to NH_4^+ with linear near-Nernstian response of 51 mV decade^{-1} change over the concentration range 10^{-1}–10^{-4} M NH_4^+ [21]. This electrode responds poorly to Li^+ ion activity and the selectivity for NH_4^+ over K^+ and Na^+ was reported to be superior to glass electrodes sensitive to ammonium ion. The nonactin electrode can be used in an enzyme system and will be discussed later. Valinomycin-based potassium ion-selective electrodes have been studied [22].

Gas electrodes, prepared by placing a gas-permeable membrane over a housing which contains a pH electrode (or Pt electrode in the case of O_2) and internal filling solution, are available for many species that either participate in or are produced from enzyme reactions, such as O_2, CO_2 and NH_3. The CO_2 electrodes from Instrumentation Laboratories, Radiometer and Orion, and the NH_3 electrode from Orion, work well, but have longer response times than desirable, due to the presence of the gas membrane. If this membrane could be omitted, a much improved gas electrode, and hence enzyme electrode, would result. This can be done very easily in the case of O_2 by simply using a flat Pt electrode (Beckman) which is polarized at a potential of -0.6 V vs. S.C.E. This electrode is covered with immobilized enzyme and works very well [23,24]. For measurements of CO_2 and NH_3 it is recommended that the new air-gap electrode, developed by Ruzicka and Hansen [25] be used. In this electrode the hydrophobic gas-permeable membrane generally employed in gas electrodes is in this new sensor replaced by an air gap which separates the electrolyte layer from the sample solution. By avoiding the membrane construction, and utilizing a very thin, hydrophilic layer of electrolyte at the surface of the indicator electrode, a very high speed of response is obtained. Furthermore, the lifetime of the sensor is substantially increased since the electrode does not come into direct physical contact with the sample solutions; thus there are no interferences from surfactants, particulate matter or organic solvents. The electrolyte layer can be easily renewed or even changed according to the requirements of a particular analysis, so that the same electrode can be used for measurements of a variety of gases. For a CO_2 electrode, for example, a layer of 0.01 M $NaHCO_3$ + surfactant (Victawet 12, Stauffer Chem. Co.) is placed on the glass electrode, for NH_3 a layer of 0.01 M NH_4Cl + surfactant. The gas liberated from solution diffuses to the surface where it interacts with the surface layer, effecting a change in pH that is measured

$$NH_4^+ \rightleftharpoons NH_3 + H^+ \tag{5}$$

Finally, in addition to these ion-selective electrodes, there is another large class of electrodes, the inert metal electrodes, i.e., Pt or Au, which could be used to follow enzyme reactions in which an electroactive product or reactant is involved. These electrodes have been well characterized in the literature, and are available from companies such as Beckman Instrument Co. (Fullerton, Calif.).

In summary, many commercially available (or home-made) electrodes have been used and can be used in the construction of enzyme electrodes: (1) glass electrodes —H^+ and monovalent cation; (2) gas — NH_3, CO_2 and O_2; (3) solid and "solid" liquid membrane electrodes — NH_4^+, S^-, CN^- and I^-, and (4) Pt electrodes.

Two basic techniques may be used in assay with enzyme electrodes, kinetic and equilibrium. In the kinetic method, the rate of change in the potential or current at the electrode is measured within the first 10—30 s of response. This change, ΔE or $\Delta i/\Delta t$, is then related to the content of substance to be assayed. In the equilibrium method, the steady-state current or potential, i or E, is measured, and related to the concentration. This relationship is logarithmic in the case of potential (E) and linear in the case of current (i). The rate method is faster, and eliminates the possibility of error due to side reactions which may consume the electroactive substance.

(D) ASSAY OF ENZYMES USING ION-SELECTIVE ELECTRODES

Now let us consider the use of ion-selective electrodes in the monitoring of enzyme activity, then proceed to the self-contained biochemical electrode systems in which the enzyme or substrate has been placed on the sensor itself. A good review on electrochemical methods of monitoring enzymatic reactions has been presented by Christian [26] that includes some mention of use of electrodes.

Probably the most common electrochemical method for the assay of enzymes which produce or consume an acid is to follow the pH change of the reaction mixture as a measure of the activity of the enzyme using the most selective of all electrodes, the low-sodium error glass electrode. This method is not generally employed directly since the activity of a given enzyme is affected by changes in pH. Instead a "pH stat" method is used in which the pH of the assay mixture is maintained by the addition of an acid or base. The rate of the addition of reagent gives the reaction velocity.

The activity of an enzyme in a system in which oxygen is consumed can be determined using an oxygen electrode. The electrode is a gold cathode separated by an epoxy casting from a silver anode. The inner sensor body is housed in a plastic casing and comes in contact with the assay solution only through the membrane. When oxygen diffuses through the membrane it is electrically reduced at the cathode by an applied potential of 0.8 V. This reaction causes a current to flow between the anode and cathode which is proportional to the partial pressure of oxygen in the sample. The rate of uptake of oxygen can be related to the activity of the enzyme or the concentration of substrate in the assay mixture. Good correlation between glucose values determined in blood by a measurement of oxygen uptake with those obtained by standard chemical tests was found by Kadish and Hall [27] and by Makino and Koono [28].

Ion-selective electrodes have been used to determine the activity of rhodanase and cholinesterase. In the rhodanase system, a cyanide-selective electrode followed the decrease of cyanide ion during the reaction

$$CN^- + S_2O_3^{2-} \xrightarrow{\text{rhodanase}} SCN^- + SO_3^{2-} \tag{6}$$

which is catalyzed by rhodanase [29]. Results obtained by this method are comparable to those obtained by spectrophotometric procedures. The method was easily adapted to automated systems as shown by Guilbault and co-workers [29]. A solution method for assay of rhodanase using a CN^- electrode was described by Llenado and Rechnitz [30]. The cholinesterase assay was performed using a sulfide-selective electrode to monitor the amount of thiocholine released under the influence of cholinesterase.

$$\text{acetylthiocholine} + H_2O \xrightarrow{\text{cholinesterase}} \text{thiocholine} + CH_3COOH \tag{7}$$

The amount of thiocholine released is proportional to the activity of cholinesterase [31]. Llenado and Rechnitz used systems very similar to those described above for the assay of β-glucosidase [32,33], rhodanase [32], and glucose oxidase [32]. An ion-selective electrode for cyanide was used to follow the production of cyanide ion in the assay of β-glucosidase

$$\text{amygdalin} + H_2O \xrightarrow{\beta\text{-glucosidase}} \text{benzaldehyde} + 2 \text{ glucose} + HCN \tag{8}$$

and the consumption of cyanide in the rhodanase system as was

12

described previously [29]. An iodide-selective electrode was used with the glucose oxidase assay to measure the decrease in iodide concentration resulting from oxidation of iodide to iodine by hydrogen peroxide. An autoanalyzer system was used to monitor the three enzymes.

$$\beta\text{-D-glucose} + H_2O + O_2 \xrightarrow[\text{oxidase}]{\text{glucose}} \text{D-gluconic acid} + H_2O_2 \qquad (9)$$

$$H_2O_2 + 2H^+ + 2I^- \xrightarrow{\text{Mo(VI)}} 2H_2O + I_2 \qquad (10)$$

This same reaction sequence for assay of glucose oxidase was subsequently turned around by Llenado and Rechnitz [34] and used for the assay of glucose. A soluble glucose oxidase was used, again in the Autoanalyzer system. However, the use of the catalyzed iodide oxidation in conjunction with an iodide electrode in the assay of glucose was shown to be subject to many interferences in blood by Guilbault and co-workers [35]. Reducing agents, such as ascorbic acid, tyrosine and uric acid interfere strongly, and glucose measurements in samples of blood containing them can only be made after vigorous pretreatment of the sample.

A cyanide ion-selective electrode was used by Gutknecht and Guilbault [36] for monitoring the removal of cyanide from aqueous systems. The enzyme β-cyanoalanine synthase (injectase) was used in an immobilized form

$$CN^- + \text{cysteine} \rightarrow HS^- + \beta\text{-cyanoalanine} \qquad (11)$$

One of the first examples of the use of ion-selective electrodes in enzymic analysis was the report of Guilbault et al. [37] who described the use of a monovalent cation glass electrode for following the urea—urease, glutamine—glutaminase, asparagine—asparaginase and D- and L-amino acids—amino acid oxidase reactions. In all cases the NH_4^+ produced was measured

$$\text{substrate} \xrightarrow{\text{enzyme}} NH_4^+ \qquad (12)$$

In order to overcome the interferences of Na^+ and K^+ ions on the use of such a glass electrode, Montalvo [38] constructed an ammonium "ion-specific" electrode by coupling a hydrophonic ammonia-permeable membrane to a Beckman cation electrode. In this electrode the ammonia produced from the urea—urease reaction diffuses through the ammonia membrane, where it reacts with tris buffer forming NH_4^+ which is then sensed by the cation electrode. Response times of about 20 min were required for the enzyme reaction. Moreover, a constant flow of buffer must be pumped over the cation electrode making the electrode totally impractical.

This same specificity for NH_4^+, and elimination of response to Na^+ and K^+ can be accomplished either by use of the Orion NH_3 electrode directly at pH 7 or 7.5 as described by Anfalt et al. [39] (response time, 2 min) or the air-gap electrode as used by Hansen and Ruzicka [40] (response time, <1 min, at pH 11 after a 2 min incubation time at lower pH). In this latter paper the ammonia, catalytically released from urea in the presence of soluble urease, is volatilized and sensed by a glass electrode covered with a film of electrolyte. Since the electrode never actually touches the sample solution, the problems caused by the presence of proteins, blood cells and species which reportedly block the pores in the membranes of other gas electrodes, are avoided.

Baum [1] has reported an electrometric method for the determination of cholinesterase (ChE) activity based on the use of a liquid membrane ion-selective electrode which responds to the concentration of acetylcholine remaining during an enzymatic catalyzed hydrolysis

$$\text{acetylcholine} \xrightarrow{\text{ChE}} \text{choline} + \text{acetic acid} \tag{13}$$

The electrode, sold by Corning (Corning, N.Y.) is perhaps the only selective non-enzymatic organic species measuring electrode that has been described in the literature, having a selectivity for acetylcholine over choline which is similar in structure. As the acetylcholine reacts, a drop in potential is measured, and equated to the activity of enzyme. Baum later combined this sensitive electrode with immobilized cholinesterase, present in the form of a disc which was simply added to the solution to be assayed [41]. Immobilized cholinesterase discs were used extensively for 64 days with good results.

Guilbault and Gibson [42] proposed methods for the assay of cholinesterase activity, based on the linear rate of change of pH of a weak tris buffer at a pH around the pK value, measured with a glass electrode. Two methods were proposed, one a static semi-automatic method capable of doing up to 12 assays per hour, the second an automatic flow approach which does 40 assays per hour.

Larson et al. [164] described kinetic methods for assay of deaminase enzymes using the air-gap electrode. The rate of NH_3 pro-

14

duced by action of urease, arginase and amino acid oxidase was used to assay for the activity of these enzymes.

2. Preparation and properties of enzyme electrodes

In Sect. 1 we considered the basic principles of the enzyme electrode and the two basic parts of such an electrode, the enzyme and the ion-selective electrode sensor. Let us now discuss the experimental parameters involved in the construction of enzyme electrodes and the observed characteristics and limitations of these electrodes.

(A) IMMOBILIZATION METHODS

(1) General

Cost is a factor in enzymic analysis, because continuous or semi-continuous analysis requires large amounts of unrecoverable enzymes. Immobilized enzymes can be used continuously and expensive enzymes can be recovered, thus resolving the problem of cost. Also, most solubilized enzymes are very unstable, often must be stored at low temperatures, and then only retain activity for a short period of time. In cytochemical systems most, if not all, of the enzymes are attached to cell surfaces or trapped within cell membranes. Immobilization of enzyme puts them in a more natural environment with the result that they are usually most stable and efficient.

Within the last six years a new technology based on enzyme immobilization has rapidly emerged. Five methods have been used for the preparation of water-insoluble derivatives of enzymes:

(a) microencapsulation within thin-wall spheres;

(b) adsorption on inert carriers;

(c) covalent crosslinking by bifunctional reagents into macroscopic particles;

(d) physical entrapment in gel lattices; and

(e) covalent binding to water-insoluble matrices.

Let us now consider some of the different methods of enzyme immobilization and the procedures followed to immobilize enzymes used in electrodes. For more details see one of several review articles [43—54].

(2) Immobilization techniques

a. Microencapsulation. Microencapsulation within thin wall spheres, the newest approach to enzyme insolubilization, was first introduced only several years ago by Chang [55—60]. The thin wall of the spheres is semipermeable such that enzymes are physically prevented from diffusing out of the microcapsule while reactants and products can readily permeate the encapsulating membrane.

Microencapsulation is performed by depositing polymer around emulsified aqueous droplets, either by interfacial coacervation or interfacial polycondensation. Such a technique has not yet been applied to the preparation of electrodes but could be. Hence it will not be considered further here.

The main disadvantage of the use of spheres is that many of the interfacial polymerization procedures cause enzyme deactivation. Rony [61] has circumvented this problem by first producing hollow fibers instead of spheres and then placing the enzyme in the fiber and sealing the ends. Hollow fibers have the added advantage of:

(a) being easy to fill with many types of catalyst;

(b) allowing recovery of the trapped enzyme; and

(c) being mass produced, thus commercially available.

b. Adsorption. Adsorption of enzymes on insoluble supports results from ionic, polar, hydrogen bonding, hydrophobic or π-electron interactions. In early works, adsorption to inorganic carriers was used as a technique for the characterization of enzymes, and since then the characterization of the adsorption phenomenon has been the major concern of most studies. Early works on protein adsorption have been reviewed [52,62]. A few of the adsorbents that have been used are glass [62], quartz [63], charcoal [64], silica gel [65], alumina [66], ion exchange resins [67,68], and bentonite [69].

The major advantage of this method of immobilization is its extreme simplicity. Attachment is by simple exposure and conditions are mild. But the disadvantage is a very serious one. Adsorbed enzymes are easily desorbed by factors depending on pH, solvent, substrate, and temperature effects.

c. Covalent crosslinking by bifunctional reagents. Bifunctional reagents have been used to insolubilize enzymes and other proteins by intermolecular crosslinking, with the concomitant formation of

16

macroscopic particles [49,50,52]. Bifunctional reagents can be divided into two classes, "homo" bifunctional and "hetero" bifunctional, depending on whether the reagent possesses two identical or different functional groups.

A few "homo" bifunctional reagents are glutaraldehyde [70—73] bisdiazobenzidine-2,2'-disulfonic acid [74,75], 4,4'-difluoro-3,3'-dinitrodiphenyl sulfone [76], diphenyl-4,4'-dithiocyanate-2,2'-disulfonic acid [77], 1,5-difluoro-2,4-dinitrobenzene [78], and phenol-2,4-disulfonyl chloride [79]. A few "hetero" bifunctional reagents are toluene-2-isocyanate-4-isothiocyanate [80], trichloro-s-triazine [81,82], and 3-methoxydiphenyl methane-4,4'-diisocyanate [80]. The advantages of this method are: (a) simplicity and (b) chemical binding of the enzyme enabling control of the physical properties and particle size of the final product. The major disadvantage is that many enzymes are sensitive to the coupling reagents such that they lose activity in the process. This method is thus limited in its applicability, but one such reagent, glutaraldehyde has been applied with very good success to the manufacture of enzyme electrodes [23, 24,39].

CHO
|
(CH₂)₃   Glutaraldehyde
|
CHO

d. Inclusion in gel lattices. An enzyme can be occluded with a gel matrix by carrying out a polymerization in an aqueous solution containing the enzyme. The polymer used most often consists of acrylamide as monomer and *N,N'*-methylenebisacrylamide as crosslinking agent. The gel can be formed as a thin film where it is to be used or it can be formed in a block and mechanically dispersed into the desired particle size to be used in solution, in a column, etc. Polyacrylamide gels were first prepared by Bernfeld and Wan [83] and later used by many others [84—89]. Other types of gels that have been used are starch gel stabilized onto polyurethane foam [90—92] and silastic [93,94].

Inclusion of enzymes in a gel has the advantage of mild reaction conditions without significant alteration of the enzyme. It can, therefore, theoretically be applied to any enzyme.

There are several disadvantages, however, which should be mentioned. Because there is a broad distribution of pore size in poly-acrylamide-type polymers, there is a continuous leakage of the occluded enzyme. Free radicals generated during the polymerizing process may affect the enzyme such that it loses some or most of its activity. Also, there will probably be diffusion control of the substrates and products and the reaction is limited to small substrates which can diffuse readily into the gel (i.e., glucose, urea, amino acids). Stability is about 3 weeks or a hundred assays maximum.

e. Covalent binding to water-insoluble matrices. A new technology, based on the immobilization of enzymes by covalent coupling to insoluble carriers, has rapidly emerged within the last six years. Because covalent coupling places enzymes in a more natural environment, they usually function efficiently, have increased stability, and have the added advantage of immobilization which is not reversed by pH, ionic strength, substrate, solvents, or temperature. It is little wonder that covalent binding of enzymes to insoluble carriers is now the most widely used method of enzyme insolubilization.

Carriers are chosen by their properties of solubility, functional groups, mechanical stability, surface area, swelling, and hydrophobic or hydrophilic nature. Essentially three types of carriers have been employed: inorganics, natural polymers, and synthetic polymers. A

TABLE 2

Some common carriers used for covalent binding of enzymes

Carrier	Ref.
Porous glass	95, 96
Polyacrylamide	97, 98
Polyacrylic acid	99, 100
Polyaspartic acid	101
Polyglutamic acid	99
Polystyrene	102
Nylon	103, 104
Cellulose	105, 106
Sephadex	108, 109
Ethylene maleic anhydride co-polymer	101, 110
Agarose	111, 112
Sepharose	107, 113
Carboxymethyl cellulose	101, 105

few of the more common carriers which have been used are given in Table 2. These carriers are usually activated by transformation into various derivatives.

Binding is carried out by way of functional groups on the enzyme which are not essential for its catalytic activity. Even though a particular amino acid residue may be necessary for catalytic activity, it is usually present several times in non-essential positions. Thus random chemical reaction will denature only a portion of the molecules present. Because binding must be carried out under conditions which do not cause denaturation, it is necessary to understand the effects of chemical modification of enzymes on their activity. Several reviews have been written on this subject [114—116].

The amino acid residues suitable for covalent binding are: (a) α- and ϵ-amino groups; (b) the phenol ring of tyrosine; (c) β- and γ-carboxyl groups; (d) the sulfhydryl group of cysteine; (e) the hydroxyl group of serine; and (f) the imidazole group of histidine. Of

Diazonium coupling

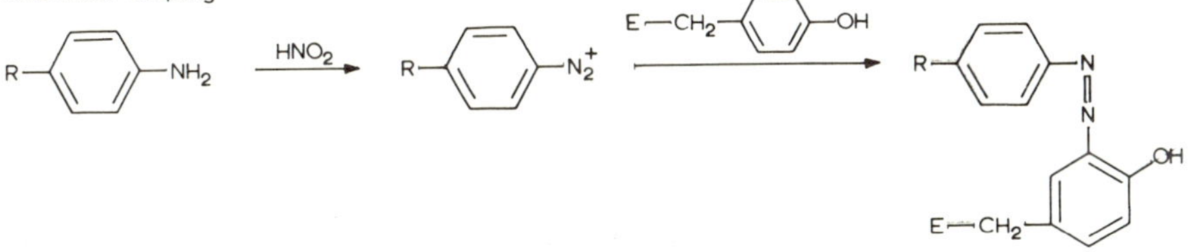

(14)

Isothiocyano coupling

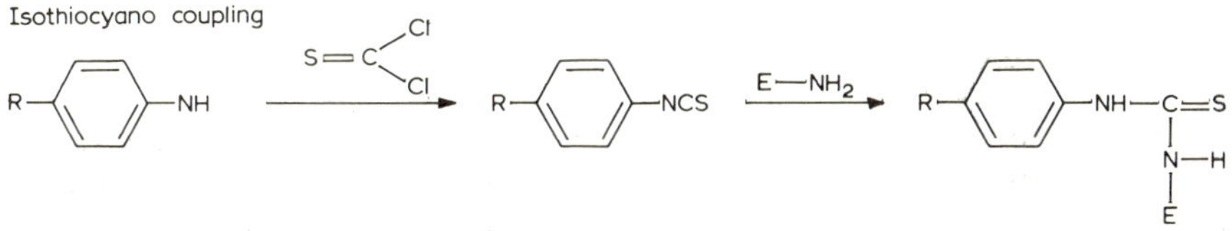

(15)

Hydrazide coupling

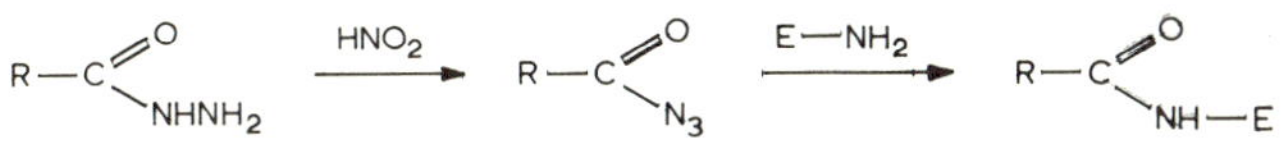

(16)

Amide formation

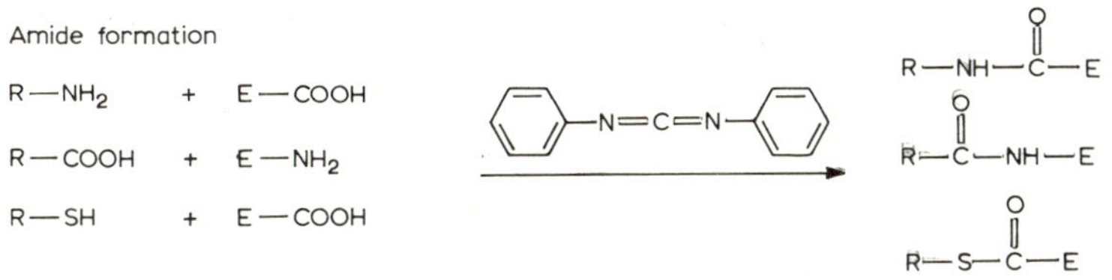

R—NH$_2$ + E—COOH

R—COOH + E—NH$_2$

R—SH + E—COOH

(17)

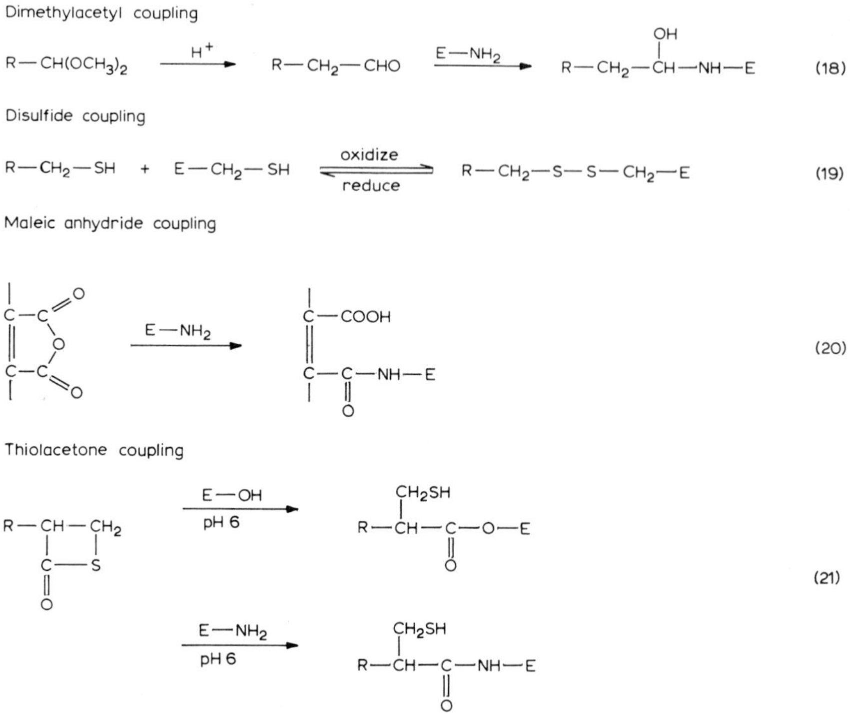

Dimethylacetyl coupling

$$R-CH(OCH_3)_2 \xrightarrow{H^+} R-CH_2-CHO \xrightarrow{E-NH_2} R-CH_2-\overset{\displaystyle OH}{\underset{\displaystyle |}{CH}}-NH-E \tag{18}$$

Disulfide coupling

$$R-CH_2-SH \;+\; E-CH_2-SH \underset{\text{reduce}}{\overset{\text{oxidize}}{\rightleftharpoons}} R-CH_2-S-S-CH_2-E \tag{19}$$

Maleic anhydride coupling

$$\tag{20}$$

Thiolacetone coupling

$$\tag{21}$$

Fig. 1. Common coupling reactions. R refers to the solid phase, E to enzyme. (From ref. 117.)

these the most widely used are the first three. There are a large number of methods of covalently coupling enzymes to water-insoluble carriers. Several of the more common methods are shown in Fig. 1.

(3) Properties of immobilized enzymes

Many properties of immobilized enzymes are different from those of their solubilized counterparts. We have already mentioned effects on the kinetics brought about by diffusion control and heterogeneous catalysis. One of the most important properties is an increased long-term stability, and temperature stability. In general, stability can be increased by the proper choice of coupling procedure and insoluble carrier. Another property which has been observed is a change in

20

reactivity. Immobilization may alter kinetic constants due to a change in activation energy, pH profile, Michaelis constants, or even specificity.

If the matrix is ionic, the microenvironment of the active site may be altered by the electrostatic field. The pH within the matrix will be different to that in the external solution. If substrates are also charged, the apparent Michaelis constant may be altered. Covalent coupling itself may bring about these effects as well as a change in specificity due to an alteration of the enzyme's net charge, nearest neighbor effects on the active site region, perturbations of intramolecular interactions, and conformational changes.

Covalent binding of enzymes to electrically neutral carriers, by way of non-active site residues, has been shown to have no effect on their kinetic behavior towards low molecular weight substrates [52, 74,118]. Detailed discussions of the effects of immobilization on enzyme properties can be found elsewhere [44, 50, 110, 119].

(B) CONSTRUCTION OF ENZYME ELECTRODES

We mentioned above that there are 4 steps to follow in the construction of an enzyme electrode. We shall consider each of these factors in more detail.

(1) Step 1

Pick the enzyme that reacts with the substance to be determined. From standard references books on enzymology, such as Biochemists Handbook [120] find an enzyme system which is suitable for your determination. In the ideal case, this will involve the use of the primary function of the enzyme, i.e., the main substrate—enzyme reaction. For example, for a glucose electrode, glucose oxidase would be used, for a urea electrode, urease, for a L-glutamic acid electrode, L-glutamate dehydrogenase. In other cases, this might necessitate using an enzyme that acts on the compound of interest as a secondary substrate, i.e., urease for N-methyl urea or malic dehydrogenase for acetic acid [121]. Of course, this latter case will introduce more interferences and less selectivity into the assay.

Note, in some cases, there are several enzymes which act on the substrate of interest, via different reactions. For example, L-tyrosine could be determined using L-tyrosine decarboxylase and measuring the CO_2 liberated [122], or using L-amino acid oxidase using a Pt electrode [123] or an NH_4^+ electrode [124,125]. The latter enzyme, although less selective, can be obtained commercially in high purity, the former is available in low purity and would have to be purified before use. Hence, the scientific capabilities of one's laboratory might dictate the choice of enzyme.

(2) Step 2

Obtain the enzyme. Having found the enzyme to be used for your application, check the catalogs of commercial suppliers, Boehringer, Calbiochem, Sigma, Worthington, etc., to see if the enzyme can be purchased, and its purity. The latter may or may not pose a problem. Many enzymes are stable in an impure state, i.e., jack bean urease, glucose oxidase from the food industry (General Mills), and can be used satisfactorily in a pseudo "immobilized" state, i.e. as a liquid covered with a dialysis membrane, for up to a week. In other cases, the impure enzyme has too low an activity to be useful in the low purity state without further purification (many of the decarboxylases available from Sigma). In the latter case, the enzyme must be purified, which although not difficult, would involve further work, and possible assistance from others. In still other cases, one might find that the enzyme that one wants to use is not available commercially. In this case there are 2 possibilities,

(1) contact a large biochemical supply house and inquire whether it will isolate and purify the enzyme you want. Many will, for a suitable fee;

(2) look up the enzyme in the literature or standard biochemistry—enzymology reference books, obtain the isolation and purification methods used, and perform the purification yourself. This we have done ourselves, in many cases, with excellent results, and in most cases the techniques are simple enough to be carried out by a person with reasonable scientific training. Frequently the results are well worth the effort.

(3) Step 3

Immobilize the enzyme. A simple rule of thumb to follow is that the better the enzyme is immobilized, the more stable it will be, and hence, the longer it will be useful and the more assays will be possible from one batch. Let us now consider the various possibilities, and the characteristics of the product.

a. Commercially available immobilized enzyme. This is the ideal case and is the first choice, if possible. There are a number of enzymes available in the immobilized form — most of these are fine products, and as good or better than most scientists can do themselves. The author has personally used the products of Boehringer and Aldrich with good success. Of the enzymes available that are likely to be of most use to the reader one could mention urease (Boehringer), glucose oxidase (Boehringer or Aldrich), ribonuclease (Boehringer) and uricase (Aldrich). Furthermore, Corning has offered to sell almost any enzyme bound to glass, under certain conditions, and the reader is invited to write for details.

b. Soluble "immobilized" enzymes. The second choice available, which is the easiest for the novice if choice (a) is not possible, is to use the soluble enzyme in construction of the electrode. A thick paste of the enzyme powder is made with a little water (1—2 μl); this paste is spread over the surface of the electrode, and the layer is covered with a dialysis membrane 20—25 μm thick cellophane (Will Scientific, Inc., or Arthur H. Thomas, U.S.A., about 100 μ pore size). Such soluble enzyme electrodes are stable for up to about a week, if kept in a 5—10°C refrigerator, between use. Electrodes with the more crude enzymes, urease or glucose oxidase mentioned above, might be stable for longer periods of time.

c. Physically entrapped enzymes. In ease of preparation this is the next choice. Many enzymes have been physically bound in polyacrylamide gels, which are crosslinked polymers with the enzyme trapped inside. A typical preparation is mentioned in ref. 88 and below, and similar preparations can be effected by anyone with a minimum of effort. The stability of the final product depends on the degree of care taken and the control of experimental conditions, as was carefully pointed out by Guilbault and Montalvo [88], but can be as long as 3—4 weeks or about 50—100 determinations.

d. Chemically bound enzymes. These are the most difficult to prepare, although preparation can be effected by anyone who has had a year's course in organic chemistry. The products are most stable, and can be used for 200—1000 assays, and stored at room temperature for over a year between assays [126]. The best actual method involved will depend on the individual enzyme, as will be discussed in other parts of this book. In the author's experience, the polyacrylacid diazo coupling [123,126] and the glutaraldehyde methods [23,24] have yielded extremely satisfactory results. The covalent binding to polyacrylamide crosslinked polymer, used by Boehringer in its commercial preparations, is also quite satisfactory.

Reactive intermediates for direct coupling of enzymes via only 1—2 steps are available from Corning (Corning, N.Y.), Aldrich (Milwaukee, Wisc.) and Koch-Light (England). These are recommended to anyone interested in making chemically-bound enzymes. The glutaraldehyde method is also quite simple to effect (glutaraldehyde is available from Sigma, St. Louis).

(4) Step 4

Place the enzyme around the appropriate electrode. In order to develop an electrode for the substrate of interest, one must have as the base sensor an electrode that responds to either one of the reactants, A or B in eqn. (22)

$$A + B \xrightarrow{\text{enzyme}} C + D \tag{22}$$

or to one of the products, C or D. The sensor can be a gas electrode (to measure all O_2-consuming reactions, NH_3 or CO_2 liberating enzymes), a glass electrode (to follow H^+ changes in reactions that liberate acid, or NH_4^+ producing enzymes), a Pt electrode (to follow all enzyme reactions involving electroactive species or O_2) or some other ion-selective electrode (i.e., the CN^- electrode for amygdalin, an NH_4^+ antibiotic electrode for deaminase enzymes, an I^- electrode for oxidative enzymes coupled with the I^--I_2 indicator reaction, the S^{2-} electrode for cholinesterase substrates, etc.). In most cases, the limiting factor in the design of an enzyme electrode will be the availability of a sensor to monitor the reaction. Of course, there are other possibilities for monitoring enzyme reactions; for example, a thermistor could be covered with enzyme and the temperature change resulting from the enzyme reaction monitored [127]. Considerable research is being performed in this area, but as of this time no satisfactory systems have been developed.

24

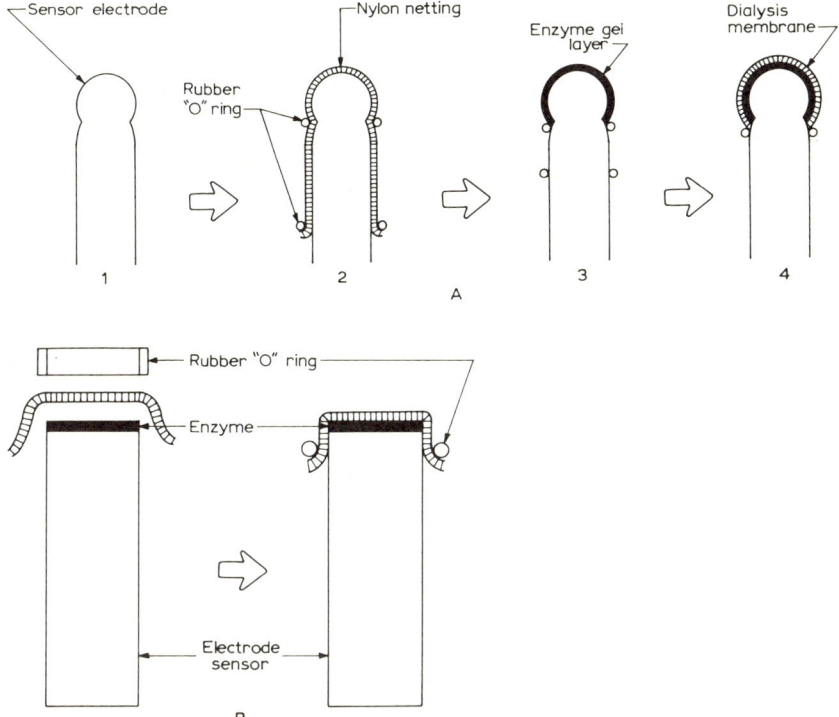

Fig. 2. Preparation of enzyme electrodes. A, Physically entrapped enzyme electrodes; B, "soluble" or chemically bound enzyme electrodes.

Assuming that a sensor is available, and the enzyme has been obtained and immobilized, let us now describe the preparation of typical enzyme electrodes (see Fig. 2).

a. Type A — dialysis membrane electrode. Turn the ion-selective electrode upside down and place 10—15 units of the soluble enzyme, physically entrapped enzyme or chemically bound enzyme (after immobilization of the enzyme the preparation should be freeze dried to form a powder) onto the electrode. Take a piece of cellophane dialysis membrane (20—25 μm thick, obtained from either Will Scientific or Arthur H. Thomas, U.S.A.) which has been cut into a circular piece with a diameter about twice the size of the electrode sensor, and wrap the cellophane around the electrode, taking care that the powder is evenly spread over the surface of the electrode in a thin layer (this might be as conveniently done by placing a thick paste of the enzyme in water onto the tip of the flat electrode while it is held upside down (Fig. 2B) and coating the enzyme onto the surface with a spatula). Place a rubber "O" ring, with a diameter that fits the electrode body snugly, around the cellophane (Fig. 2B) and gently push it onto the electrode body so that the cellophane—enzyme layer on the bottom of the electrode is held tight and flat (Fig. 2B). Place the electrode in a buffer solution for a few hours or overnight to allow penetration of buffer into the enzyme layer and permit loss of entrapped air. Store the electrode in buffer between use [123,126, 128—130].

b. Type B — physical entrapment onto surface. Holding the electrode sensor upside down (Fig. 2A) cover it with a thin nylon net (about 90 μm thick — a sheer nylon stocking obtained from any ladies shop is satisfactory) which is secured with a rubber "O" ring in the same manner as above. This serves as a support for the enzyme gel solution. Prepare the enzyme gel solution by mixing 0.1 g of enzyme (purity about 10—50 units mg^{-1}) with 1.0 ml of gel solution — 1.15 g of *N,N'*-methylenebisacrylamide (Eastman Organics, Rochester), 6.06 g of acrylamide monomer (Eastman), 5.5 g of potassium persulfate, and 5.5 mg of riboflavin in 50 ml of water. Gently pour the enzyme gel solution onto the nylon net in a thin film, making sure all the pores of the net are saturated; one ml of this solution should be enough for several electrodes. Place the electrode in a water-jacketed cell at 0—5°C and remove oxygen, which inhibits the polymerization, by purging with N_2 before and during polymerization. Complete the polymerization by irradiating with a 150-W Westinghouse projector spot light for 1 h. At the end of this time the enzyme layer should be dry and hard. Place a piece of dialysis membrane over the outside of the nylon net for further protection, and secure with a second rubber "O" ring. Soak the electrode in buffer solution overnight and store in buffer between use [88,126,131—135].

c. Type C — direct polymerization onto the membrane. This can be effected by a direct attachment of the enzyme to the surface of the electrode, if glass, by the Corning technique which has been discussed elsewhere, or by direct chemical attachment on the electrode surface, as was done by Anfält et al. [39] in the case of the Orion

26

NH$_3$ electrode. In the latter study membranes were prepared by dropping 0.1 ml of soluble urease solution (0.5 unit) onto the surface of the gas diffusion membrane of the electrode. The membrane was set aside for 12 h at 4°C to allow evaporation of the solvent, and glutaraldehyde solution was then added dropwise (2.5% in phosphate buffer, pH 6.2). The membrane was set aside for a further 1.5 h at 4°C and was then rinsed carefully with water in order to remove free enzyme and buffer. Note, some activity was lost over a 20 day period indicating insufficient enzyme was used [39]. At least 5—10 units of urease would have been better in this case (again in 0.1 ml of solution).

Of the three types of electrode described above, Type A is the author's preference for ease of preparation (once the bound enzyme is obtained) and for long-term stability (if the enzyme is chemically bound). Type B electrodes are not difficult to prepare, but are time consuming, requiring 1 h of polymerization time per electrode, and have a maximum stability of only about 3—4 weeks or 50—100 assays. Type C electrode with direct attachment to the electrode is not recommended because it essentially commits the electrode sensor to that one enzyme electrode. In Types A and B electrodes, one can easily replace the enzyme layer, when it is no longer useful, with a new layer — the sensor can be reused until its lifetime is exhausted.

(C) PERFORMANCE CHARACTERISTICS OF ELECTRODES

Now that the electrode has been made, let us consider some of the factors that affect the response and stability of the electrode.

(1) Stability

Some of the factors that affect the stability of enzyme electrodes are listed in Table 3. The stability of an enzyme electrode is a difficult term to define, since an enzyme can lose some activity resulting in a shift of the calibration curve downwards. Yet, if the slope remains constant, as is frequently the case, the electrode is still useful, needing only calibration daily. This is seldom a problem since all who use electrodes of any type, i.e., glass electrode, reset the pH or potential of their electrode at least once a day. This should be done with the enzyme electrode also using serum (i.e., Monitrol, Dade, Miami). Another problem in the definition of stability is that many

TABLE 3

Factors that affect the stability of enzyme electrodes

---

1. Type of entrapment
   (a) Soluble + dialysis membrane — 1 week or 25—50 assays
   (b) Physical — 3—4 weeks, 50—100 assays
   (c) Chemical — 4—14 months, 200—10000 assays

2. Content of enzyme in gel and purity

3. Optimum conditions of enzyme

4. Stability of base sensor

---

workers measure the potential of their electrode occasionally over a long period of time, and report this data as the stability. This may mean that the electrode was used 1 time a day or week, 10 times a day or 100 times a day. Naturally, the more the use, the shorter will be the overall lifetime. The first factor that affects stability is the type of entrapment used. As a general rule, a "soluble" electrode is useful for about 1 week or 25—50 assays, provided the electrode is kept refrigerated between uses. The physically entrapped "polyacrylamide" electrodes are good for about 3 weeks or 50—100 assays, depending crucially on the degree of care exercised in the preparation of the polymer. The chemical enzyme can be kept indefinitely if not used very much, even at room temperature (see Table 3 and Fig. 3) as long as 14 months for glucose oxidase, greater than 4—6 months for L-amino acid oxidase or uricase. One can expect to get about 200—10000 assays per each electrode, again depending on how good a synthesis is effected. Although the electrode can be stored at room temperature, it is recommended that all electrodes be kept in a refrigerator and covered with a dialysis membrane to prevent the action of bacteria which tend to feed on the enzyme, destroying its activity. The dialysis membrane (M. Wt. exclusion about 1500, pore size about 100 μ) prevents the enzyme from getting out and bacteria getting in. The stability of the physically entrapped enzyme varies greatly with experimental conditions and a thorough study of these factors was made by Guilbault and Montalvo [88].

To determine the effect of physical immobilization parameters on the stability of the urea electrode, a series of enzyme electrodes was prepared while varying one immobilization parameter and maintaining all of the other parameters constant. To determine the stabil-

28

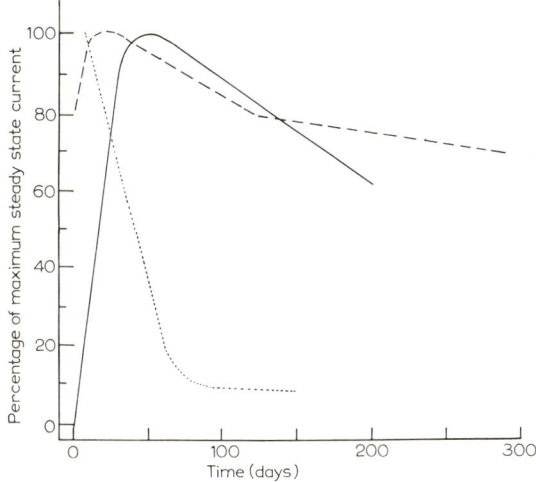

Fig. 3. Long-term stability of glucose electrodes. (- - - - -) Type 3 electrode; (————) Type 2 electrode; (· · · · · ·) Type 1 electrode. (Ref. 126 by permission of authors.)

ity of the immobilized urease coating on the surface of the cation electrode, the steady-state potential was obtained for a given urea substrate concentration at periodic time intervals. If the steady-state potential is constant within a certain period of time, no loss of activity of the immobilized enzyme has occurred. All stability data reported were obtained with the electrode stored at $25°C$ in tris buffer between measurements.

The maximum stability that could be achieved with the physically entrapped enzyme electrode was obtained with the following immobilization parameters: photopolymerizing for one hour at $28°C$ with a No. 1 150-W photoflood lamp, a gel-layer thickness of 350 μ, and an enzyme concentration in the gel of 175 mg cm^{-3} gel. The slope of the stability curve, $\Delta mV/\Delta t$, shows that the measured stability depends on the substrate concentration used in the stability measurements. When the urea concentration is high enough so that the steady-state response is independent of the substrate concentration, $\Delta mV/\Delta t$ was 0.2 mV day^{-1} over a 14-day period. At lower substrate concentrations, as for example, 1×10^{-3} M urea, the steady-state response is first order in urea concentration and a much smaller loss in activity was obtained, 0.05 mV day^{-1} over a 14-day period.

Since 1×10^{-3} M urea represents the upper limit of substrate concentration which can be measured with the enzyme electrode, the steady-state response falls by only 0.7 mV during 14 days operation at $25°$C. After 14 days, the loss in activity was much greater for both substrate concentrations.

To study the effect of the activity of immobilized urease on enzyme gel stability, physically entrapped enzyme electrodes were prepared with activity of enzyme from 375 to 3500 Sumner unit g^{-1} of enzyme. No appreciable change in stability occurred with this relatively large change in enzyme activity. On the other hand, highly purified urease is known to be very unstable in solution. A similar trend in stability would be expected with immobilized urease.

Greater stability with the enzyme electrode was always obtained when the gel solution was less than two days old. Gel solutions were stored without added polymerization catalysts when the storage period was greater than two days. The solutions were always stored in the dark at room temperature. The stability of the urease electrode was studied as a function of enzyme gel-layer thickness in the range 30—350 μ. The stability increased with increased thickness of the enzyme gel layer, but response time increased.

Several experiments were run to determine quantitatively the effect of photopolymerization light intensity and photopolymerization time on enzyme electrode stability. When the high intensity photoflood lamp is substituted with a 60-W domestic lamp, the loss in activity rises from 0.2 to 4.2 mV day^{-1} for 8.33×10^{-2} M urea. A similar loss in activity for the electrode was obtained when only the photopolymerization time was reduced from 1 h to 15 min.

To study the effect of photopolymerization temperature and water content of the gel layer during photopolymerization on the electrode stability, a series of enzyme electrodes were prepared with photopolymerization temperatures ranging from $4°$C to $43°$C. The water content of the gel layer over the electrode surface was also varied when the photopolymerization temperature was changed; this is because the rate of evaporation of water for the thin enzyme gel layer varies directly with temperature. When the photopolymerization temperature and water content of the gel were varied to study electrode stability, the other immobilization parameters were adjusted to give maximum stability. The stability, measured with 8.33×10^{-2} M urea, showed a loss of only 0.2 mV day^{-1} at $28°$C photopolymerization temperature; upon lowering the immobilization or photopolymerization temperature to 6°C, the loss in electrode activity is much higher, 3.7 mV day^{-1}. At 6°C the rate of evaporation of water from the enzyme gel layer during photopolymerization is so slow that the gel layer is still damp to the touch after immobilization is complete. This large loss in activity is due to leaching of enzyme from the gel layer. Enzyme which had leached out of the gel layer could easily be detected in the buffer solution used to store the electrode. At 28°C the rate of evaporation of water from the gel layer is sufficiently rapid so that when the polymerization is complete, the electrode is dry to the touch. The enzyme electrode is now more stable because a less porous polymer is formed. At higher polymerization temperature, such as 43°C, the resulting electrode is unstable. Therefore, maximum stability is obtained with this enzyme electrode when the photopolymerization temperature is 25°C—28°C.

To determine the effect of a film of cellophane (dialysis membrane) on enzyme electrode stability, an electrode was made by placing a thin film of cellophane over the enzyme gel layer. The cellophane was permeable to the urea substrate but not the high molecular weight enzyme. Polymerization parameters were the same as those used to obtain the maximum stability for the cellophaneless electrode. Enzyme electrode stability, measured with either 8.33 × 10^{-2} or 1 × 10^{-3} M urea, showed no measurable loss in activity for 21 days (electrode stored between measurements in tris buffer at 25°C). After 21 days, the electrode began to lose activity. The increased stability of this electrode over the membraneless electrode is apparently due to the cellophane which prevents any enzyme from leaching out of the enzyme gel layer.

Another factor that affects the apparent stability of all electrodes, especially the "soluble" and physically entrapped electrodes, is the content of enzyme in the reaction layer. As will be shown later, a certain amount of enzyme is required to yield a Nernstian calibration curve. In many cases it is advantageous to add more enzyme, say twice as much; in this case more enzymic activity can be lost yet a linear Nernstian plot will still be obtained.

Still another factor that will affect the stability of an electrode is the choice of operating conditions. An example of this is the comparison of the results obtained by Rechnitz and Llenado [136,137] and those of Mascini and Liberti [130] for the amygdalin electrode. Amygdalin is cleaved by β-glucosidase to give CN$^-$ ions, which are sensed by a CN$^-$ ion-selective electrode. Since this electrode responds best to free CN⁻ ions, obtainable only at pH's >10, Rechnitz and Llenado used this high pH for operation of their electrode. Even though the enzyme was physically bound it lost activity continually and showed a lifetime of only a few days. It is known that almost all enzymes will lose activity at pH's <3 and >9, and undoubtedly this was one contributant to the poor stability. Mascini and Liberti used only a soluble enzyme at a pH of 7 and found not only better stability (1 week which is all that can be expected from a soluble enzyme) but also faster response times. Another reason for the poor stability of Rechnitz and Llenado [137] is the "sausage" polymerization these authors tried, in which large pieces of physically entrapped enzyme are made and slices cut for each assay. From our own experience and others with such a technique, the sausage obtained is like roast beef placed in an oven for 30 min — it is well done on the outside and raw on the inside. The reader is advised not to attempt such a large scale entrapment, but should prepare individual small batches of polyacrylamide enzyme gels.

A thorough comparison of the stability of the three types of "immobilized" electrodes, soluble (Type 1), physically entrapped (Type 2), and chemically bound (Type 3) was shown by Guilbault and Lubrano [126] for glucose oxidase (Fig. 3). The long-term stabilities of Types 1, 2 and 3 electrodes were studied by testing the response of each type of electrode to 5×10^{-3} M glucose in phosphate buffer, pH 6.6, at least once a week for several months. When not in use the electrodes were stored in phosphate buffer at 25°C. The results, shown in Fig. 3, show that the long-term stability decreased in the order: chemically bound > physically bound > solubilized. Not only did Type 1 electrode response decrease drastically with time, it also decreased with each determination of glucose. This is a serious problem and as a result, Type 1 electrode is of little use analytically, except with frequent calibration and use of a large excess of enzyme. This problem is not encountered with Types 2 and 3 electrodes consisting of immobilized enzyme. The activity of these 2 electrodes actually increases for the first 20—40 days before beginning to decrease — this is probably due to the establishment of diffusion channels in the matrix over a period of time with concomitant increase in apparent activity until the channel formation ceases and only denaturation is observed, or, it could be due to changes in the conformation of the fraction of enzyme immobilized in a non-active conformation to the more stable and preferred conformation. Immobiliza-

32

tion in an unfavorable conformation can be due to pH, temperature or stirring effects during the immobilization process. The decrease in response is due to a decrease in activity of the enzyme layer because of slow denaturation, and possibly also slow irreversible inhibition. The Types 2 and 3 electrodes eventually reach a stability change of −0.25 and −0.08% of maximum response per day, respectively. The physically bound enzyme lost half its activity in 7 months, but the chemically bound had lost only 30% of its activity in 400 days (13 months). Of course, this stability would have been much less if the electrodes had been subjected to considerable use each day — actually about 200 assays for the Type 2 electrode and almost 1000 for the Type 3 enzyme electrode are possible.

Still another factor affecting the stability of some enzyme electrodes is the leaching out of a loosely bound co-factor from the active site, a co-factor which is needed for the enzymic activity. Such was found by Guilbault and Hrabankova [125] in the case of D-amino acid oxidase in a polyacrylamide membrane. The bond between protein and coenzyme (flavine adenine dinucleotide, FAD) is very weak in D-amino acid oxidase, and FAD is easily removed by dialysis against buffer without FAD. Without FAD in the solution used to store the electrode all activity is lost in 1 day; using a 4×10^{-4} M solution of FAD in tris buffer, pH 8.0, to store the electrode between use, resulted in a 3-week stability, with little loss in activity.

Finally, the stability of the enzyme electrode will depend on the stability of the base sensor. This, in most cases, is not the limiting factor in the stability, the sensor having a longer stability than the immobilized enzyme. This factor should be considered however, in use of some of the shorter lifetime electrodes, such as the liquid membrane electrodes.

(2) Response time*

There are many factors that affect the speed of response of an enzyme electrode, and these are listed in Table 4. To obtain a response, (1) the substrate must diffuse through solution to the membrane surface, (2) diffuse through the membrane and react with enzyme at the active site, and (3) the products formed must then diffuse to the electrode surface where they are measured. Let us consider each of

---

* A faster response is defined as a decrease in response time.

TABLE 4

Factors affecting the response time of enzyme electrode

---

1. Stirring rate of solution
2. Concentration of substrate $10^{-1} > 10^{-3} > 10^{-5}$
3. Concentration of enzyme
4. pH optimum
5. Temperature (most effect on rate)
6. Thickness of membrane
7. Dialysis membrane

---

these factors in detail, and see how the response time can be optimized.

a. Rate of diffusion of the substrate. A mathematical model describing this effect can be derived as was done by Blaedel et al. [138], but in simplest, practical terms the rate of substrate diffusion will depend on the stirring rate of the solution as was shown experimentally by Mascini and Liberti [130] for the amygdalin electrode and described in Fig. 4. In an unstirred solution the substrate gets to the membrane surface, albeit slowly, so that long response times are observed. At high stirring rates the substrate quickly diffuses to the membrane surface where it can react. The difference can be as much as a decrease of response time from 10 min to 1—2 min, or less. With rapid stirring for the urea electrode [88,133] a response time less than 30 s was achieved. Of importance also is the relationship of stirring rate to the equilibrium potential observed. As shown in Fig. 4, the potential shifts as a function of stirring rate due to the changes in the amount of substrate brought to the electrode surface and the degree of its reactivity. Hence, for fast response time and steady reproducible values it is recommended that a fast stirring of the solution be effected, yet with a constant stirring rate (i.e., set the speed on your stirrer and use this same setting for all readings).

b. Reaction with enzyme in membrane. The rate of reaction will depend, according to the Michaelis—Menten equation,

$$V = \frac{k_3 [E] [S]}{Km + [S]}$$

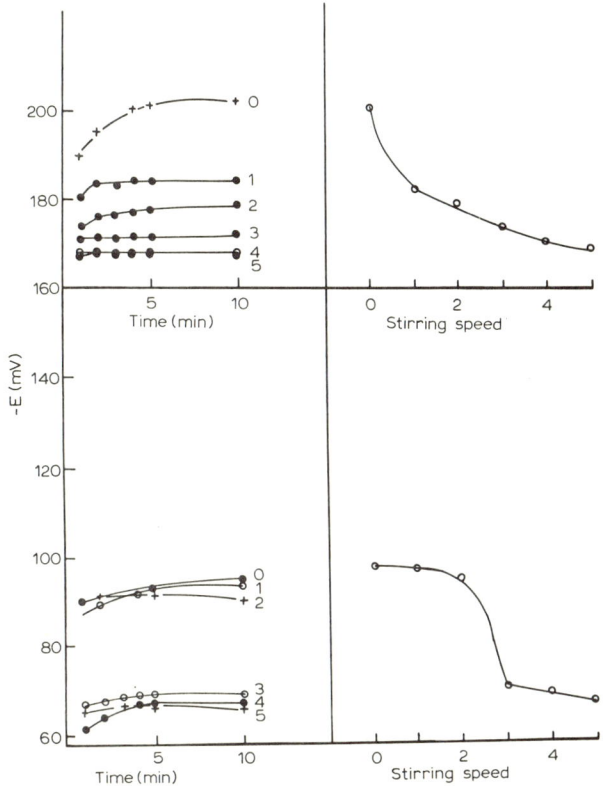

Fig. 4. Stirring effect. Response time and equilibrium value with different stirring speed. The number 0 corresponds to unstirred solutions and numbers 1–5 increasing stirring speed and are arbitrary numbers. [Amygdalin] = (a) 10^{-2} M, (b) 10^{-4} M; enzyme amount = 1 mg. (Ref. 130 by permission of authors.)

on the activity of enzyme and factors that affect it, i.e., pH, temperature, inhibitors, and on the concentration of substrate. The equilibrium potential obtained, however, should be dependent only on the substrate concentration and the temperature (since this term appears in the Nernst equation). The response rate will also depend on the thickness of the membrane layer in which reaction occurs and on the size of the dialysis membrane used to cover the enzyme layer, if one is used. Let us consider each of these factors separately.

(3) Factors that affect response time

a. Effect of substrate. Two typical examples of the effect of substrate concentration on response rate are shown in Figs. 5 and 6. Figure 5 shows the response of a β-glucosidase membrane electrode to amygdalin at various concentrations [130] and Fig. 6, the response of a glucose oxidase membrane electrode to glucose [126]. In both cases, the rate of reaction increases (as indicated by the increased inflection of the E-time of i-time curve) as the substrate concentration increases and a faster response time is observed, i.e., 1 min for 10^{-1} M amygdalin and 5 min for 10^{-4} M amygdalin. As an alternative to waiting until an equilibrium potential or current is reached, the rate of change in the current or potential (Δi or $\Delta E/\Delta t$) can be measured and equated to the concentration of substrate. This was done by Guilbault and Lubrano in the case of the glucose electrode [126], Fig. 6, and a result for glucose is obtained in 12 s.

b. Effect of enzyme concentration. The activity of enzyme in the gel will have 2 effects on an enzyme electrode: (1) it will ensure that a Nernstian calibration plot is obtained as will be discussed in (3) below, and (2) it will affect the speed of response of the electrode. However, this effect is a tricky one, in as much as an increase in the

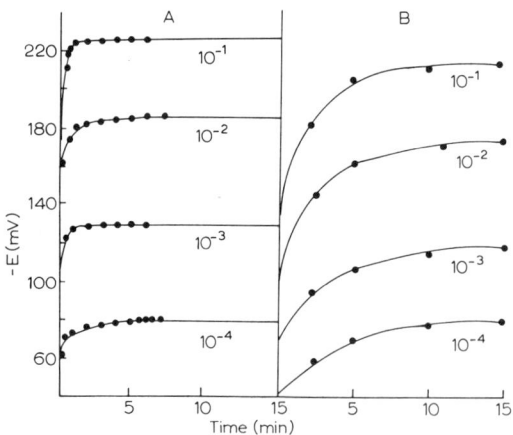

Fig. 5. Amygdalin response—time curves for an electrode containing 1 mg of β-glucosidase immobilized by a dialysis paper. A, At pH 7; B, at pH 10. (Ref. 130 by permission of authors.)

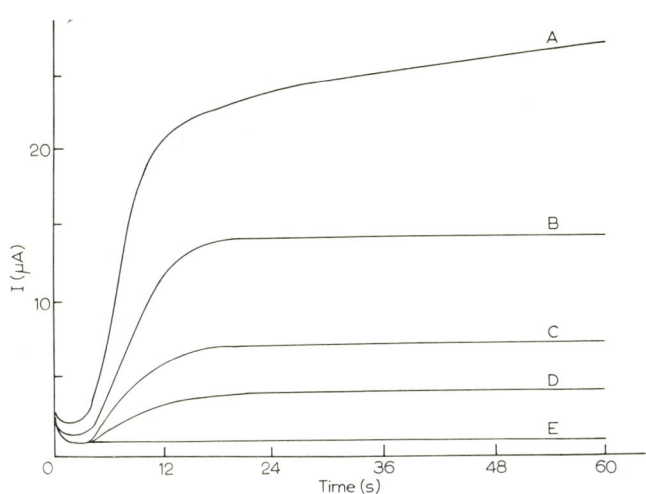

Fig. 6. Family of current—time curves for the glucose electrode poised at 0.6 V. Glucose solutions are in phosphate buffer, pH 6.0; ionic strength 0.1. A = 2.0 $\times$ 10^{-2} M; B = 1.0 $\times$ 10^{-2} M; C = 5.0 $\times$ 10^{-3} M; D = 2.5 $\times$ 10^{-3} M; E = 5.0 $\times$ 10^{-4} M. (From ref. 126 by permission of authors.)

amount of enzyme also affects the thickness of the membrane. This is demonstrated in Fig. 7, taken from the results of Mascini and Liberti [130], for the amygdalin electrode. As the amount of en-

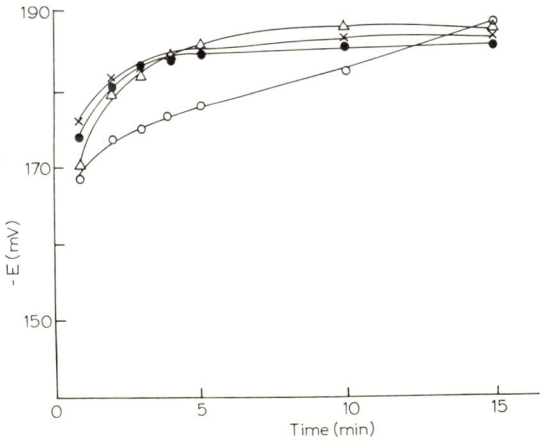

Fig. 7. Response—time curves with different amounts of enzyme. pH 7; [Amygdalin] = 10^{-3} M. ($\bullet$) 0.1 mg enzyme; ($\times$) 1 mg enzyme; ($\triangle$) 2.5 mg enzyme; ($\circ$) 5 mg enzyme. (Ref. 130 by permission of authors.)

zyme is increased from 0.1 to 2.5 mg of β-glucosidase a shorter time is observed, yet when 5 mg of enzyme was used the response time became much longer. This latter effect is due to a further thickening of the membrane layer by the use of more weight of enzyme, resulting in an increase in the time required for the substrate to diffuse through the membrane. If one weight of enzyme had been chosen and the activity of enzyme increased at constant mass, a steady increase in the rate of response would be observed, and then a gradual leveling off in response time. Hence, for best results it is recommended that as active an enzyme be used as possible, to ensure rapid kinetics, in as thin a membrane as obtainable.

c. *Effect of pH*. Every enzyme will have a maximum pH at which it is most active, and a certain range of pH in which it demonstrates some reactivity. The immobilized enzyme will have a different pH range from the range of the soluble enzyme because of its environment, as was discussed above. The pH range for immobilized glucose oxidase is about 5.8—8.0 [126] (solution enzyme 5—7), β-glucosidase about 5—8 [130]. Hence, for fastest responses one should work at pH optimum. This is not always possible, however, since the sensor electrode might not respond optimally at the pH of the enzyme reaction. Thus, a compromise is generally necessary between these 2 factors. However, one should be careful not to be trapped into forcing the enzyme system to conform with the requirements of the sensor, as was done by Rechnitz and Llenado in the case of the amygdalin electrode [136,137]. These authors tried a pH of 10, which has been shown to be optimum for the electrode sensor, the CN^- electrode. Longer response times were obtained at pH 10 (Fig. 5) compared to pH 7, since the enzyme has very little activity at this pH. Furthermore, the enzyme rapidly loses activity at high pH's (>9—10) and Rechnitz and Llenado found their immobilized enzyme electrode very unstable, changing in response curve downwards every day. Yet by working at pH 7, Mascini and Liberti [130] found their soluble enzyme electrode still useful after one week. Similar effects are noted in other studies; Guilbault and Shu [139], for example, found the response time of a glutamine electrode decreased from pH 6 to pH 5, the latter being optimum for the enzyme reaction.

Further examples of making the electrode conform to the enzyme system, instead of vice versa, are the results of Anfalt et al. [39] and Guilbault and Tarp [140] on the NH_3 sensor for the product of the urea—urease reaction. Although at the optimum pH for this enzyme reaction, 7—8.5, there is very little free NH_3 present to be sensed by a gas-type electrode, and one would predict poor results for urea assay, both groups found that the sensitivity of the sensor was more than sufficient for each measurement at these low pH's. This is partly ascribed to the fact that there is a build-up of larger amounts of product in the reaction layer than in solution, and hence, an increase in sensitivity is obtained for the sensor, which sits close to the enzyme layer.

d. Effect of temperature. One would predict a dual effect of temperature, an increase in the rate of reaction resulting in a faster response time, and also a shift in the equilibrium potential by virtue of the temperature coefficient in the Nernst and van't Hoff equations.

This was demonstrated by Guilbault and Lubrano [126] for the glucose electrode, in which the effect of temperature on the electrode response was studied from $10°C$ to $50°C$. Linear plots of the log rate and log total current vs. $1/T$ were observed as predicted by the van 't Hoff ($\ln K = \ln C - (\Delta H/RT)$, K being the equilibrium constant), and the Arrhenius ($\ln k = \ln A - (E_a/RT)$, k being the rate constant) equations. Practically, this means that the temperature of the enzyme electrode should be carefully controlled for best results, although the effect of temperature is most pronounced on reaction rate measurement. Similar effects were noted by Guilbault and Lubrano for amino acid oxidase [123].

Guilbault and Hrabankova [125] found that the response of the D-amino acid oxidase electrode to D-methionine showed only very small effects at increasing temperatures ($25°C—40°C$) although the theoretical Nernstian slope is $61.74 \text{ mV decade}^{-1}$ at $37°C$. Similarly, Papariello et al. [141] found that although the response time of his penicillin electrode was somewhat more rapid at $37°C$ than at $25°C$, no great improvement was observed.

Hence, the user of enzyme electrodes is advised to control the temperature if he is making kinetic measurements, but not to bother in making equilibrium measurements; simply use room temperature or about $25°C$ for convenience.

e. Thickness of the membrane. The time required to reach a steady-state potential or current reading is strongly dependent on the gel layer thickness. This is due to an effect on the rate of diffusion of the substrate through the membrane to the active sites of the enzyme, and on the rate of diffusion of the products through the membrane to the electrode sensor where they are measured. Guilbault and Montalvo [88] observed that the time interval for 98% of the steady-state response was about 26 s with a 60 μ thick net of urease and about 59 s with a 350 μ net for 8.33×10^{-2} M urea and an enzyme concentration of 175 mg cm^{-3} of gel. Similarly, Anfält et al. [39] in the case of a urea electrode with glutaraldehyde-bound enzyme, and Mascini and Liberti [130], using a β-glucosidase amygdalin electrode, observed an increase in response time in going from thin to thick membranes. Thus it is recommended that as thin a membrane as possible be used for best results; this can be achieved using a highly active enzyme.

f. Effect of dialysis membrane. In most cases, it is advantageous to use a dialysis membrane to cover the electrode as was previously pointed out. This membrane serves to protect the enzyme and prolong the stability of the electrode. Guilbault and Montalvo [88]

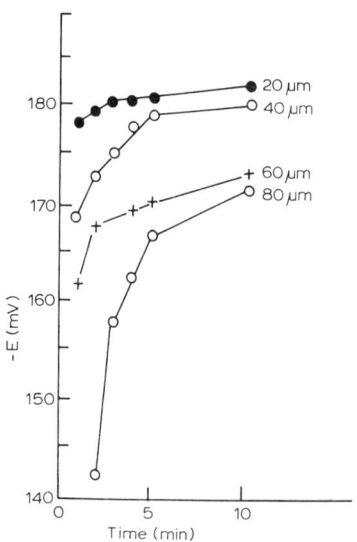

Fig. 8. Effect of thickness of dialysis paper on response time and on equilibrium values. [Amygdalin] = 10^{-2} M; enzyme amount = 1 mg. (Ref. 130 by permission of authors.)

noted that the cellophane coatings had little effect on the response time of the urea electrode.

A thorough study of the effect of the thickness of the dialysis membrane on the response rate was made by Mascini and Liberti [130] for the amygdalin electrode. Their results are shown in Fig. 8, indicating that the response time and the equilibrium values are altered by varying the thickness of the membrane. A thin 20 μm membrane (Arthur H. Thomas) or a 25 μm membrane (Will Scientific) will have essentially no effect on the response time and are recommended for use. Additional thicknesses of membrane will cause an increase in the time required for response, however.

g. Electrode base sensor. The final factor that will affect the speed of response is the electrode sensor itself, and how fast it will reach a potential or current proportional to the amount of product or reactant it sensed.

In most cases of enzyme electrodes, in which rapid stirring is used to minimize factor (1) and a thin membrane of highly active enzyme is used under optimum conditions to minimize factor (2), the determining factor will be the response time of the sensor.

Guilbault and Montalvo [88] observed a response time of 26 s with a 60 μ thick enzyme layer of urease in a urea electrode, compared to a response time of 23 s for an uncoated cation electrode. Anfält et al. [39] observed response times of their urea probe of 30 s—1 min, almost the same as those for the uncoated NH_3 gas electrode. Mascini and Liberti [130] noted a one-minute response time for their CN^- coated amygdalin electrode, at high substrate concentration (10^{-1} M), quite similar to the uncoated CN^- electrode. At low substrate concentrations (10^{-2}—10^{-5}) factor (2) becomes the rate-limiting factor on the response time.

(4) Other characteristics

a. Shape of response curve. The concentration of enzyme in the membrane will have two effects on the electrode response. The first, on the response time, was discussed above. The second is on the shape of the response curve. This is best indicated in Fig. 9, which is taken from the work of Guilbault and Montalvo [88] on the urea electrode.

To study the effect of enzyme concentration on the enzyme gel

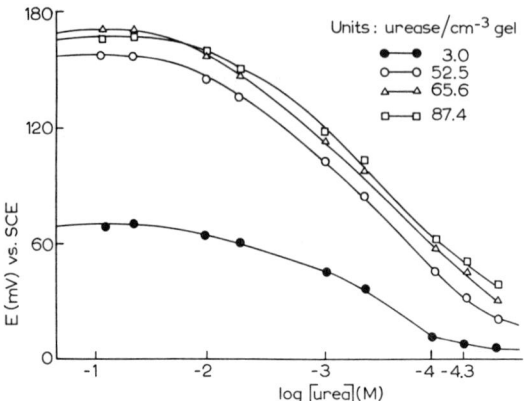

Fig. 9. Effect of enzyme concentration on electrode response; $250\,\mu$ netting. (Ref. 88 by permission of author.)

layer activity, gels were prepared with enzyme concentrations ranging from 3 to 110 units of urease cm^{-3} of gel. The steady-state response of each enzyme-coated electrode when dipped in urea solutions from 5×10^{-5} to 1.6×10^{-1} M was measured. The results are shown in Fig. 9. The slope of each curve increases with the amount of enzyme in the gel layer on the electrode until, with larger enzyme concentrations, only a small increase in activity of the gel membrane is obtained. Figure 10 is a plot of the steady-state response

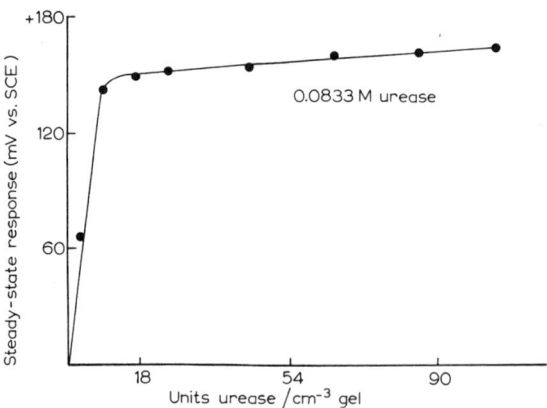

Fig. 10. Dependence of gel-layer activity on enzyme concentration, $350\,\mu$ netting. (Ref. 88 by permission of author.)

42

for 8.33×10^{-2} M urea against the amount of urease in the gel membrane. There is a rapid increase in response or activity up to 7.5 units of urease cm^{-3} of gel. Above 7.5 units of urease cm^{-3} of gel a large increase in enzyme concentration gives only a small increase in activity of the enzyme gel membrane on the cation electrode. Opttimum enzyme concentration in the gel, considering only the economy of enzyme, is obtained at about 7.5 units of urease cm^{-3} of gel.

A series of urease electrodes was prepared with the same urease concentration (65.6 units of urease cm^{-3} of gel) but with different gel compositions to determine if the activity of the gel layer depends upon the gel composition. With a 350 μ gel layer over the cation electrode, variation of the gel per cent from 5 to 17.6 at constant monomer : cross-linking ratio gave less than a 2% difference in response with 8.33×10^{-2} M urea. Variation of the per cent cross-linking material from 5 to 19 at constant gel concentration gave likewise a very small difference in response.

With a urease concentration of 65.6 units cm^{-3} of gel, the steady-state response to 8.33×10^{-2} M urea decreased by only 2% upon decreasing the gel layer thickness from 350 to 60 μ. However, the response time decreased significantly.

Similar results are obtained in the case of every enzyme electrode. The amount of enzyme to be used depends on the individual electrode, and the curve, similar to Fig. 10, that is observed. However, as a rule of thumb 10—20 units per membrane will generally be sufficient to give an excellent response curve. In the case of the more unstable "soluble" and physically entrapped enzyme electrodes, an excess of enzyme should be used (i.e., 50 units cm^{-3} of gel as shown in Fig. 10) so that a loss of enzyme will not affect the potentials observed. Likewise, purified enzymes should be used (at least 1 unit mg^{-1}) so as to keep the thickness of the membrane to a minimum for fast response rates.

b. Wash-time of the electrode. Because there is a build-up of product in the enzyme membrane, every type of enzyme electrode described requires a washing after use in order to return to the base line potential. This wash-time varies from only 20 s in the case of urease and the air-gap electrode [140] to as long as 10 min for urease with a pH electrode sensor [132]. The wash-time will increase with the thickness of the enzyme membrane, as expected, and also with the nature of enzyme and the base sensor. This latter effect is due to the charge on the enzyme, which in the case of urease, for example, is negative, with a resulting attraction for the positive NH_4^+ ions formed. The glass, likewise, has an attraction for these ions, resulting in longer wash-times being required.

The electrode can be washed in an automatic electrode washer as described by Montalvo and Guilbault [142], by simply placing the electrode in water, or by rinsing it under a distilled water tap. The type of washing will depend on the type of electrode. Soluble and physically entrapped enzymes must be washed gently since they can be easily washed out of their matrix. The more sturdy chemically bound enzymes can be washed under a water tap [140]. The latter, of course, yields a much quicker return to base line. Oxygen based electrodes [23,24] will show a quick return to base line if simply placed in a fresh buffer solution. Other electrodes, such as the Pt based amperometric enzyme electrodes [123,126,129,143], have to be pretreated before first use by applying a potential (+0.6 V vs. SCE) until the anodic current decays to a low value; after each run the electrode is washed in a stirring phosphate buffer solution until the current decays to a low value (0.5—1 min) indicating the removal of unreacted substrate and reaction products.

c. Range of substrate determinable. All enzyme electrodes sense substrate in the range of 10^{-2}—10^{-4} M, with some electrodes useful to as high as 10^{-1} M or as low as 10^{-5} M. In all cases, curves similar to those in Fig. 9 are obtained, approximately Nernstian in the linear range with a slope of close to 59.1 mV decade^{-1}. All curves level off at high substrate concentration, as predicted by the Michaelis—Menten equation, which states that the reaction becomes independent of substrate at high substrate concentration. A leveling off of the curve at low substrate concentration is also observed; this is due to the limit of detection of the electrode sensor used.

(5) Effect of interferences

Any enzyme electrode will only be as good as the selectivity it possesses. Hence, a consideration of the possible interferences is now in order. Interferences fall into 2 categories: (a) interferences in the electrode sensor, and (b) interferences with the enzyme itself.

a. Interferences in the electrode sensor. The electrode used to sense the products of, or reactants in, the enzyme reaction should be one that will not react with other substances present in the sample to be assayed, if possible.

The urea electrode probes originally described by Guilbault and Montalvo [88,133] could not be used for assay of urea in blood or urine because the cation glass sensor used to measure the NH_4^+ produced also responds to Na^+ and K^+ present in blood or urine.

A cell using a glass electrode (Beckman Electrode 39137 or 39047) as the reference electrode was tried by Guilbault and Hrabankova [134] in an attempt to eliminate the effect described above. Calibration curves for urea were found to be the same as when a cell with the SCE reference electrode was used. Also, the interferences of monovalent cations in solution are smaller with this sensor because both electrodes are sensitive to these ions. However, concentrations of Na^+ and K^+ higher than 10^{-4} M, considerably decreased the electrode response. This effect could be explained as a decrease of activity coefficients in the presence of other ions or as a decrease of enzymic reaction rate caused by a higher ionic strength. Combining the cell with the uncoated glass reference electrode with a cation exchanger, the determination of urea in blood and urine is possible with a deviation of less than 3%.

A still further improvement was described by Guilbault and Nagy [135], in which a solid antibiotic nonactin electrode was used as sensor. This electrode has a selectivity of NH_4^+/K^+ of 6.5/1 and NH_4^+/Na^+ of $7.5 \times 10^2/1$, thus partially eliminating any response to these ions by the sensor. By using a three electrode system (a chemically-bound urease over a nonactin solid electrode vs. an uncoated nonactin electrode and a calomel electrode as reference) and dilution to a constant interference level of K^+ (Na^+ does not interfere because of the high selectivity coefficient of the sensor), urea in blood was assayed with an accuracy of 2—3% or better. In this procedure a standard calibration plot of E vs. urea was prepared in the presence of KCl, at its highest level present in blood. Before each run the blood sample to be assayed was brought to the same potential as that observed in the KCl buffer solution by addition of more KCl to the sample using the uncoated electrode vs. SCE. The potential of the urease sensor vs. SCE was then read and the urea determined.

Anfält et al. [39] used an Orion NH_3 electrode, which is a glass electrode with a layer of NH_4Cl and a gas-permeable membrane over the outside, to sense the NH_3 produced from the urea—urease reaction. In theory, the use of the gas NH_3 electrode, which senses only NH_3 and not K^+, Na^+ or other ions or substrates, should provide the desired selectivity of a sensor. However, at the pH's optimum for the urea—urease reaction (7—8.5) the free NH_3 present is quite small compared to NH_4^+ ions. Yet, since the enzyme layer is placed directly onto the gas electrode there was sufficient NH_3 produced at the electrode surface to give nice linear calibration plots of pH's at 7, 7.4, 8 and 9, with slopes close to 59.1 mV decade^{-1} (actually the slope at pH 7 was 69.5 mV decade^{-1}, higher than Nernstian).

In a similar type study, Guilbault and Tarp [140] showed that the air-gap electrode for NH_3 developed by Ruzicka and Hansen [25] could be used to provide a totally specific urea assay directly at pH 8.5. The air-gap electrode is designed on a principle similar to the gas electrodes, except minus the gas membrane, which usually causes many practical problems such as long response times and clogging of the membrane pores by proteins and other substances in blood, thus tremendously decreasing the useful lifetime. In the air gap, NH_3, generated in solution by the enzyme reaction at pH 8.5, diffuses out of solution to the surface of the glass electrode which is coated with a layer of 10^{-3} M NH_4Cl and surfactant. A change in pH is observed due to the reaction

$$NH_3 + H^+ \rightleftharpoons NH_4^+ \tag{23}$$

which is monitored, and is proportional to the log urea concentration. By calibration of the response curve vs. Monitrol II (freeze dried blood serum from Dade, Miami, Fla.) the urea content of 30 blood samples from Rigshospitalet in Copenhagen were assayed with an accuracy of about 2%. The method requires only 2—4 min per assay, and using a chemically-bound urease over 500 assays per electrode are possible.

In the Pt electrode devices described for measuring oxidative enzyme systems, two approaches have been taken: to measure the peroxide produced by monitoring the total current change [126,143, 144], or the rate of change in the current [126,143] at +0.5 V vs. SCE, or to measure the uptake of oxygen by the enzymic reaction [23,24] by measuring the reduction of O_2 at —0.6 V vs. SCE. In either system, compounds present that are either oxidized or reduced at a Pt electrode at ±0.6 V vs. SCE would interfere. Clark [144] eliminated this interference by using a second uncoated Pt electrode, held at the same potential as the enzyme electrode, to measure any

46

compounds present in blood. His system, as modified, forms the basis for instruments to measure glucose sold commercially by Yellowsprings Instrument Co. (Yellowsprings, Ohio, U.S.A.) and Radiometer (Copenhagen). However, the normal values of such interferences (ascorbic acid or uric acid are examples of oxidizable compounds) are generally <5% of the glucose signal. Another way to eliminate the effect is to measure very quickly (~20 s) the rate of change in the current as was done by Guilbault and Lubrano [126, 143], who found that assays of over 200 blood samples could be performed with an accuracy and precision of better than 2%. Still better selectivity is obtained using the measurement of O_2 at -0.6 V since all other compounds present in blood that consume O_2 can be subtracted out before measurement with the enzyme sensor [23,24].

In other electrodes, such as the CN^- solid precipitate electrode used by Rechnitz and Llenado [136,137] and Mascini and Liberti [130] for amygdalin any ions capable of forming insoluble silver salts will interfere because of formation of a precipitate on the electrode surface. Also, substances capable of reducing silver ion will interfere, as will heavy metal ions and transition metal ions capable of forming cyanide complexes. In the use of an iodide sensor to measure glucose, Nagy et al. [35] found several types of interferences making this type of measurement approach very limited: interferences at the iodide electrode (thiocyanate, sulfide, CN^-, and Ag^+) and interferences from oxidizable compounds present in blood, such as uric acid, tyrosine, ascorbic acid and Fe(II), which compete in the oxidation of iodide to iodine in the peroxide—peroxidase system. These compounds had to be removed by sample pretreatment.

Finally, in the glass electrode systems for pH measurement as described by Papariello et al. [141,145] and Mosbach and co-workers [132], any acidic or basic components present would interfere in the measurement. However, by adjustment to a definite pH before initiation of the enzyme reaction, and assuming only the enzyme reaction will give rise to a pH change, these effects can be minimized or eliminated.

b. Interferences with the enzyme reaction. Such interferences fall into 2 classes: (1) substrates that can catalyze the reaction in addition to the compound to be measured, (2) substances that either activate or inhibit the enzyme.

In the case of some enzymes, such as urease, the only substrate that reacts at a reasonable rate is urea — hence, the urease coated electrode is specific for urea [88,133]. Uricase, likewise, acts almost specifically on uric acid [23]. Others, like penicillinase [132,141, 145], react with a number of substrates: ampicillin, naficillin, Penicillin G, Penicillin V, cyclibillin and dicloxacillin. All of these can be determined with a penicillinase electrode.

Similarly, D-amino acid oxidase [125] and L-amino acid oxidase [123,124] are less selective in their responses: the former in an electrode yields a good response to D-phenylalanine, D-alanine, D-valine, D-methionine, D-leucine, D-norleucine, and D-isoleucine; the latter for L-leucine, L-tyrosine, L-phenylalanine, L-tryptophan, and L-methionine. Alcohol oxidase [129,146] responds to methanol, ethanol and allyl alcohol. Hence, in use of electrodes of these enzymes, either a separation must be used if 2 or more substrates are present, or the total determined. In the case of L-amino acid assay, an attractive alternative exists, the use of decarboxylative enzymes [122] which act specifically on different amino acids. Such are known for L-tyrosine, L-phenylalanine, L-tryptophan, and others.

Glucose oxidase acts on a number of sugars [35]: glucose and 2-deoxy glucose are the main substrates, but cellubiose and maltose also react, probably due to the presence of other hydrolytic enzymes in the glucose oxidase preparation.

The activity of the enzyme can be adversely affected by the presence of certain compounds, called inhibitors. These are generally heavy metal ions, such as Ag^+, Hg^{2+}, and Cu^{2+}, and sulfhydryl reacting organic compounds, such as p-chloromercuribenzoate and phenyl mercury(II) acetate (due to their reaction with the free S—H groups present at the active site of many enzymes, especially the oxidases) [126]. One important point to realize, however, is that the immobilized enzyme is much less susceptible to inhibitors, especially weak or reversible inhibitors, due to the protection of the immobilization matrix. Thus by using the enzyme in an immobilized form, most of one's worries about inhibitors are eliminated. From personal experiences in the design and use of almost 20 different enzyme electrodes, I have not experienced a case of enzyme inhibition interfering with an assay. However, one should always be aware of this problem, especially in assaying solutions containing heavy metal ions, and especially pesticides.

3. Analytical applications of enzyme electrodes

Having now considered the methods of preparation of enzyme electrodes and their response characteristics, let us now review the current literature to see what electrodes have been designed and used.

(A) ELECTRODES FOR ASSAY OF SUBSTRATES USING IMMOBILIZED ENZYMES

(1) Glucose electrodes

The first report of an "enzyme electrode" was that of Clark and Lyons [147], who proposed that glucose could be determined amperometrically using soluble glucose oxidase held between cuprophane membranes. The oxygen uptake was measured with an O_2 electrode

$$\text{glucose} + O_2 + H_2O \xrightarrow{\text{G.O.}} H_2O_2 + \text{gluconic acid} \tag{24}$$

The term "enzyme electrode" was introduced by Updike and Hicks [131], who coated an oxygen electrode with a layer of physically entrapped glucose oxidase in a polyacrylamide gel. The decrease in oxygen pressure was equivalent to the glucose content in blood and plasma. A response time of less than a minute was observed. An instrument based on this type of approach, i.e. an immobilized glucose oxidase column and a Pt electrode is now marketed by Leeds and Northrup Instrument Co (Pennsylvania). Such a system was described by Weibel et al. [148] for the assay of glucose. These authors found that complete conversion of glucose could be accomplished in less than 60 s with immobilized glucose oxidase columns containing 400—600 μl of porous glass.

Clark [144] proposed measuring the hydrogen peroxide produced in the enzymic reaction with a Pt electrode. An instrument based on this concept, using glucose oxidase held on a filter trap, is now sold by Yellowsprings Instrument Co. (Yellowsprings, Ohio). Two platinum electrodes are used, one to compensate for any electro-oxidizable compounds in the sample, such as ascorbic acid, the second to monitor the enzyme reaction. A similar instrument is also marketed now by Radiometer (Copenhagen, Denmark).

Williams et al. [149] used quinone as the hydrogen acceptor in place of oxygen, and described enzyme electrodes for blood glucose

$$\text{glucose} + \text{quinone} + H_2O \xrightarrow{\text{G.O.}} \text{gluconic acid} + \text{hydroquinone} \qquad (25)$$

$$\text{hydroquinone} \xrightarrow{\text{Pt}} \text{quinone} + 2H^+ + 2e^- \quad (E = 0.4 \text{ V vs. SCE}) \qquad (26)$$

Using glucose oxidase trapped in a porous or gelled layer and covered with a dialysis membrane over a Pt electrode, glucose could be determined by monitoring the electro-oxidation of quinone. About 3—10 min were required to obtain a steady-state current.

Guilbault and Lubrano [126,143] described a simple, stable, rapid reading electrode for glucose. The electrode consists of a metallic sensing layer (Pt or Pt-glass [150]) covered by a thin film of immobilized glucose oxidase held in place by means of cellophane. When poised at the correct potential, the current produced is proportional to the glucose concentration. The time of measurement using this amperometric approach is less than 12 s using a kinetic method. The electrode is stable for over a year when stored at room temperature with only a 0.1% change from maximum response per day. The enzyme electrode determination of blood sugar compares favorably with commonly used methods with respect to accuracy, precision and stability, and the only reagent needed for assay is a buffer solution. An instrument based on this system is marketed by Owens (Illinois, Toledo, Ohio).

Nagy et al. [35] described a self-contained electrode for glucose based on an iodide membrane sensor

$$\text{glucose} + O_2 \xrightarrow[\text{oxidase}]{\text{glucose}} \text{gluconic acid} + H_2O_2 \qquad (27)$$

$$H_2O_2 + 2I^- + 2H^+ \xrightarrow{\text{peroxidase}} 2H_2O + I_2 \qquad (28)$$

The highly selective iodide sensor monitors the 'ocal decrease in the iodide activity at the electrode surface. The assay of glucose was performed in a flow stream and at a stationary electrode. Pretreatment of the blood sample was required to remove interfering reducing agents, such as ascorbic acid, tyrosine and uric acid.

Nilsson et al. [132] described the use of conventional hydrogen ion glass electrodes for the preparation of enzyme-pH electrodes by either entrapping the enzymes within polyacrylamide gels around the glass electrode or as a liquid layer trapped within a cellophane membrane. In an assay of glucose, based on a measurement of the gluconic acid produced, the pH response was almost linear from 10^{-1} to 10^{-3} M with a ΔpH of about 0.85 decade^{-1}. Electrodes of this type were also constructed for urea and penicillin (see below). The ionic strength and pH were controlled using a weak (10^{-3} M) phosphate buffer, pH 6.9, and 0.1 M Na_2SO_4.

(2) Urea electrodes

Guilbault and Montalvo [88,133] prepared several electrodes for urea by physically entrapping urease in a polyacrylamide gel held over the surface of a monovalent cation electrode by cellophane film.

$$\text{urea} + 2H_2O \xrightarrow{\text{urease}} 2NH_4^+ + 2HCO_3^- \tag{29}$$

The urea diffuses into the urease layer where it is converted to ammonium ions, which are sensed by the cation electrode. The electrode could be used for up to 3 weeks with no loss of activity and responded to urea in the concentration range 5×10^{-5}—1.6×10^{-1} M with a response time of about 35 s.

Because sodium and potassium ions interfered in the measurement, Guilbault and Hrabankova [134] used an uncoated NH_4^+ ion electrode as reference electrode to the urease-coated NH_4^+ electrode, and added ion exchange resin in attempts to develop a urea electrode useful for assay of blood and urine. Good precision and accuracy were obtained.

In attempts to improve the selectivity of the urea determination, Guilbault and Nagy [135] used a silicone rubber based nonactin ammonium ion-selective electrode as the sensor for the NH_4^+ ions liberated in the urease reaction. The selectivity coefficients of this electrode were 6.5 for NH_4^+/K^+; 5×10^2 for NH_4^+/Na^+ and much higher for other cations. The reaction layer of the electrode was made of urease enzyme chemically immobilized on polyacrylic gel. A still further improvement was described by Guilbault et al. [128] using a three-electrode system, which allowed dilution to a constant interference level. Analysis of blood serum showed good agreement with spectrophotometric methods, and the enzyme electrode was stable for 4 months at 4°C.

Still further improvement in the selectivity of this type of electrode was obtained by Anfält et al. [39] who polymerized urease directly onto the surface of an Orion ammonia gas electrode probe by means of glutaraldehyde. Sufficient NH_3 was produced in the enzyme reaction layer even at pH's as low as 7—8 to allow direct assay of urea in the presence of large amounts of Na^+ and K^+. A response time of 2—4 min was observed.

Guilbault and Tarp [140] described a still better, total interference-free, direct reading electrode, for urea, using the air-gap electrode of Ruzicka and Hansen [25]. A thin layer of urease chemically bound to polyacrylic acid was used, at a solution pH of 8.5, where good enzyme activity was still obtained, yet where sufficient NH_3 is liberated to yield a sensitive measurement with the air-gap NH_3 electrode. The urea diffuses into the gel; the NH_3 produced diffuses out of solution to the surface of the air-gap electrode where it is measured. A linear range of 3×10^{-2}—5×10^{-5} M was obtained with a slope of 0.75 pH unit decade^{-1}. The electrode could be used continuously with good serum analysis for up to one month (at least 500 samples) with a precision and accuracy of better than 2%. The response time was 2—4 min at pH 8.5, and the electrode was washed under a water tap for 5—10 s after each measurement. Absolutely no interference from any levels of substances commonly present in blood was observed (Na^+, K^+, NH_4^+, ascorbic acid, etc.).

Guilbault and Stakbro [151] described a new concept in the use of immobilized enzymes, the enzyme stirrer, a self-contained unit in which the enzyme is attached directly to the stirring bar. This stirrer was used for the assay of urea in blood with excellent precision and accuracy. One unit was used for about 500 assays.

A urea electrode using physically entrapped urease and a glass electrode to measure the pH change in solution was described by Mosbach et al. [132]. The response time of the electrode to urea was about 7—10 min and had a linear range from 5×10^{-5} to 10^{-2} M with a change of about 0.8 pH unit decade^{-1}. The electrode could be kept at room temperature for about 2—3 weeks. The ionic strength and pH were controlled using a weak (10^{-3} M) tris buffer and 0.1 M NaCl.

Still another possibility for a urea electrode is the use of a CO_2 sensor to measure the second product of the urea—urease reaction, HCO_3^-. Guilbault and Shu [122] evaluated the use of the CO_2 sensor and found that a urea electrode, prepared by coupling a layer of urease covered with a dialysis net to a CO_2 electrode, had a linear range of 10^{-4}—10^{-1} M, a response time of about 1—3 min and a slight response to only acetic acid. Na^+ and K^+ ions had no interference.

(3) Amino acid electrodes

The CO_2 sensor was evaluated by Guilbault and Shu for response to tyrosine when coupled with tyrosine decarboxylase held in an immobilized form by a dialysis membrane [122]. A linear range of $2.5 \times 10^{-4}-10^{-2}$ M was observed with a slightly faster response time than observed with the urea electrode mentioned above. A slope of 55 mV decade^{-1} was obtained, compared to 57 mV decade^{-1} for the urea electrode.

Enzyme electrodes for the determination of 1-amino acids were developed by Guilbault and Hrabankova [124] who placed an immobilized layer of L-amino acid oxidase over a monovalent cation electrode to detect the ammonium ion formed in the enzyme catalyzed oxidation of the amino acid. These electrodes are stable for about 2 weeks, and have a 1—2 min response time.

Two different kinds of enzyme electrodes were prepared by Guilbault and Nagy for the determination of L-phenylalanine [152]. One of the electrodes used a dual enzyme reaction layer — L-amino acid oxidase with horseradish peroxidase — in a polyacrylamide gel over an iodide-selective electrode. The electrode responds to a decrease in the activity of iodide at the electrode surface due to the enzymatic reaction and subsequent oxidation of iodide.

$$\text{L-phenylalanine} \xrightarrow{\text{L-amino acid oxidase}} H_2O_2 \tag{30}$$

$$H_2O_2 + 2H^+ + I^- \xrightarrow[\text{peroxidase}]{\text{horseradish}} I_2 + H_2O \tag{31}$$

The other electrode was prepared using a silicone rubber based nonactin type ammonium ion-selective electrode covered with L-amino acid oxidase in a polyacrylic gel. The same principle of diffusion of substrate into the gel layer, enzymatic reaction and detection of the released ammonium ion applied to this system. Linear calibration plots were also obtained for L-leucine and L-methionine in the range $10^{-4}-10^{-3}$ M.

Electrodes specific for D-amino acids, which are oxidatively catalyzed by D-amino acid oxidase were reported by Guilbault and Hrabankova [125]. The NH_4^+ ion produced is monitored with a cation electrode.

$$\text{D-amino acid} + O_2 \xrightarrow{\text{oxidase}} NH_4^+ + HCO_3^- \tag{32}$$

The stability of these electrodes could be maintained for 21 days if they are stored in a buffered flavine adenine dinucleotide (FAD) solution, since the FAD is weakly bound to the active site of the enzyme and is needed for its activity. Electrode probes suitable for the assay of D-phenylalanine, D-alanine, D-valine, D-methionine, D-leucine, D-norleucine, and D-isoleucine were developed. An electrode for asparagine was also developed using asparaginase as the catalyst [125]; no co-factor was necessary.

Guilbault and Shu [139] described an enzyme electrode for glutamine, prepared by entrapping glutaminase on a nylon net between a layer of cellophane and a cation electrode. The electrode responds to glutamine over the concentration range 10^{-1}—10^{-4} M with a response time of only 1—2 min. Guilbault and Lubrano [123] prepared an electrode for L-amino acids by coupling chemically-bound L-amino acid oxidase to a Pt electrode which senses the peroxide produced in the enzyme reaction

$$\text{L-amino acid} + O_2 + H_2O \rightarrow R\text{—COCOOH} + NH_3 + H_2O_2 \qquad (33)$$

The time of measurement is less than 12 s using a kinetic measurement of the rate of increase in current per unit time and the only reagent required is a phosphate buffer. The L-amino acids cysteine, leucine, tyrosine, phenylalanine, tryptophan and methionine were assayed.

(4) Alcohol electrodes

Alcohol oxidase catalyzes the oxidation of lower primary aliphatic alcohols.

$$RCH_2OH + O_2 \xrightarrow[\text{oxidase}]{\text{alcohol}} RCHO + H_2O_2 \qquad (34)$$

The hydrogen peroxide produced in these reactions or the oxygen consumed may be determined amperometrically with a platinum electrode as in the determination of glucose above. Guilbault and Lubrano [129] used the alcohol oxidase obtained from Basidiomycete to determine the ethanol concentration of 1 ml samples over the range 0—10 mg/100 ml, with an average relative error of 3.2% in the 0.5—7.5 mg/100 ml range. This procedure, based on a measurement of the peroxide produced, should be adequate for clinical determinations of blood ethanol since normal blood from individuals who have not ingested ethanol ranges from 40—50 mg/100 ml. Methanol is a serious interference in the procedure since the alcohol oxidase is more active for methanol than ethanol. However, the concentration of methanol in blood is negligible compared to that of ethanol.

Still better results were obtained by Guilbault and Nanjo [153] based on a measurement of the oxygen consumption in the enzymatic reaction at —0.6 V vs. SCE. Since the peroxide produced reacts with the aldehyde to produce acid, which is in turn enzymatically oxidized, the oxygen measurement provided a more selective and sensitive electrode for alcohols. Furthermore, by measuring oxygen consumption the electrode became more selective for ethanol than methanol, by over 3 orders of magnitude.

(5) Uric acid electrode

A self-contained rapid reading electrode for uric acid was described by Guilbault and Nanjo [23]. The electrode was prepared by placing a layer of glutaraldehyde bound uricase over the tip of a Beckman Pt electrode, the enzyme was then covered for support with a thin layer of dialysis membrane. The decrease in the level of dissolved oxygen in solution due to the enzymic reaction

$$\text{uric acid} + O_2 \xrightarrow{\text{uricase}} \text{allantoin } H_2O_2 + H_2O \tag{35}$$

was measured at an applied potential of —0.6 V vs. SCE. The current observed is proportional to the level of uric acid at concentrations of 10^{-5}—10^{-1} M. By measuring the initial rate of change in current an assay can be performed in less than 30 s. Further studies indicated the electrode could be used for the assay of glucose and amino acids [24].

It was found that the peroxide produced in the reaction could not be monitored at +0.6 V vs. SCE as described in the method of Guilbault and Lubrano for glucose [126,143], amino acids [123] and alcohols [129], since (1) the polarographic curves for peroxide and uric acid are too close to be separated at any pH useful for the enzyme reaction, and (2) an allantoin—peroxide complex is the product of the oxidation of uric acid, not free peroxide. Additionally, the oxygen uptake method was found to be more sensitive allowing the assay of lower concentrations of substrates.

The use of a Pt electrode rather than the Clark-type oxygen electrodes eliminates all the problems associated with gas membrane electrodes, namely slow response and blockage of the membrane by substances present in blood.

(6) Lactic acid electrode

Williams et al. [149] used ferricyanide as a hydrogen acceptor for lactic acid, and described an enzyme electrode for lactate based on the following reaction

$$\text{lactate} + Fe(CN)_6^{3-} \xrightarrow{\text{LDH}} \text{pyruvate} + 2Fe(CN)_6^{4-} \tag{36}$$

$$2Fe(CN)_6^{4-} \xrightarrow{Pt} 2Fe(CN)_6^{3-} + 2e^- \tag{37}$$

By monitoring the electro-oxidation of the ferrocyanide produced at +0.4 V vs. SCE at a Pt electrode covered with a porous or gelled layer of lactate dehydrogenase and a dialysis membrane, a current was produced proportional to the concentration of lactic acid. About 3—10 min were required for measurement. Because of the low Km value of this enzyme ($Km = 1.2 \times 10^{-3}$ M), it was necessary to dilute the sample with buffered ferricyanide. A linear plot was obtained over the range 10^{-4}—10^{-3} M.

(7) Amygdalin electrode

An electrode specific for amygdalin based on a solid-state cyanide electrode was reported by Rechnitz and Llenado [136,137]. The enzyme β-glucosidase immobilized in acrylamide gel, was used

$$\text{amygdalin} \xrightarrow{\beta\text{-glucosidase}} HCN + 2C_6H_{12}O_6 + \text{benzaldehyde} \tag{38}$$

A linear range of 5×10^{-3}—10^{-5} M was reported, with a slope of about 40 mV decade^{-1} and a response time of about 10 min at concentrations of 10^{-2} and 10^{-3} M amygdalin and 30 min at 10^{-4}—10^{-5} M. The electrode rapidly lost activity indicating an incomplete physical entrapment had been effected.

One reason for this long response time and poor stability was the high pH used (10.4), a pH at which the enzyme has low activity and is denatured. This was recognized by Mascini and Liberti [130] who improved the response time and other electrode characteristics by working at a pH of 7. The electrode was prepared by spreading the enzyme directly onto the membrane surface and covering it with a thin dialysis membrane. Since the enzyme was not immobilized a stability of less than a week was obtained and the response time was only about 1—2 min at 10^{-1}—10^{-3} M and 6 min at 10^{-4} M. Furthermore, a linear calibration was obtained from 10^{-1}—10^{-4} M with a slope of 53 mV decade^{-1} (compared to about 40 mV decade^{-1} at pH 10).

(8) Penicillin electrode

The first attempt at the design of a penicillin electrode was made by Papariello et al. [141]. The electrode was prepared by immobilizing penicillin β-lactamase (penicillinase) in a thin membrane of polyacrylamide gel molded around, and in intimate contact with, a glass (H$^+$) electrode. The increase in hydrogen ion from the penicilloic acid liberated from penicillin is measured

$$\text{penicillin} \xrightarrow{\text{penicillinase}} \text{penicilloic acid} \tag{39}$$

The response time of the electrode was very fast (<30 s) and had a slope of 52 mV decade^{-1} over the range 5×10^{-2}—10^{-4} M for sodium ampicillin. The reproducibility of the electrode was very poor, probably because no attempt was made to control the ionic strength and pH.

Mosbach and co-workers [132] prepared a penicillin electrode in a similar fashion, using polyacrylamide entrapped penicillinase on a glass (H$^+$) electrode, yet controlled the ionic strength and pH by using a weak 0.005 M phosphate buffer, pH 6.8, and 0.1 M NaCl. Good results were obtained, in comparison to the results of Papariello et al.; the calibration plot was linear from 10^{-2} to 10^{-3} M with a ΔpH of 1.4 and as little as 5×10^{-4} M sodium penicillin could be determined. The electrode could be stored for 3 weeks and the average deviation was $\pm 2\%$ with a response time of about 2—4 min.

In a later paper [145] Cullen et al. described an improved electrode for assay of penicillins. The authors claimed that a membrane must be placed between the glass electrode and the enzyme (penicillinase) layer to produce a useful electrode. Another layer of dialysis membrane over the enzyme protected its activity. Using constant ionic strength media good results were claimed for the assay of several penicillins.

(9) Acetic, formic and succinic acid electrodes

Nanjo and Guilbault [153] found that acetic acid can be assayed with the alcohol oxidase electrode developed for ethanol. In addition to acetic acid, formic and lactic acids were catalytically oxidized in the presence of the enzyme to CO_2, thus permitting the assay of any of these acids in the absence of the others. The reactivity to lactic acid could be eliminated by adding lactate dehydrogenase, however.

Nanjo et al. [154] described an interesting electrode for succinic acid, using the mitochondria of a rat, known to contain high amounts of succinate dehydrogenase. As little as 10^{-5} M succinate was assayable, based on a measurement of the O_2 uptake in the enzymatic reaction with a platinum electrode held at -0.6 V vs. SCE.

(10) Phosphate ion electrode

An enzyme electrode for phosphate ion was described by Guilbault and Nanjo [155]. A dual enzyme system was used

$$\text{glucose-6-phosphate} \xrightarrow[\text{phosphatase}]{\text{alkaline}} \text{glucose} + PO_4^{3-} \tag{40}$$

$$\text{glucose} + O_2 \xrightarrow[\text{oxidase}]{\text{glucose}} \text{gluconic acid} + H_2O_2 \tag{41}$$

The competitive inhibition by phosphate ion added causes a decrease in the consumption of dissolved oxygen by the second reaction (eqn. (41)), which can be detected amperometrically by the base electrode, a platinum disc electrode, poised at -0.6 V vs. SCE.

Besides being able to assay phosphate ion at concentrations of 10^{-2}–10^{-4} M, the dual enzyme electrode was also found useful for the assay of other oxyacids, such as arsenate, tungstate, molybdate and borate. None of the common ions present in water (i.e. halides, NO_3^-, SO_4^{2-}, etc.) interfered.

(11) Nitrate ion electrode

Hussein and Guilbault [156,157] have described the use of nitrate and nitrite reductases from *Escherichia coli* for the specific assay of nitrate (and nitrite) ions. The enzymes degrade nitrate to NH_4^+ which was sensed with an antibiotic NH_4^+ ion electrode. Assay of nitrate ion in the range 10^{-2}–10^{-4} M is possible in the presence of most other common anions.

58

(B) SUBSTRATE ELECTRODES FOR THE ASSAY OF ENZYMES

(1) Urease electrode

Attempts have been made to determine the activity of enzymes using "immobilized" substrates. Such electrodes have two limitations that the enzyme electrodes above do not have: (1) the substrate, unlike the enzyme, is used up in an assay; hence, a limiting factor will be the amount of substrate available; (2) the enzyme, since it is not consumed, will continually act on the substrate, to produce product; hence, the analysis must involve a kinetic rather than an equilibrium method.

Montalvo [158,159] designed an electrode for urease by continually passing a layer of soluble urea between the tip of an NH_4^+ cation electrode and a dialysis membrane. Urea diffuses through the membrane and is hydrolyzed by urease in the dilute aqueous solutions. The ammonium ion diffuses back through the membrane to the cation electrode where it is sensed. Although an interesting approach, it is one that lacks practicality.

(2) Cholinesterase electrode

In another study an enzyme-sensing electrode system for serum cholinesterase was prepared by coupling a pH-sensing electrode to a thin polymer membrane with a low molecular weight cut-off [160]. The electrode system utilized two thin-layer solutions to form a micro-electrochemical cell. One layer contained the serum to be assayed, the second the acetylcholine substrate which had been stabilized to balance the non-enzymatic decay of substrate by using a high molecular weight buffer. A pseudo-linear curve was obtained from $10-70$ units ml^{-1} of cholinesterase and assays could be performed in $1.5-4.5$ min.

A more practical approach to a cholinesterase substrate electrode was proposed by Gibson and Guilbault [161], who prepared the insoluble reineckate salt of acetylcholine, and placed this on the tip of a pH electrode covered with a nylon net permeable to enzyme for support. The enzyme diffuses into the substrate layer producing acid which is then sensed by the pH electrode

$$\text{acetylcholine reineckate} \xrightarrow{\text{ChE}} \text{acetic acid} + \text{choline-reineckate} \qquad (42)$$

It is believed by the author that such substrate electrodes do have a future, although a limited one, and will become generally accepted only if preparable in simple, self-contained systems, such as those developed for the enzyme electrodes.

(C) INSOLUBILIZED CO-ENZYMES IN ELECTRODES

There are a large number of enzyme systems that require a co-factor, such as NAD, FAD, etc., to function. The dehydrogenases fall into this category. In the fabrication of electrodes for substrates utilizing enzymes that need a co-factor, a large saving in cost could be achieved if the co-factor could be immobilized and included in the electrode. If co-factor activity is retained, the range of enzymes useful in electrodes would be greatly extended.

The first attempt at this was described by Davies and Mosbach [162] who incorporated co-enzymically active dextran-bound NAD^+ into an enzyme electrode with soluble glutamate dehydrogenase and soluble lactate dehydrogenase

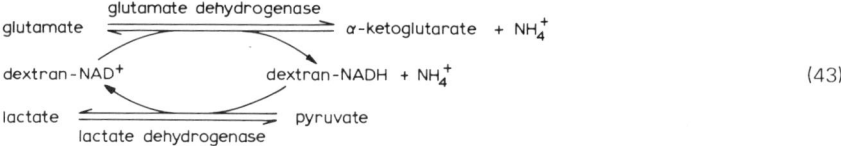

$$(43)$$

The electrode was used to determine glutamate concentrations in the range 10^{-4}—10^{-3} M according to the scheme in eqn. (43). The generation of NH_4^+ in the vicinity of the NH_4^+ sensitive electrode caused a millivolt deflection which was proportional to the log of the glutamate concentration. The dextran-NADH formed in the reaction was reconverted to dextran-NAD^+ by the lactate dehydrogenase catalyzed reaction. The enzyme electrode was also used to measure pyruvate concentrations in the range 2×10^{-5}—8×10^{-4} M.

The main disadvantage of this system is that although the co-factor is insolubilized the enzymes are soluble. If both could be insolubilized indeed a useful electrode would result, although the theoretical feasibility of both co-enzyme and enzyme being insoluble, and yet able to react, is questionable and has yet to be resolved.

(D) COMPARISON OF ELECTRODES

(1) Enzyme electrodes for substrates

We have considered all the various types of electrodes that have been described for substrates. Let us now discuss these various electrodes, compare them, and see which designs are best. First, it can be said that none of these electrodes are available as self-contained units that can be purchased separately. There are already three instruments on the market for glucose (Yellowsprings Instrument Co., Radiometer and Leeds and Northrup) as we have mentioned, and soon we can expect others for glucose and urea (Owens-Illinois, Toledo), that use an enzyme electrode as an integral part of the instrument. But there are no plans at present for sale of individual electrodes. However, these are easy to make and use as we have seen.

Many electrodes have been described for urea based on pH [132], cation (NH_4^+) [88,128,133,135], CO_2 [122] and NH_3 [39,140, 151] electrode probes. From a practical point of view, the best is the NH_3 electrode based probe [140,151] since it eliminates any interference problems from Na^+ and K^+ in solution [88,128,133,135]. With it, and a chemically-bound enzyme, over 500 assays per electrode are possible in 1—4 min/assay [140,151]. The CO_2 based sensor [122] is also quite selective, although slower in response, and has the disadvantage of necessity for compensation for the CO_2 present in blood before assay. The use of a glass (pH) electrode as probe for urea [132] is a sound idea, except that long times are required to get a stable pH change (5—10 min) and the electrode has a more limited range of concentrations in which it is useful. I believe all urea electrodes of the future will be based on some type of NH_3 sensor.

Undoubtedly, the best electrodes for glucose are those based on the use of a Pt sensor. Those based on the iodide electrode are definitely not useful, being subject to many interferences, are slow in response and have a narrow linear range [35]. The glass (pH) electrode based sensor [132], though interesting, does have a long response time, and a long wash-time, and is not as generally useful as the Pt based sensor. However, the initial work of Mosbach and co-workers [132] does represent a new vista of electrodes, in that these authors have shown it is possible to measure any acid-producing enzyme reaction with good accuracy. The concept should prove useful in the numerous enzyme systems in which a change in pH is observed.

Much credit in the field of enzyme electrodes is due to Leland Clark, one of the pioneers in this field, who described the first "enzyme" electrode, the glucose electrode, using a Pt sensor and the soluble enzyme [144,147]. This concept has led to enzyme electrode systems marketed by Yellowsprings Instrument Co. and Radiometer. He has also used the polarographic H_2O_2 monitoring system for other electrodes, for example alcohols [146], and amino acids [163], and has provided stimulation to others around him. The electrode of Williams et al. [149], based on measurement of the oxidation of hydroquinone formed when quinone replaces O_2 as the hydrogen acceptor is interesting, but only theoretically so.

The Pt based sensors, based on peroxide measurement [126,143, 144,147] and on measurement of O_2 consumption [24] are the most sound and generally useful and may form the basis for future instruments, although the use of an O_2 gas electrode with immobilized glucose oxidase [131,148] has attracted some attention by instrument builders (an instrument by Leeds and Northrup for glucose is based on such a design).

The Pt electrode [123], the gaseous CO_2 electrode [122], the cation (NH_4^+) electrode [124,125,139], the I^- electrode [152], the antibiotic NH_4^+ electrode [152] and the Pt based O_2 electrode [24] have been tried as base sensors for measurement of amino acids. Of these I believe the electrodes of the future will be based on the CO_2 or NH_3 electrodes, either regular gas type or air-gap, with the use of specific decarboxylases or lyases to build electrodes for the various amino acids. An alternative will be the Pt based sensors, with measurement of O_2, with L-amino acid oxidase as a general type probe. Using either the air-gap or the Pt sensors a fast response over a wide range is possible.

The Pt electrode based sensor for lactic acid electrode [149] is of much too limited a usefulness and will not attract any attention in its current design. However, by a direct measurement of O_2 uptake with a platinum electrode [23,24] I believe useful electrodes for all dehydrogenase enzyme systems might be feasible.

For alcohols, the Pt based sensors will be used, with the best results probably attained with the O_2 measurement at -0.6 V vs. SCE [153]. The electrode is fast in response, and more sensitive than the H_2O_2 measurement system at $+0.6$ V vs. SCE.

Penicillin electrodes based on the pH sensor will continue to attract attention, and hopefully the excellent preliminary work of Mosbach and Papariello and co-workers [132,145] will be followed up by others. The uric acid electrode based on O_2 measurement using a Pt electrode without a membrane is a fine electrode and should be widely studied [23]. Alternatively, a CO_2 sensor could be used to measure the other product of this reaction. For amygdalin the electrode of Mascini and Liberti [130] is the best; further studies with chemically-bound enzymes should give a good final product, although measurements of amygdalin are of very limited analytical usefulness.

Other electrodes in the near future will probably be a creatinine electrode using creatinase with measurement of the NH_3 liberated with an air-gap electrode or Orion gas electrode; a L-phenylalanine electrode employing a lyase to selectively cleave this amino acid to NH_3 which could be sensed by the NH_3 electrode; and a cholesterol electrode using cholesterol oxidase and the Pt sensor. The number of new sensors is limited only by the imagination.

(2) Substrate electrodes for enzymes

This is a field in which only a little work has been done, much of it of little practical use. As was discussed above, the problems with a substrate electrode are: (1) the substrate is consumed in the assay, and hence the electrode would have to be renewed periodically; (2) the enzyme will continually act on substrate to produce product, hence, a kinetic assay of the reaction must be made in a short time; (3) the substrate, if immobilized, must be tailored such that it still fits into the active site of the enzyme.

The most obvious solution to the problem is to use a sensor electrode over which fresh substrate is continually passed. However, this approach leads to a large expense (large supply of substrate) and also a final "electrode" that is anything but simple, with pumps and regulators to control substrate flow. This approach was taken by Montalvo in a urea electrode for urease [158,159] and by Guilbault et al. [29,42] and Crochet and Montalvo [160] for a cholinesterase sensing electrode.

The electrode for urease has a thin layer of urea solution trapped between the glass electrode and a dialysis membrane. Upon placing the electrode, which has urea continually pumped about the surface, in a solution containing urease, the urea diffuses out of the membrane into solution, reacts with urease, which cannot diffuse through the membrane, forming ammonium ion. This ion diffuses back through the membrane, where it is sensed by the cation glass electrode. The electrode has a slow response time, yet a curve was obtained of potential vs. urease concentration [158,159].

Guilbault and co-workers [29] described a cholinesterase assay system using a sulfide electrode and acetylthiocholine chloride as substrate. The thiol produced after reaction with cholinesterase was measured by the sulfide sensor and a potential produced was a function of the enzyme concentration

$$\text{acetylthiocholine chloride} \xrightarrow{\text{ChE}} \text{thiocholine chloride} \qquad (44)$$

The response time was about 90 s. The sulfide electrode, however, is not a very reliable one for continuous operation. Guilbault and Gibson [42] found the use of a glass electrode to monitor the acid liberated in the reaction

$$\text{acetylcholine} \xrightarrow{\text{ChE}} \text{choline + acetic acid} \qquad (45)$$

was a far better way to assay for cholinesterase and could do up to 60 assays per hour in a Technicon type apparatus.

Crochet and Montalvo [160] also described a cholinesterase sensing electrode, prepared by coupling a pH sensing electrode to a thin dialysis membrane separating two solutions. A layer of cholinesterase to be assayed is passed over the inside layer closest to the electrode and a layer of acetylcholine and polyethylenimine buffer (to react with free acid produced by non-enzymic hydrolysis) was passed over the outside layer. The acetic acid, produced when the acetylcholine diffuses from the outer layer to the inner layer and reacts with cholinesterase, is sensed by the glass electrode. The potential change, ΔE min^{-1}, is a function of the cholinesterase concentration. The electrode has a response time of 1.5—5 min, and requires 2 pumps, one for enzyme injection, the second for a continuous substrate layer.

Although interesting, all of the above work has little practical value, since an enzymic assay can be performed by other techniques with simpler equipment, cheaper and faster. If, however, the substrate could be immobilized over a sensor electrode, just like the enzyme is in enzyme electrodes, and the enzyme could react with it yielding products that can be quickly sensed, then a practical electrode would result. The first attempts at such a system were made by Gibson and Guilbault [161], who placed substrate, immobilized as either an insoluble salt (acetylcholine reineckate) or on an ion exchange resin (acetylcholine$^+$ or acetylthiocholine$^+$ resin$^-$). The enzyme reacts with the substrate, producing acid which is sensed by a pH electrode or $-$SH which is sensed by a sulfide electrode, and the rate of potential change, ΔE min^{-1}, is equated to enzyme concentration.

If any future work on substrate electrodes follows, it will have to be based on "immobilized" substrate to be of any practical use, and of advantage over existing methods. However, because of the necessity of periodic replacement of the substrate membrane, this field will not blossom as widely as the enzyme electrode area.

4. The future of enzyme electrodes

The future of enzyme electrodes is bright. Many electrodes have already been developed, and it is quite likely we shall see many more described in the near future. The existence of hundreds of enzyme systems and many good sensing probes makes it quite possible that electrodes can be produced for most of the inorganic and organic substances of importance in the areas of clinical, environmental, medical, and agricultural chemistry, to name but a few. The electrodes are so simple to prepare, that it is hoped that this descriptive chapter will provide the stimulus to others to try to develop new sensors.

A major problem in this area so far has been the lack of commercial products available to all, based on the ideas so far developed. Instead of marketing self-contained electrodes which many would want to buy and use, most companies are working towards total instruments, based on the "enzyme electrode" as the central ingredient (i.e., Yellowsprings Instrument Co., Radiometer, Leeds and Northrup, Owens-Illinois). It is the author's hope that at least one company will introduce a line of individual electrodes. In the meantime, the concepts are so simple, that the reader is encouraged to make his own probes.

5. Acknowledgement

The financial assistance of the National Science Foundation (Grant No. GP-31518) and the National Institutes of Health (Grant

No. 1 R01 GM17268) in support of the experimental work of this author in the area of enzyme and substrate electrodes is gratefully acknowledged.

References

1 G. Baum, Anal. Biochem., 39 (1971) 65; 42 (1971) 487.
2 E.W. White, F. McCapra and G.F. Field, J. Amer. Chem. Soc., 85 (1963) 337.
3 S.P. Colowick and N.O. Kaplan (Eds.), Methods of Enzymology, Academic Press, New York, 1957, p. 107.
4 G.G. Guilbault, Handbook of Enzymic Methods of Analysis, Marcel Dekker, New York, 1976.
5 M. Cremer, Z. Biol., 47 (1906) 562.
6 F. Haber and Z. Klemasiewicz, Z. Phys. Chem., 67 (1909) 385.
7 G. Karremas and G. Eisenman, Bull. Math. Biophys., 24 (1962) 413.
8 O.K. Stephanova, M.M. Shultz, E.A. Matarova and B.P. Nicolsky, Vestn. Leningrad. Univ., 4 (1963) 93.
9 E. Pungor, Anal. Chem., 39 (13) (1967) 28A.
10 G.A. Rechnitz, Chem. Eng. News, 45 (1967) 146.
11 R.A. Durst (Ed.), Ion-Selective Electrodes, Special Publication 314, National Bureau of Standards, Washington, D.C., 1969.
12 G.J. Moody and J.D.R. Thomas, Talanta, 19 (1972) 623.
13 G.J. Moody, R.B. Oke and J.D.R. Thomas, Analyst (London), 95 (1970) 910.
14 G.H. Griffiths, G.J. Moody and J.D.R. Thomas, Analyst (London), 97 (1972) 420.
15 J.E.W. Davies, G.J. Moody and J.D.R. Thomas, Analyst (London), 97 (1972) 87.
16 J. Pick, K. Toth, E. Pungor, M. Vasak and W. Simon, Anal. Chim. Acta, 64 (1973) 477.
17 R.M. Garrels, in G. Eisenmann (Ed.), Glass Electrodes for Hydrogen and Other Cations, Marcel Dekker, New York, 1967, p. 344.
18 R.N. Khuri, in R.A. Durst (Ed.), Ion-Selective Electrodes, Special Publication 314, National Bureau of Standards, Washington, D.C., 1969, p. 287.
19 D. Ammann, E. Pratsch and W. Simon, Anal. Lett., 5 (1972) 843.
20 L. Pioda, M. Wachter, R. Dohner and W. Simon, Helv. Chim. Acta, 50 (1967) 1373.
21 G.G. Guilbault and G. Nagy, Anal. Chem., 45 (1973) 417.
22 E. Eyal and G.A. Rechnitz, Anal. Chem., 43 (1971) 1090.
23 G.G. Guilbault and M. Nanjo, Anal. Chem., 46 (1974) 1769.
24 G.G. Guilbault and M. Nanjo, Anal. Chim. Acta, 73 (1974) 367.
25 J. Ruzicka and E.H. Hansen, Anal. Chim. Acta, 69 (1974) 129.
26 G.D. Christian, Advan. Biomed. Eng. Med. Phys., 4 (1971) 95.
27 A.H. Kadish and D.A. Hall, Clin. Chem., 9 (1965) 869.
28 Y. Makino and K. Koono, Rinsho Byori, 15 (1967) 391.
29 W.R. Hussein, L.H. vonStorp and G.G. Guilbault, Anal. Chim. Acta, 61 (1972) 89.

30 R.A. Llenado and G.A. Rechnitz, Anal. Chem., 44 (1972) 1366.
31 L.H. vonStorp and G.G. Guilbault, Anal. Chim. Acta, 62 (1972) 425.
32 R.A. Llenado and G.A. Rechnitz, Anal. Chem., 45 (1973) 826.
33 R.A. Llenado and G.A. Rechnitz, Anal. Chem., 44 (1972) 468.
34 R.A. Llenado and G.A. Rechnitz, Anal. Chem., 45 (1973) 2165.
35 G. Nagy, L.H. vonStorp and G.G. Guilbault, Anal. Chim. Acta, 66 (1973) 443.
36 W.F. Gutknecht and G.G. Guilbault, Environ. Lett., 2 (1971) 51.
37 G.G. Guilbault, R.K. Smith and J.G. Montalvo, Anal. Chem., 41 (1969) 600.
38 J. Montalvo, Anal. Chim. Acta, 65 (1973) 189.
39 T. Anfält, A. Granelli and D. Jagner, Anal. Lett., 6 (1973) 969.
40 E. Hansen and J. Ruzicka, Anal. Chim. Acta, 72 (1974) 353.
41 G. Baum, F. Ward and S. Yaverbaun, Clin. Chim. Acta, 36 (1972) 406.
42 G.G. Guilbault and K. Gibson, Anal. Chim. Acta, 76 (1975) 245.
43 P.V. Sundaram, A. Tweedale and K.J. Laidler, Can. J. Chem., 48 (1970) 1498.
44 G.J.H. Melrose, Rev. Pure Appl. Chem., 21 (1971) 83.
45 E. Katchalski, in P. Desnuelle (Ed.), Structure and Function of Related Proteolytic Enzymes, Academic Press, New York, 1970, p. 198.
46 K. Mosbach, Sci. Amer., 224 (1971) 26.
47 J. Gryszkiewicz, Folia Biol. (Warsaw), 19 (1971) 119.
48 A.N. Emery and C.A. Kent, Birmingham Univ. Chem. Eng., 21 (1970) 71.
49 L. Goldstein, in D. Perlman (Ed.), Fermentation Advances, Academic Press, New York, 1969, p. 391.
50 L. Goldstein and E. Katchalski, Fresenius Z. Anal. Chem., 243 (1968) 375.
51 A.H. Sehon, Symp. Ser. Immunobiol. Stand., 4 (1967) 51.
52 I.H. Silman and E. Katchalski, Ann. Rev. Biochem., 35 (1966) 873.
53 G. Manecke, Naturwissenschaften, 51 (1964) 25.
54 I. Chibata and T. Tosa, Kagaku To Seibutsu, 7 (1969) 147.
55 T.M.S. Chang, Science, 146 (1964) 524.
56 T.M.S. Chang, Ph.D. Thesis, McGill University, Montreal, 1965.
57 T.M.S. Chang, F.C. MacIntosh and S.G. Mason, Can. J. Physiol. Pharmacol., 44 (1966) 115.
58 T.M.S. Chang, Sci. J., (July 1967) 62.
59 T.M.S. Chang, Sci. Tools, 16 (1969) 35.
60 T.M.S. Chang, Biochem. Biophys. Res. Commun., 44 (1961) 1531.
61 P.R. Rony, Biotechnol. Bioeng., 13 (1971) 431.
62 J.P. Hammel and B.S. Anderson, Arch. Biochem. Biophys., 112 (1965) 443.
63 G. Lindau and B. Rhodius, Z. Physik. Chem., Abt. A, 172 (1935) 321.
64 E.S. Vorobeva and O.M. Poltorak, Vestn. Mosk. Univ., Khim., 21 (1966) 17.
65 M.G. Goldfeld, E.S. Vorobeva and O.M. Poltorak, Zh. Fiz. Khim., 40 (1966) 2594.
66 E.F. Gale and H.M.R. Eppe, Biochem. J., 38 (1944) 232.
67 A.Y. Nikolayev, Biokhimiya, 27 (1962) 843.
68 T. Tosa, T. Mori, N. Fuse and I. Chibata, Biotechnol. Bioeng., 9 (1967) 603.
69 G. Hamoir, Experimentia, 2 (1946) 257.
70 E.F. Jansen and A.C. Olsen, Arch. Biochem. Biophys., 129 (1969) 221.

71 R. Haynes and K.A. Walsh, Biochem. Biophys. Res. Commun., 36 (1969) 235.
72 A.F.S.A. Haboeb, Arch. Biochem. Biophys., 119 (1967) 264.
73 F.M. Richards, Annu. Rev. Biochem., 32 (1963) 268.
74 I.H. Silman, M. Albu-Weissenberg and E. Katchalski, Biopolymers, 4 (1966) 441.
75 R. Goldman, H.L. Silman, S.R.Caplan, O. Kedem and E. Katchalski, Science, 150 (1965) 758.
76 F.J. Wold, Biol. Chem., 236 (1961) 106.
77 G. Manecke and G. Gunzel, Naturwissenschaften, 54 (1967) 647.
78 H. Zahn and H. Meienhofer, Makromol. Chem., 26 (1958) 126, 153.
79 D.J. Herzig, A.W. Rees and R.A. Day, Biopolymers, 2 (1964) 349.
80 H.F. Schick and S.J. Singer, J. Biol. Chem., 236 (1961) 2447.
81 G. Kay and E.M. Crook, Nature (London), 216 (1967) 514.
82 B.P. Surinov and S.E. Manoylov, Biokhimiya, 31 (1966) 387.
83 P. Bernfeld and J. Wan, Science, 142 (1963) 678.
84 P. van Duijn, E. Pascoe and M. van der Ploog, J. Histochem. Cytochem., 15 (1967) 631.
85 W.N. Arnold, Arch. Biochem. Biophys., 113 (1966) 451.
86 G.P. Hicks and S.J. Updike, Anal. Chem., 38 (1966) 726.
87 G.R. Penzer and G.K. Radda, Nature (London), 213 (1967) 251.
88 G.G. Guilbault and J.G. Montalvo, J. Amer. Chem. Soc., 92 (1970) 2533.
89 K. Mosbach, Acta Chem. Scand., 24 (1970) 2084.
90 E.K. Bauman, L.H. Goodson, G.G. Guilbault and D.N. Kramer, Anal. Chem., 37 (1965) 1378.
91 G.G. Guilbault and D.N. Kramer, Anal. Chem., 37 (1965) 1675.
92 F.L. Aldrich, V.R. Usdin and B.M. Vasta, U.S. Army Report DA-18-108-405 CML-828, 1963.
93 S.N. Pennington, H.D. Brown, A.B. Patel and C.O. Knowles, Biochim. Biophys. Acta, 167 (1968) 479.
94 H.D. Brown, A.B. Patel and S.K. Chattopadhyay, J. Biomed. Mater. Res., 2 (1968) 231.
95 H.H. Weetall, Nature (London), 223 (1969) 959.
96 H.H. Weetall, Science, 166 (1969) 615.
97 J.K. Inman and H.M. Dintzis, Biochemistry, 8 (1969) 4074.
98 S.A. Barker, P.J. Somers, R. Epton and J.V. McLaren, Carbohyd. Res., 14 (1970) 287.
99 R.P. Patel, D.V. Lopiekes, S.R. Brown and S. Price, Biopolymers, 5 (1967) 577.
100 B.F. Erlander, M.F. Isambert and A.M. Michelson, Biochem. Biophys. Res. Commun., 40 (1970) 70.
101 A.B. Patel, S.N. Pennington and H.D. Brown, Biochim. Biophys. Acta, 178 (1969) 26.
102 W.E. Hornby, H. Filippuson and A. McDonald, FEBS Lett., 9 (1970) 8.
103 W.E. Hornby and H. Filippuson, Biochim. Biophys. Acta, 220 (1970) 343.
104 D.J. Inman and W.E. Hornby, Biochem. J., 129 (1972) 255.
105 K.P. Wheller, B.A. Edwards and R. Whittam, Biochim. Biophys. Acta, 191 (1969) 187.

106 G.R. Craven and V. Gupta, Proc. Nat. Acad. Sci. U.S.A., 67 (1970) 1329.
107 K. Mosbach and B. Mattiasson, Acta Chem. Scand., 24 (1970) 2093.
108 D. Gabel, P. Vretbald, R. Axen and J. Porath, Biochim. Biophys. Acta, 214 (1970) 561.
109 R. Axen and J. Porath, Nature (London), 210 (1966) 367.
110 L. Goldstein, Methods Enzymol., 19 (1970) 935.
111 M.L. Green and G. Crutchfield, Biochem. J., 115 (1969) 183.
112 D. Gabel and B. Hofsten, Eur. J. Biochem., 15 (1970) 410.
113 G. Kay and M.D. Lilly, Biochim. Biophys. Acta, 198 (1970) 276.
114 B.L. Vallee and J.F. Riordam, Ann. Rev. Biochem., 38 (1969) 733.
115 J. Sir Ram, M. Bier and P.H. Maurer, Advan. Enzymol., 24 (1962) 105.
116 H. Fraenkel-Conrat, in P.D. Boyer, H. Lardy and K. Myrback (Eds.), The Enzymes, Vol. 1, Academic Press, New York, p. 589.
117 G. Lubrano, "Amperometric Enzyme Electrodes." Ch. 4, p. 34, Thesis submitted to Louisiana State University in New Orleans for Ph.D. thesis (1973).
118 A. Bar Eli and E. Katchalski, J. Biol. Chem., 238 (1963) 1690.
119 E. Katchalski, in A.S. Hofman (Ed.), Solid Phase Proteins: Their Preparation, Properties and Application, Battelle Seattle Research Center, 1971, p. 1.
120 C. Long (Ed.), Biochemists Handbook, Van Nostrand, Princeton, 1961.
121 G.G. Guilbault, R. McQueen and S. Sadar, Anal. Chim. Acta, 45 (1969) 1.
122 G.G. Guilbault and F. Shu, Anal. Chem., 44 (1972) 2161.
123 G.G. Guilbault and G.J. Lubrano, Anal. Chim. Acta, 69 (1974) 183.
124 G.G. Guilbault and E. Hrabankova, Anal. Lett., 3 (1970) 53.
125 G.G. Guilbault and E. Hrabankova, Anal. Chim. Acta, 56 (1971) 285.
126 G.G. Guilbault and G.J. Lubrano, Anal. Chim. Acta, 64 (1973) 439.
127 A. Johansson, J. Lundberg, B. Mattiasson and K. Mosbach, Biochim. Biophys. Acta, 304 (1973) 217.
128 G.G. Guilbault, G. Nagy and S.S. Kuan, Anal. Chim. Acta, 67 (1973) 195.
129 G.G. Guilbault and G.J. Lubrano, Anal. Chim. Acta, 69 (1974) 189.
130 M. Mascini and A. Liberti, Anal. Chim. Acta, 68 (1974) 177.
131 S.J. Updike and G.P. Hicks, Nature (London), 214 (1967) 986.
132 H. Nilsson, A. Akerlund and K. Mosbach, Biochim. Biophys. Acta, 320 (1973) 529.
133 G.G. Guilbault and J.G. Montalvo, J. Amer. Chem. Soc., 91 (1969) 2164.
134 G.G. Guilbault and E. Hrabankova, Anal. Chim. Acta, 52 (1970) 287.
135 G.G. Guilbault and G. Nagy, Anal. Chem., 45 (1973) 417.
136 G.A. Rechnitz and R. Llenado, Anal. Chem., 43 (1971) 283.
137 G.A. Rechnitz and R. Llenado, Anal. Chem., 43 (1971) 1457.
138 W.J. Blaedel, T.R. Kissel and R.C. Bogaslaski, Anal. Chem., 44 (1972) 2030.
139 G.G. Guilbault and F.R. Shu, Anal. Chim. Acta, 56 (1971) 333.
140 G.G. Guilbault and M. Tarp, Anal. Chim. Acta, 73 (1974) 355.
141 G.J. Papariello, A.K. Mukherji and C.M. Shearer, Anal. Chem., 45 (1973) 790.
142 J.M. Montalvo and G.G. Guilbault, Anal. Chem., 41 (1969) 1897.
143 G.G. Guilbault and G.J. Lubrano, Anal. Chim. Acta, 60 (1972) 254.
144 L.C. Clark, U.S. Patent 3,539,455 (1970).

145 L.F. Cullin, J.F. Rusling, A. Schleifer and G.V. Papariello, Anal. Chem., 46 (1974) 1955.
146 L. Clark, Biotechnol. Bioeng. Symp. No. 3 (1972) 377.
147 L.C. Clark and C. Lyons, Ann. N.Y. Acad. Sci., 102 (1962) 29.
148 M.K. Weibel, W. Dritschilo, H. Bright and A. Humphrey, Anal. Biochem., 52 (1973) 402.
149 D.L. Williams, A.R. Doig and A. Korosi, Anal. Chem., 42 (1970) 118.
150 G.G. Guilbault, G.J. Lubrano and D. Gray, Anal. Chem., 45 (1973) 2255.
151 G.G. Guilbault and W. Stakbro, Anal. Chim. Acta, 76(1) (1975) 237.
152 G.G. Guilbault and G. Nagy, Anal. Lett., 6 (1973) 301.
153 M. Nanjo and G.G. Guilbault, Anal. Chim. Acta, 75(2) (1975) 169.
154 M. Nanjo, T. Billedeaux and G.G. Guilbault, unpublished results.
155 G.G. Guilbault and M. Nanjo, Anal. Chim. Acta, 78 (1975) 69.
156 W.R. Hussein and G.G. Guilbault, Anal. Chim. Acta, 72 (1974) 381.
157 W.R. Hussein and G.G. Guilbault, Anal. Chim. Acta, 76 (1975) 183.
158 J.G. Montalvo, Anal. Biochem., 38 (1970) 359.
159 J.G. Montalvo, Anal. Chem., 42 (1969) 2093.
160 K.L. Crochet and J.G. Montalvo, Anal. Chim. Acta, 66 (1973) 259.
161 K. Gibson and G.G. Guilbault, Insolubilized Substrates for Assay of Cholinesterase. Unpublished Results, The Technical University of Denmark, 1974.
162 P. Davies and K. Mosbach, Biochim. Biophys. Acta, (1974).
163 L. Clark, Proc. Int. Union Physiological Sciences, Vol. 9, 1971.
164 N. Larsen, E. Hansen, G.G. Guilbault, Anal. Chim. Acta, 79 (1975) 155.

Chapter II

Molecular fluorescence spectroscopy

G.G. GUILBAULT

1. Introduction to luminescence

(A) HISTORY OF LUMINESCENCE

Luminescence is one of the oldest and most established analytical techniques, having been first observed by Monardes in 1565 from an extract of *Ligirium nephiticiem*. Sir David Brewster noted the red emission from chlorophyll in 1833, and Sir G.G. Stokes described the mechanism of the absorption and emission process in 1852. Stokes also named fluorescence after the mineral fluorspar (Latin fluo = to flow + spar = a rock), which exhibits a blue—white fluorescence.

Phosphorescence dates back to the early 1500's, being so named after the Greek word for "light bearing". In fact, the element phosphorus was named from this same Greek word in 1669 since it was found to produce a bright light in a dark room.

Luminescence is one of the most active research fields in science today, as evidenced by the increasing number of papers, reviews, and monographs published each year. Fluorescence, phosphorescence, chemiluminescence, and atomic fluorescence provide some of the most sensitive and selective methods of chemical analysis.

Some of the better general references in luminescence spectroscopy are listed at the end of this chapter. Books by Guilbault [1], Hercules [2], Passwater [3], Phillips and Elevitch [4], Udenfriend [5,6], White and Argauer [7], Konstantinova-Shlesinger [8], and Pringsheim [9] are worth reading. Excellent chapters by Weissler and White on fluorescence appear in the Handbook of Analytical Chemistry [10] and in Scott's Standard Methods of Chemical Analysis [11]. Workers in fluorescence should also consult the reviews [12] by White in Analytical Chemistry every two years and receive the free pamphlets published monthly by the American Instrument Company [13] and G.K. Turner Associates [14].

(B) THEORY OF LUMINESCENCE

When a quantum of light impinges on a molecule it is absorbed in about 10^{-5} s and an electronic transition can take place to a higher electronic state (Fig. 1). This absorption of radiation is highly specific, and radiation of a particular energy is absorbed only by a characteristic structure.

In the ground state of most molecules each orbital electron in the lower energy levels is paired with another electron whose spin is opposite to its own spin. Such a state is called a singlet state, S. When the molecule absorbs radiation the electron is raised to an upper excited singlet state, S_1, S_2, and so forth. These singlet transitions are

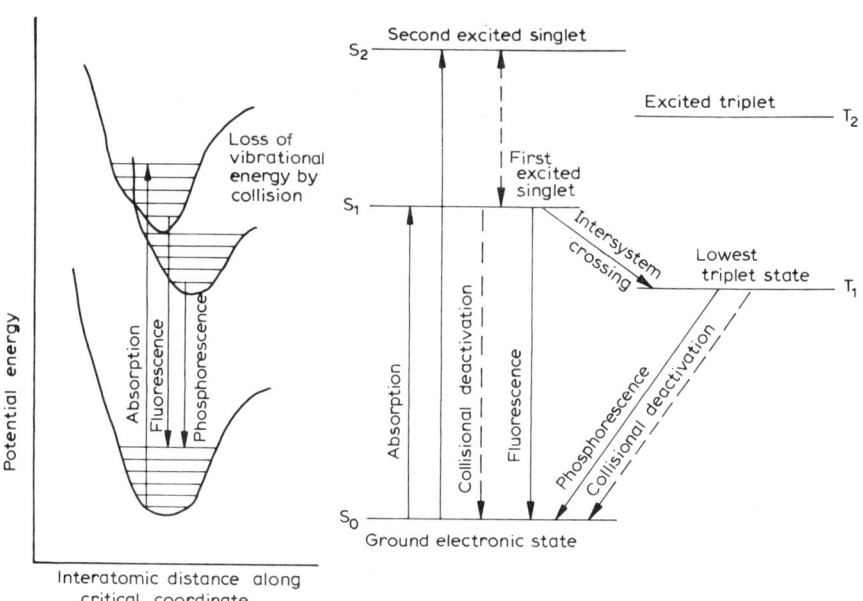

Fig. 1. Schematic energy-level diagram for a diatomic molecule.

responsible for the visible and UV absorption spectra observed for molecules. The absorption transitions usually originate in the lowest vibrational level of the ground electronic state.

During the time the molecule can spend in the excited state, 10^{-4}—10^{-8} s, some energy in excess of the lowest vibrational energy level is rapidly dissipated. The lowest vibrational level ($v = 0$) of the excited singlet state S, is attained. If all the excess energy is not further dissipated by collisions with other molecules, the electron returns to the ground electronic state with the emission of energy. This phenomenon is called fluorescence. Because some energy is lost in the brief period before emission can occur, the emitted energy (fluorescence) is of longer wavelength than the energy that was absorbed.

The phenomenon of phosphorescence involves an intersystem crossing or transition from the singlet to the triplet state. A triplet state results when the spin of one electron changes so that the spins are the same or unpaired. The transition from the lowest triplet state, T, to the singlet ground state is quantum mechanically "not allowed", consequently, transition times of 10^{-4}—10 s are observed. Hence, a characteristic feature of phosphorescence is an afterglow, i.e., emission which continues after the exciting source is removed. Because of the relatively long lifetime of the triplet state, molecules in this state are much more susceptible to radiationless deactivation processes, and only those substances dissolved in a rigid medium phosphoresce.

(C) TYPES OF FLUORESCENCE AND EMISSION PROCESSES

The fluorescence normally observed in solutions is called Stokes fluorescence. This is the re-emission of less energetic photons, which have a longer wavelength (lower frequency) than the absorbed photons. If thermal energy is added to an excited state or a compound has many highly populated vibrational energy levels, emission at shorter wavelengths than those of absorption occurs. This is anti-Stokes fluorescence, often observed in dilute gases at high temperatures. A common example is the green emission from copper-activated cadmium sulfide excited by red light.

Resonance fluorescence is the re-emission of photons possessing the same energy as the absorbed photons. This type of fluorescence is never observed in solution because of solvent interactions, but it does occur in gases and crystals. It is also the basis of atomic fluorescence. Atomic fluorescence spectroscopy is an excellent technique for the assay of many elements.

If an electron is excited by an absorbed photon of energy to a higher vibrational level with no electronic transition, energy is entirely conserved and a photon of the same energy is re-emitted within 10^{-15} s as the electron returns to its original state. The emitted light has the same wavelength as the exciting light since the absorbed and emitted photons are of the same energy. The emitted light is referred to as Rayleigh scattering and occurs at all wavelengths. Its intensity, however, varies as the fourth power of the wavelength, so its effect can be minimized by working at longer wavelengths. It is a problem when the intensity of fluorescence is low in comparison with the exciting radiation and when the absorption and fluorescence spectra of a substance are close together.

Another form of scattering emission related to Rayleigh scattering is the Raman effect. Raman scatter appears in fluorescence spectra at higher and lower wavelengths (the former being more common) than the Rayleigh-scatter peak, and these Raman bands are satellites of the Rayleigh-scatter peak with a constant frequency difference from the exciting radiation. These bands are due to vibrational energy being added to, or subtracted from, this excitation photon. The Raman bands are much weaker than the Rayleigh-scatter peak but become significant when high intensity sources are used. The relationship between the fluorescence band, Rayleigh scatter, and Raman scatter is shown in Fig. 2.

(D) EXCITATION SPECTRUM

Any fluorescent molecule has two characteristic spectra: the excitation spectrum (the relative efficiency of different wavelengths of exciting radiation to cause fluorescence) and the emission spectrum (the relative intensity of radiation emitted at various wavelengths).

The shape of the excitation spectrum should be identical to that of the absorption spectrum of the molecule and independent of the wavelength at which fluorescence is measured. This is seldom the case, however, the differences being due to instrumental artifacts. Examination of the excitation spectrum indicates the positions of the absorption spectrum that give rise to fluorescence emission. The excitation spectrum of the Al chelate of Acid Alizarin Garnet R (Fig.

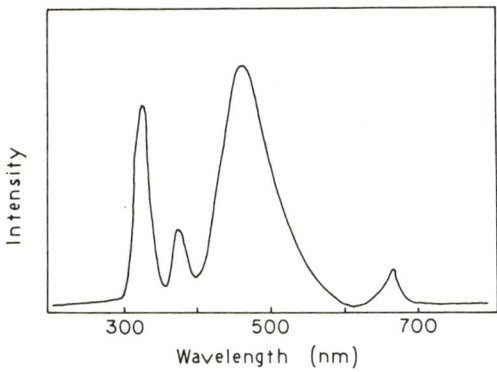

Fig. 2. Fluorescence spectra of quinine sulfate in 0.1 N sulfuric acid (λ_{ex} = 320 nm). Peaks: 320 nm, Rayleigh scatter; 360 nm, Raman scatter of water; 450 nm, quinine fluorescence; 640 nm, second-order Rayleigh scatter; 720 nm, second-order Raman scatter.

3), for example, indicates peaks at 350, 430 and 470 nm. The absorption spectrum exhibits peaks at 270, 350 and 480. To obtain the true or "corrected" spectra of this compound the apparent curve would have to be corrected for changes with frequency of (a) the photomultiplier, (b) the band width of the monochromator and (c) the changing transmission of the monochromator.

A general rule of thumb is that the longest wavelength peak in the excitation spectra is chosen for excitation of the sample. This minimizes possible decomposition caused by the lower wavelength higher energy radiation.

(E) EMISSION SPECTRUM

The emission or fluorescence spectrum of a compound results from the re-emission of radiation absorbed by that molecule. The quantum efficiency and the shape of the emission spectrum are independent of the wavelength of exciting radiation. If the exciting radiation is at a wavelength different from the wavelength of the absorption peak, less radiant energy will be absorbed and hence less will be emitted. The emission spectrum of the Al Acid Alizarin Garnet R complex indicates a fluorescence peak at 580 nm (Curve C, Fig. 3).

Each absorption band to the first electronic state will have a corresponding emission or fluorescence band. These 2 bands or spectra

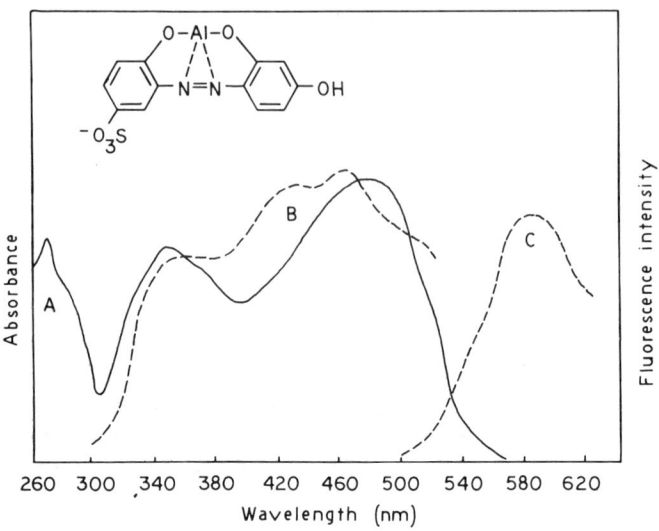

Fig. 3. Aluminum complex with Acid Alizarin Garnet R (0.008%). Curve A, the absorption spectrum; Curve B, the fluorescence excitation spectrum; and Curve C, the fluorescence emission spectrum.

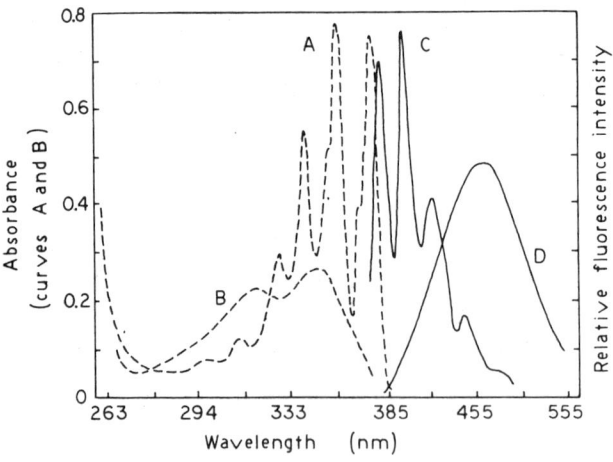

Fig. 4. Absorption and fluorescence spectra of anthracene (in ethanol) and quinine (in 0.1 N sulfuric acid). Curve A, anthracene absorption; Curve B, quinine absorption; Curve C, anthracene fluorescence; Curve D, quinine fluorescence.

will be approximately mirror images of each other. In fact, this mirror image principle is useful in distinguishing whether an absorption band is another vibrational band in the first excited state or a higher electronic level. Additional fluorescence peaks other than the mirror image of the absorption spectrum indicate scatter or the presence of impurities. Rayleigh and Tyndall scatter could be observed in the emission spectrum at the same wavelength as the excitation wavelength, and also at twice this value (second-order grating effect). At very dilute solutions one may also observe Raman scatter. The wider the fluorescence band is, the more complex and less symmetrical the compound.

Figure 4 shows the absorption and emission spectra of anthracene and quinine. Four major absorption peaks are observed in the anthracene spectrum — all correspond to transitions from S_0 to S_1^*, but denote transitions to different vibrational levels. Four major emission peaks, each a mirror image of the peaks in the absorption spectrum, are likewise observed. For quinine 2 excitation peaks are observed, one at 250 nm corresponding to a $S_0 \rightarrow S_2^*$ transition, and a second at 350 nm corresponding to a $S_0 \rightarrow S_1^*$ transition. Only one emission peak, corresponding to the $S_1 \rightarrow S_0$ transition, is observed.

The fact that some compounds possess several excitation and/or emission peaks is of analytical usefulness. If 2 compounds have overlapping excitation bands, as in the case of anthracene and quinine, both could be excited together, then differentiated by their emission spectra. Quinine could be measured at a λ_{em} of 450 nm, whereas anthracene could be monitored at a λ_{em} of 400 nm. Similarly, if 2 compounds emit radiation at the same wavelength, they could still be measured together in the same solution, if they had different, non-overlapping, excitation peaks. This, in fact, is one of the major advantages of fluorescence over absorption spectroscopy.

Any portion of the spectrum where absorption occurs may produce fluorescence since emission almost always takes place from the lowest vibrational level of the first excited singlet state in solution regardless of which vibrational level or which state the molecule is originally excited. The fluorescence peak will be at the same wavelength regardless of the excitation wavelength; however, the intensity of the fluorescence will vary with the relative strength of the absorption (or the sum total of all the absorptions).

A physical constant that is characteristic of luminescent molecules is the difference between the wavelengths of the excitation and emission maxima. This constant is called the Stokes shift and indicates the energy dissipated during the lifetime of the excited state before return to the ground state

$$\text{Stokes shift} = 10^7 \left(\frac{1}{\lambda_{ex}} - \frac{1}{\lambda_{em}} \right) \qquad (1)$$

where λ_{ex} and λ_{em} are the corrected maximum wavelengths for excitation and emission, and are expressed in nanometers. The Stokes shift is of interest to analytical chemists since the emission wavelength can be greatly shifted by varying the form of the molecule being excited. The fluorescence-maximum shift of 5-hydroxyindole from 330 nm at pH 7 to 550 nm in strong acid occurs with no change in the excitation peak (295 nm) and is due to excited-state protonation.

The Stokes shifts for various molecular species of 3-hydroxypyridine are listed in Table 1. The Stokes shifts for the cations undergoing excited-state ionizations are much higher than those of the neutral ionic species, indicating that processes other than light absorption and emission are involved, and that energy is dissipated in bringing about ionization in the excited state.

(F) FLUORESCENCE QUANTUM EFFICIENCY AND LIFETIME OF THE EXCITED STATE

Every molecule possesses a characteristic property which is described by a number called the quantum efficiency, Φ. This is the ratio of total energy emitted per quanta of energy absorbed.

$$\Phi = \frac{\text{number of quanta emitted}}{\text{number of quanta absorbed}} = \text{quantum yield} \qquad (2)$$

TABLE 1

Stokes shifts for 3-hydroxypyridine [a]

Species	pH optimum	Stokes shift (cm^{-1})
Normal cation	8 N HCl	5750
Excited-state cation	3 N HCl	9830
Dipolar ion	pH 6	6240

[a] Reprinted from ref. 15, by courtesy of publisher.

The higher the value of Φ, the greater the fluorescence of a compound. A non-fluorescent molecule is one whose quantum efficiency is zero or so close to zero that the fluorescence is not measurable. All energy absorbed by such a molecule is rapidly lost by collisional deactivation. Some typical quantum yields of various substances are given in Table 2.

The value of Φ can be determined by measuring the fluorescence of a dilute solution (F_1) of a substance whose quantum efficiency is known, Φ_1, such as quinine sulfate. The fluorescence and absorbance of a solution of the substance whose Φ is to be determined is then measured and the quantum efficiency is calculated as follows

$$\Phi_2 = \Phi_1 \cdot \frac{F_2}{F_1} \cdot \frac{A_1}{A_2} \cdot \frac{q_1}{q_2} \tag{3}$$

where A is the absorbance and q is the relative photon output of the source at the excitation wavelength (taken directly from the curve). A value of Φ, of 0.51 at 10^{-3} M or 0.55 at infinite dilution is taken for Φ_1.

The fluorescence lifetime of most organic molecules is in the nanosecond region. The fluorescence lifetime or day time τ, refers to the mean lifetime of the excited state; the probability of finding a given molecule that has been excited still in the excited state after time t is $e^{-t/\tau}$. The general equation relating the fluorescence intensity, I, and

TABLE 2

Quantum yields of various substances [a]

Compound	Solvent	Quantum yield
Fluorescein	Water, pH 7	0.65
	0.1 N NaOH	0.92
Rhodamine B	Ethanol	0.97
Riboflavin	Water; pH 7	0.26
Anthracene	Benzene	0.29
Naphthalene	Alcohol	0.12
Phenol	Water	0.22
Chlorophyll a	Ethanol, methanol	0.23
Chlorophyll b	Methanol	0.10

[a] Selected values from ref. 16.

the lifetime, τ, is

$$I = I_0 e^{-t/\tau} \tag{4}$$

where, I = fluorescence intensity at time t, I_0 = maximum fluorescence intensity during excitation, t = time after removing source of excitation, and τ = average lifetime of excited state.

The average lifetimes of the excited state of some typical compounds are given in Table 3.

(G) RELATION BETWEEN FLUORESCENCE INTENSITY AND CONCENTRATION

The basic equation defining the relationship of fluorescence to concentration is

$$F = \Phi I_0 (1 - e^{-\epsilon b c}) \tag{5}$$

where Φ is the quantum efficiency, I_0 is the incident radiant power, ϵ is the molar absorptivity, b is the path length of the cell, and c is the molar concentration.

The basic fluorescence intensity—concentration equation indicates that there are 3 major factors other than concentration that affect the fluorescence intensity:

(1) the quantum efficiency, Φ. The greater Φ, the greater will be the fluorescence. This was discussed above;

(2) the intensity of incident radiation, I_0. Hence, theoretically, the more intense source will yield the greatest fluorescence. In actual

TABLE 3

Average lifetime of the excited state (τ) of some compounds [a]

Compound	τ (ns)
Anthracene	4.26
Chlorophyll	30.0
Fluorescein anion	4.9
NADH	4.5
Quinine	19.0
Resorcinol	1.7

[a] Selected values from ref. 17.

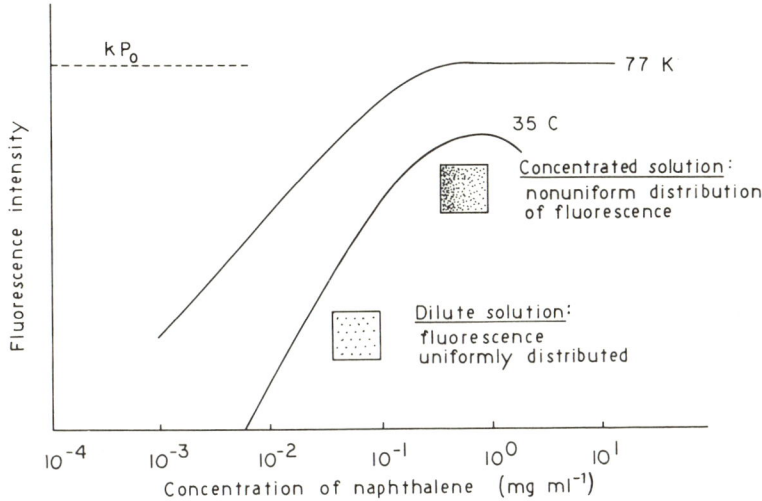

Fig. 5. Dependence of fluorescence upon concentration of fluorophor and temperature.

practice, a very intense source can cause photodecomposition of the sample. Hence, one compromises on a source of moderate intensity, i.e., a Hg or Xe lamp is used;

(3) the molar absorptivity of the compound, ϵ. For a molecule to emit radiation it must first absorb radiation. Hence, the higher the molar absorptivity, the better will be the fluorescence intensity of the compound. It is for this reason that saturated non-aromatic compounds are non-fluorescent.

For very dilute solutions, the equation reduces to one comparable to Beer's law in spectrophotometry.

$$F = K\Phi I_0 \epsilon b c \tag{6}$$

Thus a plot of fluorescence vs. concentration should be linear at low concentrations, then reach a maximum at higher concentrations (Fig. 5). At high concentrations quenching becomes so great that the fluorescence intensity decreases (inner cell effect). The linearity of fluorescence as a function of concentration holds over a very wide range of concentration (Fig. 5). Measurements down to 0.00001 μg ml^{-1} are feasible and linearity extends up to 100 μg ml^{-1} or higher.

(H) INTRODUCTION TO EXPERIMENTATION

The fundamental principles of fluorescence measurement are illustrated by the following simplified schematic representation of a filter fluorometer (Fig. 6). The desired narrow band of wavelengths of exciting radiation are selected by a filter (called the primary filter) placed between the radiation source and the sample. The wavelength of fluorescence radiant energy to be measured is selected by a second optical filter (called the secondary filter) placed between the sample and a photodetector located at a 90° angle from the incident optical path. The output of the photodetector, a current which is proportional to the intensity of the fluorescent energy, is amplified to give a reading on a meter or a recorder. In a spectrofluorometer, the filters are replaced by prism or grating monochromators, and an $x—y$ recorder is used to display the excitation and emission spectra. Further details on fluorescence apparatus will be discussed below.

(I) USE OF LUMINESCENCE

Fluorescence was noted in solutions and minerals in the early 1800's. Sir David Brewster observed the red emission from chlorophyll in 1833 and in 1852 Sir G.G. Stokes suggested the name fluorescence for the emission process, naming the phenomenon after the mineral fluorspa which produces a blue—white fluorescence. The emission from minerals, such as barite, was observed early in the

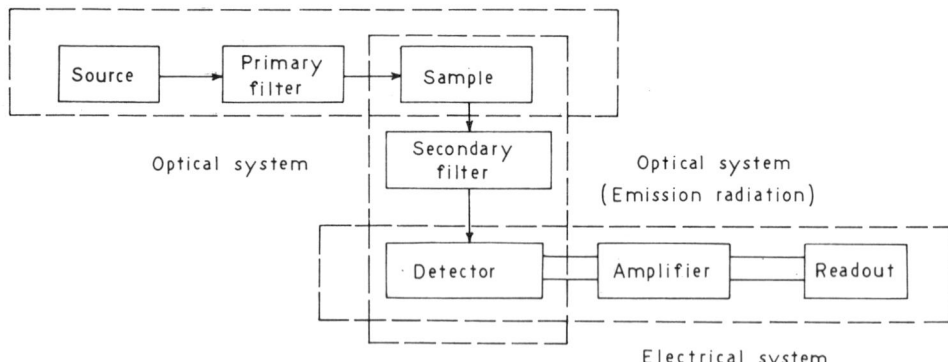

Fig. 6. Schematic diagram of the optical components of a typical fluorometer.

TABLE 4

Applications of luminescence

CLINICAL PATHOLOGY

Electrolytes	Calcium, magnesium, inorganic sulfate, inorganic phosphate
Steroids	Corticosteroids, estrogens, progesterone, androgens, testosterone, bile acids
Lipids	Lipoproteins, phospholipids, cholesterol, triglycerides
Proteins	Serum albumin, protein electrophoresis
Amino Acids and Metabolites	Tryptophan, serotonin, phenylalanine, tyrosine, catecholamines, 3-o-methylcatecholamines, homovanillic acid, DOPA, tyramine and 3-methoxytyramine, histidine and histamine, creatine, kynurenic acid, xanthurenic acid
Immunology	Fluorescent antibodies, fluorescent antigens, blood typing
Enzymes	Dehydrogenases, transaminases, phosphatases, proteases, lipases, creatine kinase, LDH-isoenzymes, peroxidases
Drugs	Barbiturates, salicylates, quinidine, LSD, tetracyclines
Metabolites	Blood glucose, porphyrins, carboxylic acids and ketones
Other	BUN, ammonia, hippuric acid, hematin iron

INORGANIC

Metals — Anions	Cyanide, fluoride, sulfate, silicate, iodide
Cations	Aluminum, arsenic, beryllium, boron, cadmium, cerium, calcium, gallium, iron, lithium, magnesium, rare earths, selenium, silicon, tin, tungsten, uranium, zinc, zirconium

AGRICULTURAL CHEMISTRY

Inorganic	As noted above, especially selenium, magnesium, boron, fluorides, aluminum, tin
Tracing Techniques	Insecticide and pesticide spray coverage studies; residue evaluations
Natural Products	Gibberellic acid, chlorophylls, pigments
Vitamins	A, B_1, B_2, B_6, C, D, and E
Proteins	Protein in milk

TABLE 4 (continued)

PUBLIC HEALTH	
Pollution control	Insecticide aerial drift studies, water pollution studies, spent sulfite liquor
Bacteriology	Identification and counting of bacteria
Metal Poisoning	Beryllium, boron, lead, uranium, cadmium
Immunology	Fluorescent antibody control
Screening Programs	P.K.U., histidemia

15th century and was named phosphorescence from the Greek, "light bearing".

Fluorescence, phosphorescence and chemiluminescence provide some of the most sensitive and selective methods of analysis for many compounds. Some typical examples of analysis in clinical pathology, inorganic analysis, agricultural chemistry and public health are listed in Table 4.

(J) PRACTICAL CONSIDERATIONS

(1) Advantages of fluorescence

Molecular emission (fluorescence and phosphorescence) is a particularly important analytical technique because of its extreme sensitivity and good specificity. Fluorometric methods can detect concentrations of substances as low as one part in ten billion, a sensitivity 1000 times greater than most spectrophotometric methods. The main reason for this increased sensitivity is that in fluorescence the emitted radiation is measured directly, and can be increased or decreased by altering the intensity of the exciting radiant energy. An increase in signal over a zero background signal is measured in fluorometric methods. In spectrophotometric methods the analogous quantity, absorbed radiation, is measured indirectly as the difference between the incident and the transmitted beams. This small decrease in intensity of a very large signal is measured in spectrophotometry with a corresponding large loss in sensitivity.

The specificity of fluorescence is the result of two main factors.

One, there are fewer fluorescent compounds than absorbing ones because all fluorescent compounds must necessarily absorb radiation, but not all those compounds that absorb radiation emit. Second, there are two wavelengths used in fluorometry compared to one used in spectrophotometry. Two compounds that absorb radiation at the same wavelength will probably not emit at the same wavelength. The difference between the excitation and emission peaks ranges from 10–280 nm.

Materials that possess native fluorescence, those that can be converted to fluorescent compounds (fluorophors), and those that extinguish the fluorescence of other compounds can all be determined quantitatively by fluorometry.

(2) Limitations of fluorescence

The principal disadvantage of fluorescence as an analytical tool is its serious dependence on the environment (temperature, pH, ionic strength, etc.). The UV light used for excitation may cause photochemical changes or destruction of the fluorescent compound giving a gradual decrease in the intensity reading. Fluorescence is not usually suited for the determination of the major constituents of a sample because for larger amounts, the accuracy is considerably less than that attainable by gravimetric or volumetric methods.

Quenching, the reduction of fluorescence by a competing deactivating process resulting from the specific interaction between a fluorophor and another substance present in the system, is also frequently a problem. The general mechanism for the quenching process can be denoted as

$$M + h\gamma \rightarrow M^* \text{ (light absorption)} \tag{7}$$

$$M^* \rightarrow M + h\gamma \text{ (fluorescence emission)} \tag{8}$$

$$M^* + Q \rightarrow Q^* + M \text{ (quenching)} \tag{9}$$

$$Q^* \rightarrow Q + \text{Energy} \tag{10}$$

Four common types of quenching are observed in luminescence processes: temperature, oxygen, concentration and impurity quenching. One of the most notorious quenchers is dissolved oxygen, which causes a reduction in fluorescence intensity and a complete destruction of phosphorescent intensity. Small amounts of iodide and nitrogen oxides are very effective quenchers and interfere. Small amounts of highly absorbing substances like dichromate interfere by robbing the fluorescent species of the light available for excitation. For this reason most workers prefer not to wash their cuvettes with dichromate cleaning solution.

a. Temperature quenching. As the temperature is increased, the fluorescence decreases. This is illustrated in Fig. 7, which shows the effect of temperature on the fluorescence of four common substances. The degree of temperature dependence varies from compound to compound. Tryptophan, quinine, and indoleacetic acid are compounds whose fluorescence varies greatly with temperature.

Temperature effects on luminescence are a type of excited-state quenching by encounter. The fluorescence changes are nearly those of molecular activity with temperature, which suggests that increasing temperature increases molecular motion and collisions, and hence robs the molecule of energy. The change in fluorescence is

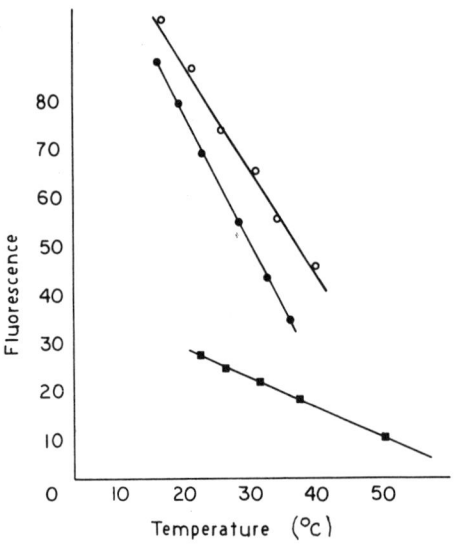

Fig. 7. Variations in the fluorescence intensity of several compounds as a function of temperature. All compounds were dissolved in 0.1 M phosphate buffer, pH 7.0, except quinine. ●, Tryptophan or indoleacetic acid; ○, indoleacetic acid in buffer saturated with benzene; ■, quinine in 0.1 N sulfuric acid.

normally 1% per 1°C; however, in some compounds, such as tryptophan or Rhodamine B, it can be as high as 5%.

In a practical sense, temperature control should be exercised for maximum precision and accuracy. For some instruments, such as the Aminco—Bowman spectrophotofluorometer, temperature-control accessories are available. In cases where the sample must be stored under refrigeration, care must be exercised to bring the sample back to room temperature before measurement. This may pose problems of photodecomposition or change in sample concentration due to evaporation of the solvent that should be considered.

Temperature exerts other effects than quenching on fluorescence. The total fluorescence at the normal excitation, or fluorescence maxima, is lowered due to a dissipation of energy by vibrational energy transitions. Also higher temperatures produce more band maxima due to the increased population of higher energy vibrational levels. For best results all standards and samples should be kept at the same temperature during measurements of fluorescence.

b. Oxygen quenching. Oxygen, present in solutions at a concentration of 10^{-3} M, normally reduces the fluorescence of a typical compound by 20%. Unsubstituted aromatics are still more severely affected by oxygen. Oxygen must be completely removed for phosphorimetry.

We shall investigate the causes of oxygen interference later. Let us say now that oxygen quenching is a type of excited-state quenching, and it is possible to measure the dissolved-oxygen content by its quenching. The ratio of the observed fluorescence in aerated and non-aerated solutions is given by the ratio L_0/L. A few of these ratios

TABLE 5

Ratio L_0/L for several organic compounds

Compound	Ratio L_0/L
Anthracene	1.25
Benzene	2.4
Carbazole	1.9
Naphthalene	6.5
Pyrene	1.0
Toluene	3.0

are presented in Table 5. A value of 1 means that the fluorescence in a deaerated solution, L_0, is the same as that in an aerated solution, L, and hence oxygen has no effect on the luminescence.

The analytical sensitivity can be increased by oxygen removal. This can be accomplished by bubbling an inert gas, such as nitrogen, through the solution for 5—10 min or, better, by a freeze-thaw cycle. It is important that all solvents for use in phosphorimetry be purchased degassed and stored under nitrogen.

c. Concentration quenching. Absorption causes many problems during a fluorometric assay, just as fluorescence causes a problem when the absorbance of a solution is measured.

In order for fluorescence to be observed absorption must occur. As we have already seen, the fluorescence intensity is proportional to the molar absorptivity: the more highly absorbing the substance, the greater its fluorescence. But when the absorption is too large, no light can pass through to cause excitation. Thus, at low concentrations, when the absorbance is less than about 0.05, there is a linear relationship between fluorescence and concentration (Fig. 8). At intermediate concentrations the light is not evenly distributed along the path of light. The portion of the solution nearest the light source absorbs so much radiation that less and less is available for the rest of the solution. As a result, considerable excitation occurs at the front

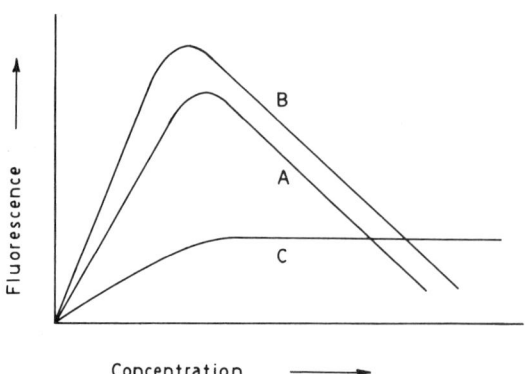

Fig. 8. Relationship between fluorescence and concentration.
A: End-on detector
B: Right angle detector
C: Surface detector

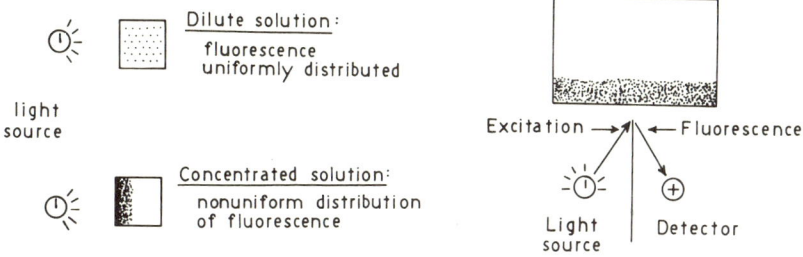

Fig. 9. Effect of concentration on fluorescence.

of the solution (Fig. 9), but less and less occurs throughout the rest of the cell. This type of concentration quenching causes a fluorescence loss that is called the inner-cell effect.

When fluorescence is measured at the surface (Fig. 9), the fluorescence increases linearly and then levels off as predicted. This is explained by the fact that even in concentrated solutions only the surface is observed, and hence quenching is not important. However, in solution, quenching does become important; hence the fluorescence decreases at high concentrations.

Hence an unknown sample should always be tested for concentration quenching. Consider curve A in Fig. 8. There are two possible concentrations for each relative fluorescence value, one on each side of the maximum. To know which is correct, the analyst must dilute his sample and read again. If the fluorescence increases, then the previous reading was on the negative slope, and an incorrect value was obtained. In this case the solution should be diluted and read again. If the fluorescence decreases, then a correct value is obtained. All fluorescence methods should incorporate a linear curve with checks for concentration quenching.

Concentration quenching can be reduced instrumentally by using wide slits so that little sample remains between the source, the fluorescing region of the sample and the detector. A special cell with thick walls is also helpful.

An important form of concentration quenching involves dimer or polymer formation, called excimer quenching. The excimer has a different electron orientation and a longer emission wavelength than the monomer. Hence if an instrument monitors the fluorescence at one wavelength, the emission at the longer wavelength will go undetected. So the observed fluorescence will decrease with increasing concentration.

   d. Impurity quenching. Many researchers have become completely disenchanted with fluorescence as an analytical tool because of the phenomenon of impurity quenching. Fluorometry is generally considered a specific and sensitive tool, not subject to chemical interference, since most measurements are made in dilute solutions that contain only trace amounts of impurities. When these impurities are present at moderate concentrations, interferences result. This interference can be in the form of the inner-cell effect, collisional quenching, energy transfer, charge transfer, or the heavy-atom effect. To get round this problem the analyst could try to reduce the interference by dilution or by separation techniques. If an interference is known to be present, it should be added to the standards used in preparing the calibration curve.

(K) TYPES OF LUMINESCENCE

   The various types of luminescence can be classified according to the means by which energy is supplied to excite the luminescent molecule.

   When molecules are excited by interaction with photons of electromagnetic radiation the form of luminescence is called *photoluminescence*. If the release of electromagnetic energy is immediate or from the singlet state the process is called *fluorescence*, whereas *phosphorescence* is a delayed release of energy from the triplet state. Some molecules exhibit a *delayed fluorescence* which might incorrectly be assumed to be phosphorescence. This results from two intersystem crossings, first from the singlet to the triplet, then from the triplet to the singlet.

   If the excitation energy is obtained from the chemical energy of reaction the process is *chemiluminescence*. In *bioluminescence* the electromagnetic energy is released by organisms.

   Triboluminescence (Greek tribo, to rub) is produced as a release of energy when certain crystals, such as sugar, are broken. The energy stored on crystal formation is released in this process.

   Other types of luminescence — *cathodoluminescence*, resulting from a release of energy produced by exposure to cathode rays or *thermal luminescence*, which occurs when a material existing in high vibrational energy levels emits energy at a temperature below red heat, after being exposed to small amounts of thermal energy — are much less commonly encountered.

2. Instrumentation

(A) COMPONENT PARTS

The basic components of any instrument designed to measure luminescence (Fig. 10) are: light source, wavelength selectors, sample compartment and detector system. The instrumentation is the same as that used in spectrophotometry, with two exceptions, (1) the detector is rotated 90° to the incident light path, and (2) a second wavelength selector is placed in front of the detector. Thus any good spectrophotometer can be adapted to fluorescence work at a small additional cost.

In fluorescence the detector is placed at 90° to the path of incident light so that little of the incident radiation will strike the detector. Only light emitted from the sample reaches the detector, so

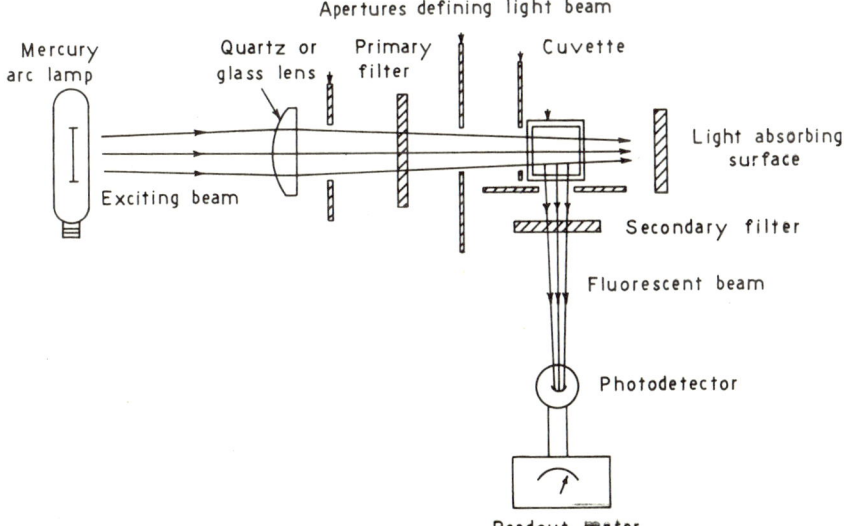

Fig. 10. Schematic diagram of the optical components of a typical fluorometer.

that this device will register zero signal when no luminescence occurs. Then, an increase in signal indicates an emission from the sample. This is the major reason for the greater sensitivity of luminescence over spectrophotometric methods.

The second wavelength selector is placed before the detector to remove all radiation except that emitted by the sample. This provides another degree of specificity to the analysis. A discussion of each of the components used in fluorescence instrumentation will help the reader to be aware of how these parameters can be optimized for maximum performance.

There are two principal types of instrument: filter fluorometers and spectrofluorometers. The filter instrument uses filters to select the desired excitation and emission wavelengths by absorbing or reflecting unwanted radiation. We shall discuss the various types of instruments available commercially in each of these classes later in this chapter.

(1) Light sources

a. *Mercury-vapor lamp.* The mercury-vapor lamp is widely used in filter fluorometers because of its intense light emission and stability. Since an elaborate power supply is not needed, the lamp is economical. The most commonly used wavelength in the mercury-vapor lamp is the resonance line at 365—366 nm, although other useful lines are present at 254, 302, 313, 405, 436, 546, 577, and 579 nm. A list of the relative intensities of lines of the mercury-vapor lamp is presented in Table 6.

TABLE 6

Relative intensity of spectral lines in a mercury-vapor lamp

Line (nm)	Relative intensity	Line (nm)	Relative intensity
253.7	1	366.3	1.4×10^{-3}
296.5	6.0×10^{-3}	404.7	8.9×10^{-3}
302.2	1.1×10^{-2}	435.8	1.7×10^{-2}
312.2	7.1×10^{-3}	546.1	1.2×10^{-2}
313.2	1.1×10^{-2}	577.0	1.7×10^{-3}
365.0	8.9×10^{-3}	579.0	1.8×10^{-3}
365.5	2.1×10^{-3}		

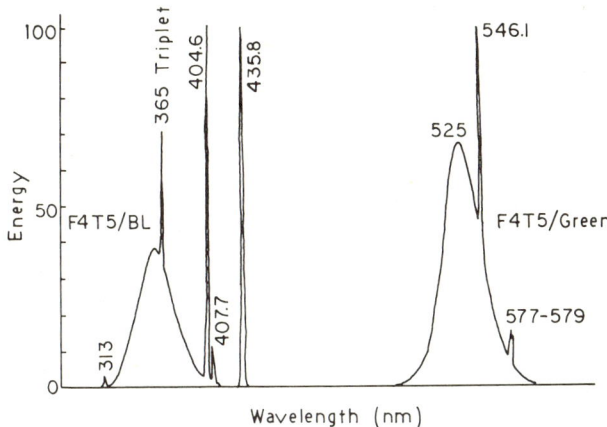

Fig. 11. Energy spectra of mercury-phosphor lamps.

A low-pressure mercury-vapor lamp can be modified to provide energy at wavelengths other than the resonance lines. The inner surface of the lamp is painted with a thin layer of crystalline phosphors that absorb the mercury-vapor resonance radiation and generate a broad band at longer wavelengths. The Aminco and Turner filter fluorometers, for example, use a 4-W low-pressure mercury-phosphor

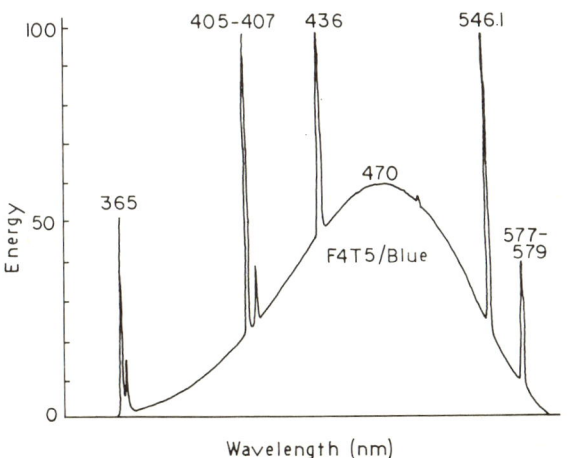

Fig. 12. Energy spectrum of the F4T5 blue-coated mercury lamp.

lamp with a broad emission band having its peak at 360–365 nm. A green lamp with a 525-nm peak, a blue lamp with 405- and 436-nm peaks (Fig. 11), and a blue lamp with 405-, 436-, 546-, and 577-nm peaks (Fig. 12) are available.

b. Xenon lamp. Most grating instruments such as the Aminco Bowman SPF, use a high-pressure xenon lamp (commonly 150-W). The xenon lamp has a good continuum (Fig. 13), better in the UV than that of the tungsten lamp, but it does not have the intensity that the mercury lamp has at its resonance lines (at those lines the mercury lamp has twice the intensity of the xenon lamp). Moreover, it is necessary to use an expensive d.c. converter and stabilizer with this lamp. A mercury lamp cannot be used with a scanning spectro-fluorometer, however, because the excitation spectrum obtained would simply be that of the mercury resonance lines superimposed on the sample's excitation spectrum. A lamp with a smooth continuum is needed in the UV region also.

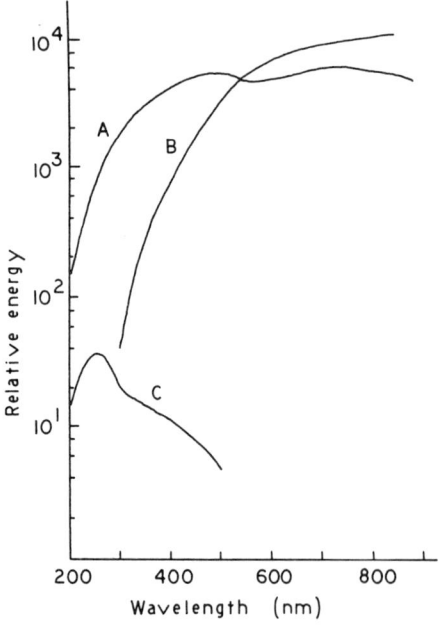

Fig. 13. Relative output of various lamps. A, xenon; B, tungsten; C, deuterium.

94

The usefulness of the xenon lamp is limited by the regulation of the power supply, which directly affects the stability and life of the lamp. Any fluctuations in the power supply will appear in the lamp. "Arc wander", the variation in the position of the arc passing between the electrodes of the xenon lamp, is frequently a problem. Better lamp construction and greater stability in the power supply have helped to minimize arc wander. Arc wander will cause a slight shift in the excitation spectrum and hence must be eliminated.

The ellipsoidal condensing system accessory to the Aminco Bowman SPF is much more sensitive to arc instability. But the excitation wavelength remains constant due to the entrance-slit arrangement.

c. *Hazards of gas-vapor lamps.* Gas-ionization lamps are filled with high-pressure gases, 5 atm at room temperature or 20 atm at operating temperatures. Since the lamp can easily explode on receiving a shock, several layers of cloth should be wrapped around it when handled. Furthermore, the gas-vapor lamps produce UV radiation, which can severely burn the retina of the eye. A pair of UV filter glasses should be worn when working with these lamps, and one should never look directly at a gas-ionization lamp.

Gas-ionization lamps also convert air into ozone and nitrogen oxides, which are toxic. Ozone is produced by UV radiation of 180—210-nm wavelength; in fact the production of ozone is a good measure of the UV output of the xenon lamp. Ozone is easily detected by its characteristic odor.

The fluorometer lamp should be vented to a hood or exhaust to remove ozone. If venting is impossible, the ozone should be catalytically converted to molecular oxygen by passing the exhaust through a molecular sieve (10X) or platinum dust.

(2) Monochromators

The purpose of the monochromator is to isolate narrow bands of electromagnetic radiation from the source. It is a wavelength selector. There are two types of monochromators used in luminescence equipment — a filter or a grating. In fact, instruments are classified as filter fluorometers (non-scanning) or grating spectrofluorometers (scanning). Filters are not used in instruments because they give their greatest dispersion in the UV not the visible where most measure-

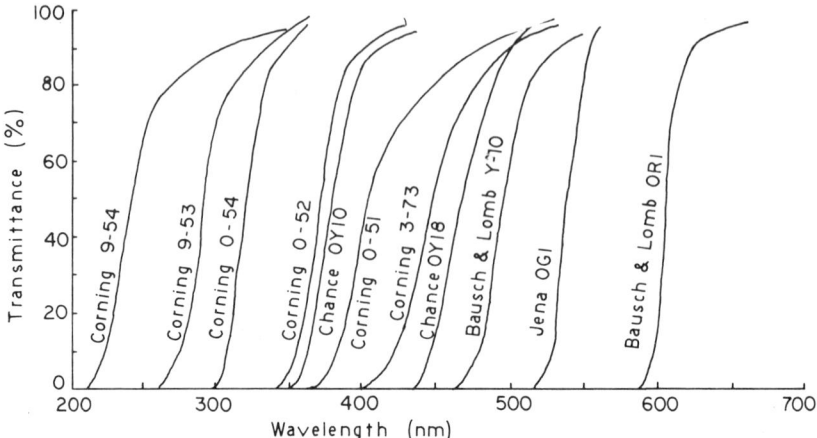

Fig. 14. Transmittance characteristics of secondary filters made by the Corning Glass Works (U.S.A.), the Chance Pilkington Optical Works (England), Bausch and Lomb (U.S.A.), and Jena Glaswerke (Germany).

ments are made and to obtain adequate sensitivity a large prism would be needed. This is expensive.

a. Filters. There are three types of filters.

(1) Neutral tint. These give a nearly constant transmission over a

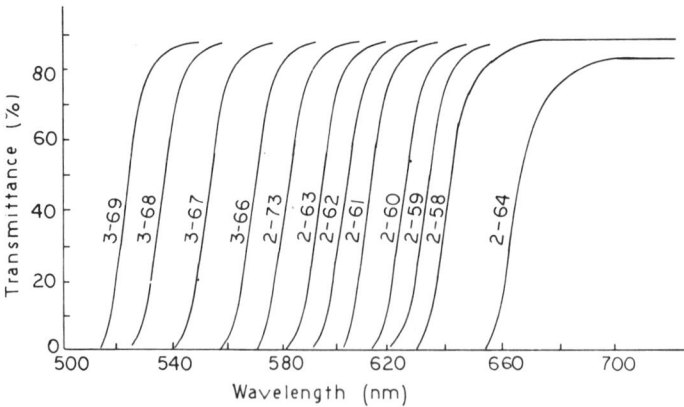

Fig. 15. Sharp cut-off filters.

96

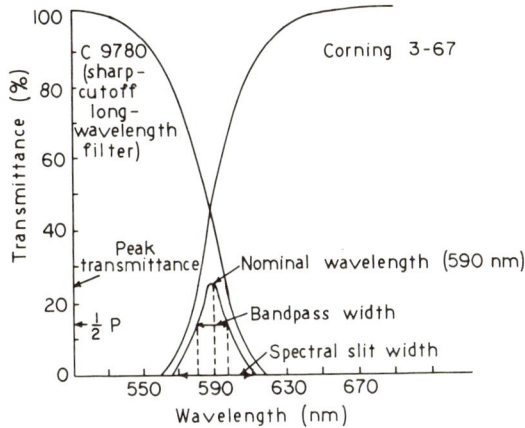

Fig. 16. Spectral transmittance characteristics of a composite glass absorption filter and its components.

wide range. They are designed to decrease the intensity of the fluorescence signal uniformly and are used with strongly fluorescing compounds.

(2) Cut-off filters. This type of filter is used to cut off stray or unwanted radiation since it produces a sharp cut-off in the spectrum with complete transmission on one side of the cut-off with little or no transmission on the other. Some typical cut-off filters are shown in Figs. 14 and 15.

(3) Bandpass filters. These are composite filters constructed from

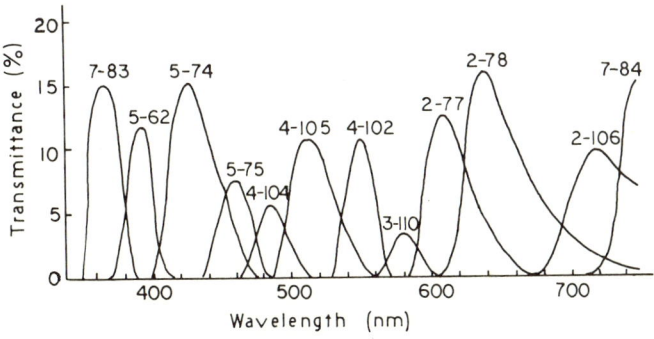

Fig. 17. Transmittance spectra of some glass filters (Corning Glass Works).

sets of cut-off filters. One part consists of a long-wavelength sharp cut-off filter (blue and green series), and the other is a short-wavelength cut-off filter (red and yellow series). Figure 16 illustrates the use of two such filters in composing a glass absorption filter with a nominal wavelength of 590 nm. Other such filters are available for wavelengths of 360—740 nm, as illustrated in Fig. 17. Normally these filters have relatively wide bandpass widths (width of the band at one-half the peak transmittance) with peak transmittances of about 25%.

There are several typical filters that are useful in conjunction with the mercury-vapor lamp (Tables 7 and 8). Many other filters are available for isolating almost any spectral region, and the reader is advised to consult Eastman Kodak, Turner, or Aminco catalogs for the wide selection available.

The choice of the proper filter is empirical. For practical purposes, for example, excitation at the strongest mercury line at 365 nm will give the greatest sensitivity, if the compound absorbs there, even though this might not be the wavelength of maximum absorption. Hence, a filter peaking at 365 nm, rather than one peaking at, say, 390 nm, might be used, even though the λ_{max} were 390 nm.

Another potential source of error is the fluorescence of filters themselves. A fluorescent secondary filter, such as the Corning 22, will cause a positive error in a turbid solution. And very little turbidity can cause quite large errors when improper filters are used.

b. Gratings. A grating consists of a large number of parallel lines or grooves ruled at extremely close intervals (e.g., 30000 lines per inch) on a highly polished surface, such as aluminum. A master grating is used as a mold in the production of replica gratings. A film of parting compound is applied to the master, the film is aluminized, the crevices are filled with epoxy resin, and an optical flat is bonded by

TABLE 7

Wratten filters useful with mercury-vapor lamps

Filter No.	Principal lines transmitted (nm)
18A	365
22	577 and 579
50	436
74	546

TABLE 8

Primary (activation) filters

Wavelength	Filter No.	Comments
254	110—810 (7—54)	Passes also the 313-, 365-, and part of the 405-nm lines, both of which are relatively minor in the required 110—851 lamp; quartz cuvettes (110—802) must be used.
254	110—810 (7—54) + 110—815	Provides pure 254-nm activation when measurement in the near ultraviolet is required; quartz cuvettes (110—802) and 110—851 lamp required.
325	110—810 (7—54) + Wratten 34A (unmounted) 110—836	Gelatin held in place by the Corning filter; used with the standard (110—850) lamp to provide a narrow band of activating light peaking at 325 nm; with the 110—851 lamp it isolates the 313-nm line. In neither case is the activating light very intense, the former being generally best, but it is suitable for a great deal of work. Pyrex cuvettes are generally satisfactory.
365	110—811 (7—60) or 110—834 (7—37)	The general-purpose primary filter supplied with the Turner fluorometers; normally used with the standard (110—850) lamp and Pyrex cuvettes; 110—834 recommended for solid samples or paper-chromatogram door.
405	110—812 (405)	Should be used with the dark Corning 7—51 glass away from the light as the Wratten 2C is slightly fluorescent and must be blocked; used with either 110—850 or 110—851 lamp (about two-fold gain in sensitivity with latter) and Pyrex cuvettes.
405 + 436	110—813 (47B)	Seldom used alone, but for some applications where either 405 or 436 may be used provides increased sensitivity by providing both; like the 405 filter, it is normally used with 110—850 lamp, but the 110—851 lamp provides increased sensitivity; used with Pyrex cuvettes.
436	110—816 (2A) + 110—813 (47B)	The 2A is placed nearest the lamp and eliminates the 405-nm line. Other comments as for the 405.

TABLE 8 (continued)

Wavelength	Filter No.	Comments
470	110—827 (3) + 110—831 (48)	Used only with the 110—853 blue lamp.
546	110—814 (1—60) + 110—822 (58) or 110—832 (546)	The 1—60 is placed nearest the lamp. Other comments as for the 405. The 110—832 is recommended for tracer work with Rhodamine B or Pontacyl Brilliant Pink B or for best results with paper-chromatogram door.

the epoxy to the aluminum replica of the master grating pattern. When the epoxy hardens, the replica grating, completely anodized, is separated from the master.

Two factors are important in discussing the characteristics of a grating: blaze angle and the number of lines or grooves in the grating. The basic equation for a grating is

$$n\lambda = 2d \sin \theta \tag{11}$$

where n is the order of diffraction, d is the distance between adjacent grooves, θ is the angle of reflectance, and λ is the wavelength of radiation.

When radiant energy strikes the grating so that the angles of incidence and diffraction are equal but opposite in sign, then $n\lambda = 0$. This is the zero-order, which corresponds to spectral reflection in a reflection grating. When $n = 1$, the diffraction is of the first-order; when $n = 2$, it is of the second-order, etc. (Fig. 18).

The resolving power R of a grating is given by the formula

$$R = \frac{\lambda}{\Delta\lambda} = mN \tag{12}$$

where $\Delta\lambda$ is the wavelength difference between two lines that are just barely distinguishable, λ is their average wavelength, N is the number of lines or grooves in the grating, and m is the length of the grating in centimeters.

For example, a grating with 600 lines per millimeter and 15 cm long would have a theoretical resolving power of 90000 (15 cm × 6000 lines per centimeter). At 5400 Å this grating would yield a

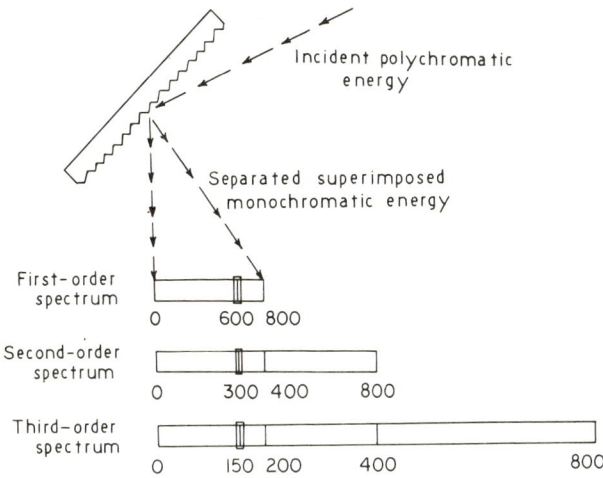

Fig. 18. Overlap of first-, second-, and third-order spectra from a reflection grating.

theoretical $\Delta\lambda$ of 0.06 Å:

$$\Delta\lambda = \frac{\lambda}{R} = \frac{5400}{90000} = 0.06 \text{ Å}.$$

Gratings with resolving powers of 500,000 have been ruled. The greater the number of lines, the greater the resolution. The more lines per unit length, the greater the dispersion in the first-order. The Aminco SPF-1000 has a grating of 30000 in the first order, or a theoretical resolution of 0.02 nm at the 577-nm line of mercury. The observed resolution is only about 0.7 nm, however, due to the influence of other factors in the design of the monochromator.

The blaze is the second concept that is important to understand in selecting a grating. The blaze is the wavelength at which the maximum output of the grating is concentrated, and the blaze angle is the angle at which the grooves are ruled in the grating. If a grating is blazed at 500 nm, its maximum output is 500 nm. The efficiency of a grating drops off rapidly as the wavelength differs from the wavelength for which the grating is blazed. The first-order grating efficiency drops off more rapidly on the short wavelength side than on the longer wavelength side. The bandpass of the efficiency curve for

TABLE 9

Typical efficiencies and bandpass widths of gratings used in spectrofluorometry

Parameter	Grating blaze (nm) [a]		
	300	500	750
First-order blaze (nm)	300	500	750
Second-order blaze (nm)	150	250	375
First-order efficiency — first-order blaze (%)	85	90	90
Second-order efficiency — second-order blaze (%)	60	80	75
First-order bandpass (nm)	200—600	335—1000	500—1500
Second-order bandpass (nm)	100—255	170—750	250—1125

[a] All gratings with 600 lines per millimeter.

a grating extends from two-thirds of the blaze wavelength to twice the blaze wavelength.

In the Aminco Bowman SPF, for example, the excitation grating is blazed at 300 nm and the emission grating is blazed at 500 nm. Thus the former can be used from 200 to 600 nm, the latter from 335 to 1000 nm. For maximum efficiency, however, a grating blazed at 750 nm should be used for scanning the near-IR region. The typical efficiencies and bandpass widths of gratings used in spectrofluorometry are listed in Table 9.

The advantages of gratings are the following: (1) gratings have uniform resolution and linear dispersion at all wavelengths; (2) up to 80% of the incident radiation can be directed into the first-order by blazing the grating; (3) gratings are less expensive than prisms; (4) all wavelengths can be dispersed.

The major disadvantage of the grating is that several orders of spectra are passed. This can be offset by using filters in the optical path. Thus to observe the 600-nm spectral line without interference from the second-order 300-nm spectral line, a filter cutting off all radiation below 400 nm should be used.

(3) Cell compartment

Pyrex cells are useful for measurements above 320 nm (which compose 95% of all common analyses). Only below 320 nm are

102

quartz or fused-silica cells required. Hence for all practical purposes the large additional cost of quartz or fused silica (Supersil, etc.) is not justified.

Below 320 nm, where quartz is necessary, the researcher should consider the properties of the different kinds of quartz. Corning quartz possesses a lower native fluorescence than do other varieties. However, even this quartz possesses sufficient fluorescence to be detected on the spectrophotofluorometer at high sensitivity (λ_{ex} = 265 and 330 nm; λ_{em} = 500 nm). Fused silica (e.g., Supersil) is preferred over fused quartz. Care should also be given to the selection of sample cuvettes, especially with respect to scratches and surface flaws.

(4) Cell configuration

A few words on cell configuration are desirable at this time. One could use any cell configuration from 0 to 180° for measurement of fluorescence. A 30 or 45° configuration works as well as a 90° one. However, the lowest backgrounds from incident radiation are obtained at 90°, and, for this reason, this configuration is most commonly used in all instruments.

When concentration quenching is encountered in the conventional 90° configuration and the sample cannot be diluted, the worker can try one of the following: (a) use a front-surface configuration with a solid-sample accessory or (b) use a microcell to eliminate all regions of the sample solution that are not both fluorescing and observed by the detector.

(5) Slits

The most important parameter in determining the resolution of an instrument is the slit width. The distribution of energy as a function of wavelength for the light passing through the exit slit of a monochromator can be represented as an isosceles triangle if the entrance and exit slits are of equal width (Fig. 19). The middle wavelength (peak transmittance) is called the nominal wavelength and is the value read on the dial of the instrument. The bandpass is the bandwidth at one-half the peak transmittance and is essentially the width of the exit slit. The spectral slit width is twice the bandpass and is the total width of the base line. Within the bandpass width is con-

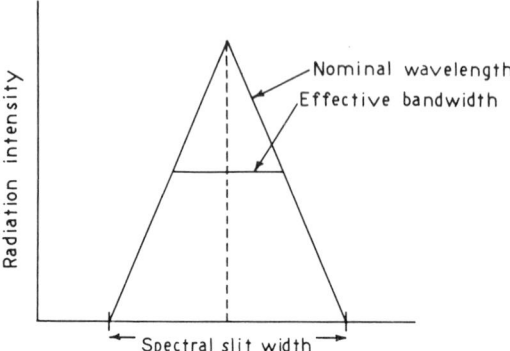

Fig. 19. Distribution of radiant energy emerging from a slit as a function of wavelength.

tained three-quarters of the transmitted radiant energy. In a grating instrument the bandpass for a given slit is constant through the spectrum and depends on the ruling of the grating.

Figure 20 illustrates the effect of slit width on the spectral isolation of the monochromatic 546-nm mercury line in the Beckman DU instrument. At a slit opening of 0.1 mm the bandpass is 3.4 nm. This means that two peaks closer than approximately 6 nm could not be

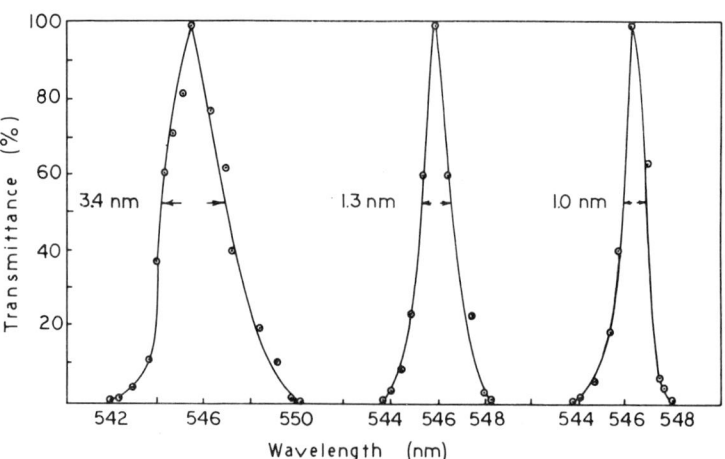

Fig. 20. Spectral isolation with various slit widths. Beckman Model DU spectrophotometer. After Cary and Beckman [18].

104

resolved. At a slit width of 0.02 nm the bandpass is 1.3 nm, but at 0.01 nm it is only 1.0 nm. Hence there is no linear relationship between bandpass and slit width, but it can be said that better resolution can be obtained by decreasing the slit width. However, decreasing the slit width decreases sensitivity since the intensity of light emerging from the monochromator decreases. This is why most good fluorometers use a photomultiplier, which is a more sensitive device for measuring the emergent radiation.

There are three types of slit: fixed, unilateral, and bilateral. Fixed slits are slots cut in an opaque material. Some instruments, such as the Aminco Bowman SPF, use a series of interchangeable fixed slits to provide reproducible settings. Unilateral slits are made from two beveled blades or jaws, with one jaw movable, and allow continuous variation through a limited range. The disadvantage of the unilateral slit is that the center of the spectral line is shifted as the slit width is altered. Bilateral slits have two blades that move symmetrically to maintain a constant center line. This is the type used in the Perkin–Elmer MPF-2A. Slits should be cleaned periodically with cellophane or similar material.

Resolution is normally expressed in spectral bandpass, since this takes into account the spread of wavelength leaving the exit slit. The bandpass is related to the dispersion D, the focal length F, and the slit width W by the equation

$$BP = \frac{W}{FD} \tag{13}$$

(6) Detectors

The light emitted by a sample in the UV and visible region of the spectrum is best measured with a photomultiplier tube. This detector multiplies the light signal up to a million times and presents a signal output. The photomultiplier combines photocathode emission with multiple cascade stages of electron amplification to achieve a large amplification of primary photocurrent within the envelope of the phototube itself, with retention of linear response. As shown in Fig. 21, the photomultiplier is constructed so that the primary photoelectrons from the cathode are attracted and accelerated to the first dynode with considerable energy. Each dynode consists of a plate coated with a substance having a small force of attraction for the

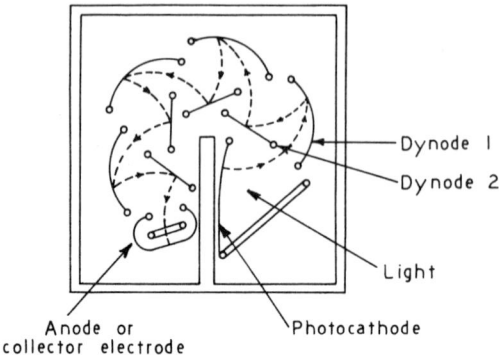

Fig. 21. Schematic diagram of photomultiplier tube. Broken lines are the paths traveled by the secondary electrons as they are focused by each succeeding dynode's field in turn.

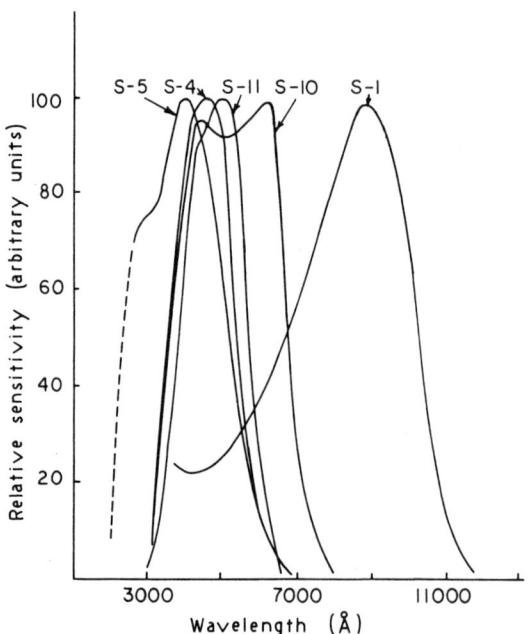

Fig. 22. Spectral response curves of some commercial photoemissive surfaces. All curves have been adjusted so that the wavelength of maximum sensitivity is 100 ordinate units.

106

escaping electrons. The impinging high-energy electrons strike with enough energy to dislodge and eject 2—5 secondary electrons. These electrons are accelerated to the second dynode by an additional positive potential, and so on. The process can be repeated 10—20 times in stages of potential from 20 to 110 V. This requires a high-voltage, low-current power supply that is very stable.

The photomultiplier surface contains a thin layer of one or more elements possessing a low ionization potential. As a result, valence electrons are easily released when struck by photons. The alkali metals are commonly used and are plated on an Ag/Ag_2O cathode. Photomultiplier tubes can be made to respond to different wavelengths by varying the elements of the photosensitive surface. The tubes are classified by "S" ratings from S-1 to S-22 (see Fig. 22). The response of the human eye most closely resembles the S-4 or S-11 curve.

Because of manufacturing problems, it is difficult to get two photomultiplier tubes with the same response characteristics. The researcher is advised to select carefully and to buy tubes with known sensitivity and spectral response.

There are several sources of noise in a photomultiplier tube, two of which are shot noise, which arises from extraneous electron emission, and dark-current noise, which arises from the thermionic emission in the dynodes.

For best operation and low noise, it is recommended that the photocathode be kept in the dark at all times. Sensitivity can be increased by lowering the noise. The signal-to-noise ratio (S/N) should be at least 2, but preferably 10. Thus, lowering the noise by cooling the photomultiplier could increase sensitivity.

(B) BASIC INSTRUMENTS

As we have already mentioned, all instruments available for the measurement of luminescence fall into two classes: filter instruments and grating instruments.

(1) Filter instruments

a. General remarks. Filter fluorometers are inexpensive, very sensitive, and simple in design. The instrument uses a mercury lamp, a primary filter to pass only certain wavelengths to excite the sample, a

TABLE 10

Comparison of common filter fluorometers

Fluorometer [a]	Lamp	Lowest detectable concentration (at $S/N = 1$) of quinine sulfate ($\mu g\ ml^{-1}$)	Cost [b] ($)	Remarks
Aminco	Hg	0.0002	1300	Photomultiplier detector; single beam; temperature control; seven scales; quartz optics.
Baird-Atomic Fluorimet	Hg	0.001	750	Solid state.
Beckman ratio	Hg (phosphor-coated sleeve)	0.0010	1200	Photomultiplier detector; double beam, ratio recording; Vycor optics.
Coleman 12C	Hg arc	0.003	570	Phototube detector; single beam; one scale; glass optics.
Farrand ratio	Hg	0.0001	1800	1P28 Photomultiplier, optical balance.
Turner 110, 111	Hg	0.003	1300 (110) 1800 (111)	Photomultiplier detector; double beam; optical balance; temperature control; quartz optics; four scales; 110 = null balance; 111 = recording.

[a] All instruments listed use a glass-filter monochromator.
[b] Approximate, subject to market fluctuations.

108

secondary filter to pass only the fluorescence emission, and not the incident energy, strong radiation, and light scatter, to the detector, which is usually a photomultiplier. The filter fluorometer is more sensitive than the grating instrument and can do almost anything the latter can do, except scan. Table 10 lists the characteristics of some commonly available filter fluorometers.

To use a filter instrument one must first determine the wavelengths of excitation and emission in order to choose the filters to use for maximum sensitivity. To do this, one first runs the absorption spectrum of the compound on any spectrophotometer. Since the excitation spectrum should match the absorption spectrum, one need simply pick a primary filter to peak around the λ_{max} of the compound. Then one places a dilute solution (10^{-5} M) of the compound to be assayed into the fluorometer and tries several secondary filters (realizing that the emission peak is usually separated from the excitation peak by 20—150 nm) until a maximum signal is obtained.

b. Commercial instruments. Commercial instruments differ from each other in the component parts and in the manner of assembly. Some instruments like the Coleman Model 12C, the Photovolt Model

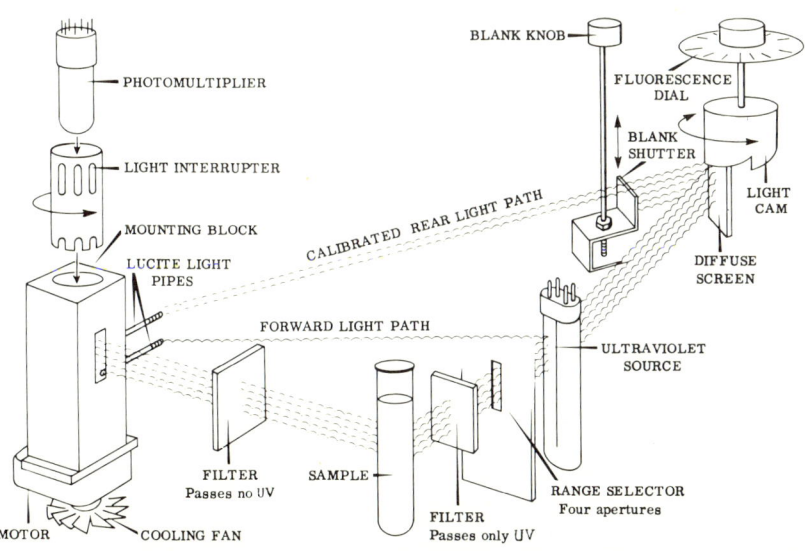

Fig. 23. Optical system of the Turner Model 110 fluorometer.

540, the Farrand, and the Aminco possess a single-beam circuit, whereas the Klett, the Lumetron, the Turner, the Beckman, and the Hilger Spekker instruments utilize some sort of a null-indicating circuit for eliminating lamp fluctuations (a type of double-beam circuit). Klett is one of the few instruments still utilizing a barrier-layer-cell detector.

G.K. Turner Associates markets two good filter fluorometers that are quite widely used. The 110 is a null-balance instrument; the Model 111 has direct output to a recorder and is more versatile, yet more expensive. Figure 23 shows the optical system for the Model 110. The Turner has a calibrated rear light path to compensate for fluctuations in the light source.

A true ratio fluorometer, such as the Beckman shown in Fig. 24, places the reference solution in one beam while the sample solution is irradiated by the second beam. This makes the fluorometer insensitive to the temperature changes that affect other instruments. A special lamp irradiates each solution alternately, and a discriminator circuit presents the ratio signal to the meter.

The Coleman Model 12C fluorometer has been used in a number of laboratories for about 35 years. It has a blue-sensitive phototube

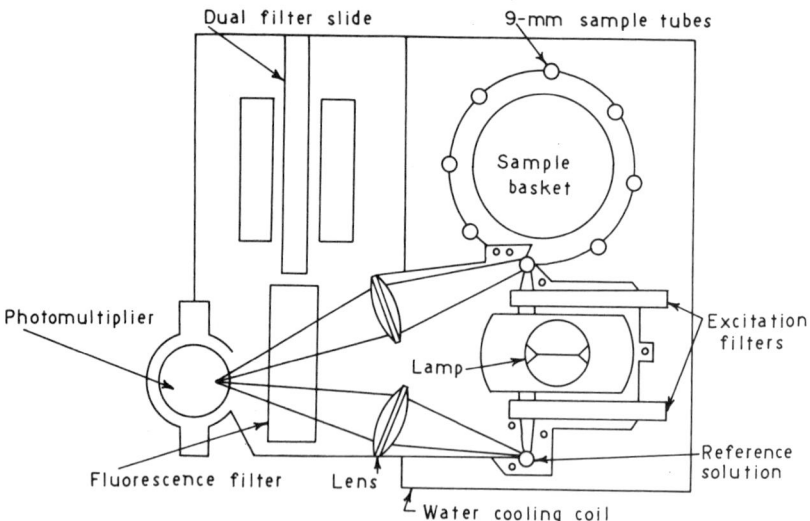

Fig. 24. Optical diagram of the Beckman ratio fluorometer. The lamp alternately illuminates the reference and sample solutions.

110

as a detector. The use of a phototube limits its sensitivity, however.

Aminco offers a good filter fluorometer; it is available in solid state and has seven linearly calibrated scales.

(2) Grating instruments

a. General remarks. A grating instrument generally uses a xenon lamp as the source of radiation, two gratings to disperse and select the desired excitation and fluorescence energy, and a photomultiplier as the detector. Various slits in the focal planes of the gratings determine the bandpass and the intensity of the energy striking the sample or the detector.

The grating spectrofluorometer is more versatile than the filter instrument and can be used for luminescence research projects. A comparison of some spectrofluorometers is presented in Table 11. In this table the various characteristics of each instrument are compared: lamp, monochromator, slit widths, lines on the grating, lowest detectable concentration of quinine sulfate, resolution, and various characteristics of each. This table is intended to give an unbiased comparison according to manufacturer's specifications.

b. Commercial instruments. One of the most popular spectro-fluorometers is the Aminco—Bowman spectrophotofluorometer (SPF), an adaptation of an instrument developed by Dr. Robert L. Bowman. The components of the spectrofluorometer are diagrammed in Fig. 25. The xenon source covers the range of excitation wavelengths from 200 to 1500 nm. A blue- or red-sensitive photomultiplier tube is used to measure the spectral signal. Two gratings are used for monochromaticity, and the resolution of the instrument is quite good (<1 nm).

The Baird-Atomic, Inc., spectrofluorometer is a two-grating monochromator based on a design originally suggested by Bowman et al. [19]. The use of two excitation and two emission gratings gives the instrument good resolution and low background scatter.

The Baird-Atomic Fluoripoint spectrofluorometer is a low cost instrument that has two grating monochromators, a Hg or Xe source, a sample compartment with four position sample turret, and a solid state photomultiplier power supply and amplifier.

The Farrand Model MK-1 spectrofluorometer is also based on a design originally suggested by Bowman et al. [19]. Two grating monochromators (220—650 nm) are used along with a 1P21 photo-

TABLE 11

Comparison of spectrofluorometers

Instrument	Lamp	Monochromator (slits)	Lowest detectable concentration (at $S/N = 1$) of quinine sulfate (p.p.b.)	Resolution (nm)	Remarks
Aminco					
Aminco—Bowman [a]	Xenon (150 W)	Grating (1—30 nm) 15,000 lines per inch; (30,000 optional)	0.2	1 [b]	1P21 Photomultiplier; seven scales; temperature control.
SPF-1000	Xenon	Grating 30,000 lines per inch (0.1—4 nm)	0.5	0.2	Photomultiplier; corrected spectra, instrument $25000
SPF-125	Hg	Grating 15,000 lines per inch (1.5—44 nm)	0.005	1.5	1P21 Photomultiplier; seven scales; solid state.
Baird-Atomic					
Fluorispec SF/1	Xenon (150 W)	Dual grating (2—32 nm)	0.1	1.6	1P28 Photomultiplier or 7102 for IR; four scales at 1 and 1/10 sec time constants.
Fluoripoint	Hg or Xe	Grating (10 or 20 nm)	0.05	10	EM 1 9771B Photomultiplier; seven scales.
Farrand					
MK-1	Xenon (150 W)	Grating (0.5—20 nm)	0.1	0.5	1P28 Photomultiplier; ten scanning speeds

Perkin—Elmer

MPF-2A	Xenon (150 W)	Grating (1—40 nm) 14,400 lines per inch	0.05 [c]	1	R-106 Photomultiplier; ratio recording; temperature control; three scan speeds; variable slits.
MPF-3	Xenon (150 W)	Grating (1—40 nm) 14,400 lines per inch	0.005	1	Same as 2A except six scan speeds; $13800 with constant quanta accessory.
204	Xenon (150 W)	Grating (10 nm) 14,400 lines per inch	0.05	10	Meter readout with thirty-six scales; $4600 with recorder.

Turner

210	Xenon (75 W)	Grating (0.5—25 nm) 30,000 lines per inch	2 [d]	0.1 [e]	Corrected spectra; double beam; temperature control; can be used as spectrophotometer.
430	Xenon (150 W)	Grating (15 nm)	0.005	15	R-136 Photomultiplier.

[a] Cost of basic instrument about $5000; corrected-spectra instrument, $9000; ratio-recording attachment, $3000.
[b] Resolution of 0.5 nm optional.
[c] At 10-nm bandpass.
[d] At 15-nm bandpass.
[e] Readability.

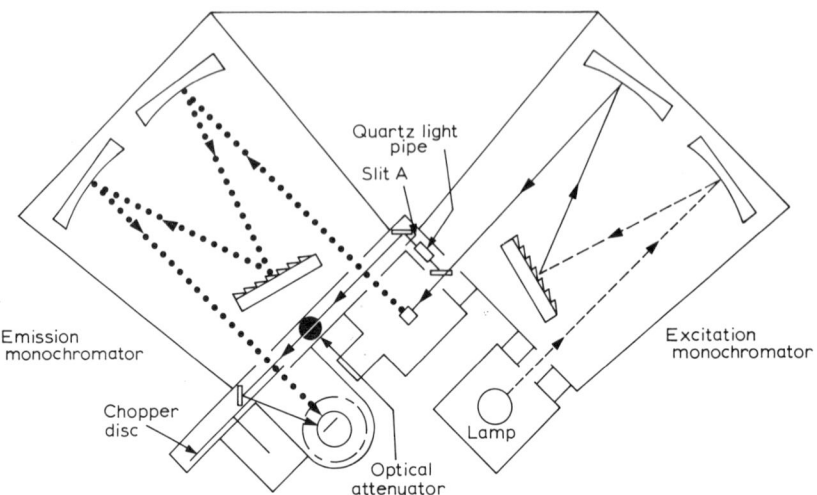

Fig. 25. Optical system of Aminco—Bowman SPF.

multiplier tube. An attachment that uses a lead sulfide phototube is available for measurements in the IR region.

The Perkin—Elmer MPF-3 instrument has two monochromators with good resolution and also slits that are variably adjustable from 1 to 40 nm. The instrument has a ratio-recording system to compensate for drifts in the light source.

Turner Instrument Company now markets a new, low cost (~$4000) grating instrument, the Model 430. The instrument has a sensitivity of 5 parts per trillion with a 15-nm bandwidth. A jacketed cell holder provides temperature control, and a range selector has precision ranges of 1, 3, 10, 30, 100, 300 and 1000.

3. Practical considerations

(A) CUVETTES

(1) Emission from cuvettes

Parker [20] has shown that various types of quartz cuvette emit an appreciable amount of fluorescence (Fig. 26). Fused quartz has an extremely high background fluorescence; synthetic silica has a

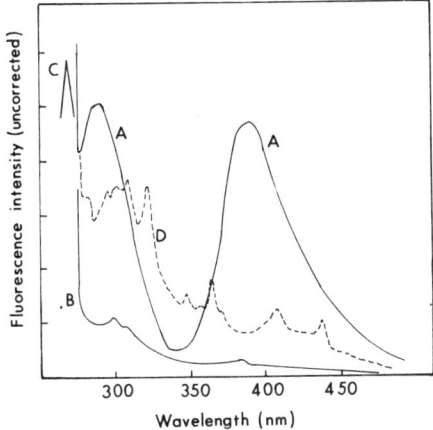

Fig. 26. Interference from scattered light and fluorescence of cuvettes. Specpure cyclohexane excited by light from a Bausch and Lomb monochromator set at 250 nm. Curve A, in fused-quartz cuvette, with filters (sensitivity 1000); Curve B, in synthetic silica cuvette, with filters (sensitivity 1000); Curve C, as for Curve B, but sensitivity 500 to show intensity of Raman band; Curve D, in synthetic silica cuvette, without filters (sensitivity 1000). Reprinted from ref. 20 by courtesy of C.A. Parker.

much lower background and is preferred for studies in the UV. As we have already mentioned, for studies in the visible region, which constitute about 80% of all measurements, pyrex or glass cells are sufficient for good results, and the large additional cost of quartz or silica is not necessary.

Fused-quartz cuvettes also emit in some regions of the visible. For example, Parker and Joyce [21] noted phosphorescence in the 370—430-nm region. Synthetic silica cells show a much lower background.

(2) Care of cells

a. Causes of cell deterioration. Cell deterioration is caused by films, etching, and contamination. Films are deposited by solvent evaporation; strong wetting agents or inadequate cleaning. Etching is due to continued use of strong alkalis or concentrated mineral acids, either in the sample or in the cleaning solution. It is also due to weak alkaline solutions left in the cell for long periods. Contamination is caused by solutions being allowed to evaporate in the cell, leaving deposits of salts, organic material, etc.

b. Effects of cell deterioration. The principal effects of cell deterioration are reduced transmission and light scattering caused by etching of the cell windows. Films and particulate matter also decrease transmission and may contaminate the sample.

c. Detection. Cell deterioration may be evident from poor base lines. The extent of cell deterioration may be judged by the following specifications: at 220, 240, and 270 nm (silica cells) or 320 nm (Vycor or Pyrex cells) the transmittance of a new cell filled with distilled water will be at least 70% of transmittance with air alone in the cell space.

d. Prevention. The following procedures should be observed for proper cell maintenance.

(1) Always start cleaning by rinsing the cell thoroughly with distilled water (aqueous samples) or a suitable non-fluorescent, purified organic solvent.

(2) Clean the cell with a mild agent as soon as possible after each use. Mild inorganic detergents (Calgonite) may be used if it is certain that they produce true solutions and do not contain particulate matter.

(3) For hard-to-remove deposits use a solution of 50% 3 N hydrochloric acid and 50% ethanol.

(4) Whenever possible, rinse the cell with the sample solution before filling.

(5) Remember, that if a reagent is not of spectrograde purity, it may leave a deposit on the cell window after evaporation.

(6) Never blow the cell dry with air. It is better to speed evaporation of the solvent with the aid of vacuum.

(7) Never use any brush or instrument that might scratch the sides of the cell.

(8) Never use alkalis, abrasives, etching materials, or hot concentrated acids.

(9) Never use ultrasonic devices to clean cells.

(B) SOLVENTS

It cannot be too strongly emphasized that pure solvents must be used in fluorometric work in order to obtain meaningful results. It is

116

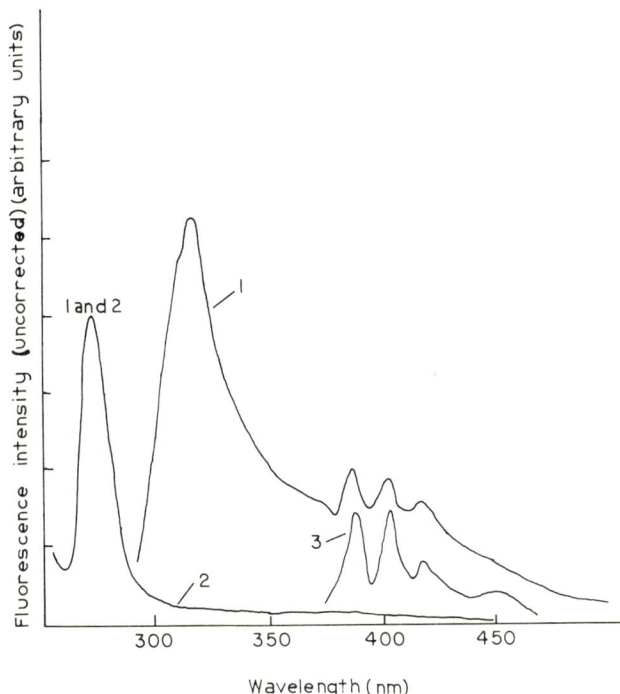

Fig. 27. Fluorescence of trace impurities in ethanol excited by light of 250-nm wavelength: Curve 1, absolute ethanol as received; Curve 2, ethanol purified by distillation; Curve 3, purified ethanol with added anthracene (0.0003 μg/ml). The peak at 270 nm in Curves 1 and 2 is the Raman emission. Reprinted from ref. 20 by courtesy of C.A. Parker.

not sufficient merely to demonstrate that the solvent or reagent is not fluorescent in itself. Non-luminescent impurities can act as quenchers. Traces of peroxides in diethyl ether can significantly reduce the luminescence of nitrogen heterocyclics due to a charge-transfer reaction [22]. This occurs even when the peroxide level is low. The same type of excited-state reaction occurs between nitrogen heterocyclics and aliphatic alcohols, which are frequently present as solvent contaminants.

Generally, most commercial solvents must be purified before use in fluorescence. Methyl and ethyl alcohols and dimethylformamide should be distilled before use. Distillation of absolute ethanol from potassium hydroxide has been found to be satisfactory. Figure 27 shows the spectra obtained from commercial absolute ethanol (Curve 1) and ethanol purified by distillation (Curve 2). The anthracene present as impurity (Curve 3) is removed by distillation.

n-Butanol, benzene, heptane, ethylene dichloride, ether and acetone have been purified for fluorescence assay by shaking with 0.1 M hydrochloric acid, 0.1 M sodium hydroxide, and then water. Purification can also be effected by passing solvents over silica gel.

Harleco reagents for fluorescence studies have been marketed by the Hartmann—Leddon Company (Philadelphia) for about 10 years. Alcohols, water chlorinated solvents, hydrocarbons, and acids with very low background fluorescences are available. Good solvents for fluorescence are those marketed as "gas-chromatography grade". These solvents have a very low concentration of impurities because of the sensitivities of gas-chromatography methods.

Frequently contamination not originally present in the solvent can result from packaging and from contact with rubber, glass, and cork stoppers, grease and filter papers, which contain fluorescent contaminants or quenchers. Boron can be extracted from pyrex glass, and aluminum, calcium, and silica from glass. Many reagents (e.g., hydrochloric acid) contain aluminum in appreciable concentrations.

Many synthetic detergents are fluorescent. The luminescent properties of a detergent should be checked before use. Inorganic cleaning agents, such as Calgonite, are good for fluorometry. Dichromate cleaning solutions should never be used because even traces of it absorb significant radiation and will interfere with an assay. When necessary, hot nitric acid can be used to clean cuvettes.

(C) SOURCES AND DETECTORS

Xenon lamps are used in all commercial grating instruments, mercury lamps in all filter instruments. Safety precautions that should be followed using these lamps were presented previously in Sect. 2.

Most fluorometers and spectrophotofluorometers today use a photomultiplier tube as the detector. The most common photomultiplier tube is the 1P21 (300—600 nm), but investigators should be familiar with other tubes that can be alternatively used, such as the 1P28 (200—600 nm), or the 7102 (500—1000 nm) for IR studies. As we mentioned above, no two photomultiplier tubes, even from the same manufacturer, will have the same characteristics. Care in order-

118

ing and evaluating photomultipliers should be exercised. In no case should the response characteristics of a photomultiplier tube given by a manufacturer be accepted and used in correcting spectra. Photodetectors should be kept in the dark at all times to obtain low noise and good operation.

(D) TEMPERATURE

Fluorescence is quite sensitive to changes in temperature, as we have already pointed out. Some compounds, like indoleacetic acid, undergo a 50% change in fluorescence in the $20-30°C$ temperature region.

Heat radiating from the light source used for excitation is the major cause of temperature instability. In some instruments the temperature of the solution in the cell compartment will rise as much as $6-10°$ in a few minutes. Under these circumstances it would be impossible to measure the fluorescence of a compound like indoleacetic acid, which has a high temperature coefficient.

The cell compartment should be kept as close to room temperature as possible. Some instruments, such as the Aminco—Bowman, have cooling attachments to maintain a constant temperature. Circulating water or air also help to maintain room temperature in the sample cell. A constant-temperature room does not always help. Also, samples that must be carried through a procedure involving heating or cooling should be allowed to come to room temperature before assay.

(E) FILTER VS. GRATING INSTRUMENTS

We discussed previously the differences between filter and grating instruments, and the features of various instruments. The choice of a filter or grating instrument will mainly depend on the budget of one's institution. A good filter instrument costs about $1000; a good grating instrument about $5000. More money than this will buy various attachments, such as corrected-spectra, temperature control, and variable slits. Actually a good filter instrument will enable one to do anything a grating instrument will do, except scan and resolve closely absorbing or emitting species. It is durable, efficient, and more sensitive than the grating instrument.

The grating instrument will allow one to achieve good resolution and to choose the wavelengths of maximum excitation and emission for analysis. The spectra obtained will not be the true spectra, however, because of differences in the intensity of the spectral source, the blaze of the monochromator, and differences in the detector response with differences in wavelength. This will be discussed later. Filters are an essential part of the filter fluorometer, but they can be quite useful in grating instruments. The proper filter placed between the emission monochromator and the phototube can often eliminate many difficulties caused by scatter and second-order spectra.

(F) CARE OF FILTERS

Filters are very delicate and should be handled with care. Handle filters only by the edges or extreme corners. Keep filters flat and dry — moisture tends to cloud them. Never wash the filter with water under any circumstances. If water should come in contact with the gelatin at the edges of the filter, this will cause the filter to swell. The glasses will separate so air can enter the filter. If the filter becomes so dirty that it cannot be cleaned by simply rubbing, a piece of lens cleaning paper or soft cloth moistened with a solvent should be used. Be careful, however that the cloth and glass are free of grit, which might scratch the glass, and be sure the solvent does not touch the cemented edges of the filter. Dry, cool storage is desirable for filters; a desiccated sealed container is desirable. Filters should never be used at temperatures higher than 120—130°C.

(G) SELECTION OF THE EXCITATION AND EMISSION WAVELENGTHS

In choosing the best wavelength for excitation of the molecule to be studied, several factors should be considered. The wavelength should be near a strong intensity point of the instrument's radiation source and also at a strong absorption band of the compound. The excitation wavelength should also be reasonably separated from the emission band. Since scatter is usually observed, there should be at least 20—30 nm separating the two peaks, and preferably 50 nm. If the compound possesses two or more peaks in the excitation spectrum, all of good intensity, the one at longest wavelengths should, generally, be chosen. This will ensure minimum photodecomposition of the compound.

If absorbing interferences are present in the sample, one might wish to choose a wavelength for excitation different from the maximum wavelength. If a grating instrument is available, the emission peaks are found by running the emission spectrum at the wavelength of maximum excitation. The latter is determined by setting the emission monochromator at various wavelengths from 350 to 600 nm in 50-nm increments and manually scanning the excitation monochromator. When a maximum is observed in the excitation spectrum, the emission monochromator is scanned until a maximum is observed. A permanent excitation spectrum, uncorrected of course, is then obtained by using an x—y recorder at the wavelength of maximum emission. Then the emission spectrum is obtained by recording at the wavelength of maximum excitation.

If only a filter fluorometer is available, the absorption spectrum of the compound is obtained with any good spectrophotometer. The λ_{max} should be the λ_{ex} of the compound. One then chooses a filter that best fits the λ_{ex} of the compound and one of the lines of the mercury lamp that is used in the filter fluorometer. Since the lamp will have a maximum intensity at these spectral lines, it is often best to pick a filter to match this line rather than the λ_{ex} of the compound, provided the compound exhibits appreciable absorption at this wavelength. The 254- or 365-nm lines of the lamp would be best for the excitation of, for example, quinine sulfate (λ_{ex} 250 and 350 nm).

Then one chooses a filter for selection of the emission wavelength by trying different secondary filters at 50-nm intervals, starting at λ_{ex} + 50 nm. Care should be taken to choose a secondary filter that is compatible with the primary filter. When one has found a filter that gives a good signal, one could try other filters that peak in the same area to attempt to optimize the signal.

Next, let us consider the emission spectrum of the compound. The wavelength chosen to measure emission should be separated by at least 20—50 nm from the excitation peak to eliminate scatter. The fluorescence spectrum should closely resemble a mirror image of the absorption spectrum. In many cases frequencies of vibration excitation in the lowest excited singlet are within 10—20% of those in the ground state, which is the factor responsible for the mirror-image relationship between absorption and fluorescence spectra. The presence of such a relationship between two spectra may thus be taken to indicate a large change in the configuration of the molecule caused by photoexcitation.

Fewer peaks are sometimes observed in the emission spectra than in the excitation spectra; this is because emission frequently occurs from the first excited state, whereas absorption can take place to the first, second and sometimes the third excited singlet. Thus quinine sulfate has peaks at 250 and 350 nm but only one emission peak at 450 nm. If the intensity of the emission peak is less than that of the excitation peak, photodecomposition should be suspected. Any other peaks in the excitation spectrum are either scatter or Raman peaks. The Raman peaks in water appear at 248, 313, 365, and 436 nm. In the spectrum of quinine sulfate scatter would be observed at either 250 or 350 nm, depending on the wavelength used for excitation. Second-order scatter would appear at 500 and 700 nm.

(H) LIGHT SCATTERING

The term "scattered light" refers to light which is of the same or longer wavelength as that of the exciting beam and which emerges from the cuvette. The scatter that occurs at the wavelength of excitation is composed of Rayleigh scattering from the solvent, Tyndall scattering from colloidal particles, and scatter from the surface of the container. There is another type of scatter, called Raman scatter, which occurs at wavelengths longer than the exciting radiation. It is a physical property of the pure solvent and differs from λ_{ex} by a con-

TABLE 12

Raman bands for several solvents corresponding to the various mercury lines [a]

Solvent	Wavelength (nm) of Raman band produced by excitation at				
	248	313	365	405	436
Carbon tetrachloride	—	320	375	418	450
Chloroform	—	346	410	461	502
Cyclohexane	267	344	408	458	499
Ethanol	267	344	409	459	500
Water	271	350	416	469	511

[a] After Parker [23].

122

stant frequency. The Raman peaks of several common solvents are listed in Table 12. At high sensitivities all types of scatter appear, but Rayleigh and Tyndall scatter are more intense than Raman scatter. Hence, in water solution, excitation at 313 nm will produce Rayleigh peaks at 313 and 626 nm (second-order) and a Raman peak at 350 nm. When the fluorescence and excitation wavelengths are close together, the distortion due to scattering severely limits sensitivity.

Scatter can be diminished by the use of a secondary filter. With radiation at 365 nm one would use a secondary cut-off filter that has zero transmission below 420 nm. This would eliminate both Rayleigh and Raman scatter.

Price et al. [24] showed that the scatter components, Raman and Rayleigh, differ from fluorescence in their degree of polarization and can be removed by polarized filters. However, such filters also reduce signal intensity and limit sensitivity. Chen and Bowman [25] used the quartz Polarcoat polarizing filter to eliminate scattered light. Such filters have a higher transmission in the UV [26]. Chen [27] pointed out that scatter produced by the UV source can be decreased by a single polarized filter in the excitation beam. He found that the use of two polarizers in the excitation beam was not more effective.

(I) PHOTODECOMPOSITION

Photodecomposition was discussed earlier. It becomes an extremely important problem in dilute solutions; in concentrated solutions the amount of substance decomposed is minor in comparison with the total amount. To minimize photodecomposition it is important that the fluorometer has a shutter, so that the solution is only irradiated during the short period of measurement. The measurement should be made as rapidly as possible. If labile compounds are under study, a weaker source and a more sensitive photodetector should be used; for example, proteins which are photosensitive [28], should be handled in this manner.

Photodecomposition occurs locally in a 1-cm cuvette, and often diffusivity allows the solution to recover and re-equilibrate. Re-equilibration could also be accomplished by stirring or inverting the cuvette. Photodecomposition becomes more of a problem with microcuvettes.

(J) STANDARDIZATION

Because of differences in lamp intensity and photomultiplier sensitivity, as well as other instrumental variables, it is impossible to obtain exactly the same reading each and every day on a fluorometer. Hence the instrument should be standardized and set to a constant sensitivity each day. Generally, a solution of a stable fluorophor of definite known concentration is used to set the instrument to a given level. One of the most widely accepted standards is quinine sulfate, which has a λ_{ex} of 350 nm and a λ_{em} of 450 nm. This compound is particularly good for assays that use the 365-nm mercury line. A solution of quinine sulfate, say 10^{-5} M, is prepared and is used to set the instrument to a preset level (e.g., 80 on a particular scale) every day.

Quinine sulfate in 0.1 N sulfuric acid, although a good standard, has certain limitations:

(1) It is useful as a standard only for substances that are close to it in fluorescent properties (350 and 450 nm). In other regions other standards, such as indole, phenols, and fluorescein, have been used.

(2) It is light-sensitive, particularly at concentrations less than 1 μg ml^{-1}, and should be kept refrigerated in a black bottle.

(3) It will appear to decrease in concentration at low concentrations (<50 μg ml^{-1}) due to adsorption onto glass surfaces. For best results a more concentrated stock solution of quinine sulfate should be stored and dilutions made when necessary.

A set of six fluorescence standards useful in the standardization of any fluorometer is sold by Perkin—Elmer (Norwalk, Conn.). The set consists of six fluorescent compounds dissolved in a plastic matrix: (a) a mixture of anthracene and naphthalene, (b) orilene, (c) p-terphenyl, (d) tetraphenylbutadiene, (e) a proprietary mixture, and (f) Rhodamine B. By using any one of these six standards, any region can be standardized. Since the samples are solid (they look like plastic blocks), they are stable and can be used indefinitely without special storage. The set is recommended for those doing routine fluorescence measurements.

Alternatively, the Raman spectrum of the solvent could be used to standardize instrument sensitivity daily [23]. Since the Raman peak is always located near the excitation maximum, this is a more versatile method of standardization. The Raman peak appears on the spectrum as the internal standard of sensitivity.

124

(K) CALIBRATION OF WAVELENGTH

Original calibration is performed at the factory, but every spectro-fluorometer must be calibrated periodically to ascertain that the reading on the wavelength dial is the true wavelength. Differences could cause loss in sensitivity and error in an assay.

In calibration the difference between the apparent wavelength and the known wavelength for a series of mercury emission lines is used as a test of wavelength accuracy. A low-pressure mercury-arc lamp (Pen Ray quartz lamp, UV Products, Inc., San Gabriel, Calif., or equivalent) is placed in the sample cell holder of the instrument. The 12 mercury lines at 253.65, 296.73, 302.15, 313.16, 334.15, 366.33, 404.66, 407.78, 435.84, 546.07, 576.96, and 597.07 nm are used. The second-order lines at 626.32, 668.3, 732.66, and 809.32 nm can also be used for calibration if desired. For each line the position of the wavelength dial is adjusted to give maximum signal, and the wavelength reading is recorded. The difference between the observed reading and the true reading represents the correction factor that must be applied to the reading on the dial to give the true value. To compensate for dial backlash, always adjust the dial to the peak reading from the same side. In scanning instruments turn the dial to the peak in the same direction as the dial is turned by the scan motor.

To adjust the excitation monochromator once the emission mono-chromator has been adjusted, place in the cuvette a suspension of Ludox (an aqueous suspension of colloidal silica, E.I. du Pont de Nemours, Wilmington, Del.). Place the Pen Ray lamp in the standard position and adjust the excitation wavelength dial setting until a maximum meter signal is produced at each of the previously deter-mined wavelength settings for the emission monochromator. The correction factors for each of the excitation wavelengths are deter-mined for future use.

In many cases the errors obtained in wavelength reading are size-able. For example, Hercules [29] found an error of 15—20 nm in the fluorescence maxima of 1-naphthol. He pointed out that the pen-and-ink recorders can also be responsible for spectral errors unless they are checked at various recording speeds.

(L) CORRECTION OF SPECTRA

The excitation and emission curves recorded from most spectro-fluorometers are only approximate curves and do not represent the true spectra. If one simply wishes to report the maximum wavelengths of excitation and emission for a compound in describing an analytical procedure, one need only calibrate one's instrument as described in the preceding section and need not worry about what the true spectrum looks like. If, however, one wants to obtain the true spectrum of a compound for publication, one must correct the spectrum. This will not be discussed here, but the reader is invited to see refs. 1 and 30—37 for more information.

4. Structural and environmental effects

In order to effectively utilize luminescence as an analytical tool it is necessary that every researcher know the basics of the effects of structure and the sample environment on the emission process. Under structural effects we shall briefly consider what types of compounds fluoresce, and how we might increase the total emission by changes in structure. Under environmental effects we shall study how pH, the solvent, other ions and other factors change the luminescence.

(A) STRUCTURAL EFFECTS

Fluorescence phenomena are not generally sensitive to the finer details of molecular structure. Among the huge number of known organic compounds, only a small fraction exhibits intense luminescence. Therefore, the mere fact that a molecule fluoresces can in itself provide significant information regarding its structural features. Aromatic hydrocarbons possessing large conjugated systems often exhibit fluorescence. In these systems π electrons, which are less strongly held than sigma electrons, can be promoted to π^* antibonding orbitals by absorption of electromagnetic radiation of fairly low energy without extensive disruption of bonding (Fig. 28). The $\pi \rightarrow \pi^*$ (K band) has a high extinction coefficient and hence is likely to occur.

Characteristically, fluorescence is the light emission from a π, π^* singlet. Fluorescence is less likely from an n, π^* state. The extinction coefficient for the $n \rightarrow \pi^*$ transition is low, hence, less likely to occur. Secondly, the characteristics of the resulting n, π^* state are such that other energy-dissipating processes compete successfully

126

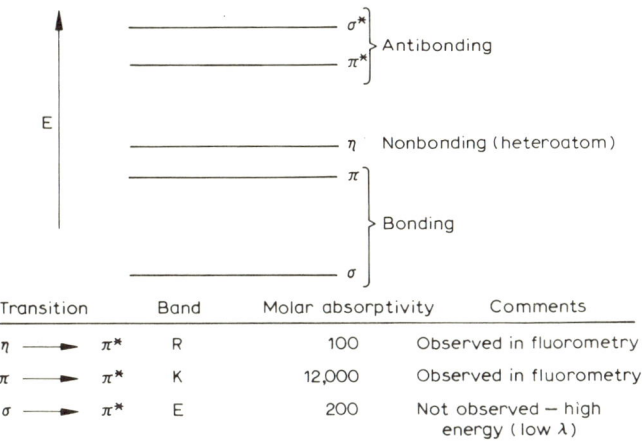

Transition	Band	Molar absorptivity	Comments
$\eta \longrightarrow \pi^*$	R	100	Observed in fluorometry
$\pi \longrightarrow \pi^*$	K	12,000	Observed in fluorometry
$\sigma \longrightarrow \pi^*$	E	200	Not observed — high energy (low λ)

Fig. 28. Types of transitions involved in the absorption process.

against the emission of light. Other processes ($\sigma \to \sigma^*$, $\sigma \to \pi^*$) never lead to fluorescence since they are of such high energy that bond breakage occurs before emission.

Only a relatively few aliphatic and saturated cyclic compounds fluoresce or phosphoresce. This is because the electrons in aliphatic compounds are normally tightly bound and participate in sigma bonding. When aliphatics and saturated cyclics absorb UV energy, photodecomposition usually results. Aliphatic aldehydes and ketones, where the carbonyl oxygen can be excited to an antibonding π^* orbital, do fluoresce, though weakly. Some highly conjugated aliphatic compounds such as Vitamin A, are fluorescent enough in the native state that direct analysis of these compounds is possible.

Because they possess a large number of π electrons, aromatic compounds possess a high fluorescence probability. As the degree of conjugation increases, the intensity of fluorescence increases and a bathochromic shift (shift to higher wavelengths) is observed (Table 13).

Certain ring structures normally produce intense fluoresence, such

TABLE 13

Effect of conjugation on fluorescence

Compound	Φ_F	Excitation maximum	Fluorescence maximum
Benzene	0.11	204	278
Naphthalene	0.29	286	321
Anthracene	0.46	375	400
Naphthacene	0.60	390	480
Pentacene	0.52	580	640

as quinone, pyrrol, fluorescein, furan, etc. Heterocyclic compounds usually do not fluoresce in non-polar solvents, but many do in polar solvents or acid solutions. Some typical heterocyclics possessing native fluorescence are the indoles, quinolines, coumarins and resorufin derivatives.

The nature of substituent groups (especially chromophoric groups) plays an important role in the nature and extent of fluorescence by a molecule. Fluorescence yields (intensities) and energies of aromatic and heterocyclic hydrocarbons are usually altered by ring substitution. Unfortunately, we cannot often make many broad generalizations. Substituent effects upon the chemical and physical properties of organic molecules in their ground electronic states constitute a lively area of investigation at present. Furthermore, only little is known about the influence of substituents on the behavior of excited states. Both effects must be understood before generalizations concerning the effect of various substituent groups can be made.

In general, the substitution of certain electron-donating functional groups for hydrogen, such as $-F$, $-OH$, $-OCH_3$, $-NH_2$, $-NHCH_3$ or $-N(CH_3)_2$, usually produces a more intensely fluorescing compound, while electron-withdrawing groups such as halides (except F), $-NO_2$ or $-COOH$ often yield a compound of decreased fluorescence intensity (Table 14).

The only exception to the general rule of substitution, is cyanide, which is *meta*-directing, yet which always causes an increase in intensity. In all cases, except that of phenol, the ionized hydroxy group is more fluorescent than the $-OH$ group. Certainly the electronic configuration of a molecule changes when a proton is extracted from a functional group. The changes, however, may not necessarily be of the same order of magnitude in both the ground

TABLE 14

Fluorescence of monosubstituted benzenes

Compound	Substituent	λ_{ex}	λ_{em}	Rel. intensity
Ortho—para directing				
Benzene	H	269	291	1
Aniline	NH_2	290	345	46
Dimethylaniline	$N(CH_3)_2$	297	363	114
Fluorobenzene	F	269	285	13
Chlorobenzene	Cl	281	294	0.02
Bromobenzene	Br	—	None	0
Iodobenzene	F	—	None	0
Phenol	OH	279	302	112
Meta—directing				
Benzoic acid	COOH	—	—	0
Nitrobenzene	NO_2	—	—	0
Benzaldehyde	CHO	—	—	0
Benzonitrile	CN	287	294	45

and excited electronic states. In many cases, such as the fluoresceins, the presence of a carboxylic acid functional group can act to increase or decrease fluorescence depending on whether or not the group is ionized.

In many cases two compounds with almost identical structures will exhibit vastly different fluorescent properties. Consider, for example, phenolphthalein(I) which is non-fluorescent and fluorescein(II) which is highly fluorescent. This indicates that ring closure is conducive to fluorescence. Structural rigidity reduces vibrational amplitudes that promote radiationless losses, and is a prime factor in fluorescence capability.

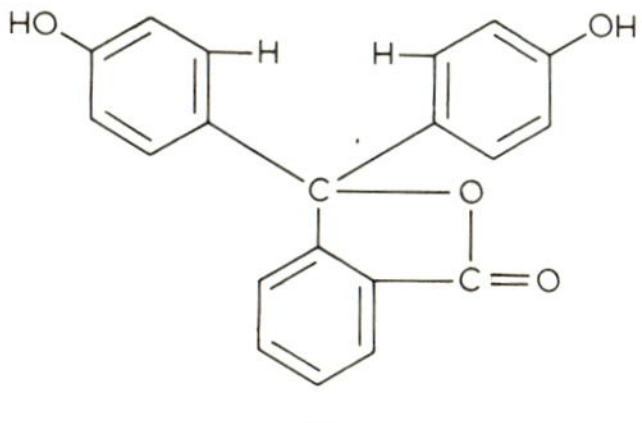

(I) 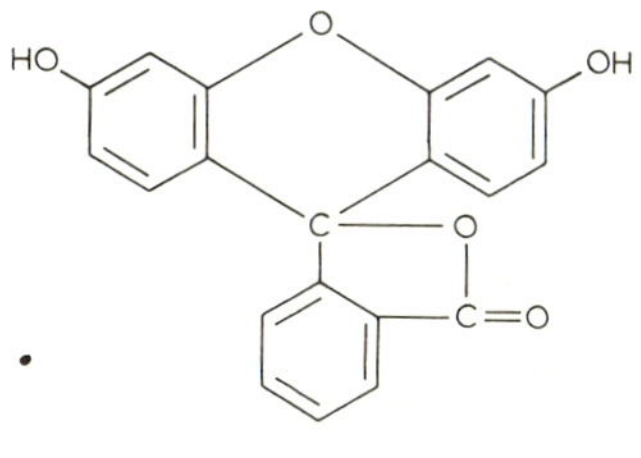   (II)

TABLE 15

Luminescence of dianions of fluorescein dyes at 77 K [a]

Substituent and number	Φ_F	Φ_P/Φ_F
None	0.83	0
Cl, 2	0.79	0
Br, 1	0.60	0.13
Br, 2	0.29	0.21
I, 1	0.15	0.67
I, 2	0.054	1.05

[a] Selected data from ref. 38.

The introduction of heavy atoms (e.g., halogens) decreases the fluorescence in favor of phosphorescence. Thus fluoro- or chloro-fluorescein is fluorescent, but iodofluorescein is phosphorescent but not fluorescent (Table 15).

As one traverses the substituent series Cl, Br, I, phosphorescence is increasingly favored and the quantum efficiency ratio of phosphorescence to fluorescence (ϕ_P/ϕ_F) increases while ϕ_F decreases. These trends can be rationalized by saying that heavy-halogen substitution increases the rates of $S_1^* \rightarrow T_1^*$ intersystem crossing and $T_1^* \rightarrow S_0$ phosphorescence [39]. This occurrence is commonly termed the "heavy-atom effect".

Heteroatom substitution greatly affects the luminescence of aromatic compounds. The major reason for this is that most heteroatoms possess at least one "lone" pair of non-bonding (n) electrons. Absorption of radiation results in an $n \rightarrow \pi^*$ transition. This $n \rightarrow \pi^*$ transition is responsible for many of the differences between the luminescence properties of heterocyclic compounds and those of aromatic hydrocarbons. Because population of the (n, π^*) singlet is much less probable ($\epsilon \sim 10^2$) than population of the (π, π^*) singlet ($\epsilon > 10^4$) and the energy separation between the first excited singlet and lowest triplet is very low for (n, π^*) states, a $S_1^* \rightarrow T_1^*$ intersystem crossing occurs with a greatly enhanced ϕ_P/ϕ_F being observed for heteroatom compounds.

Similarly, most aromatic carbonyl compounds also possess lowest-excited-singlet states of (n, π^*) character, so that intersystem crossing to the triplet state is very efficient. Hence a large number of aromatic aldehydes and ketones exhibit fairly intense phosphorescence, but no fluorescence.

130

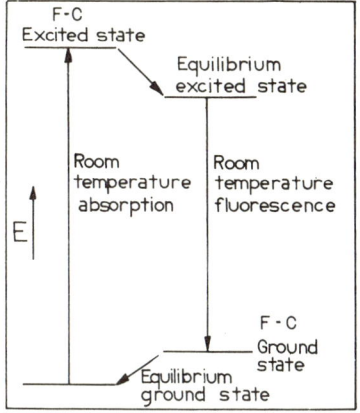

Fig. 29. Schematic representation of equilibrium and Franck—Condon electronic states.

(B) ENVIRONMENTAL EFFECTS

Environmental factors can strongly influence the fluorescence of polyatomic molecules. Molecular environment constitutes an important parameter which can be used by the analyst to increase the sensitivity and selectivity of fluorometry. A large number of environmental effects are of importance. We will only discuss a few of these: pH, heavy atoms, metal ions, oxygen and temperature. It cannot be too strongly emphasized that pure solvents must be used in fluorometry to obtain good results. It is not generally sufficient to demonstrate that the solvent does not itself fluoresce since non-luminescent impurities can act as quenchers.

(1) Solvent effects

An electronic transition must occur while the photon is in the vicinity of the absorbing molecule in about 10^{-18} s. This is $10^3 - 10^4$ times faster than the rate of bond stretching, so that an electronic transition occurs before any change in interatomic distance will occur. This observation is known as the Franck—Condon principle, and is pictured in Fig. 29. The molecule, unchanged except for the electronic transition, finds itself in a metastable, excited state, called the Franck—Condon excited state. The molecule then readjusts to its

TABLE 16

Electrostatic solvent effects on anthracene fluorescence

Solvent	v_F (cm^{-1})	λ_F (nm)	$\lambda_F - \lambda_A$
(Vapor)	27380	365	180
Hexane	26517	377	156
Methanol	26461	378	177
Dioxane	26220	381	280
Toluene	26170	382	228
Benzene	26116	383	262
Chlorobenzene	26064	384	279
Acetonitrile	25972	385	285
Formamide	25508	392	347
N-Methylformamide	25112	398	368

new environment by a solvent reorientation in about 10^{-12} s, and reaches the equilibrium excited state. Emission then occurs from the equilibrium excited state to the Frank—Condon ground state in about 10^{-8} s. The molecule than reorients itself to its environment and reaches the equilibrium ground state.

Thus, in solutions, the molecule enters into one excited state and leaves from another excited state, each state having different energy levels. Therefore, the wavelengths of absorption and emission are different. The fact that the molecule is excited to the Franck—Condon excited state first, is the principal reason that the fluorescence spectrum is more subject to different solvent effects than the absorption spectrum.

In most polar molecules the excited state is more polar than the ground state. Hence an increase in the polarity of the solvent produces a greater stabilization of the excited rather than the ground state. Consequently, a shift in both the absorbance and fluorescence spectra to lower energy or higher wavelength is observed (Table 16 and Fig. 30). This type of behavior is observed for almost all cases, even when the solute and solvent are not polar, because of the induced dipole in the excited state. The magnitude of the change, however, is not as great.

Concerning intensity, electrostatic solvent—solute interactions do not produce significant variations in fluorescent yields if both solute and solvent are non-polar. Some solvents may appear to be better than others because of solvent quenching.

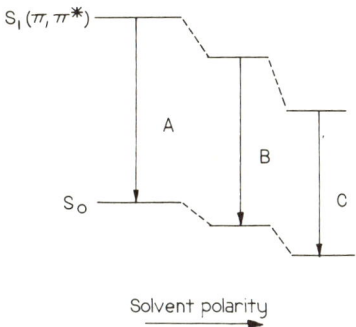

Fig. 30. Schematic representation of the red shift in $\pi^* \to \pi$ fluorescence spectra as the polarity of the solvent is increased. Solvent A is less polar than solvent B, which is in turn, less polar than solvent C.

$$^1A^* + S \rightleftharpoons [A^- S^+] \to {}^3A^* + S \qquad\qquad (14)$$
$$\underset{A}{\overset{\text{radiationless}}{\rightsquigarrow}}$$

For polar solutes or solvents, electrostatic intensity perturbations are relative to those produced by hydrogen bonding or other specific interactions.

(2) Effect of external heavy atoms and paramagnetic ions

The effect of an external heavy atom (iodide as solvent, for example) or of a paramagnetic ion (Cu^{2+}, Fe^{3+} in solution) is to increase the rate of intersystem crossing ($S_1^* \to T_1^*$) and also the rate of the $T_1^* \to S_0$ transition [40]. The net result is an increase in the phosphorescence/fluorescence ratio (ϕ_P/ϕ_F). The effect is believed due to the formation of weak charge-transfer complexes of the substance A and the heavy atom S.

$$A^* + S \to [A \to S] \to {}^3A^* \overset{h\nu}{\longrightarrow} A \qquad\qquad (15)$$

A shift in the spectra to longer wavelengths is frequently observed. This effect can be used to distinguish n, π^* and π, π^* triplets in car-

TABLE 17

External heavy-atom effect on phosphorescence intensity [a]

Compound	Concentration $(\mu g\ ml^{-1})$	P/P_0 [b]
Naphthalene	13	0.1
Anthracene	18	0.9
Triphenylene	2.3	0.05
Naphthacene	6.0	4.6
1,2-Benzanthracene	23	1.4
3,4-Benzpyrene	25	3.5
1,2-Benzfluorene	22	13
2,3-Benzfluorene	22	25

[a] From ref. 41.
[b] Ratio of phosphorescence intensity in ethanol—ethyl iodide (5/1 v/v) to that in ethanol.

bonyls. Phosphorescence from the n, π^* state is much less susceptible to heavy atom enhancement than the π, π^* state. The effect of a heavy atom (ethyl iodide) on the phosphorescence intensity of various solutes is given in Table 17.

(3) Effect of pH

Most molecules that contain an ionizable hydroxy group will exhibit an increase in fluorescence intensity as the pH is increased — one of the exceptions to this is phenol, as was previously mentioned. Most hydroxy compounds also possess a lower pK_a in the excited state than they do in ground state due to a stabilization of the ϕ—O^{-*} species.

The acid—base properties of 2-naphthol serve as an excellent example of excited state dissociation. The pK_a for ground state 2-naphthol is about 9.5. The 2-naphthol molecule exhibits a single fluorescence peak at 359 nm, whereas, the 2-naphtholate anion exhibits a fluorescence peak at 429 nm. The large energy separation of these two fluorescence peaks makes it easy to measure the fluorescence of either molecular 2-naphthol or ionic 2-naphtholate, without interference from the other species. In the pH region where molecular 2-naphthol predominates the overall process may be represented

134

as:

$$C_{10}H_7OH + h\nu_M \qquad C_{10}H_7O^- + h\nu_I$$

$$\uparrow (2) \qquad\qquad \uparrow (5)$$

$$C_{10}H_7OH + h\nu \xrightarrow{(1)} C_{10}H_7OH^* \underset{(4)}{\rightleftharpoons} C_{10}H_7O^{-*} + H^+ \qquad (16)$$

$$(3) \qquad\qquad (6)$$

$$\downarrow \qquad\qquad\qquad \downarrow$$

$$C_{10}H_7OH \cdot \qquad\qquad C_{10}H_7O^-$$

(1) Absorption of radiant energy to produce the excited *molecule*. Remember the pH is less than 9.5 and so 2-naphthol predominates.

(2) Deactivation of the excited molecule by molecular fluorescence (359 nm).

(3) Radiationless deactivation of the excited molecule.

(4) Dissociation of the *excited* molecule producing a proton and an excited anion.

(5) Deactivation of the excited anion by ionic fluorescence (429 nm).

(6) Radiationless deactivation of the excited anion.

In the pH range below 9.5 molecular 2-naphthol predominates in the ground state and so it may be possible to observe an excited state dissociation reaction by measuring the fluorescence of 2-naphtholate ion (429 nm) while exciting 2-naphthol (solution of pH less than 9.5). The excited state acidity of 2-naphthol is found to be 3.1, over 6 orders of magnitude different from the ground state. This means that if the fluorescence of 2-naphthol is measured at pH's >4, a large increase in sensitivity results.

Brønsted acidity differences between the ground and the lowest excited singlet state of organic molecules are large, commonly ranging from 4 to 9 pK units. Some compound classes, especially phenols, thiols and aromatic amines, become much stronger acids on excitation, whereas others (nitrogen and sulfur heterocyclics, carboxylic acids, aldehydes and ketones) with the lowest (π, π^*) singlets become more basic (Table 18).

Thus one may use pH as a parameter in fluorometric analysis to reduce interference by extraneous solutes in a mixture or to obtain the most strongly fluorescent species for analysis. Since excited-state pK_a values and proton transfer rates are, in some cases, rather sensitive to changes in molecular structure [41,45,46] analytical utilization of excited-state proton-transfer reactions may assist in increasing the selectivity and versatility of analytical procedures.

TABLE 18

Excited-state acidities for some aromatic compounds [a]

Compound	pK_a		
	Ground	Singlet	Triplet
Phenol	10.0	4.0	8.5
4-Methoxyphenol	10.2	5.6	8.6
2-Naphthol	9.5	3.1	7.7
1-Naphthoic acid	3.7	~11	4.6
2-Naphthoic acid	4.2	~11	4.2
Quinolinium ion	5.1	10.5	5.8
Acridinium ion	5.5	10.6	5.6
2-Naphthylammonium ion	4.1	~−2	3.1

[a] Data from refs. 42—44.

(4) Hydrogen bonding

Hydrogen bonding interactions of substituted aromatic molecules with the solvent or with other solutes can greatly affect their fluorescence behavior. The reader is directed to a review on the effects of hydrogen bonding on the fluorescence of organic molecules [47] for a more detailed explanation than that presented here.

We assume that a solute molecule A can hydrogen-bond with a molecule or solvent (or other solute) B. Formation of an excited-state hydrogen-bonded complex between the two can occur by excitation of such a species already present in the ground state.

$$A-B \xrightarrow{h\nu} (A-B)^* \tag{17}$$

Alternatively, during the lifetime of an uncomplexed, excited A^* molecule hydrogen bonding with B can occur.

$$A \xrightarrow{h\nu} A^*; A^* + B \to (A-B)^* \tag{18}$$

In the former case it is obvious that both the absorption and fluorescence spectra of A will be affected by hydrogen bonding with B. However, when hydrogen bonding takes place only after excitation, only the fluorescence spectrum of A will be perturbed by the interaction.

136

TABLE 19

Fluorescence yield for 5-hydroxyquinoline in various solvents [a]

Solvent	$-\Delta H$ [b] (kcal mol^{-1})	Φ_F	
		298 K	77 K
Isopentane	—	0.30	0.27
Acetonitrile	3.5	0.24	—
Sulfolane	3.5	0.21	—
Dioxane	4.4	0.19	—
Diethyl ether	5.1	0.12	0.24
Dimethylformamide	5.3	0.09	—
Tetrahydrofuran	5.5	0.09	—
Dimethyl sulfoxide	6.4	0.07	0.20

[a] From ref. 48.
[b] Enthalpy of hydrogen-bond formation between phenol and solvent; data from ref. 49.

It is often (but not always) observed that excited-state hydrogen bonding reduces the ϕ_F of A. An interesting example of this effect is 5-hydroxyquinoline. In hydrogen-bond-accepting solvents (acetonitrile, dioxane, DMSO, ether, etc.) it is found [48] that, as the enthalpy of formation of hydrogen bonds between a phenol and the solvent increases, ϕ_F decreases in room-temperature solutions (Table 19).

Unlike most nitrogen heterocyclics, 5-hydroxyquinoline exhibits very inefficient $S_1^* \to T_1^*$ intersystem crossing in any solvent. The influence of solvent on ϕ_F for this compound must therefore be attributed to increased $S_1^* \to S_0$ internal conversion induced by hydrogen bonding with the solvent [48]. Thus, as a general rule, the analyst performing fluorometric or phosphorometric analyses of molecules containing such functional groups as —OH, —CO$_2$H, —NR$_2$, or —SH should, when feasible, choose solvents that will not hydrogen-bond strongly with the substituent.

The influence of hydrogen bonding on the fluorescence of organic molecules is not limited to interactions with the solvent. As already noted, the fluorescence yield of 5-hydroxyquinoline can be correlated with the ability of the solvent to act as a hydrogen-bond acceptor. It therefore might seem reasonable to expect similar correlations for

8-hydroxyquinoline, particularly since the absorption spectra of 5- and 8-hydroxyquinoline are quite similar.

However, the ϕ_F's for 8-hydroxyquinoline are generally about 100 times smaller than that for 5-hydroxyquinoline in the same solvent, despite the fact that the absorption spectra of the two compounds are almost identical. Second, there is no obvious correlation between the hydrogen-bond-accepting ability of a solvent and the fluorescence yield of 8-hydroxyquinoline in that solvent. To understand these differences we must note that 8-hydroxyquinoline can engage in intramolecular (III) and intermolecular hydrogen bonding, whereas only the latter is possible for 5-hydroxyquinoline. One may conclude [48] that both intermolecular and intramolecular hydrogen

(III)

bonding affect the fluorescence of 8-hydroxyquinoline by increasing the rate constant for $S_1^* \to S_0$ internal conversion; this effect has also been noted in other systems exhibiting intramolecular hydrogen bonding. In a molecule like 8-hydroxyquinoline competition between intramolecular and intermolecular hydrogen bonding can have important effects. In a solvent that is sufficiently effective as a hydrogen bonder to disrupt the intramolecular hydrogen bond, but does not form an extremely strong intermolecular hydrogen bond with the phenolic hydroxyl, the fluorescence efficiency of a molecule like 8-hydroxyquinoline should be maximal. Thus, whenever fluorescence or phosphorescence assays are designed for a molecule capable of intramolecular hydrogen bonding, such competition should be taken into consideration in defining solvent and temperature conditions.

The general effects of excited-state hydrogen bonding on the luminescence of nitrogen heterocyclics are dependent on whether the lowest excited singlet is (n, π^*) or (π, π^*). It is normally observed that the excited (π, π^*) singlets of nitrogen heterocyclics are much

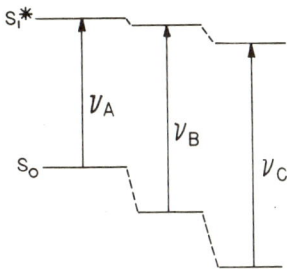

Increasing solvent hydrogen-bond donation

Fig. 31. Schematic representation of the blue shift in $n \to \pi^*$ absorption spectrum as the hydrogen-bonding power of the solvent is increased. Hydrogen-donating power increases in the order A < B < C.

more strongly basic than the ground state; one therefore expects the excited state to hydrogen-bond more strongly with protic solvents than the ground state. Thus for a nitrogen heterocyclic with a low-lying (π, π^*) excited singlet the frequencies of both absorption and fluorescence should shift to lower frequency as the hydrogen-bond-donating ability of the solvent increases, but the magnitude of the fluorescence shifts should be greater than those in absorption. In contrast (n, π^*) excited singlets of nitrogen heterocyclics are obviously much less basic than the ground state and should therefore be less susceptible to hydrogen bonding interactions. Thus, while energies of $n \to \pi^*$ absorption spectra should increase dramatically with increasing hydrogen-bond-donor power of the solvent, $\pi^* \to n$ fluorescence spectra should be virtually insensitive to this property of the solvent (Fig. 31).

It is often observed that aromatic carbonyl compounds and nitrogen heterocyclics fluoresce very weakly, or not at all, in non-polar, aprotic solvents, but that their fluorescence yields increase sharply on the addition of hydrogen-bonding solvents. In such compounds the lowest excited state is usually (n, π^*) in aprotic solvents, but the lowest (π, π^*) singlet often is not very much more energetic than the (n, π^*) singlet. On increasing the "proticity" of the solvent, the (π, π^*) states shift to lower energies, relative to the ground state, while the energies of (n, π^*) singlets increase as noted in the preceding paragraph. Therefore, it is possible for the opposing energy shifts (n, π^*) and (π, π^*) singlets, on addition of polar, hydrogen-bonding solvents, to be sufficiently large for a molecule whose lowest singlet is (n, π^*) in non-polar media to be (π, π^*) in hydroxylic solvents. Since fluorescence from (π, π^*) singlets is usually more efficient than that from (n, π^*) singlets, addition of hydroxylic solvents often produces increased fluorescence from hydrocarbon solutions of aldehydes, ketones, and heterocyclics [50,51]. For example, addition of n-propanol to glassy solutions of 3,4-benzoquinoline enhances ϕ_F by a factor of as much as 35 while simultaneously effecting a significant decrease in ϕ_P and an increase in the phosphorescence radiative lifetime [52]. Such observations are often noted for heterocyclics in which the lowest (n, π^*) and (π, π^*) singlets do not interchange as the proticity of the solvent is increased.

(5) Effect of temperature

As was previously mentioned, a decrease in temperature usually produces an increase in fluorescence intensity. The reason for this observation is that competing radiationless processes cannot compete as effectively at low temperatures as they can at higher temperatures. In fact, at very low temperatures one finds the vibrational structure of the molecule fully resolved.

A blue-shift (shift to shorter wavelengths) is also observed at low temperatures because the Franck—Condon excited state is frozen in and is the same as the equilibrium excited state. The result is a large ΔE and hence a longer wavelength.

There is one pronounced effect of temperature, and this is in the process known as delayed fluorescence. Consider, for example, the energy-level diagram in Fig. 32, in which T_1^* lies just below S_1^*. At sufficiently high temperatures intersystem crossing from S_1^* to T_1^* may be followed by thermal excitation of the triplet back to S_1^*. In that case, the fluorescence of the molecule will exhibit two spectrally identical components, one of which has a "normal" decay time while the other has a lifetime slightly shorter than that of T_1^*. The fluorescence intensity will increase with temperature. Such "delayed fluo-

140

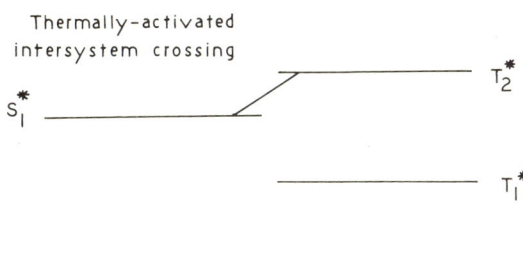

Fig. 32. Energy-level diagram for a molecule that exhibits thermally activated delayed fluorescence.

rescence", thermally activated, is exhibited by a number of molecules (e.g., anthraquinone).

(6) Effect of oxygen

Perhaps the most notorious quencher of fluorescence and phosphorescence is molecular oxygen. In liquid solutions quenching of excited-singlet states of organic molecules by dissolved O_2 molecule, which has a very large diffusion coefficient, especially in polar solvents, is so efficient that it can be a very serious problem.

It has long been known that some organic molecules are much more susceptible than others to fluorescence quenching by O_2. While the ground state of O_2 is a triplet, it is no longer believed that "paramagnetic" effects play a generally important role in O_2 quenching. Among the mechanisms that have recently been proposed to account for the quenching of excited singlet states by O_2 the two most important are [53]:

Enhanced intersystem crossing in $^1A^*$: $^1A^* + {}^3O_2 \rightarrow {}^3A^* + {}^3O_2$ (19)

Enhanced internal conversion in $^1A^*$: $^1A^* + {}^3O_2 \rightarrow {}^1A + {}^3O_2$ (20)

By a process of elimination, enhanced $S_1^* \rightarrow T_1^*$ intersystem crossing in the presence of oxygen remains the only plausible general oxygen-quenching process [54].

Affirmative evidence for reaction (20) has recently been acquired by laser-flash spectroscopy [54]. When oxygen-containing solutions of pyrene are flashed, the yield of pyrene triplet states is significantly larger than it is when O_2 is absent. It is inferred that the basic process is charge transfer in nature with the efficiency of intersystem crossing in $^1A^*$ being greatly enhanced when it forms an encounter complex with O_2.

That the presence of O_2 appears to enhance the efficiency of $S_1^* \rightarrow T_1^*$ intersystem crossing in many fluorescent molecules does not mean that dissolved oxygen can enhance the sensitivity of phosphorometry, for in fact O_2 is also a very effective triplet quencher [54]. Because the triplet has a much longer lifetime than an excited singlet, it is much more susceptible to collisional quenching processes involving impurities such as O_2 (or impurities produced by photodecomposition of the solute). Consequently, phosphorescence is rarely observed in liquid solution unless great care is exercised in the purification of solutes and solvent, and then only when exhaustive vacuum-degassing procedures are followed [55].

The effect of oxygen on fluorescence varies from compound to compound. Many substituted aromatics and some heterocyclics are almost insensitive to oxygen; the fluorescence from unsubstituted aromatic and aliphatic aldehydes and ketones is very sensitive to oxygen. In most cases, only a 20% decrease in intensity is observed, so it is not worth the trouble to eliminate oxygen from the solution in practical fluorescence measurements.

(7) Fluorescence quenching by metal ions

The presence of metal ions can influence the luminescence characteristics of organic molecules, even when the ions do not form stable complexes with the ground state of the solute of interest. Paramagnetic transition metal ions generally produce the largest effects, suggesting that the paramagnetic species increases the rate constants for intersystem crossing in the organic molecule. Diamagnetic, non-transition metal ions are usually poor quenchers and hence form good fluorescent chelates. It has, however, become evident that, for a given fluorescent solute, it is not generally possible to correlate the extent of quenching produced by metal ions with their paramagnetic moments. For example, Mn^{2+}, which possesses five unpaired electrons, is generally an inefficient quencher [56,57]. It is clear that one cannot rationalize all available information simply by assuming

142

that metals quench excited states by paramagnetic enhancement of spin—orbit coupling.

(8) Effect of solute concentration on fluorescence

It is usually found that the fluorescence intensity of a given solute increases linearly with increasing concentration at relatively low concentrations. At higher concentrations the fluorescence intensity may reach a limiting value and even decrease with further increases in concentration. Several processes are responsible for these "concentration-quenching" effects:

(1) Because the low-frequency tail of the absorption spectrum of a solute often overlaps the high-frequency end of its fluorescence spectrum, fluorescence from an $^1A^*$ molecule can be reabsorbed by a ground-state molecule of the same solute. The probability of such an event increases with increasing solute concentration. "Self-absorption" distorts the shape of the fluorescence spectrum, since only the higher frequencies in the spectrum are reabsorbed. Self-absorption ultimately reduces the fluorescence intensity, unless ϕ_F for $^1A^*$ is equal to unity. The importance of self-absorption errors can often be reduced by using "front-surface" excitation [58].

(2) Many aromatic molecules (especially those with functional groups capable of hydrogen bonding) form dimers or higher aggregates in solution, particularly in non-polar and non-hydrogen-bonding solvents (hydrocarbons, CCl_4, CS_2, etc.). This tendency will of course be greater at high solute concentrations. Often the dimers are less strongly fluorescent than the monomer. Also, since the energy of S_1^* for the dimer is invariably lower than that for the monomer, the dimer can quench monomer emission by radiative, or long-range non-radiative, energy transfer [59].

(3) We have already referred to the propensity of excited singlets $^1A^*$ to form excimers with ground-state solute molecules [60].

$$^1A^* + A \rightarrow {}^1(A^*A) \tag{21}$$

The excimer has its own characteristic emission spectrum, which is red-shifted relative to the monomer spectrum. Many aromatic compounds show excimer emission at concentrations well below those required for the formation of ground-state dimers. For example, aromatic hydrocarbons form excimers at concentrations in the order of 10^{-3} M.

In principle, concentration quenching is not an important analytical difficulty since it can easily be remedied by diluting the solution. However, numerous published studies show that in practice concentration quenching occurs and is not properly compensated for.

5. Phosphorescence

(A) GENERAL

Lewis and Kasha [61] identified the phenomenon of phosphorescence as a transition from a metastable triplet state and hinted that molecules might be identified by their emission spectra in 1944. The design and construction of a phosphorimeter were described by Freed and Salmre [62] in 1958. These authors reported the luminescence characteristics of indole, 5-hydroxytryptamine, tryptophan, and reserpine. They found the sensitivities of analysis for these compounds to be 10 times greater with phosphorimetry than with fluorometry; the sensitivities were limited mainly by the presence of a high solvent background and residual quartz luminescence.

In 1957 Keirs et al. [63] critically evaluated a number of analytical methods based on the phenomenon of phosphorescence. Among the techniques that were demonstrated to be of analytical importance for improving the selectivity of analysis were excitation resolution, emission resolution, and phosphoroscopic resolution. A synthetic mixture of 4-nitrobiphenyl, benzaldehyde, and benzophenone (decay times of 0.08, 0.006 and 0.006 s, respectively) was resolved into a single fast-decaying component and two slower decaying species. Benzaldehyde and benzophenone were resolved by measurement of the phosphorescence emission at two different wavelengths (emission resolution) using simultaneous equations. Acetophenone and benzophenone (decay times of 0.008 and 0.006 s, respectively) were resolved phosphoroscopically with the aid of two simultaneous equations. Triphenylamine and diphenylamine were resolved by selective excitation at two different wavelengths (excitation resolution) using simultaneous equations. Simultaneous equations are necessary if, and only if, instrumental resolution is not completely effective. The instrumentation included a high-pressure mercury arc with filters for excitation resolution, a Becquerel phosphoroscope with resolution times of 0.001—0.02 s, and either a spectrograph for the photographic recording of spectra or a photomultiplier for the measurement of total emission during phosphoroscopic resolution.

The possibilities of chemical analysis by phosphorimetry were reviewed in 1962 by Parker and Hatchard [64]. A modified spectrofluorometer was used for the measurement of weak phosphorescence spectra. Two quartz-prism monochromators, two separate choppers driven by synchronous motors, a photomultiplier for emission measurement, and a photomultiplier for monitoring the exciting radiation comprised the instrumental components. Rapidly decaying phosphors having ratios of phosphorescence intensity to fluorescence intensity as small as 10^{-5} were studied. Emission spectra were corrected for monochromator—photodetector characteristics. The areas of the emission spectra were proportional to the respective fluorescence and phosphorescence quantum efficiencies. A number of fluorescence and phosphorescence quantum efficiencies were tabulated for aromatic compounds at 77 K. It was concluded that all major applications of phosphorimetry would probably require measurements in a rigid solution at low temperatures, which led Parker and Hatchard to believe that phosphorimetry would be applicable only when fluorometry at room temperature was insensitive or nonspecific. Parker and Hatchard also investigated the phosphorescence emission of several compounds at room temperature and found weak but measurable emission. In spite of this lack of sensitivity, it was still possible to measure low concentrations of some compounds.

Freed and Vise [65] constructed a spectrophosphorimeter to determine the phosphorescence excitation and emission spectra of *N*-acetyl-L-tyrosine ethyl ester, whose spectrum was similar to that of tryptophan. They used a solvent of water—methanol—ethanol in a ratio of 5 : 11 : 4, v/v/v. Chips of fused quartz were added to the sample cells to prevent cracking of the solvent glasses at 77 K. An internal standard of benzyl alcohol was also employed to account for such instrumental instabilities as source drift and sample positioning. Freed and Vise concluded that solvent purity was the chief limitation in phosphorimetry.

In 1963 Winefordner and Latz [66] described the construction of a spectrophosphorimeter consisting of an unfiltered 150-W mercury arc for excitation, a rotating-can phosphoroscope, and a grating monochromator for the measurement of phosphorescence emission spectra. Latz [67] has reported a thorough study of the procedural factors that influence phosphorimetry as a means of chemical analysis.

Phosphorimetry has been of limited value in the past because of marginal precision and accuracy, of solvent limitations and difficulties and time of sampling. In 1968 Hollifield and Winefordner [68] described a rotating-sample-cell method for increasing the precision of low-temperature phosphorescence measurements. In conventional phosphorescence analysis the sample cell is stationary; the achievable relative standard deviation is in the 10—20% range. With a rotating sample cell, the achievable relative standard deviation was about 1—2%, representing an improvement by a factor of 10. Moreover, cracking of the sample on freezing was less of a problem with the rotated cell. This observation was probably one of the most important discoveries in the field, since it changed phosphorescence to a more quantitative technique.

Other significant advances include the use of a more stable source power supply and the use of aqueous solvents. Aqueous solutions (20 μl) can be placed in a quartz capillary tube by capillary action and no cell cracking is observed on cooling to 77 K.

(B) THEORETICAL CONSIDERATIONS

Each energy level can be occupied by two electrons which must have opposite spins, designated as plus and minus. If all the electrons are "paired" in this way, the system is in the singlet state. However, if the atom or molecule has two unpaired electrons, both having the same spin, it is then in a triplet state. The lowest energy level available, the "ground state", is a singlet state. If the absorption of energy causes one of the electrons to be raised to a higher vacant level, without change of spin, the result is an excited singlet state. If a change in spin occurs, the result is an excited triplet state.

There are a variety of possible electron-energy transitions for a molecule, accompanied by an absorption or emission of light. If a pair of π electrons are excited to a higher π level, an antibonding state designated as π^*, the resulting state is a π, π^* singlet if no change in spin has occurred, but a π, π^* triplet if the spin has flipped over to the opposite sign.

The light emission from a π, π^* singlet is fluorescence. If the excited state is a π, π^* triplet, the higher improbability of a spin-flipping transition back to the ground state ($T_1^* \to S_0$) causes the light emis-

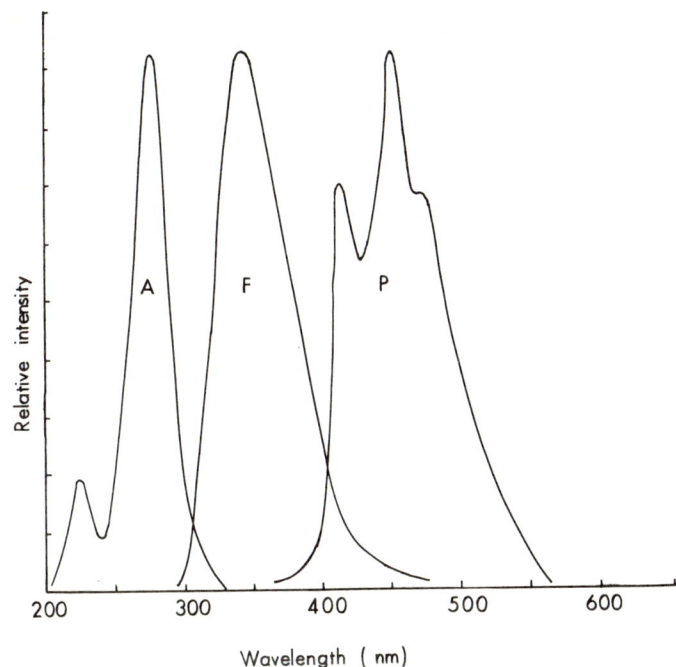

Fig. 33. Absorption (A), fluorescence (F), and phosphorescence (P) spectra of tryptophan.

sion to be greatly delayed and the result is phosphorescence. Also, because more energy is lost in the process, the wavelength of phosphorescence is shifted to longer wavelengths than fluorescence (Fig. 33).

The greatest difference between fluorescence and phosphorescence is the afterglow observed in phosphorescence. When one cuts off the excitation radiation, fluorescence ceases, but phosphorescence persists for some time. The duration of the afterglow is expressed as its half-life or mean lifetime. A second characteristic of phosphorescence is that the excited state is paramagnetic. This can be shown by magnetic susceptibility measurements [69] and electron-spin-resonance experiments [70]. Another distinction of phosphorescence is that the lifetime of the excited state and the intensity of emission are extremely sensitive to heavy atoms and paramagnetic ions.

TABLE 20

Substituent effects upon luminescence of naphthalenes and fluoresceins in EPA

Compound	Φ_P/Φ_F	Φ_F	λ_F (nm)	λ_P (nm)
Naphthalene	0.093			
1-Chloronaphthalene	5.2			
1-Bromonaphthalene	6.4			
1-Iodonaphthalene	>1000			
Fluorescein	0	0.83	527	
2',7'-Dichlorofluorescein	0	0.79	538	
4',5'-Dibromofluorescein	0.21	0.29	540	650
4',5'-Diiodofluorescein	1.05	0.054	544	667

(C) STRUCTURAL EFFECTS

The effect of halogen substitution on the luminescence of aromatic hydrocarbons is of considerable importance to chemists. One generally observes that as the substituent series F, Cl, Br, I is traversed, phosphorescence is increasingly favored relative to fluorescence. This effect is illustrated in Table 20 which shows the ratio of the quantum efficiency of phosphorescence to fluorescence, Φ_P/Φ_F, for a number of substituted naphthalenes and fluoresceins. The ratio Φ_P/Φ_F increases across the series F → I. The trends for yields of luminescence and excited-state lifetimes in haloaromatics can be rationalized only if one postulates that heavy-halogen substitution increases the rate of intersystem crossing from a singlet to a triplet state (radiationless process) and subsequent triplet to singlet transition (a radiative process). Halogen substitution must therefore increase the extent of spin orbit coupling in aromatic systems, the increase being larger for heavier halogens. This perturbation is commonly termed the "heavy-atom effect". In the case of the substituted naphthalenes the increase in the ratio of phosphorescence to fluorescence is primarily due to an increase of probability of singlet to triplet transitions. However, in the fluorescein series the primary effect is the result of an increased non-radiative transition from the excited state back to the ground state.

Since fluorescein and dichlorofluorescein are highly fluorescent, but do not phosphoresce, whereas the corresponding bromo- and iodo-substituted derivatives are weakly fluorescent, but do phosphoresce, a study of the luminescent properties of the fluorescein family provides an interesting introduction to some of the structural factors affecting luminescence. Paramagnetic ions also effect a transition from the singlet to the triplet with a resulting increase in phosphorescence intensity. Metal chelates of such ions are generally phosphorescent.

(D) INSTRUMENTATION

(1) General

The instrumentation used to study phosphorescence is very similar to that used for fluorescence. To see phosphorescence in the presence of fluorescence we must take advantage of the slight time difference involved between the absorption and emission of radiant energy. This is accomplished with a mechanical device called a phosphoroscope (Fig. 34) which modulates the radiation from the light source incident on the sample and simultaneously modulates the luminescence radiation from the sample which is incident on the photodetector. The modulation is periodic and out of phase so that no incident exciting or luminescent radiation reaches the photodetector during one phase, whereas, only long decaying luminescence (phosphorescence) radiation reaches the photodetector during the other phase. Therefore, the main function of the phosphoroscope is to allow measurement of phosphorescence in the presence of fluorescence and scattered radiation.

At room temperature the energy of the triplet state is readily lost by a collisional deactivation process involving the solvent, and phosphorescence is not observed. At reduced temperatures a solidified sample does not lose energy readily and phosphorescence can be easily observed. The usual procedure in phosphorescent studies is to place the samples in small quartz tubes which are then placed in liquid nitrogen (77 K) and held in a quartz Dewar flask. A typical quartz Dewar flask is shown in Fig. 34. The incident radiation energy passes through the unsilvered part of the Dewar flask, and luminescence is observed through the same part of the flask at right angles to the incident beam. The sample cell is immersed directly into the coolant, which is usually liquid nitrogen.

The solvents used in phosphorimetry at liquid nitrogen temperatures must form clear, rigid glasses, must have good solubility characteristics for the compounds to be studied, must be readily available

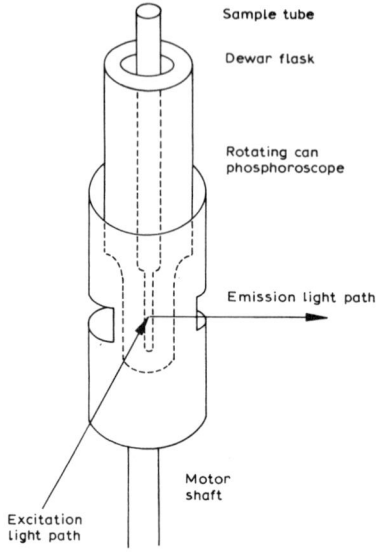

Sample tube

Dewar flask

Rotating can
phosphoroscope

Emission light path

Motor
shaft

Excitation
Light path

Fig. 34. Schematic diagram of a rotating can phosphoroscope.

and inexpensive, and must neither absorb strongly nor luminesce greatly in the spectra regions of interest. The most commonly mixed solvent is EPA (a 5 : 5 : 2 v/v/v mixture of diethylether, isopentane, and ethanol).

Phosphorescence attachments which can be attached to a commercially available fluorometer are available for the Aminco—Bowman SPF, the Baird Atomic Fluirispec, the Farrand MK1 spectrofluorometer, and the Aminco filter fluorometer. The last (Accessory D2-63019,) is the cheapest and most readily available. A typical spectrophosphorimeter is pictured in Fig. 35. Source light passes through a rotating shutter to excite the sample to phosphoresce; emitted light alternately passes to the detector.

(2) The phosphoroscope

The phosphoroscope is a mechanical device used (a) to modulate the radiation from the light source incident on the sample, and (b) to modulate the luminescence radiating from the sample and incident on the photodetector. The modulation is periodic and out of phase,

150

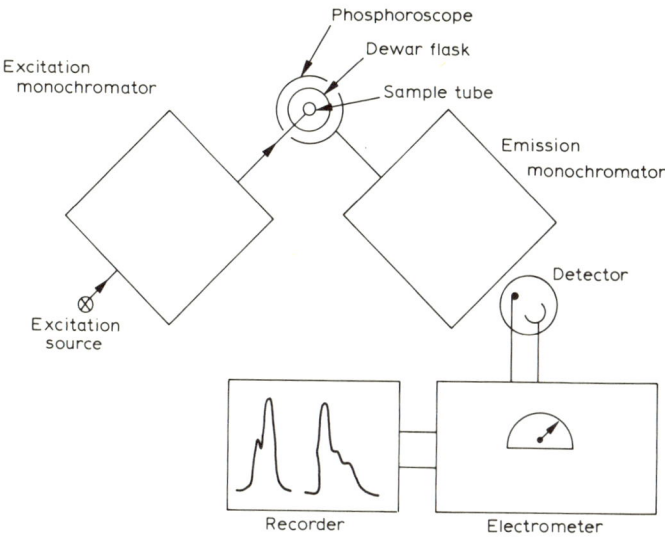

Fig. 35. Schematic diagram of a spectrophosphorimeter.

so that no incident exciting or luminescence radiation reaches the photodetector during one phase, and only long-decaying luminescence (phosphorescence) radiation reaches the photodetector during the second phase. Therefore, the main purpose of the phosphoroscope is to allow measurement of phosphorescence in the presence of incident-light scattering and fluorescence.

There are two major types of mechanical phosphoroscopes. The rotating-can phosphoroscope (see Fig. 34) was first described by Lewis and Kasha and is used in the Aminco—Bowman spectrophotofluorometer. It is a hollow cylinder with one or more slits equally spaced in the circumference. As the can is turned by a motor, radiation from the excitation monochromator is alternately allowed to strike the sample, and the light emitted by the sample is alternately, but out of phase with excitation, allowed to reach the emission-monochromator entrance slit. The fluorescence and exciting radiation decay rapidly after the exciting radiation is terminated. Hence only the long-decaying phosphorescence will remain when the phosphoroscope has turned from the point at which excitation is stopped to the point where the luminescence of the sample is measured.

The other major type of mechanical phosphoroscope (Fig. 36) is

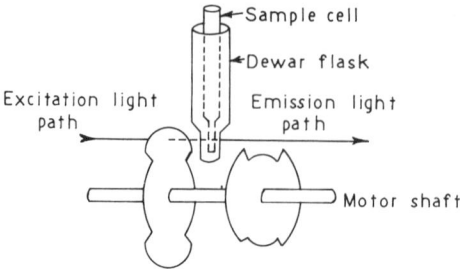

Fig. 36. Schematic diagram of a Becquerel disk phosphoroscope.

the Becquerel, or rotating-disk type. This type of phosphoroscope is currently being used in several of the compensated-luminescence instruments because it is more versatile than the rotating-can type. It

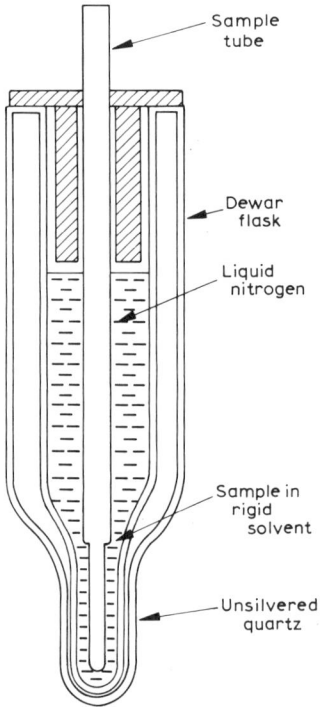

Fig. 37. Schematic diagram of sampling system used in phosphorimetry.

152

consists of two disks with notches cut in them at equal intervals and mounted on a common axis turned by a motor. Once again excitation occurs out of phase with measurement of luminescence.

(3) The sample cell

In phosphorimetry the sample must be frozen; this limits the internal dimensions of the sample cell. Too small a cell diameter will prevent the excitation energy from entering the sample due to reflection; too large a cell diameter leads to cracking of the frozen solution and a high degree of scattering. Generally, small quartz tubes 1 mm and 3 mm are used. These are placed in liquid nitrogen (77 K) and held in a quartz Dewar flask. The crystalline nature of the frozen sample requires precise positioning and repositioning of the cell. For good precision (1%) the sample should be rotated as described by Hollifield and Winefordner [68].

The cells should be cleaned with successive rinses of fuming nitric acid, distilled water, and finally with the solvent and sample solution to be measured. The small cells are emptied by means of a polyethylene tube connected to a water aspirator.

A typical quartz Dewar flask that was used in the phosphorescence attachment to the Aminco—Bowman spectrophotofluorometer is pictured in Fig. 37. Light passes through the unsilvered part of the Dewar flask, and luminescence is observed through a similar part of the flask at right angles. The sample cell is immersed directly into the coolant which is usually liquid nitrogen. This immersion technique has the advantage of simplicity and speed of operation. However, it also has several disadvantages. First, the exciting radiation and the luminescence radiation must pass through three quartz layers, which results in considerable light loss. Second, the light paths must pass through the thermostating medium, limiting the choice of coolants to transparent, non-luminescent liquids. Third, as the coolant in the sample-viewing area warms up, convection of the coolant results in a change in its refractive index and a resultant flicker in the exciting radiation and the luminescence radiation. Fourth, ice crystals and foreign objects, such as dust, in the flask tend to produce nucleation sites, causing bubbling of the coolant, which again produces a flicker noise.

These disadvantages could be minimized by using a conduction-cooling device. In this system the sample cell is cooled by contact with a copper rod immersed in some coolant. Thermal gradients are reduced by the high thermal conductivity of the copper, and quartz surfaces are minimized. Unfortunately, it is difficult to obtain good thermal contact between the sample cell and the copper rod. Also, it is difficult to thermally insulate the cooling apparatus; this results in fogging of the viewing area, making quantitative measurements difficult. Hoerman and Manciewicz [71] have successfully applied conduction cooling to the study of calcified tissue. They formed potassium bromide pellets of particles of bone, enamel, and dentin and cooled the pellets to 93 K by conduction using a liquid nitrogen reservoir.

At the present time immersion cooling is the only readily available method for quantitative analysis. Unfortunately, the method of positioning the sample cell in the Dewar flask as well as arc wander of the xenon source limit the precision of phosphorimetric measurements to about ±5—10% relative standard deviation with the Aminco phosphoroscope attachment. With rotation of the sample, however, 1% precision is attainable [68].

(4) Solvents

Of all the possible coolants (e.g., liquid nitrogen, air, oxygen, rare gases, nitrous oxide, and other boiling liquids, as well as liquid/solid mixtures) only liquid nitrogen is sufficiently pure to be transparent and non-luminescent at all wavelengths between 200 and 800 nm and safe, convenient, and inexpensive to use.

The solvents used in phosphorimetry at liquid nitrogen temperature must form clear, rigid glasses, must have good solubility characteristics for the compounds to be studied, must be readily available and inexpensive, and must neither absorb strongly nor luminesce greatly in the spectral regions of interest.

Some of the solvents used for phosphorimetric analysis are classified as acidic, alcoholic, aqueous, basic, ether, halides, hydrocarbon, and plastics or solid matrices. Ethanol has been combined with a variety of other solvents to form good mixed solvents, the most popular of which is EPA, a 5 : 5 : 2 (v/v/v) mixture of ether, isopentane and ethanol (sold by the American Instrument Co., Silver Spring, Md.).

Solvents, even the so-called spectrograde ones, are generally not sufficiently pure to be used without further purification. Purification in some cases may be as simple as several extractions with dilute acid and base. In any event, the procedure with the least number of transfers and least exposure to the atmosphere and to glassware is likely to be best. Absolute ethanol as received from the supplier must be purified. This is accomplished using a 5-ft. vacuum-jacketed distillation column of 1 in. diam., packed with 3/32 in. glass helices.

Hydrocarbons, such as hexane, isopentane, and heptane, can also be purified by distillation. Alternatively, the hydrocarbons can be dried over anhydrous sodium sulfate or sodium ribbon and then passed through a 2-ft. column of 200-mesh silica gel activated for 12 h at 350°C prior ro use [72].

Lukasiewicz et al. [73—75] have described another improvement to allow the measurement of aqueous or predominately aqueous solutions at 77 K by means of phosphorimetry: an open quartz capillary cell as the sample cell. Sample solutions (20 μl) fill the cell by capillary action, and the sample cell is then rotated. As a result of the open capillary cell, these cells with predominately (or pure) aqueous solutions do not crack when rapidly cooled to liquid N_2 temperatures or when warmed back to room temperature by use of a heat gun. The major advantages of using aqueous or aqueous-organic mixture solutions in phosphorimetry are:

(1) Most biochemical species are more soluble in aqueous solutions than in the solvents needed to produce clear, rigid glasses at liquid nitrogen temperatures.

(2) Solution conditions, e.g., pH, ionic strength, etc., can be varied for optimal analytical results.

(3) Water is easier to obtain in higher phosphorimetric purity than most organic solvents.

(4) Contamination problems are less when water can be used to clean glassware, cells, and equipment.

(5) Commercial equipment

Phosphorescence attachments are available to commercial instruments such as the Aminco—Bowman spectrophotofluorometer, the Baird-Atomic Fluorispec, and the Aminco filter fluorometer. The latter is not recommended since it has given very poor results in our laboratory. The attachments to the Aminco—Bowman instrument and the Baird-Atomic Fluorispec consist of a phosphoroscope can, a quartz Dewar flask, and quartz sample tubes. Both have been used in our laboratory and work well. It is recommended that the quartz sample tube be rotated as suggested by Winefordner [68] for optimum precision (1%). It is surprising that no commercial equipment yet uses this simple expedient.

(E) ANALYTICAL APPLICATIONS

(1) Practical aspects of measurement

a. Instrument calibration. In phosphorescence the instrument sensitivity should be adjusted daily as described for fluorometry in Sect. 3. The sensitivity must be periodically adjusted so that the overall measurement system has the same sensitivity as that used in all previous studies. As an example, Moye [76] used a standard solution of toluene in ethanol as a reference standard to adjust instrument sensitivity.

b. Procedure for measurement. The excitation and emission monochromators are usually adjusted to the peak excitation and emission wavelengths unless interferences that prevent this are present. The instrument sensitivity is then checked as described above and set to the proper range for measurements to be made. The Dewar flask is cleaned and filled with liquid nitrogen. The sample tube is cleaned and filled with sample, and the sample-tube holder is aligned in the Dewar flask. If the sample solution cracks or forms a snow on cooling, the sample should be discarded and a new sample solution placed in the sample tube. (Note: with the rotating-sample approach of Winefordner [68] cracked glasses and snows can be tolerated.) The signal is determined from the read-out device, and this reading is used to determine the concentration of the sample by use of an analytical curve of read-out signal vs. concentration of sample. The standard analytical curve should be prepared by using standard solutions of the sample and measured under experimental conditions identical with those used for the unknown sample. In plotting an analytical curve each point on the curve must be corrected for the solvent blank. It is interesting to note that most phosphorescence analytical curves are linear over a concentration range of 10000 or more. Negative curvature usually results in the 10^{-4} M and higher concentration range due to self-absorption, molecular aggregation,

156

and concentration quenching. Positive curvature near the minimum detectable sample concentration is generally a result of contaminant luminescence.

c. Quenching and enhancement of phosphorescence. In general, phosphorescence is less susceptible to quenching than is fluorescence. The use of low temperatures and rigid media in phosphorimetry diminishes the probability of non-radiative deactivation of the triplet state by diffusion-controlled processes.

The presence of heavy atoms affects both the radiative and non-radiative triplet-to-singlet transitions. McGlynn et al. [77] have studied the external-heavy-atom effect and have suggested that it might be a useful means of enhancing the phosphorescence intensity of some species. Hood and Winefordner [78] have studied the influence of ethyl iodide—ethanol as a solvent for enhancing the phosphorescence signal of naphthalene, phenanthrene, and 10 other polynuclear aromatic hydrocarbons. They found that a 5 : 1 (v/v) ethanol—ethyl iodide solvent enhanced the phosphorescence signal of 1,2-benzfluorene and 2,3-benzfluorene by 13 and 25 times, respectively, whereas the same solvent resulted in a diminution of the phosphorescence signal of naphthalene and phenanthrene by 10 and 5 times, respectively. Many of the compounds influenced most significantly by the heavy-atom effect are carcinogens (e.g., 3,4-benzpyrene, 1,2,5,6-dibenzanthracene, and 1,2-benzanthracene). Since the heavy-atom effect can enhance or depress the phosphorescence signal, an increase in the selectivity of measurement as well as increased sensitivity can be effected by properly selecting the solvent for a specific system.

It must be kept in mind that, even though the phosphorescence intensity may actually be increased due to the heavy-atom affect, there may still be a decrease in the phosphorescence signal because the heavy atom may result in such a reduction in the decay time τ that there may be a significant decrease in the phosphorescence signal due to phosphoroscope delay time (i.e., the value of α' decreases considerably).

Paramagnetic ions are also effective quenchers of phosphorescence in many instances. The acetylacetone complexes of paramagnetic transition metal ions are not phosphorescent, whereas the same complexes of diamagnetic ions are. Oxygen as a quencher of luminescence has been the subject of much debate. However, from an analytical viewpoint, oxygen does not appear to be a major quencher for samples measured in rigid media at low temperatures.

d. Analysis of mixtures of phosphorescent compounds. There are several methods of achieving selectivity in the measurement of mixtures of phosphorescent compounds. These methods are called excitation resolution, emission resolution, phosphoroscopic resolution, and time-resolved phosphorimetry. We shall discuss each of these separately.

(1) Excitation resolution. Excitation resolution is achieved by varying the wavelength of the excitation monochromator to excite each molecule separately. If the phosphorescence of a mixture is excited by a wavelength that is considerably more strongly absorbed by one component than by all others, then there is obtained predominantly the emission spectrum of this compound. Similarly, if several excitation wavelengths are employed, it becomes possible to obtain the spectra of the individual components more or less undistorted, provided the absorbance spectra of the compounds present in the mixture differ sufficiently. This technique, which has long been used in fluorometry, has been found extremely useful also in phosphorimetry.

Figure 38 gives an example. In it are reproduced the phosphores-

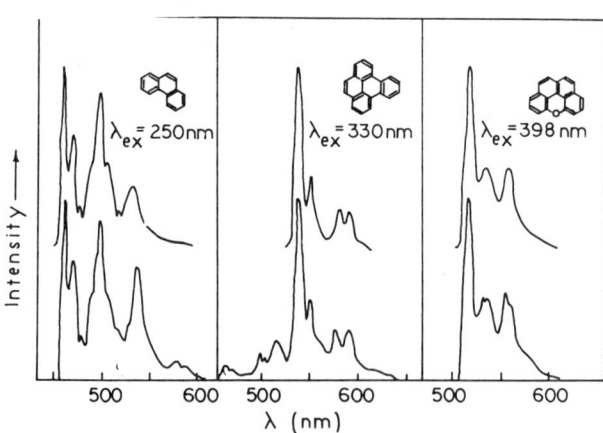

Fig. 38. Phosphorescence spectra of a mixture of 40% phenanthrene, 30% 1,2-benzpyrene, and 30% peri-(1,8,9)-naphthoxanthene on excitation with 250, 330, and 398 nm. The curves were measured on the pure compounds in each case. (From ref. 1 by permission of the author.)

cence spectra of the three-component mixture of phenanthrene, 1,2-benzpyrene, and peri-(1,8,9)-naphthoxanthene excited by three different wavelengths. For each of these wavelengths a different compound gives an intense absorption maximum, and the others give either minimal or weak absorption. Above the spectra of the mixture are shown the spectra of the pure components. As can be seen, in each case only one component appears clearly in the spectrum of the mixture.

(2) Emission resolution. Emission resolution is achieved by varying the wavelength of the emission monochromator in order to measure preferentially the luminescence of just one molecule. For example, two pesticides, methoxychlor and Kelthane, can be assayed in this way. As indicated in Fig. 39, the emission monochromator is first adjusted to 525 nm in order to measure Kelthane (B) in the presence of methoxychlor. Then the emission is set at 400 nm in order to measure methoxychlor (A).

(3) Phosphoroscopic resolution. By suitable change in the time of measurement of the luminescence radiation after termination of the exciting radiation by a shutter, the phosphoroscopic resolution of two molecules with sufficiently different decay times (at least a tenfold difference) can be performed. The slow-decaying molecule is measured at a time sufficient for at least 99% of the fast-decaying species to have decayed. Keris et al. [79] were the first to suggest phosphoroscopic resolution based on rotation speed and the geometry of a Becquerel phosphoroscope. If there is a sufficient difference in the decay times of two phosphors, it is possible to vary the phosphoroscope speed in order to measure only the slower decaying species. Once the concentration of the slower decaying species is known, the concentration of the faster decaying species can also be determined.

(4) Time-resolved phosphorimetry. St. John and Winefordner [80] have developed a new technique called time-resolved phosphorimetry (a type of phosphorescence resolution) to measure two-component mixtures of similar phosphorescent compounds. The method utilizes the difference in the decay times of the phosphors. The exponentially decaying phosphors are resolved by using a logarithmically responding instrument. The concentration of the two components can be determined from the recording of the logarithm of the phosphorescence signal due to the mixture vs. time after termination of excitation by a method analogous to that used in radioactive-isotope

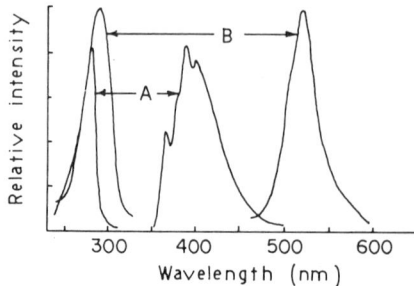

Fig. 39. Resolution of the emission of a mixture of two pesticides: (A) methoxy-chlor and (B) Kelthane. From ref. 1 by permission of the author.

analysis. Mixtures of tryptophan and tyrosine and of benzoic acid and benzaldehyde were measured by this technique with an overall relative error of about 3%. In Fig. 40, logarithmic decay curves are given for benzoic acid $(3.00 \times 10^{-4}$ M, $\tau = 2.4$ s) benzaldehyde $(6.5 \times 10^{-5}$ M, $\tau < 0.1$ s) in ethanol, and for tryptophan $(6.04 \times 10^{-6}$ M, $\tau = 6.4$ s) and tyrosine 5×10^{-6} M, $\tau = 1.4$ s) dissolved in 5×10^{-3} M sodium methoxide in ethanol.

   e. Complementary nature of fluorescence and phosphorescence. The techniques of phosphorescence and fluorescence are com-

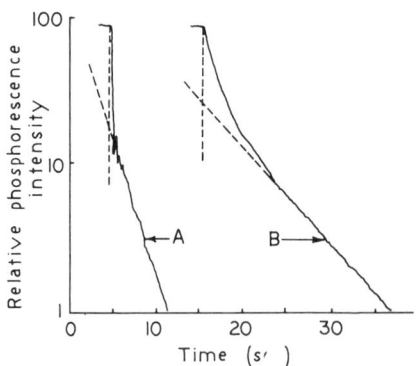

Fig. 40. Logarithmic decay curves for a mixture of (A) benzoic acid and ben-zaldehyde and (B) a mixture of tryptophan and tyrosine. Ethanol was the solvent for all solutions. (From ref. 1 by permission of the author.)

160

plementary rather than competitive. If a compound is strongly fluorescent, it will be weakly phosphorescent, and vice versa; that is, the larger ϕ_F, the lower ϕ_P. This was demonstrated earlier for the halofluoresceins. Unsubstituted fluorescein and its fluoro- and chloro-derivatives are fluorescent, and not phosphorescent. Bromofluorescein and iodofluorescein are phosphorescent, not fluorescent.

f. Improved instrumentation for phosphorimetry. Zweidinger and Winefordner [81] have shown that, by using a rotating sample cell, a more stable power supply, and a better solvent clean-up procedure, the detection limits for most organics can be lowered by more than a hundred-fold and that precision and accuracy can be increased by more than ten-fold. Furthermore, the time and effort involved in sampling and measuring can be considerably reduced. Precise, accurate, sensitive, selective, and rapid analysis could also be performed in solvents forming opaque or densely cracked glasses — no longer is it necessary that a clear solid be formed for good results (see Table

TABLE 21

Precision of phosphorescence measurements [a] for clear glasses and snows with the Varian Spinner Assembly [b]

N [c]	Nature of matrix	Stationary random orientation	Stationary aligned	Rotating
4	Clear [d]	8.7	1.3	0.8
6	Clear [d]	6.0	0.5	0.8
5	Cracked [d]	13.7	3.4	1.4
11	Clear [d]	3.1	0.9	0.3
10	Clear [d]	2.9	2.8	1.0
5	Snow [e]	3.6	1.6	0.9
10	Snow [e]	2.8	0.8	0.6
10	Snow [e]	2.7	2.4	0.7

[a] Phosphorescence measurements made on 1.6×10^{-5} M sulfanilamide solutions, which gives a signal five orders of magnitude above the phototube dark current. Precision expressed in relative standard deviation.
[b] Data from Zweidinger and Winefordner [81].
[c] Number of determinations.
[d] Ethanol solvent.
[e] Isooctane—ethanol, 4 : 1 (v/v) mixture as solvent.

21 for typical results). This increases the applicability of phosphorimetry in biological analysis.

The rotating sample cell assembly currently being used is a modification of the one described by Hollifield and Winefordner [68]; the present system is shown in Fig. 41. The rotating assembly described in detail by Zweidinger and Winefordner [81] and modified by Lukasiewicz et al. [73–75], consists primarily of a Varian A60-A high resolution NMR spinner assembly (Varian Associates, Palo Alto, Calif.) mounted on an AMINCO phosphoroscope sample compartment in place of the usual lid. The pressure cap of the spinner assembly is covered with black tape to ensure a completely light-tight sample compartment. The rotating sample cell is driven by a flow of nitrogen or air gas taken from a compressed gas tank. The sample cell

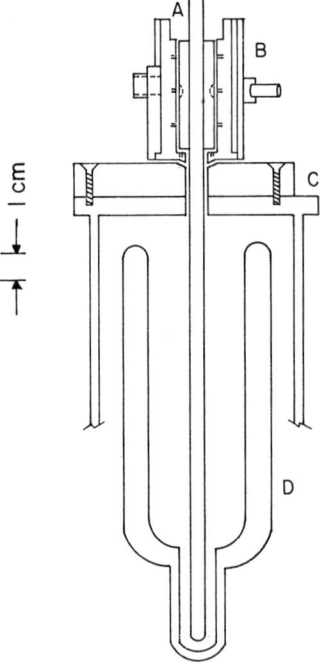

Fig. 41. Schematic diagram of rotating sample cell assembly: (A) quartz sample cell; (B) Varian (909614-04) spinner assembly for NMR; (C) Aminco light cover mount; (D) Aminco quartz Dewar flask. (From ref. 1 by permission of the author.)

162

must have a 6 mm O.D. and be about 25 cm long. The sample rotation speed is easy to maintain constant at some speed between 450 and 1400 r.p.m.; the actual speed is unimportant as long as it is maintained constant during a series of measurements, which can be assured by means of a normal two-state regulator and a rotameter flow meter to monitor the gas flow rate. The normal quartz sample cells used by Zweidinger and Winefordner [81] were 5 mm O.D. and 3 mm I.D., whereas the quartz capillary cells used by Lukasiewicz et al. [73—75] were 5.0 mm O.D. and 0.90 mm I.D. The cells were made of synthetic, high-purity, optical-grade quartz (Quartz Scientific Co., Eastlake, Ohio for the capillary tubing and Amersil, Inc., Hillside, New Jersey, for the normal tubing).

(2) Analytical determinations

Phosphorimetry has been used for analysis of agriculturals, human blood and urine, amino acids and proteins, nucleic acids, nucleotides, nucleosides, purines and pyrimidines, enzymes, pharmaceuticals, petroleum products and air pollutants. The reader is referred to ref. 1 for a complete listing of various applications. Typical examples of analyses are reported below.

Moye and Winefordner [82] reported the phosphorescence characteristics of 52 pesticides. Of the 32 pesticides that phosphoresce the authors listed the spectral characteristics, lifetimes, limits of detection and linear region. Phosphorescence was much more sensitive than fluorescence for a number of these compounds, especially the carbamates.

Hollifield and Winefordner [83] described a method for the analysis of sulfur drugs in blood. The procedure avoids the necessity of prior separation steps.

Steele and Szent-Gyorgyi [84] reported the phosphorescence characteristics of a number of purines and pyrimidines: adenine, adenosine, AMP, ADP, and ATP. The latter was very phosphorescent.

Winefordner and co-workers have described procedures for the assay of a number of pharmaceuticals; aspirin [85], procaine, phenobarbital, cocaine and chlorpromazine [86], sulfonamides [87], and anticoagulants [88].

The anticoagulants studied were dicumarol, diphenadione, phenindione, ethyl biscoumacetate, and warfarin. The authors reached the following conclusions: both fluorometry and phosphorimetry are sensitive and selective techniques; fluorometry is somewhat simpler than phosphorimetry, but it is not as selective. The anticoagulants were extracted from whole blood with the ethanol single-step deproteinization extraction method. The phosphorescence of the sample was corrected for the blank. Fluorometry was more sensitive for warfarin, whereas for dicumarol, phosphorimetry was more sensitive. Fluorescence blanks were considerably greater than those found for phosphorescence. Recoveries of doped samples were excellent, and the relative standard deviation in all cases was less than 5%.

Drushel and Sommers [89] have presented an extensive study of well over 100 aromatic petroleum compounds containing sulfur, nitrogen, or oxygen as assayed by phosphorescence. The authors stated that the techniques used for the characterization of compounds containing sulfur, nitrogen, and oxygen has usually been limited to three spectrometric methods (mass, IR and UV absorption). Luminescence techniques, according to the authors, offered significant advantages for the identification of unknown components; in particular phosphorimetry offered the largest quantity of easily obtainable physical information: excitation spectrum, emission spectrum, and easily determinable lifetime. In addition, with sufficient monochromator resolution, the predominent vibrational spacing of the ground electronic state could be determined. The authors demonstrated that phosphorimetry provided the requisite information to identify the components from a gas chromatography separation of 430—650° F petroleum fraction. They also demonstrated that phosphorescence was invaluable in the characterization of a nitrogen compound concentrate from a catalytic hydrogenation of quinoline. Compound-type identification was easily accomplished with the use of phosphorimetry.

Numerous polycyclic aromatic hydrocarbons are found in the atmosphere of industrial cities. Many of these compounds have been found to produce cancer in animals and humans. A number of thorough luminescent examinations of these compounds have been carried out by Sawicki and workers at the Air Pollution Control Center, Research Triangle Park, North Carolina.

Sawicki [90] showed that phosphorescence spectroscopy should be a valuable technique for trace analysis in a review in 1964. This technique was particularly valuable for the qualitative identification of aromatic and heterocyclic carbonyl compounds in atmospheric dust. Sawicki and Johnson [91] showed that polycyclic aromatics in

164

air can be identified by phosphorescence on thin-layer chromatography strips.

Thin-layer chromatography and phosphorimetry were used by Sawicki et al. [92] to characterize certain heterocyclic hydrocarbons. Sawicki and Pfaff [93,94] demonstrated that phosphorimetric measurements can be made directly on glass-fiber chromatograms. Limits of detection (in ng) were as low when samples were collected on glass-fiber chromatograms at low temperatures as when samples were in EPA (0.1 ml) rigid glasses or on cellulose thin-layer chromatograms. This was true for such molecules as benzo[e]pyrene, 4-hydroxyacetophenone, anthraquinone, benzo[h]quinoline, triphenylene, and 2-nitrofluorene.

An extensive study of the room-temperature and low-temperature fluorescence and phosphorescence of compounds containing the 4-nitroaniline phosphor and analogous groups has been reported by Sawicki and Pfaff [95]. The limits of detection, lifetimes, and the excitation and emission spectra for 27 phosphorescent compounds were reported in EPA rigid-glass solution. The complementary nature of fluorescence and phosphorescence was emphasized. It was noted that aromatic nitro compounds usually yield low-intensity absorption spectra, and therefore, luminescence was the technique of choice. Most 4-nitrophenylhydrazones of aliphatic aldehydes and ketones were non-fluorescent but strongly phosphorescent, whereas most 4-nitrophenylhydrazones of aromatic aldehydes and ketones were non-phosphorescent but fluorescent in solvents of low dielectric constant. It was stated that phosphorescence and room-temperature and low-temperature fluorescence were three powerful complementary analytical tools that achieved their greatest value when used together in microchemical analysis rather than separately.

The effect of solvents and the physical state on the phosphorescence emission of trypsin, ribonuclease, phenylalanine, tryptophan, and tyrosine were studied by Nag-Chaudhuri and Augenstein [96]. The spectra in several solvents were presented.

Bobrovich and Konev [97] reported the phosphorescence of amylase. Vladimirov and Litvin [98] reported a detailed study of the relation of the phosphorescence of single amino acids to protein phosphorescence. The authors studied zein, human-serum albumin, and egg albumin, and prepared synthetic mixtures of tyrosine, tryptophan, and phenylalanine in the same relative abundance as in the proteins. The phosphorescence spectra of egg albumin and serum albumin contained what appeared to be tryptophan emission; similar results were obtained in the equivalent mixtures of amino acids. The phosphorescence spectrum of zein, however, was composed of tyrosine-like emission; similar results were reported for the equivalent amino acid mixture.

Results similar to those of Vladimirov and Litvin [98] were noted by Vladimirov and Burshtein [99] who studied γ-globulin and actinomycin. In addition, a number of other similar studies have been reported [100—104]. The phosphorescence behavior of proteins has been attributed to an intramolecular-energy-transfer mechanism [105]; energy absorbed by phenylalanine is transferred to tyrosine or tryptophan residues; energy absorbed by, or transferred to, tyrosine is transferred to tryptophan residues; finally, tryptophan emission is observed. For proteins where tryptophan emission is absent the transfer mechanism stops with tyrosine, and thus tyrosine emission is observed.

Grossweiner [106] studied the phosphorescence spectra of tyrosine, tryptophan, indole, phenol, and egg albumin. He concluded that the excited spectrum of egg albumin was the sum of those from the tyrosine and tryptophan residues.

The phosphorescence of ribonuclease and insulin was investigated by Freed et al. [107]. Both of these substances gave tyrosine emission, but no tryptophan emission.

Douzou and Francq [100] compared the phosphorescence of human and horse serum albumin with the emission of phenylalanine, tyrosine, histidine, and tryptophan. Stauff and Wolf [103] studied the phosphorescence of albumin, lactoglobulin, alcohol dehydrogenase, xanthine oxidase, cytochrome c, peroxidase, and catalase. The phosphorescence was enhanced on addition of eosin.

Phosphorescence excitation and emission spectra and phosphorescence decay times and quantum efficiencies provide valuable information concerning the structure, chemical behavior, and environment of organic molecules in real, complex systems. Phosphorimetry is similar to fluorometry in methodology and areas of application, and offers a powerful complementary tool to fluorometry.

6. Determination of inorganic ions

(A) GENERAL CONSIDERATIONS

The combination of an inorganic ion with a non-fluorescent organic ligand to form a highly fluorescent metal chelate can provide a very sensitive and highly specific method for the determination of a metal ion. Procedures for the analysis of nearly thirty different metals have been devised using this basic approach. In addition, elements and ions such as cyanide, fluoride, iron, copper, and oxygen may be determined indirectly by measuring the amount of quenching of the fluorescence of a chelate or by causing the release of a ligand which can then react to form a fluorescent product as in the determination of cyanide.

CN^- + Pd-8-quinolinol-5-sulfonate

(Non-fluorescent chelate)

$\rightarrow$ 8-quinolinol-5-sulfonate + $Pd(CN)_6^{4-}$ (22)

$$\downarrow Mg^{2+}$$

(Fluorescent chelate)

The organic ligand should be an aromatic molecule containing oxygen or nitrogen, which is itself non-fluorescent. The presence of non-bonding n electrons makes it probable that the excited state will be n, π^*, which is non-fluorescent or weakly fluorescent. Upon complex formation with a metal ion, the n electrons are utilized in bonding with the metal, and thus become less accessible for excitation. The π, π^* state then results upon excitation, and strong fluorescence is observed.

The organic ligands that have most commonly been used to react with metal ions to form fluorescent chelates are the 2,2'-dihydroxyazo dyes; 8-quinolinol and its derivatives; flavanols; salicylidenes, and 2,2'-dihydroxymethines; salicylic acid; benzoin; rhodamine B, G, and S; salicylaldehydes; β-diketones; hydroxynaphthoic acid; and hydroxyanthraquinones. Other ligands have found specific applications and new compounds are continuously added to the list.

In the course of a study of azo compounds to determine the minimum structural requirements for combination of an azo compound with calcium and magnesium, for example, Diehl et al. [108] found that *o,o'*-dihydroxyazobenzene was unique in reacting with magnesium and not calcium. At pH 10 this ligand reacts with magnesium to produce a stable, orange fluorescent chelate, but forms no chromogenic or fluorescent complex with calcium. The chelate is fluorescent in water solution at pH's greater than 11, but the fluorescence

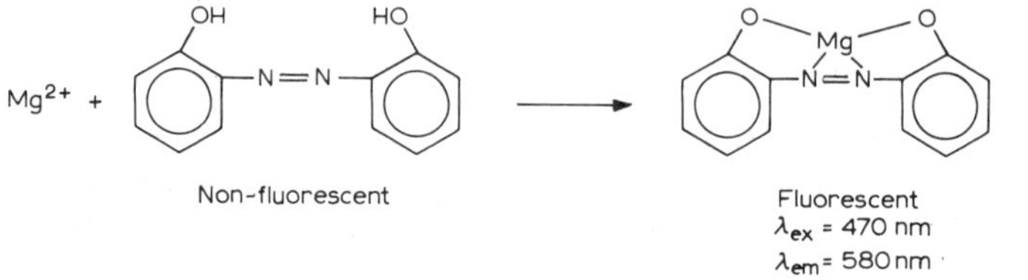

Non-fluorescent

Fluorescent
λ_{ex} = 470 nm
λ_{em} = 580 nm

(23)

intensity is increased in ethanol—water solutions at pH's of 10—11.4. At more acidic pH values protons compete more effectively with the Mg^{2+} for the hydroxy-oxygens and consequently the chelate is not formed quantitatively. At pH values greater than about 11.5, magnesium hydroxide may form in appreciable amounts.

Reference material on the analysis of inorganic ions by fluorescence can be found in books by White [109], Guilbault [110], Hercules [111], Bozhevol'nov [112], Udenfriend [113,114] and Konstantinova-Shlezinger [115]. Weissler and White [116] have

TABLE 22

Assay of inorganic ions with inorganic reagents

Ion	Reagent	Sensitivity (p.p.m.)	Refs.
As	HCl or HBr	0.15	118
Bi	HCl or HBr	0.002	119, 120
$C_2O_4^{2-}$	Ce^{4+}	8.8	121
Ce	$HClO_4$	0.1	122
Fe^{2+}	Ce^{4+}	5.6	121
I^-	Ce^{4+}	0.6	121
Os	Ce^{4+}	0.5	121
Pb	HCl, HBr, or LiCl	0.01	123
Sb	HCl or HBr	0.001	119
Se	HCl or HBr	0.06	119
Te	HCl	0.02	124
Tl	HCl or NaCl	0.01	125

written a chapter on the analysis of inorganic and organic compounds in Meites' book and have written review articles on the fluorometric analysis of inorganic substances that appear every 2 years in Analytical Chemistry [117].

(B) DIRECT ANALYSIS OF INORGANIC SUBSTANCES

Many inorganic substances either fluoresce or phosphoresce directly in the solid state. Salts of rare-earth elements and uranyl salts fluoresce in solution. Other inorganic ions fluoresce in solutions after the addition of an inorganic reagent (HCl or HBr). See ref. 110 for a complete listing of these methods.

Good sensitivity and selectivity is achievable using inorganic reagents for the assay of inorganic ions in many cases, as is shown by the data in Table 22.

(C) FLUORESCENT CHELATES AND QUENCHING REACTIONS

The formation of a highly fluorescent chelate by the combination of an ion with an organic ligand has proved to be one of the most sensitive and highly specific methods for the determination of many elements. Some of the elements for which analytical procedures have been described by this method are Al, As, Au, B, Be, Bi, Ca, Cd, Cu, Ga, Ge, Hf, Hg, Mg, Nb, Pd, Rh, Ru, S, Sb, Se, Sn, Si, Ta, Tb, Te, Th, Tl, Zn, Zr, W. In general, fluorescent chelates are formed primarily with diamagnetic ions for reasons discussed above, although methods have been proposed for some paramagnetic ions, such as Cu^+. Some anions, such as CN^- or F^-, can be determined by direct fluorophore formation or, alternatively, could be assayed by their quenching of the fluorescence of a chelate or the release of a ligand to form a fluorescent product.

Some typical fluorometric methods for the determination of inorganic substances are listed in Table 23. Most of these methods are highly selective and sensitive, and represent the best analytical methods for these substances.

In some cases many different reagents have been described for the analysis of an ion. For example, aluminum could be assayed by chelating with Alizarin Garnet R, 3-hydroxy-2-naphthoic acid, Morin, Mordant Blue 9, Pontachrome Blue Black R, 8-quinolinol and Salicylidene-o-amino phenol (SOAP) [128—130]. Of these, the best are

Alizarin Garnet R, Salicylidene-o-amino phenol, and Mordant Blue 9. This latter dye is reportedly able to detect 0.5 p.p.b. of Al. A pre-

(Mordant Blue 9)

caution should be noted on the meaning of sensitivity values given in the literature for Al on two points, the reagents and the method. All common reagents, including distilled water, contain Al in appreciable quantities. Distilled water run through Al pipes or allowed to stand in glass containers may contain as much as 32 p.p.b. Al. Ammonia, acetic acid, sodium acetate, all contain many parts per billion of Al. This must be removed in dealing with p.p.b. amounts of Al. The second point of precaution is to consider how the author obtained his sensitivity value. For example, if the reagent is added to 1 ml of an Al solution of 10 ng ml^{-1}, then diluted to 10 ml, a sensitivity of 1 ng ml^{-1} might be reported. However, a 1 ng ml^{-1} solution would not produce a measurable fluorescence. These general considerations hold true of all analysis.

Marienko and May [132] described the use of Rhodamine B as a reagent for the assay of Au. The AuCl$_4^-$ anion in 0.4 M HCl forms a complex with Rhodamine B that has a λ_{ex} of 550 nm and a λ_{em} of 575 nm. The method is reported to be 25 times more sensitive than atomic absorption.

Benzoin is a highly specific reagent for B and will detect 10 p.p.b. The complex has a λ_{ex} at 365 and a λ_{em} at 480 nm. The complex has a greater fluorescence intensity in formamide than ethanol [133— 136]. Marcantonatos et al. [137] found that the complex with dibenzoylmethane (λ_{ex} = 385, λ_{em} = 410 nm) is able to detect 0.5 p.p.b. B.

The most selective reagent for Be is 1-amino-4-hydroxy anthra-quinone [138,139]. A more sensitive, though less specific, reagent is 8-hydroxyquinaldine [140] able to detect 1 p.p.b.

A very specific reaction for CN$^-$ is that with quinones to yield a highly fluorescent product (λ_{ex} = 440, λ_{em} = 500 nm). Guilbault and Kramer [147,148] estimated as little as 0.5 μg could be detected.

170

TABLE 23

Fluorescent methods for the assay of inorganic ions

Ion	Reagent	Method [a]	Sensitivity (p.p.m.)	Refs.
Ag	Eosin + 1,10-phenanthroline	Q	0.004	126
	8-Hydroxyquinoline-5-sulfonic acid	C	0.013	127
Al	Acid Alizarin Garnet R	C	0.007	128
	Mordant Blue 9	C	0.0005	129
	Salicylidene-o-aminophenol	C	0.0003	130
As	Gutzeit Test	Ch	1.0	131
Au	Rhodamine B	C	0.02	132
B	Benzoin	C	0.01	133—136
	Dibenzoylmethane	C	0.0005	137
Be	1-Amino-4-hydroxy-anthra-quinone	C	0.2	138, 139
	8-Hydroxyquinaldine	C	0.001	140
Ca	Calcein	C	0.2	141—143
Cd	p-Tosyl-8-aminoquinoline	C	0.02	144
Ce	Sulfonaphtholazoresorcinol	C	0.05	145
CN⁻	Pd complex of 8-OH-quinoline-5-sulfonic acid	Q	0.02	146
	Quinones	C	0.01	147, 148
Co	Al-pontachrome BBR	Q	0.001	149
Cr	Triazinylstilbexone	Q	0.004	150
Cu	2-(2'-Hydroxyphenyl)benzoxazole	Q	0.1	151
Eu	Hexafluoroacetone-trioctyl phosphine oxide	C	0.0001	152
	2-Theonyltrifluoroacetone	C	0.0001	153
F⁻	Al-acid Alizarin Garnet R complex	Q	0.001	128
	Ternary complex with Zr + Calcein Blue	C	0.01	154
Ga	Rhodamine B	C	0.01	155
	Rhodamine 6 G	C	0.1	156
	Sulfonaphtholazo-resorcinol	C	0.001	157
	2,2',4'-trihydroxy-5-chloro-1,1'-azobenzene-3-sulfonic acid	C	0.001	158
Ge	Benzoin	C	2.0	159
Hf	Flavonol	C	0.1	160
	Quercetin	C	1.0	192
Hg	Rhodamine B	Q	0.1	161
In	8-Hydroxyquinoline	C	0.04	162

TABLE 23 (continued)

Ion	Reagent	Method [a]	Sensitivity (p.p.m.)	Refs.
Mg	8-Hydroxyquinoline	C	0.01	163—167
	Bis-salicylidene ethylene diamine	C	0.0002	168
Mn	8-Hydroxyquinoline-5-sulfonic acid	C	0.005	169
Mo	Carminic acid	C	0.9	170, 171
NH_4^+	Hantzsch reaction	C	0.01	172, 173
	NADH	E	0.01	174
NO_3^-	2,3-Diamino naphthalene	C	0.01	175
Ni	Al-1-(2-pyridylazo)-2-naphthol	Q	0.0003	176
PO_4^{3-}	NADPH	E	0.01	177
S^{2-}	Fluorescein mercuriacetate	C	0.00005	178
Se	3,3'-Diaminobenzidine	C	0.02	179—181
	2,3-Diaminonaphthalene	C	0.02	182—187
Sm	2-Theonyltrifluoroacetone	C	0.0001	153
Sn	8-OH-quinoline-5-sulfonic acid	C	0.005	188
Tb	EDTA-sulfosalicylic acid	C	0.006	189
W	Carminic acid	C	0.3	170, 171
Y	8-Hydroxy quinoline	C	0.02	190
Zn	2,2'-Methylene dibenzothiazole	C	0.002	191
Zr	Flavonol	C	0.1	160

[a] Key: C, chelate; Ch, chemical; E, enzymatic; Q, quenching.

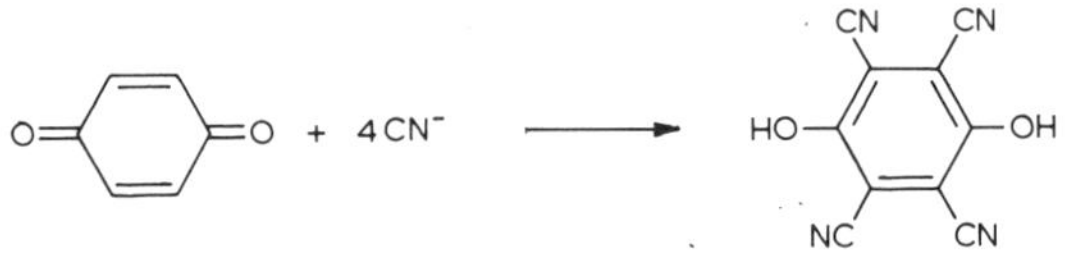

$$O={\bigcirc}=O + 4CN^- \longrightarrow HO-{\bigcirc}-OH \quad (24)$$

with CN, CN, NC, CN substituents

A very specific method for Cu is based on its quenching of the fluorescence of 2-(2-hydroxyphenyl)benzoxazole in acetone [151]. At pH > 6 the reagent possesses a green fluorescence, which decreases in the presence of Cu due to formation of a non-fluorescent chelate.

Har and West [154] have described a direct assay procedure for F^- based on the ternary complex formed with Zr and Calcein Blue. The complex ($\lambda_{ex} = 350$, $\lambda_{em} = 410$ nm) was found to have a Zr—Calcein Blue—F ratio of 1 : 1 : 1.

172

Most reagents for Al are also useful for Ga assay at lower pH, although less sensitively. Morin, 8-hydroxyquinoline, 8-hydroxyquinaldine and 5,7-dibromo-8-hydroxyquinoline have been proposed for Ga in the parts-per-billion range.

The best reagent for the analysis of Ga is Rhodamine B [155 (IVa)] or 6G (reported as 6Zh in the Soviet literature) [156]. These dyes form fluorescent chelates with Ga^{III}, Au^{III}, Tl^{III}, and Al^{III}, which are extracted by benzene from HCl solution. The optimum HCl concentration for Ga is 4 M. The ratio of Ga to Rhodamine B is 1 mole of $GaCl_4^-$ to 1 mole of RhB^+. The ring with the carboxyl is unnecessary since Acridine Red (Va) and Thiopyronine (Vb) also form complexes. The $GaCl_4^-$ probably combines with the $-NR_2Cl$ or $-NR_2$ group since fluorescein (IVb), which is similar to Rhodamine B (IVa) but lacks these groups, does not form a complex with Ga in HCl.

(IV)

a: $R = N(C_2H_5)_2$; $R' = N(C_2H_5)_2Cl$

b $R = OH$; $R' = O$

(V)

a: $X = O$; $R = NH(CH_3)$;
   $R' = NHCH_3Cl$

b: $X = S$; $R = N(C_2H_5)_2$;
   $R' = N(C_2H_5)_2Cl$

The determination of Ga by sulfonaphtholazoresorcinol is even more sensitive (0.001 p.p.m.) [157]. It is carried out in aqueous alcohol at pH 3.0, with monochloroacetic acid as buffer. The difference between Solochrome Black and sulfonaphtholazoresorcinol is that the complex formed by Solochrome Black with Ga fluoresces only after extraction from aqueous solution with an alcohol that is sparingly soluble in water, such as butyl, amyl, or hexyl alcohol.

At present an even more sensitive method has been developed for the quantitative determination of Ga. Bozhevol'nov et al. [158] studied the effect of substituents on the fluorescence properties of chelate compounds of Ga with dihydroxyazo compounds, and they found 2,2′,4′-trihydroxy-5-chloro-1,1′-azobenzene-3-sulfonic acid to be a more sensitive reagent for Ga (when used in an aqueous medium) than sulfonaphtholazoresorcinol; moreover its Ga complex is extracted with isoamyl alcohol and the fluorescence is more intense after the extraction. The fluorescence intensity of the Ga complex of this reagent is practically constant in the pH range 1.7—3.5. The fluorescence of the extracted complex is increased by a factor of 3.5 if equal volumes of isoamyl alcohol and the aqueous sample solution are used. The fluorescence intensity of solutions of this reagent is proportional to the Ga concentration, both in aqueous solutions and in isoamyl alcohol, provided the Ga concentration does not exceed 100 p.p.b. In aqueous solution the sensitivity of the reaction is 1 p.p.b. If isoamyl alcohol is used for the extraction of the samples, and if the volume ratio of isoamyl alcohol to aqueous solution is 1 : 10, Ga can be detected in amounts of 0.1 p.p.b. A detailed study of the effect of various cations and anions on the fluorescence intensity of the Ga complex showed that the fluorescence is quenched by Sn, Zr, and Pr when their amount is 100 times that of Ga, and by Cu, Fe, V, and Mo when their amount is 10 times that of Ga. Other cations do not quench even when present in amounts a thousand times larger than that of Ga. Aluminum is capable of forming a fluorescent complex, but its fluorescence is less intense. When the Ga—Al ratio is 1 : 1, the presence of Al can be ignored, and the measurements can be carried out at pH 1.7—3.5. If a ten-fold excess of Al is present, the pH of the solution should be 1.7—2.7, and in the case of a hundred-fold excess of Al the working pH range is narrower, between 1.7 and 2.2. The method of additions makes it possible to carry out the determination also in the presence of quenching impurities.

Benzoin is a highly specific reagent for B, but under proper conditions it can also be made a reagent for Ge. The fluorescence is yellow—green; as little as 2 p.p.m. is detectable [159]. Interfering ions include As, B, Be, CrO_4^{2-}, NO_2^-, and SiO_3^{2-}. Flavanol reacts with Hf to give a fluorophore with λ_{ex} 365 and λ_{em} 460 nm [160]. As little as 0.1 p.p.m. is detectable, but Zr, Al, F^-, Fe, and PO_4^{3-} interfere.

A more specific reagent is quercetin [192], which reacts with Hf in 9 M $HClO_4$ to give a green fluorophore (λ_{em} 505, λ_{ex} 340 nm). As little as 1 p.p.m. is determinable in the presence of Zn.

Alcoholic solutions of Mg^{2+}-8-hydroxyquinoline exhibit a fluorescence (λ_{ex} 420, λ_{em} 530 nm) at pH 6.5 [163]. Subsequently Schachter [164] suggested the use of the more water soluble 8-hydroxyquinoline-5-sulfonic acid for assay of Mg^{2+}. Small aliquots of serum are assayable without the need of deproteinization. Hill [165] adapted the 8-hydroxyquinoline procedure to an automated assay in serum and urine. He used an autoanalyzer system in conjunction with a Model 111 Turner fluorometer. Analyses were performed on serum dialyzates; potassium oxalate was added to prevent Ca^{2+} interference. Hill investigated the specificity of the automated method and showed that the zinc and phosphate normally found in serum and urine do not interfere. The method was shown to be highly reproducible and to yield serum values that compare almost exactly with those obtained by flame photometry. Klein and Oklander [166] utilized 8-hydroxyquinoline-5-sulfonic acid in an automated fluorometric determination of Mg in serum.

The 8-hydroxyquinoline method for Mg is now considered to be a standard clinical procedure. A modification of the Schachter procedure has appeared in Standard Methods of Clinical Chemistry [167]. More recently, Pruden et al. [193] compared serum values obtained by fluorometry with those obtained by photometry, atomic absorption, and flame emission. They reported that fluorometry and photometry gave values that were slightly higher than those obtained by flame emission or atomic absorption. It is quite likely that the higher values were the result of Ca interference and that the use of potassium oxalate, as in the automated procedure of Hill [165], would have yielded correct values. An interesting observation made by Pruden et al. [193] is that repetitive freezing and thawing of serum increases the fluorescence obtained in the 8-hydroxyquinoline assay. This results from release of interfering substances and not from increments in Mg.

Bissalicylideneethylenediamine (α,α'-(ethylenedinitrilo)di-o-cresol) forms a 1 : 1 chelate with Mg^{2+} in DMF solution and can detect 0.17 p.p.b. of Mg [168]. This is one of the most sensitive reagents for Mg. The fluorescence spectra of this chelate are shown in Fig. 42.

Pal and Ryan [169] described a fluorometric method for Mn based on its complexation of 8-hydroxyquinoline-5-sulfonic acid in concentrations as low as 5 p.p.b. The λ_{ex} was 375 and λ_{em} was 485–490 nm. The only serious interference under the reaction conditions used came from Ce.

Kirkbright et al. [170,171] have described carminic acid as a reagent for assay of Mo in steel. As little as 0.9 p.p.m. is determinable.

A fluorometric method for NH_4^+ is based on the reaction of a

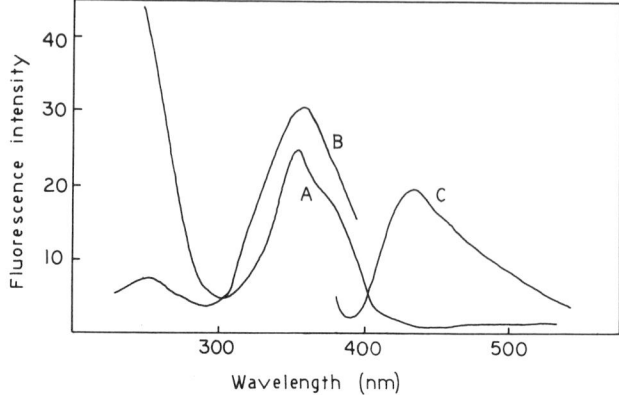

Fig. 42. Absorption (A), fluorescence excitation (B), and fluorescence emission (C) spectra of the Mg-bissalicylideneethylenediamine chelate in dimethylformamide.

formaldehyde plus diketone with NH_4^+ to give a lutidine derivative (the Hantzsch reaction) (λ_{ex} 405, λ_{em} 510 nm). From 0.01 to 0.25 p.p.m. N is determinable [172,173].

(Carminic acid)

Rubin and Knott [174] introduced an enzymatic procedure to measure NH_4^+ fluorometrically. This was achieved by utilizing the enzyme glutamic dehydrogenase, α-ketoglutarate, and reduced diphosphopyridine nucleotide, NADH. By suitable choice of conditions the reaction was made dependent on the NH_4^+ concentration and was followed by the disappearance of NADH fluorescence. The extreme sensitivity of pyridine nucleotide fluorescence permits assay of submicrogram quantities of NH_4^+. The rapidity of assay eliminates errors caused by NH_4^+ generation from biological material, which occurs during lengthy procedures of isolation or digestion.

Sawicki [175] described a fluorometric method for NO_3^- based on its reduction to nitrite with hydrazine sulfate and the subsequent

176

determination of the nitrite formed with 2,3-diaminonaphthalene. Solvent effects on the fluorescence intensity and optimum reaction conditions were discussed. The wavelengths used are λ_{ex} 364 and λ_{em} 412 nm. As little as 0.01 p.p.m.—1.13 p.p.m. is determinable.

Schenk et al. [176] described a fluorescent method for Ni^{II} based on its quenching of the fluorescence of Al-1-(2-pyridylazo)-2-naphthol in the range 10^{-9}—10^{-7} M. The method was described as far more sensitive than atomic absorption methods for Ni and has few interferences.

Lowry et al. [177] described an enzymatic method for phosphate, sensitive to 2×10^{-11} mole.

$$\text{Glycogen} + PO_4{}^{3-} \xrightarrow{\text{phosphorylase}} \text{glucose-1-phosphate} \tag{25}$$

$$\text{Glucose-1-phosphate} \rightarrow \text{glucose-6-phosphate} \tag{26}$$

$$\text{Glucose-6-phosphate} + NADP \xrightarrow{E} NADPH + \text{6-phosphogluconolactone} \tag{27}$$

The fluorescence of NADPH is measured and equated to the $PO_4{}^{3-}$ concentration.

Pal and Ryan [188] assayed Sn by the fluorescence of the 8-hydroxyquinoline-5-sulfonic acid complex at λ_{ex} 360 and λ_{em} 515 nm. As little as 5 p.p.b.—0.25 p.p.m. is assayable. Quenching of the fluorescence occurs in the presence of Fe, Cu, Hg, Ni, F^-, EDTA, citrate, oxalate, and tartrate. Traces of Al, Zn, Hf, and Zr enhance the fluorescence.

A specific spectrofluorometric method for Tb as a ternary complex with EDTA and sulfosalicylic acid has been described [189]. From 0.0064 to 3.2 p.p.m. is determinable. The λ_{ex} is 320 and the λ_{em} is 410 and 545 nm. No interference was noted from fifty-fold excesses of 33 metal ions and 14 anions.

With 8-hydroxyquinoline yttrium forms a complex that is extracted with $CHCl_3$ from pH 9.5. As little as 0.02 p.p.m. is assayable. Interference from Ce and La is observed [190].

(VI)   (VII)

Zinc can be determined with *p*-tosyl-8-aminoquinoline in concentrations of about 20 p.p.b. [144]. A more sensitive reagent is 2,2′-methylenedibenzothiazole (VI) which will detect 2 p.p.b. [191]. Picolinaldehyde-2-quinolylhydrazone (VII) will detect 26 p.p.b.

The blue—white fluorescence (λ_{ex} 390, λ_{em} 465 nm) exhibited by flavanol in the presence of Zr in H_2SO_4 solution is the basis of a method for determining as little as 0.1 p.p.m. Hafnium is the only cation that interferes; F^- and PO_4^{3-} also interfere [160].

7. Determination of organic compounds

Very sensitive and highly selective analytical procedures have been developed for the assay of hundreds of organic compounds, including amino acids, vitamins, steroids, chlorophylls and drugs. A complete listing of methods can be found in ref. 1.

Some organic compounds are naturally fluorescent, possessing structures that are rigid, coplanar and possess labile π electrons.

Fluorescein(VIII), resorufin(IX), indoxyl(X) and umbelliferone (XI), for example, are all fluorescent in the sub-nanogram region.

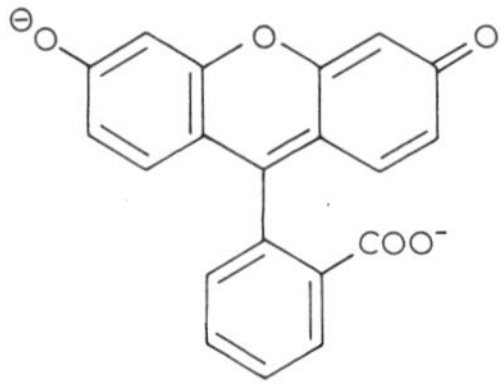

Fluorescein
(λ_{ex} = 495 nm;
λ_{em} = 535 nm)

(VIII)

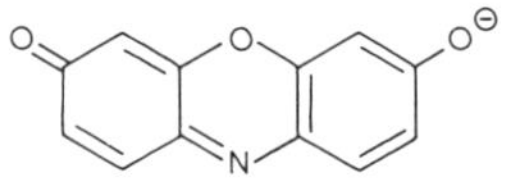

Resorufin
(λ_{ex} = 560 nm;
λ_{em} = 580 nm)

(IX)

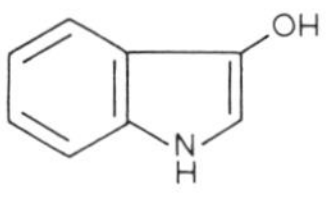

Indoxyl
(λ_{ex} = 495 nm;
λ_{em} = 570 nm)

(X)

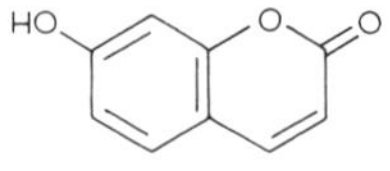

Umbelliferone
(7-Hydroxy coumarin)
(λ_{ex} = 325 nm;
λ_{em} = 440 nm)

(XI)

Amino acids, such as phenylalanine, tyrosine and tryptophan, likewise possess a native fluorescence that can be used for their assay.

178

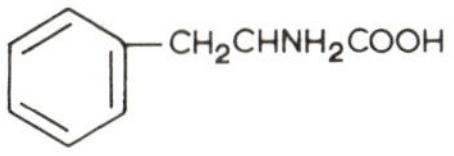

Phenylalanine
($\lambda_{ex} = 260$ nm;
$\lambda_{em} = 282$ nm)
Rel FI = 0.5

(XII)

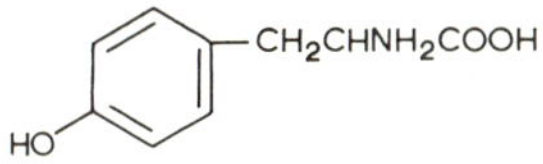

Tyrosine
($\lambda_{ex} = 275$ nm;
$\lambda_{em} = 303$ nm)
Rel FI = 9

(XIII)

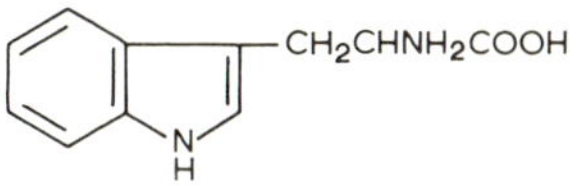

Tryptophan
($\lambda_{ex} = 287$ nm;
$\lambda_{em} = 348$ nm)
Rel FI = 100

(XIV)

The fluorescence of these amino acids demonstrates the effect of structures on the luminescence. Phenylalanine, with only a benzene ring and a $-CH_2-$ side chain is weakly fluorescent. Add a hydroxyl group as in tyrosine and the fluorescence goes up by 20; add the indole ring as in tryptophan and the relative fluorescence is 200 times better.

Fluorescence has been a valuable aid in identifying the mechanism of photosynthesis, since the different chlorophylls, a, b, c, d, etc., pheophytin, protochlorophyll, bacteriopheophytin and other photosynthesis precursors and products are fluorescent at different wavelengths. An excellent chapter on this subject can be found in Guilbault's book [1].

Other organic compounds, themselves non-fluorescent or weakly fluorescent, can be converted to good fluorophores by a simple chemical reaction.

Luminol is measured by its intense luminescence, produced via the following reaction.

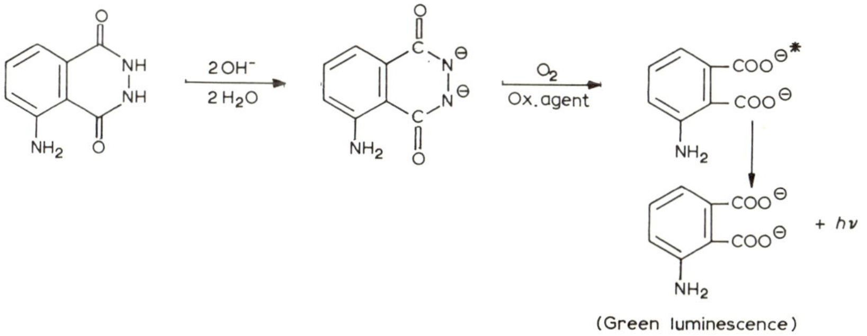

(28)

(Green luminescence)

The luminol reaction has been used for the determination of oxidizing agents, such as peroxide, and for metal ions such as Cu or

Co, which catalyze the reaction. As little as 2 p.p.b. Co or 30 p.p.b. Cu can be determined [194—195].

Acetol can be determined by a condensation with *o*-amino benzaldehyde to produce the fluorophore 3-hydroxy quinaldine [196].

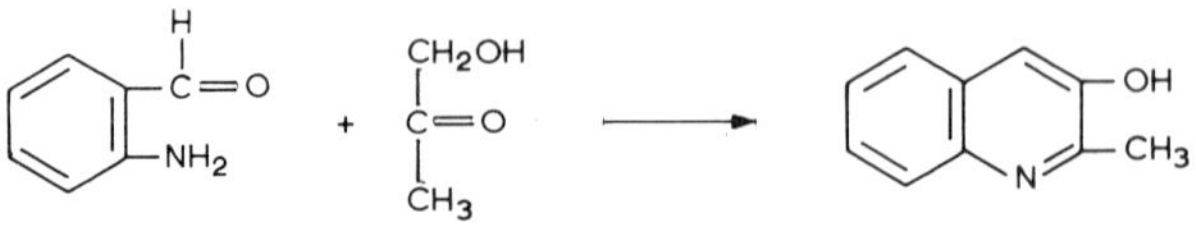

$$\lambda_{ex} = 365 \, nm; \; \lambda_{em} = 440 \, nm \qquad (29)$$

Organic acids, like malic acid, can be assayed by a condensation with resorcinol to yield umbelliferone derivatives [197].

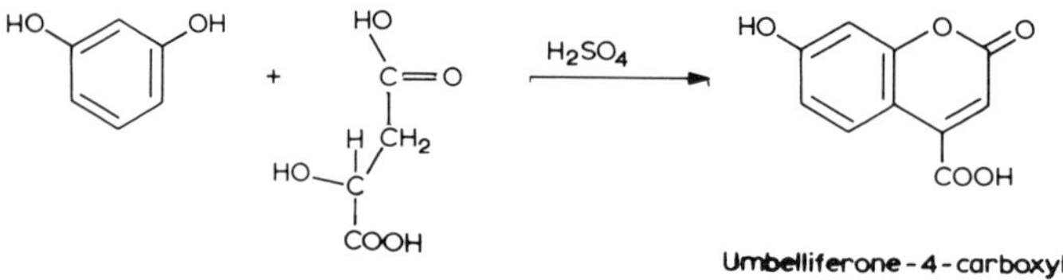

Umbelliferone-4-carboxylic acid

$$(30)$$

Some of the organic acids assayable and the relative fluorescences are listed in Table 24.

Adrenaline and dopamine [198] are similarly assayed via fluorophore formation to highly fluorescent indoxyl derivatives.

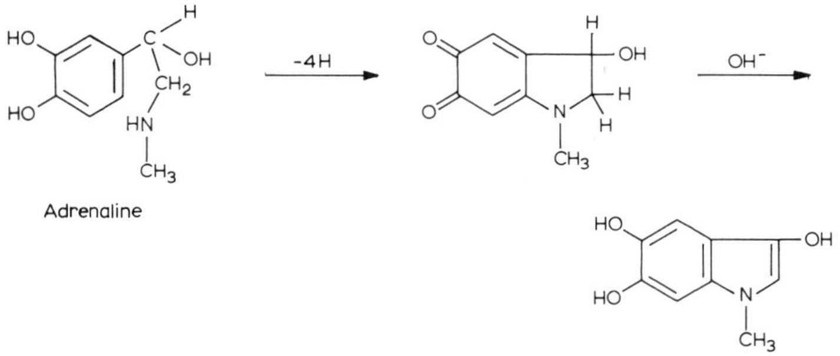

Adrenaline

$$(\lambda_{ex} = 420 \, nm; \; \lambda_{em} = 520 \, nm) \qquad (31)$$

Dopamine $\qquad (\lambda_{ex} = 345 \, nm; \; \lambda_{em} = 410 \, nm)$

$$(32)$$

180

TABLE 24

Fluorescence of resorcinol polycarboxylic acid derivatives [a]

Acid	Color of fluorescence	Relative fluorescence
Malic	Blue violet	22
Fumaric	Blue violet	24
Succinic	Yellow green	20
Isocitric	Light blue	58
Citric	Sky blue	89

[a] Ref. 196.

Some vitamines, as Vitamin A, possess a native fluorescence, and can be measured directly [199] as was mentioned above (λ_{ex} = 327 nm; λ_{em} = 510 nm for Vitamin A in the 0—10 p.p.m. range). Others, like thiamine and riboflavin, are best converted to fluorophores by simple dehydration reactions.

$$\text{Thiamine} \xrightarrow{-H} \text{Thiochrome } (\lambda_{ex} = 365 \text{ nm}; \lambda_{em} = 435 \text{ nm}) \qquad (33)$$

$$\text{Riboflavin} \xrightarrow{OH^-} \text{Lumiflavin } (\lambda_{ex} = 440 \text{ nm}; \lambda_{em} = 550 \text{ nm}) \qquad (34)$$

Vitamins D_2 and D_3 are treated with trichloroacetic acid to give fluorophores measured at 480 nm [200].

Cholesterol is commonly assayed fluorometrically by treatment with H_2SO_4 to yield a red—orange fluorophore (λ_{ex} = 546 nm; λ_{em} = 590 nm). From 0.1 to 2 μg of cholesterol are assayable [201]. Similarly, other steroids can be assayed by fluorophore formation with H_2SO_4, and the reader is referred to the chapter on Steroids in Udenfriend's book [202].

Many purines, pyrimidines, and coenzymes possess a native fluorescence in solution that can be used for their assay (Table 25). The fluorescence spectra of adenosine, AMP, ADP, and ATP are shown in Fig. 43, those of the pyridine nucleotides in Fig. 44.

The pyridine nucleotides exist in oxidized and reduced forms (Fig. 45). Reduced NAD (NADH) and NADP (NADPH) have maximum absorbance at 340 nm and a high fluorescence (λ_{ex} 340, λ_{em} 460 nm) [206]. Lowry et al. [206] have described procedures for using the native fluorescence of NADH and NADPH for their assay in tissues and have discussed the effect of solvents, pH, and trace metals on the

TABLE 25

Fluorescence characteristics of bases and derivatives [a]

Compound	Medium	Absorption maximum (nm)	Excitation maximum (nm)	Fluorescence maximum (nm)
Adenine	pH 1	263	265 [b]	380
Adenosine	5 N H_2SO_4	257	272	390
ADP	5 N H_2SO_4	257	272	390
ATP	5 N H_2SO_4	257	272	390
FAD	pH 6.6	—	365	520
FMN	HOAc-CHCl$_3$	—	365	520
Guanine	pH 1	272 (sh) [c]	275 [b]	360
	pH 11	273	275 [b]	350
Guanosine	pH 1	277 (sh) [c]	285	390
1-Methylguanine	pH 1	274 (sh) [c]	290	370
NADH	pH 8—11	340	340	460
NAD	NaOH (7 M)	—	340	460
NADPH	pH 8—11	340	340	460
NADP	NaOH (7 M)	—	340	460
Purine	pH 13	271	285	370

[a] Some data from ref. 203.
[b] Corrected values.
[c] Shoulder or point of inflection.

fluorescence. The fluorescence spectra of NADH and NADPH are shown in Fig. 44. Kaplan et al. [207] found that the oxidized pyridine nucleotides (NAD and NADP) are converted to highly fluorescent products when treated with alkali. Lowry et al. [206] reported the spectra of the NAD^+ and $NADP^+$ alkali product (Fig. 44) which are almost identical with those of NADH or NADPH.

Huff and Perlzweig [208] have shown that NAD and NADP condense with acetone in alkali solution to form highly fluorescent products. Methods were developed for these coenzymes in blood and urine.

Fluorescence has proved to be a powerful tool for the analysis of drugs and medicinal agents because of its great sensitivity and high specificity. Pharmacology requires analytical methods that can distinguish between the various drugs and their metabolic products. Since many drugs are administered at very low doses, sometimes as little as 100 μg a day, the analytical method must be highly sensitive,

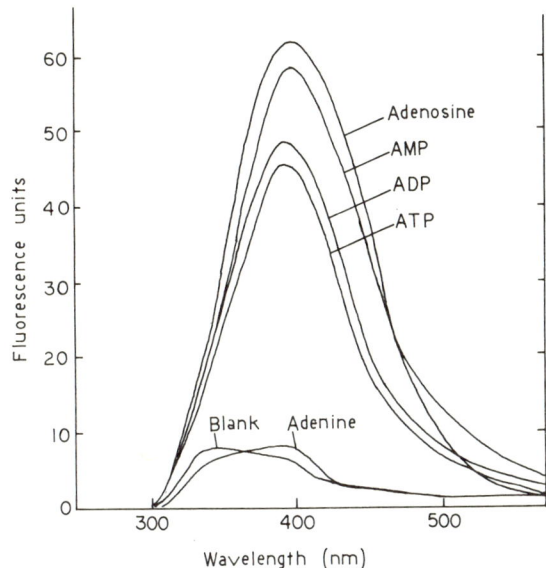

Fig. 43. Fluorescence spectra of adenine and its nucleoside and nucleotides in 7 N H_2SO_4. Excitation was at 265 nm, and the concentration was 5 μg ml^{-1} in each instance. Reprinted from ref. 204 by courtesy of the author.

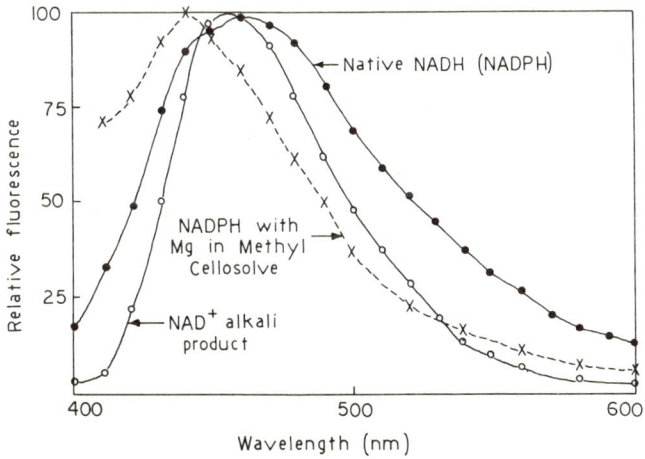

Fig. 44. Fluorescence spectra of pyridine nucleotides. Values are plotted as percentage of the maximum fluorescence. On an absolute scale NADH fluorescence would have to be reduced by a factor of about 8. Reprinted from ref. 205 by courtesy of the author.

References pp. 198—205

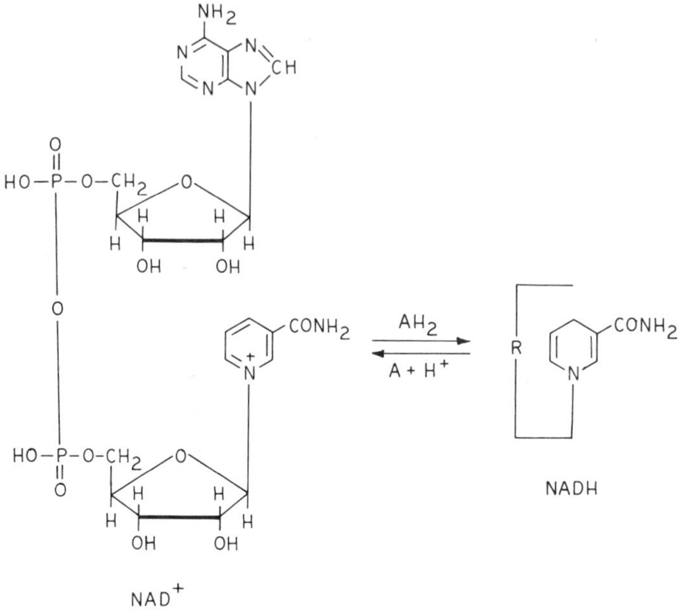

Fig. 45. Oxidized and reduced forms of nicotinamide adenine dinucleotide.

sensitive enough to pick up the small amounts excreted from the body. Fluorescence has this capability. Fluorometric methods for many of the common drugs and medicinal agents are listed in Table 26. Detailed procedures for the assay of all these drugs are not presented here, but the reader is referred to Udenfriend's Fluorescence Assay in Biology and Medicine [5,6] and Guilbault's Practical Fluorescence [1] for details.

Finally, the porphyrins, which are among the most highly fluorescent compounds in nature, are derivatives of porphine (four pyrazole rings joined into a ring system by four methane bridges). The porphyrins are excited in the visible and emit fluorescence in the red and IR. This fluorescence can be detected with standard instruments equipped with 1P21 and 1P28 photomultiplier tubes, but the use of a red-sensitive detector (RCA 7102 tube) increases the sensitivity markedly.

The naturally occurring porphyrins include hemoglobin, myoglobin, cytochromes, chlorophyll, and other pigments. Chlorophyll

184

TABLE 26

Fluorometric methods for assay of drugs [a]

Substance	Conditions	λ_{ex}	λ_{em}	Sensitivity (p.p.m.)
Actinomycin D	H_2O_2-OH^-	370	420	0.10
N-Allylnormorphine	pH 1	285	355	0.10
Antimycin	pH 7—9	350	420	0.10
Aspirin	HOAc-$CHCl_3$	280	335	0.01
Atropine	Eosin Y	365	556	1.0
Chloropromazine	pH 11	350	480	0.01
Codeine	pH 1	285	350	0.1
Digitalis	HCl-glycerol	350	465	0.1
Epinephrine	Ferricyanide	365	495	0.002
Estrogens	pH 13	490	546	0.1
LSD	pH 7	325	365	0.002
Morphine	Ferricyanide	250	440	0.1
Penicillin	2-Methoxy-6-Cl-9-(β-aminoethyl)amino-acridine	365	540	0.05
Phenobarbital	pH 13	265	440	0.5
Procaine	pH 11	275	345	0.01
Quinine	pH 1	350	450	0.002
Streptomycin	pH 13	366	445	0.10
Sulfanilamide	pH 3—10	275	350	0.10
Tetracycline	pH 11	390	515	0.02
Yohimbine	pH 1	270	360	0.01

[a] Selected values from ref. 1, Table 66.

will not be discussed but full details can be found in Chap. 13 of ref. 1.

Dhere [209] has carried out a thorough investigation on the fluorescence of porphyrins. He described and classified the fluorescence spectra of the porphyrins in a variety of solvents. The same spectra are observed in alcohol, dioxane, ammonia, and pyridine.

In acid solution only three bands appear. Pyridine protoporphyrin has a λ_{em} at 634 nm, whereas hematoporphyrin has a λ_{em} at 625 nm. Coproporphyrin, uroporphyrin, mesaporphyrin, and etioporphyrin have almost the same fluorescence characteristics as hematoporphyrin.

A survey of the luminescence characteristics of porphyrins was written by Schwartz et al. [210], who discussed factors that affect the fluorescence and quenching agents. Sharp emission bands were observed in organic solvents; as little as 10^{-7} g was determinable.

Solov'ev et al. [211,212] related the spectral characteristics of porphyrins to molecular structure. The porphyrins were found to have two emission bands with mirror-image symmetry to the two absorption bands.

Runge [213] developed a microfluorospectrophotometer that was able to detect as little as 10^{-10} g of porphyrin in tissue sections. Specific porphyrins in tissues and plasma were identified.

Martinez and Mills [214] developed a rapid procedure for the assay of total porphyrins in urine based on a combination of anion exchange and spectrophotofluorometry. After the porphyrins are eluted from the resin with 3 M HCl, their fluorescence at 646 nm is measured.

8. Assay of enzymes

Since the enzyme is a catalyst, theoretically one molecule of this material would eventually produce a sufficient change in the substrate to be measured. Hence high sensitivities can be realized in enzyme analysis. Because the concentration of enzyme is so small, it always limits the rate of reaction, and the rate can be taken as a measure of the enzyme concentration. The oxidation of glucose by oxygen to give peroxide and gluconic acid is catalyzed by glucose oxidase. The rate of peroxide production is measured by a second coupled reaction, the oxidation of a leuco dye, such as o-dianisidine, to yield a highly colored dye. When glucose, leuco dye, and oxygen are not rate limiting, the overall rate of reaction, as indicated by the rate of dye production, is proportional to the glucose oxidase activity.

Because of limitations in molar absorptivities, measurements of gas volumes, or of changes in pH, most methods previously described for measuring components in enzyme reactions are limited to reactions of reagents present at concentrations greater than 10^{-6} M. Because fluorometric methods are generally several orders of magnitude more sensitive than chromogenic ones, a large increase in the sensitivity of measurement should result. Thus much lower concentrations of reactants would be needed, and one could devise methods for substances at 10^{-9} M concentrations and lower. Moreover, fluorometric

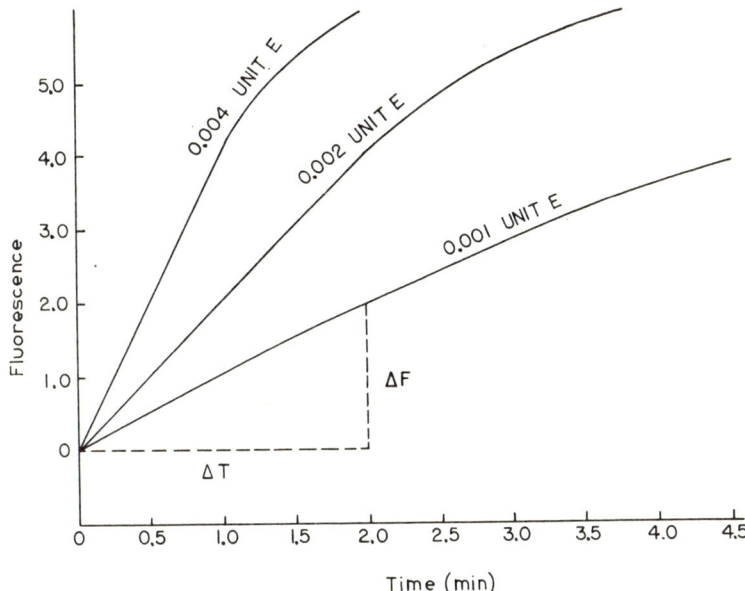

Fig. 46. Rate of increase in fluorescence with time on addition of enzyme.

methods are quite useful in biochemical work, in the localization of enzymes and related substrates (activators), within organs, and even within individual cells.

Because of their sensitivity and specificity, fluorescent methods have found increasing use in enzymology. For example, NADH and NADPH, the reduced forms of nicotinamide adenine dinucleotide (NAD) and nicotinamide adenine dinucleotide phosphate (NADP), are highly fluorescent. Thus all NAD- and NADP-dependent reactions involved in enzymatic analysis can be measured fluorometrically, with an increase of two to three orders of magnitude in sensitivity over colorimetric techniques.

In fluorometric assay methods for enzymes generally no fluorescence is initially observed. On addition of enzyme the fluorescence increases (Fig. 46). The rate of change in the fluorescence with time, $\Delta F \, min^{-1}$, is proportional to the concentration of enzyme.

Fluorescence methods have also been used extensively for the determination of hydrolytic enzymes, based on the enzyme-catalyzed hydrolysis of a non-fluorescent ester to a highly fluorescent alcohol or amine.

(1) Cholinesterase

Guilbault and Kramer have described four new fluorogenic substrates that can be used for a rapid and specific determination of cholinesterases: resorufin butyrate [215], indoxyl acetate [216], and 1- and 2-naphthyl acetate [217]. All of these are non-fluorescent, but they are hydrolyzed by cholinesterase to highly fluorescent compounds.

The substrates, resorufin acetate and butyrate, are hydrolyzed by cholinesterase, acid phosphatase, and chymotrypsin to the highly fluorescent resorufin (λ_{ex} 540, λ_{em} 580 nm). The rate of resorufin production with time is proportional to the concentration of cholinesterase from 0.0001 to 0.123 unit ml^{-1}, with a deviation of only 1.0%. By the choice of substrates, some specificity is possible. For example, resofurin acetate is not hydrolyzed by lipase nor by acetyl-cholinesterase (from bovine erythrocytes or electric eel), but resorufin butyrate is hydrolyzed by lipase. Similar concentrations of cholinesterase can be determined with a slightly lower deviation (0.9%) by using indoxyl acetate as substrate. Various authors [218] have reported the colorimetric assay of cholinesterase, based on the formation of indigo blue from indoxyl acetate. Kramer and Gelman [219] and Gehauf and Goldenson [220] have reported that indoxyl is a highly fluorescent compound and is easily air-oxidized to indigo blue. Guilbault and Kramer [216] found that indoxyl acetate could be used as a fluorogenic substrate for cholinesterase, by a proper control of experimental conditions. Indoxyl acetate is hydrolyzed first to indoxyl, which is fluorescent. Air then rapidly effects the oxidation of indoxyl to indigo white, which is twice as fluorescent as indoxyl, and then to the non-fluorescent indigo blue. An oxygen scavenger, ascorbic acid, prevents the oxidation of indoxyl, so that one can obtain a stable fluorescent product. However, in the absence of ascorbic acid below pH 7 indigo white forms and is not oxidized to indigo blue.

Guilbault et al. [221] prepared several esters as substrates for cholinesterase: the acetate, propionate, and butyrate esters of N-methylindoxyl, umbelliferone, and 4-methylumbelliferone. Comparison of these substrates with other fluorogenic esters (indoxyl acetate, indoxyl butyrate, resorufin acetate, 2-carbonaphthoxy-choline, and 2-naphthyl acetate) indicated that N-methylindoxyl acetate and butyrate were the best substrates for true and pseudo-cholinesterase, respectively. Analysis of as little as 5×10^{-5} unit ml^{-1} of cholinesterase can be performed by a direct initial-reaction-rate method in 2—3 min with an accuracy and precision of about 1.5%. All three N-methylindoxyl esters are very stable in solution, with a very low rate of spontaneous hydrolysis and a high rate of enzymatic hydrolysis. All have good K_m values, and the N-methyl-indoxyl formed is not easily air-oxidized to indigo derivatives.

(2) β-D-Galactosidase

Rotman [222] reported an extremely sensitive procedure for the fluorometric assay of β-D-galactosidase activity. The substrate was reported to be 6-hydroxyfluoran-β-D-galactopyranoside, which on enzymatic hydrolysis supposedly yielded the highly fluorescent 6-hydroxyfluoran. Further investigation by Rotman et al. [223] indicated that the substrate used in the earlier studies [222] was really fluorescein-di-(β-D-galactopyranoside), (XV), and the product liberated by galactosidase action was fluorescein.

(XV)

The use of fluorescein-di-(β-D-galactopyranoside) as a substrate for galactosidase offers many advantages when the assay is carried out under the usual laboratory conditions. Furthermore, Rotman [222] was able to demonstrate the unique applications that are made possible by the extremely high sensitivity of fluorometric assay. In this study he developed methods for measuring the activity of individual molecules of the enzyme.

(3) β-Glucuronidase

Umbelliferone (7-hydroxycoumarin (XVI)) and 4-methylumbelliferone are highly fluorescent compounds. In the body they are converted to the corresponding glucuronides, which are non-fluorescent. In the presence of β-glucuronidase the glucuronides are split to release the free fluorescent products [224].

$$(C_6H_9O_6)O \xrightarrow{\beta\text{-glucuronidase}} HO \quad + \quad C_6H_{10}O_7$$

(XVI)

The glucuronide of 4-methylumbelliferone can be prepared in excellent yield by feeding the free compound (commercially available) to rabbits and isolating the derivative from the urine. The procedure described by Mead et al. [224] gives a 20—25% yield (isolated) from an administered dose of 2.5 g.

Verity et al. [225] devised a method for β-glucuronidase assay in which the glucuronide of 1-naphthol is used as substrate. After incubation of 1-naphthyl-β-D-glucuronide at pH 4—5 with tissue extract, the solution is chilled, and made alkaline to about pH 11.0 by the addition of an appropriate volume of 0.5 N NaOH; fluorescence is then assayed. The free 1-naphthol formed during incubation is excited maximally at 340 nm and fluoresces at 460 nm. The glucuronide is also fluorescent but is maximally excited at about 310 nm and emits in the UV at 345 nm.

(4) Lipase

Guilbault and Kramer [226,227] described a rapid and simple method for the determination of lipase based on its catalysis of the hydrolysis of the non-fluorescent dibutyryl ester of fluorescein.

$$\text{dibutyrylfluorescein} \xrightarrow{\text{lipase}} \text{fluorescein} \tag{36}$$

This reaction can be monitored by measuring the rate at which the highly fluorescent fluorescein is produced with time. The concentration of enzyme can then be calculated from linear calibration plots of $\Delta F \ \text{min}^{-1}$ vs. enzyme concentration.

In a thorough study of fluorometric substrates for lipase Guilbault and Sadar [228] evaluated 12 different compounds from the aspects of stability, spontaneous hydrolysis, enzymatic hydrolysis, Michaelis constant for the enzyme—substrate complex, and total fluorescence of the final product. Optimum conditions of analysis were found for all substrates, and the lowest detectable enzyme concentration was found for each substrate. From all aspects, 4-methyl umbelliferone heptanoate was found to be the best substrate for pig-pancreas lipase, and 4-methylumbelliferone octanoate was the best for fungal lipase. As little as 1×10^{-5} unit could be determined by a direct reaction-rate method, with an accuracy and precision of about 1.5%.

Guilbault and Hieserman [229] prepared several new fluorometric substrates for the assay of lipase. A study of six N-methylindoxyl esters as substrates for lipase indicated N-methylindoxylmyristate to be best. By using this ester, from 0.0002 to 4.0 unit ml^{-1} of pig pancreas can be determined, in the presence of several other esterases, with an accuracy and precision of about 1.5%. Analysis is performed by a direct initial-reaction-rate method in 2—3 min.

(5) Phosphatase

Moss [230] described the use of the naphthyl phosphates as substrates for acid and alkaline phosphatase. The excitation maxima for 1-naphthol are at 250 and 335 nm, with peak emission at 455 nm. The corresponding phosphate ester is maximally excited at 235 and 295 nm, emission being at 365 nm. Thus, even though the ester is fluorescent, it is possible to select appropriate wavelengths to distinguish the free phenol from the ester. This is achieved at 335 nm for excitation and 455 nm for fluorescence. With 2-naphthol, excitation at 350 nm and measurement of fluorescence at 425 nm permits measurement of the free phenol in the presence of the ester. Both 1- and 2-naphthol have maximal and constant fluorescence at pH 10 and higher. It is therefore possible to follow alkaline phosphatase continuously. However, acid phosphatase necessitates alkalinization of the reaction mixture before fluorometric assay.

The fluorescein phosphates offer quite sensitive assays for phosphatase activity, but they are not very stable. Land and Jackim [231] found that the background fluorescence of a 3-O-methyl-fluorescein phosphate solution (0.1 mg ml^{-1}) at pH 8 increases seven-fold in 24 h at room temperature. Enough blank fluorescence appears on incubation at 37°C to limit the sensitivity of phosphatase assays. Land and Jackim [231] synthesized the phosphate ester of 3-

hydroxyflavone (flavone-3-phosphate) and found it to be highly stable and an excellent substrate for phosphatase. The liberated 3-hydroxyflavone is maximally excited at 360 nm and fluoresces at 510 nm. In 50% ethanol fluorescence is increased 200 times. The aluminum complex of the dye is even more highly fluorescent, with excitation at 400 nm and fluorescence at 450 nm. Picogram quantities (10^{-12} g) of *Escherichia coli* alkaline phosphatase can be assayed with this reagent.

Guilbault et al. [232] prepared umbelliferone phosphate as a fluorogenic substrate for acid and alkaline phosphatase, and compared this substrate with other substrates listed in the literature. From aspects of stability, rate of enzymatic hydrolysis, and fluorescence of product formed, umbelliferone phosphate appears to be an ideal substrate. As little as 10^{-6} unit of alkaline phosphatase and 10^{-5} unit of acid phosphatase are detectable.

Guilbault and Vaughan [233,234] described the use of Naphthol AS derivatives as substrates for phosphatase. The fluorescent properties of a series of Naphthol AS derivatives (Naphthol AS, AS-BI, AS-D, AS-GR, AS-LC, AS-MX, and AS-TR) were compared. The phosphate esters of these compounds were investigated as fluorogenic substrates for acid and alkaline phosphatase. The enzyme was determined by measuring the rate of formation of the fluorescent Naphthol AS. For example, the reaction by which the non-fluorescent Naphthol AS-BI phosphate (XVII) is converted to the fluorescent Naphthol AS-BI (XVIII): λ_{ex} 405, λ_{em} 515 nm is as follows

(37)

(XVII)   (XVIII)

The amount of enzyme is calculated from calibration plots of initial rate against the concentration of alkaline phosphatase. In acid solution Naphthol AS derivatives are non-fluorescent, but calibration curves for acid phosphatase can be obtained by quenching the en-

192

zyme reaction after a fixed time with alkali, which develops the fluorescence of the liberated Naphthol AS. The amount of acid phosphatase is calculated from plots of change in fluorescence in the fixed reaction time against enzyme concentration. Naphthol AS-BI phosphate was found to be the best substrate for both enzymes.

(B) DEHYDROGENASES AND TRANSAMINASES

(1) Methods based on the fluorescence of NADH or NADPH

The dehydrogenases are an important class of enzymes that effect the dehydrogenation of hydroxy compounds in the presence of a hydrogen acceptor or coenzyme, such as NAD or NADP. Since NADH (the reduced form of NAD) is fluorescent and NAD forms a fluorescent derivative on heating in alkali, many fluorometric methods for enzyme assay with NAD—NADH and NADP—NADPH systems have been proposed. The initial rate of formation or disappearance of NADH is measured and related to the concentration of dehydrogenases, substrate, or coenzyme. For example, lactate is converted to pyruvate at pH 9 by lactic dehydrogenase (LDH) in the presence of NAD, and the reverse reaction is effected at pH 7.

$$\text{lactate} + \text{NAD} \xrightleftharpoons{\text{LDH}} \text{pyruvate} + \text{NADH} \tag{38}$$

The rate of NADH formation at pH 9 or the rate of disappearance of this substance at pH 7 is proportional to the concentration of lactic dehydrogenase, NAD (or NADH), and substrate [235].

Fluorometric methods for hexokinase have been described by Greengard [236,237], who used a coupled enzyme system. In the first method hexokinase converts glucose to glucose-6-phosphate, which is in turn converted to 6-phosphogluconate with simultaneous formation of NADPH. Again the rate of NADPH production is proportional to the ATP or hexokinase concentration.

$$\text{ATP} + \text{glucose} \xrightarrow{\text{hexokinase}} \text{ADP} + \text{G-6-P} \tag{39}$$

$$\text{G-6-P} + \text{NADP} \xrightarrow{\text{G-6-PDH}} \text{6-phosphogluconate} + \text{NADPH} + \text{H}^+ \tag{40}$$

A second method is based on the use of kinase and NADH

$$\text{ATP} + \text{3-phosphoglycerate} \xrightarrow{\text{kinase}} \text{ADP} + \text{1,3-diphosphoglycerate} \tag{41}$$

$$1,3\text{-diphosphoglycerate} + NADH + H^+ \xrightarrow[\text{phosphate dehydrogenase}]{\text{glyceraldehyde}}$$

$$\text{glyceraldehyde-3-phosphate} + P_i + NAD^+ \tag{42}$$

The rate of NADH disappearance is proportional to the ATP concentration.

Graham and Aprison [238] have described enzymatic fluorometric methods for aspartate, glutamate, and γ-aminobutyrate in nerve tissue. The method is based on the use of glutamic-oxalacetic transaminase (GOT), which in the presence of α-ketoglutarate (α-KG) converts aspartate to oxalacetate. The oxalacetate formed was reduced to malate in the presence of NADH and malic dehydrogenase (MDH)

$$\alpha\text{-KG} + \text{aspartate} \underset{}{\overset{GOT}{\rightleftharpoons}} \text{glutamate} + \text{oxalacetate} \tag{43}$$

$$\text{oxalacetate} + NADH + H^+ \rightleftharpoons \text{malate} + NAD^+ \tag{44}$$

Either the disappearance of NADH or the increase in NAD could be measured, fluorometrically. By this method as little as 5×10^{-12} mole of aspartate, 1×10^{-11} mole of γ-aminobutyrate, and 1×10^{-10} mole of glutamate can be measured.

Laursen et al. [239,240] have developed methods for transaminases by coupling the following reactions

$$\text{L-alanine} + \alpha\text{-KG} \overset{\text{transaminase}}{\rightleftharpoons} \text{pyruvate} + \text{glutamate} \tag{45}$$

$$\text{pyruvate} + NADH \overset{LDH}{\rightleftharpoons} \text{lactate} + NAD^+ \tag{46}$$

The increase in NAD^+ is a measure of transaminase activity.

(2) Resazurin method

Guilbault and Kramer [241,242] have devised a simple and rapid fluorometric method for dehydrogenases, based on the conversion of the non-fluorescent substance resazurin (XIX) to the highly fluorescent resorufin (XX) (λ_{ex} 540, λ_{em} 580 nm) in conjunction with the NAD—NADH or NADP—NADPH systems

$$\text{substrate} + \underset{\text{(NADP)}}{NAD} \xrightarrow{\text{dehydrogenase}} \text{oxidized substrate} + \underset{\text{(NADPH)}}{NADH} \tag{47}$$

194

$$\text{NADH} + \quad \text{(XIX)} \xrightarrow[\text{or diaphorase}]{\text{phenazine methyl sulfate}} \text{NAD} + \quad \text{(XX)} \tag{48}$$

Because of the intense fluorescence of resorufin (10^{-9} M is easily detectable), a two-fold increase in sensitivity over the NADH method is obtainable. The rate of resorufin formation is proportional to the concentration of the dehydrogenase, substrate, NAD, and diaphorase.

(C) OXIDATIVE ENZYMES

(1) Peroxidase

Keston and Brandt [243] have described a fluorometric method for the analysis of as little as 10^{-8} mole l^{-1} (10^{-11} mole ml^{-1}) of hydrogen peroxide. The procedure is based on the oxidation, by hydrogen peroxide and peroxidase, of the non-fluorescent diacetyl dichlorofluorescein to the highly fluorescent dichlorofluorescein. The method is also applicable to peroxidase and other enzyme systems that produce hydrogen peroxide. Andreae [244] and Perschke and Broda [245] used scopoletin (6-methoxy-7-hydroxy-1,2-benzo-pyrone) as a substrate for peroxidase. The disappearance of scopoletin fluorescence was a measure of the peroxidase concentration.

Guilbault et al. [246] have shown that homovanillic acid (XXI) is an excellent substrate for the determination of hydrogen peroxide and peroxidase.

$$\text{(XXI)} + H_2O_2 \xrightarrow{\text{peroxidase}} \text{(XXII)} + 2H_2O \tag{49}$$

The non-fluorescent compound (XXI) is oxidized to the highly fluorescent compound (XXII) ($\lambda_{ex} = 315$, $\lambda_{em} = 425$ nm). The rate of formation of (XXII) with time is proportional to the concentration of hydrogen peroxide and peroxidase.

(2) Monoamine oxidase

Monoamine oxidase (MAO) is widely distributed in animal tissue and is localized largely in the mitochondrial fraction of the cell. It has received a great deal of attention, and many procedures are available for its assay — manometric, colorimetric, and spectrophotometric. However, the most sensitive of all procedures is a fluorometric one involving the conversion of tryptamine (XXIII) to indole acetaldehyde (XXIV) and then to indoleacetic acid (XXV). The conversion of NAD to NADH then becomes proportional to the monoamine oxidase activity [247].

(50)

The sensitivity of the method makes it particularly appropriate for studies on small amounts of tissues as are available from synaptic nerves and ganglia, blood vessels, and various portions of the brain [248]. The fluorescence is measured at a λ_{ex} of 280 and a λ_{em} of 370 nm.

(D) ENZYME CYCLING FOR PYRIDINE NUCLEOTIDES

Because of the versatility of the coenzymes NAD and NADP, almost every substance of biological interest can be measured by their use. Hundreds of papers have been published on analytical procedures using these substances and either a spectrophotometric or a fluorometric method for performing the analysis. But because of instrumental limitations, one is limited to a sensitivity of about 10^{-8} mole ml^{-1} in spectrophotometry, and 10^{-11} mole ml^{-1} with the more sensitive fluorometric methods.

Lowry et al. [249] proposed an enzyme-cycling method for measuring pyridine nucleotides, in attempts to increase this sensitivity limit by several orders of magnitude.

The nucleotide to be assayed is made to catalyze an enzymatic reaction between two substrates, which are transferred in amounts far greater than the nucleotide. Thus the measurement of the nucleotide through its catalytic effect increases sensitivity by a factor of 10^3—10^4 over a direct measurement. Coenzyme NAD is measured with lactate dehydrogenase and glutamate dehydrogenase.

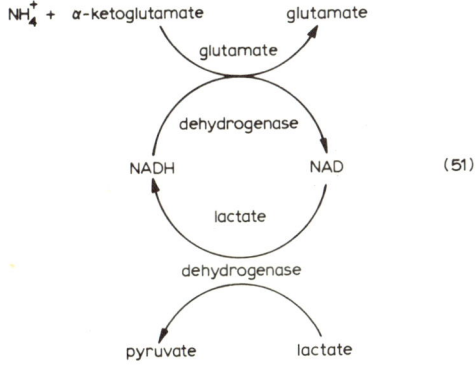

(51)

Pyruvate is produced in 2500-fold yield in 30 min and is measured in a second cycle with added NADH and lactate dehydrogenase. The rate of transformation is measured by following the change in the fluorescent NAD or NADH. Since the nucleotides are used at concentrations well below their Michaelis constants, the reaction rates are proportional to the nucleotide concentrations. The final product is again a pyridine nucleotide, so the cyclic process can be repeated with an overall multiplication factor of 10^6—10^8.

The system for NADP measurement described by Lowry et al. [249] utilizes glucose-6-phosphate dehydrogenase and glutamate dehydrogenase. Each molecule of NADP catalyzes the formation of up to 10000 molecules of 6-phosphogluconate in 30 min. The 6-phosphogluconate is then measured in a second incubation with 6-phosphogluconate dehydrogenase and extra NADP. The NADPH produced is measured fluorometrically.

Two-cycle determinations have been performed on as little as 10^{-15} mole of NADP. This detectable concentration represents the amount

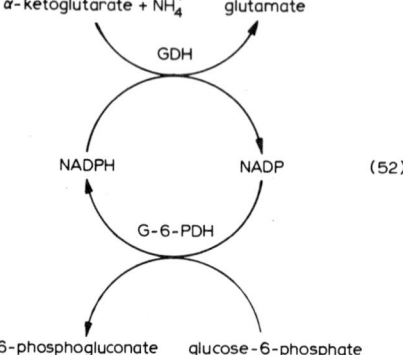

$$(52)$$

that would be formed by 1000 molecules of an enzyme with a turnover number of 10^4 per minute, if it could be coupled to a NADP reaction. In principle one could measure as little as one single enzyme molecule by the reduction of sample size or of the blank [249].

References

1 G.G. Guilbault, Practical Fluorescence. Theory, Methods and Techniques, Marcel Dekker, New York 1973.
2 D.D. Hercules (Ed.), Fluorescence and Phosphorescence Analysis, Interscience, New York, 1966.
3 R.A. Passwater, Guide to Fluorescence Literature, Plenum, New York, 1967.
4 R.E. Phillips and F.R. Elevitch, Fluorometric Techniques in Clinical Pathology, Grune and Stratton, New York, 1966, Chap. 4.
5 S. Udenfriend, Fluorescent Assay in Biology and Medicine, Vol. I, Academic Press, New York, 1966.
6 S. Udenfriend, Fluorescent Assay in Biology and Medicine, Vol. II, Academic Press, New York, 1970.
7 C.E. White and R. Argauer, Fluorescence Analysis. A Practical Approach, Marcel Dekker, New York, 1970.
8 M.A. Konstantinova-Shlesinger, Fluorometric Analysis, N. Kramer, (transl.), Davey, New York, 1965.
9 P. Pringsheim, Fluorescence and Phosphorescence, Interscience, New York, 1949.
10 A. Weissler and C.E. White, in L. Meites (Ed.), Handbook of Analytical Chemistry, McGraw-Hill, New York, 1963, Chap. 6.

11 A. Weissler and C.E. White, in F.W. Welcher (Ed.), Standard Methods of Clinical Analysis, Vol. 3A, D. Van Nostrand, Princeton, N.J., 1966, Chap. 5.

12 C.E. White, Anal. Chem., 21 (1949) 104; 22 (1950) 69; 24 (1952) 85; 26 (1954) 129; 28 (1956) 621; 30 (1958) 729; 32 (1960) 47R; 34 (1962) 82R; 36 (1964) 116R; 38 (1966) 115R; 40 (1968) 114R; 42 (1970) 57R.

13 Fluorescence News Monthly, American Instrument Co., Silver Spring, Md.

14 Traces, Monthly, G.K. Turner Associates, Palo Alto, Calif.

15 J.W. Bridges, D.S. Davies and R.T. Williams, Biochem. J., 98 (1966) 451.

16 G. Weber and F.W. Teale, Trans. Faraday Soc., 53 (1957) 646.

17 R.F. Chen, G.G. Vurek and N. Alexander, Science, 156 (1967) 949.

18 H.H. Cary and A.O. Beckman, J. Opt. Soc. Amer., 31 (1941) 682.

19 R.L. Bowman, P.A. Caulfield and S. Udenfriend, Science, 122 (1955) 32.

20 C.A. Parker, Proc. Soc. Anal. Chem., 3 (1966) 158.

21 C.A. Parker and T.A. Joyce, J. Chem. Soc., A, (1966) 821.

22 S.J. Ladner and R.S. Becker, J. Phys. Chem., 67 (1963) 2481.

23 C.A. Parker, Analyst (London), 84 (1959) 446.

24 J.M. Price, M. Kaihara and H.K. Howerton, Appl. Opt., 1 (1962) 521.

25 R.F. Chen and R.L. Bowman, Science, 147 (1965) 729.

26 M.W. McDermott and R. Novick, J. Opt. Soc. Amer., 51 (1961) 1008.

27 R.F. Chen, Anal. Biochem., 14 (1966) 497.

28 R.F. Chen, Biochem. Biophys. Res. Commun., 17 (1964) 141.

29 D.M. Hercules, Science, 125 (1957) 1242.

30 C.E. White, M. Ho and E. Weimer, Anal. Chem., 32 (1960) 438.

31 C.G. Hatchard and C.A. Parker, Proc. Roy. Soc., Ser. A, 235 (1956) 518.

32 R. Argauer and C.E. White, Anal. Chem., 36 (1964) 368.

33 C.A. Parker, Anal. Chem., 34 (1962) 502.

34 R.F. Chen, Anal. Biochem., 20 (1967) 339.

35 H.C. Borresen and C.A. Parker, Anal. Chem., 38 (1966) 1073.

36 E. Lippert, W. Noegel and I. Sieboldfalkenstein, Z. Anal. Chem., 170 (1959) 1.

37 J.W. Eastman, Appl. Opt., 5 (1966) 1125.

38 L.S. Forster and D. Dudley, J. Phys. Chem., 66 (1962) 838.

39 S.P. McGlynn, T. Azumi and M. Konoshita, Molecular Spectroscopy of the Triplet State, Prentice-Hall, Englewood Cliffs, N.J., 1969, pp. 40—43.

40 T. Medinger and F. Wilkinson, Trans. Faraday Soc., 61 (1965) 620.

41 L.V.S. Hood and J.D. Winefordner, Anal. Chem., 38 (1966) 1922.

42 G. Jackson and G. Porter, Proc. Roy. Soc., Ser. A, 260 (1961) 13.

43 A. Weller, Progr. React. Kinet., 1 (1961) 189.

44 E.L. Wehry and L.B. Rogers, J. Amer. Chem. Soc., 87 (1965) 4234.

45 N. Turro, G. Kavarnos, V. Fung, A. Lyons and T. Cole, J. Amer. Chem. Soc., 94 (1972) 1394.

46 E. Vander Donckt, Progr. React. Kinet., 5 (1970) 273.

47 E. Wehry, Fluorescence News, 6(1) (1971) 1.

48 M. Goldman and E. Wehry, Anal. Chem., 42 (1970) 1178.

49 D.P. Eyman and R.S. Drago, J. Amer. Chem. Soc., 88 (1966) 1617.

50 H. Baba, L. Goodman and P.C. Valenti, J. Amer. Chem. Soc., 88 (1966) 5410.

51 R. Rosakowitz, G.W. Byers and P. Aleermakers, J. Amer. Chem. Soc., 93 (1971) 3263.
52 J. Kropp and J. Lou, J. Phys. Chem., 75 (1971) 2690.
53 D.R. Kearns, Chem. Rev., 71 (1971) 395.
54 C.S. Parmenier and J. Rau, J. Chem. Phys., 51 (1969) 2242.
55 S. Tsai and G. Robinson, J. Chem. Phys., 49 (1968) 3184.
56 H. Linschitz and L. Pekkarninen, J. Amer. Chem. Soc., 82 (1960) 2411.
57 A.W. Varnes, R. Dodson and E. Wehry, J. Amer. Chem. Soc., 94 (1972) 946.
58 C. Parker, Photoluminescence of Solutions, Elsevier, Amsterdam, 1968, p. 226.
59 K. Rohatgi and G. Singhal, Indian J. Chem., 7 (1970) 1020.
60 E. Wehry and L. Rogers, in D. Hercules (Ed.), Fluorescence and Phosphorescence Analysis, Interscience, New York, 1966, pp. 113—118.
61 G.N. Lewis and M.J. Kasha, J. Amer. Chem. Soc., 66 (1944) 2100.
62 S. Freed and W. Salmre, Science, 128 (1958) 1341.
63 R.J. Keirs, R.D. Britt and W.E. Wentworth, Anal. Chem., 29 (1957) 202.
64 C.A. Parker and C.G. Hatchard, Analyst (London), 87 (1962) 664.
65 S. Freed and M.H. Vise, Anal. Biochem., 5 (1963) 338.
66 J.D. Winefordner and H.W. Latz, Anal. Chem., 35 (1963) 1517.
67 H.W. Latz, Ph.D. thesis, University of Florida, Gainesville, 1963.
68 H.C. Hollifield and J.D. Winefordner, Anal. Chem., 40 (1968) 1759.
69 G.N. Lewis and M. Calvin, J. Amer. Chem. Soc., 67 (1945) 1232.
70 C.A. Hitchinson and B.W. Mangun, J. Chem. Phys., 29 (1958) 952; 34 (1961) 908.
71 K.C. Hoerman and S.A. Manciewicz, Arch. Oral Biol., 9 (1964) 517.
72 W.J. Potts, J. Chem. Phys., 20 (1952) 809.
73 R.J. Lukasiewicz, J.J. Mousa and J.D. Winefordner, Anal. Chem., 44 (1972) 963.
74 R.J. Lukasiewicz, J.J. Mousa and J.D. Winefordner, Anal. Chem., 44 (1972) 1339.
75 R.J. Lukasiewicz, P.A. Royznes and J.D. Winefordner, Anal. Chem., 44 (1972) 237
76 H.A. Moye, Ph.D. thesis, University of Florida, Gainesville, 1965.
77 S.P. McGlynn, J. Daigre and F.J. Smith, J. Chem. Phys., 39 (1963) 675.
78 L.V.S. Hood and J.D. Winefordner, Anal. Chem., 38 (1966) 1922.
79 R.J. Keirs, R.D. Britt and W.E. Wentworth, Anal. Chem., 29 (1957) 202.
80 P.A. St. John and J.D. Winefordner, Anal. Chem., 39 (1967) 500.
81 R. Zweidinger and J.D. Winefordner, Anal. Chem., 42 (1970) 639.
82 H.A. Moye and J.D. Winefordner, J. Agr. Food Chem., 13 (1965) 516.
83 H.C. Hollifield and J.D. Winefordner, Talanta, 14 (1967) 103.
84 R.H. Steele and A. Szent-Gyorgyi, Proc. Nat. Acad. Sci. U.S.A., 43 (1957) 477.
85 J.D. Winefordner and H.W. Latz, Anal. Chem., 35 (1963) 1517.
86 J.D. Winefordner and M. Tin, Anal. Chim. Acta, 31 (1964) 239.
87 H.C. Hollifield and J.D. Winefordner, Anal. Chim. Acta, 36 (1966) 352.
88 H.C. Hollifield and J.D. Winefordner, Talanta, 14 (1967) 103.
89 H.V. Drushel and A.L. Sommers, Anal. Chem., 38 (1966) 10.
90 E. Sawicki, Chemist—Analyst, 53 (1964) 88.

91 E. Sawicki and H. Johnson, Microchem. J., 8 (1964) 85.
92 E. Sawicki, T.W. Stanley, J.D. Pfaff and W.L. Elbert, Anal. Chim. Acta, 31 (1964) 359.
93 E. Sawicki and J.D. Pfaff, Anal. Chim. Acta, 32 (1965) 521.
94 J.D. Pfaff and E. Sawicki, Chemist—Analyst, 54 (1965) 30.
95 E. Sawicki and J. Pfaff, Microchem. J., 12 (1967) 7.
96 J. Nag-Chaudhuri and L. Augenstein, Biopolym. Symp., No. 1, (1964) 441.
97 V.P. Bobrovich and S.V. Konev, Dokl. Akad. Nauk S.S.S.R., 155 (1964) 197.
98 I.A. Vladimirov and F.F. Litvin, Biophysics (U.S.S.R.), 5 (1960) 151.
99 I.A. Vladimirov and E.A. Burshtein, Biophysics (U.S.S.R.), 5 (1960) 445.
100 P. Douzou and J.C. Francq, J. Chim. Phys., 59 (1962) 578.
101 G.C. Barenboim, Biofizika, 7 (1962) 227.
102 E.A. Chernitskii, S.V. Konev and V.P. Bobrovich, Dokl. Akad. Nauk Belorussk. S.S.R., 7 (1963) 628.
103 J. Stauff and H. Wolf, Z. Naturforsch. B, 19 (1964) 87.
104 I.A. Vladimirov, S.L. Aksentsev and V.I. Olensov, Biofizika, 10 (1965) 614.
105 J. Nag-Chaudhuri and L. Augenstein, Biopolym. Symp., No. 1, (1964) 441.
106 L.I. Grossweiner, J. Chem. Phys., 24 (1956) 1255.
107 S. Freed, J.H. Turnbull and W. Salmre, Nature (London), 181 (1958) 1731.
108 H. Diehl, R.O. Olsen, G. Speilholtz and R. Jensen, Anal. Chem., 35 (1963) 1144.
109 C.E. White, Fluorometric Analysis. A Practical Approach, Marcel Dekker, New York, 1970, Chaps. 4—6.
110 G.G. Guilbault (Ed.), Practical Fluorescence, Marcel Dekker, New York, 1973, Chap. 6.
111 D.M. Hercules (Ed.), Fluorescence and Phosphorescence Analysis, Interscience, New York, 1966, Chap. 4.
112 E.A. Bozhevol'nov, Fluorometric Analysis of Inorganic Materials (in Russian), Khimiza, Moscow, 1960.
113 S. Udenfriend, Fluorescence Assay in Biology and Medicine, Vol. 1, Academic Press, New York, 1966, Chap. 12.
114 S. Udenfriend, Fluorescence Assay in Biology and Medicine, Vol. 2, Academic Press, New York, 1970, Chap. 16.
115 M.A. Konstantinova-Shlezinger, Fluorometric Analysis, (N. Kramer, transl.), Davey, New York, 1965, Chap. 12.
116 A. Weissler and C.E. White, in L. Meites (Ed.), Handbook of Analytical Chemistry, McGraw-Hill, New York, 1963, Chap. 6, pp. 176—196.
117 C.E. White, biennial reviews on fluorometric analysis in Analytical Chemistry, 1948—1962, and with A. Weissler, 1964—1968.
118 M.U. Belzi and I. Kushnirenko, Ref. Zh. Khim. 19GD, 1969, Abstr. No. 16G92.
119 M.U. Belzi and I. Kushnirenko, Ref. Zh. Khim. 19GD, 1969, Abstr. No. 16G67.
120 A. Solov'ev and E.A. Bozhevol'nov, Chem. Abstr., 66 (1967) 82105.
121 G.F. Kirkbright, T.S. West and C. Woodward, Anal. Chim. Acta, 36 (1966) 208.
122 N.S. Poluektov, A. Kikillov, M.A. Tishichenko and Y. Zelyukova, Zh. Anal. Khim., 22 (1967) 707.

123 M.U. Belzi and I. Kushnirenko, Ref. Zh. Khim. 19GD, 1969, (6), Abstr. No. 6G66.
124 G.F. Kirkbright, C.G. Saw and T.S. West, Analyst, 94 (1969) 457.
125 D.P. Shchernov and A.I. Ivankova, Prom. Khim. Reaktivov. Osobo Chist. Veshchestv., 8 (1967) 191.
126 M.T. El-Ghamry, R.W. Frei and G.W. Higgs, Anal. Chim. Acta, 47 (1969) 41.
127 D.E. Ryan and B.K. Pal, Anal. Chim. Acta, 44 (1969) 385.
128 W. Powell and J. Saylor, Anal. Chem., 25 (1953) 960.
129 J. de Albinati, Anales Assoc. Quim. Argentina, 53 (1965) 61; Anal. Abstr., (1966) 5432.
130 R.M. Dagnall, R. Smith and T.S. West, Talanta, 13 (1966) 609.
131 M. Haitinger, Die Fluoreszenzanalyse in der Mikrochemie, Wien-Leipzig, 1937.
132 J. Marienko and I. May, Anal. Chem., 40 (1968) 1137.
133 C.E. White, J. Chem. Educ., 28 (1951) 359.
134 C.E. White, A. Weissler and D. Busker, Anal. Chem., 19 (1947) 802.
135 C.E. White and D.E. Hoffman, Anal. Chem., 29 (1957) 1105.
136 C.A. Parker and W.J. Barnes, Analyst, 82 (1957) 606.
137 M. Marcantonatos, G. Gamba and D. Monnier, Helv. Chim. Acta, 52 (1969) 538.
138 C.E. White and C.S. Lowe, Ind. Eng. Chem., Anal. Ed., 13 (1941) 809.
139 M.H. Fletcher, C.E. White and M.S. Sheftel, Ind. Eng. Chem., Anal. Ed., 18 (1946) 179.
140 K. Motojinia, Bull. Chem. Soc. Jap., 29 (1956) 75.
141 A.B. Borle and F. Briggs, Anal. Chem., 40 (1968) 339.
142 M. Lewin, M. Wills and D. Baron, J. Clin. Pathol., 22 (1969) 222.
143 B. Fingerhut, A. Poock and H. Miller, Clin. Chem., 15 (1969) 870.
144 E.A. Bozhevol'nov, Chem. Abstr. Jap., 65 (1966) 7989.
145 C. Ti Huu, A.I. Volkova and T. Getman, Zh. Anal. Khim., 24 (1969) 688.
146 J.S. Hanker, A. Gelberg and B. Whitten, Anal. Chem., 30 (1958) 93.
147 G.G. Guilbault and D.N. Kramer, Anal. Chem., 37 (1965) 918.
148 G.G. Guilbault and D.N. Kramer, Anal. Chem., 37 (1965) 1395.
149 J. de Albinati, Anales Assoc. Quim. Argentina, 55 (1967) 61.
150 V. Temkina, E.A. Bozhevol'nov and N. Dyatlova, Zh. Anal. Khim., 22 (1967) 1830.
151 N. Iritani, T. Miyahara and I. Takahashi, Jap. Anal., 17 (1968) 1075.
152 R.P. Fisher and J.D. Winefordner, Anal. Chem., 43 (1971) 454.
153 R. Belcher, R. Perry and W.I. Stephen, Analyst (London), 94 (1969) 26.
154 T.L. Har and T.S. West, Anal. Chem., 43 (1971) 136.
155 H. Orighi and E.B. Sandell, Anal. Chim. Acta, 13 (1955) 159.
156 D.P. Shchernov, I. Solovyan and A. Drobachenko, Tezisy dokladov 6-go soveshchaniya lyuminestentu, Leningrad, 1958.
157 A. Lukin and E.A. Bozhevol'nov, J. Anal. Chem. U.S.S.R. (English transl.), 15 (1960) 45.
158 E.A. Bozhevol'nov, A. Lukin and M. Gradinarskaya, Anal. Abstr., 7 (1960) 3164.

159　N. Raju and G. Rao, Nature (London), 175 (1955) 167.
160　W.C. Alford, L. Shapiro and C.E. White, Anal. Chem., 23 (1951) 1149.
161　G. Oshima and K. Nagasawa, Chem. Pharm. Bull., 18 (1970) 687.
162　R. Bock and K. Hochstein, Z. Anal. Chim., 138 (1953) 337.
163　D. Schachter, J. Lab. Clin. Med., 54 (1959) 763.
164　D. Schachter, J. Lab. Clin. Med., 58 (1961) 495.
165　J.B. Hill, Ann. N.Y. Acad. Sci., 102 (1962) 1.
166　B. Klein and M. Oklander, Clin. Chem., 13 (1967) 26.
167　R.E. Thiers, in S. Meites (Ed.), Standard Methods of Clinical Chemistry,
　　　Vol. 5, Academic Press, New York, 1965, p. 131.
168　C.E. White and F. Cuttita, Anal. Chem., 31 (1959) 2083.
169　B.K. Pal and D. Ryan, Anal. Chim. Acta, 47 (1969) 35.
170　G.F. Kirkbright, T.S. West and C. Woodward, Talanta, 13 (1966) 1637.
171　G.F. Kirkbright, T.S. West and C. Woodward, Talanta, 13 (1966) 1645.
172　S. Belman, Anal. Chim. Acta, 29 (1965) 120.
173　V. Sardesai and H. Provido, Mikrochem. J., 14 (1969) 550.
174　M. Rubin and L. Knott, Clin. Chim. Acta, 18 (1967) 409.
175　C. Sawicki, Anal. Lett., 4 (1971) 761.
176　G. Schenk, K. Dilloway and J. Coulter, Anal. Chem., 41 (1969) 510.
177　O.H. Lowry, J.V. Passonneau and S. Schultz, Anal. Biochem., 19 (1967)
　　　300.
178　H. Axelrod, J. Cary, J. Bonelli and J. Lodge, Anal. Chem., 41 (1969) 1856.
179　E.B. Cousins, Aust. J. Exp. Biol. Med. Sci., 38 (1960) 11.
180　J.H. Watkinson, Anal. Chem., 32 (1960) 981.
181　C.A. Parker and L.G. Harvey, Analyst (London), 86 (1961) 54.
182　C.A. Parker and L.G. Harvey, Analyst (London), 87 (1962) 558.
183　W.H. Allaway and E.E. Cary, Anal. Chem., 36 (1964) 1359.
184　P.F. Lott, P. Curor, G. Moriber and J. Solga, Anal. Chem., 35 (1963) 1159.
185　W.E. Clarke, Analyst (London), 95 (1970) 65.
186　J.B. Wilkie and M. Young, J. Agr. Food Chem., 18 (1970) 946.
187　O. Olson, J. Ass. Offic. Anal. Chem., 52 (1969) 627.
188　B.K. Pal and D. Ryan, Anal. Chim. Acta, 48 (1969) 227.
189　R.M. Dagnall, R. Smith and T.S. West, Analyst (London), 92 (1967) 358.
190　M. Ichihashi, T. Shigematsu and T. Nishikawa, J. Chem. Soc. Jap., 77
　　　(1956) 1474.
191　R.R. Trenholm and D.E. Ryan, Anal. Chim. Acta, 32 (1965) 317.
192　A. Brookes and A. Townshend, Chem. Commun., 24 (1968) 1660.
193　E.L. Pruden, R. Meier and D. Plant, Clin. Chem., 12 (1966) 613.
194　E. Bovalini and M. Prazyi, Ann. Chim., 53 (1963) 1103.
195　A. Babko and M. Lukovskaya, J. Anal. Chem. U.S.S.R., 17 (1962) 47.
196　O. Bandisch and H.J. Deuel, J. Amer. Chem. Soc., 44 (1922) 1586.
197　C.E. Frohman and J.M. Orten, J. Biol. Chem., 205 (1953) 717.
198　R.J. Crout, in D. Seligson (Ed.), Standard Methods of Clinical Chemistry,
　　　Vol. 3, Academic Press, New York, 1969, p. 62.
199　N.K. De, Indian J. Med. Res., 43 (1955) 3.
200　S.W. Jones, J.B. Wilkie, W.W. Morris and L. Friedman, 138th Meeting
　　　Amer. Chem. Soc., 60C, New York (1960).
201　R.W. Albers and O.H. Lowry, Anal. Chem., 27 (1955) 1829.

202 S. Udenfriend, Fluorescent Assay in Biology and Medicine, Vol. I, op. cit. pp. 349—371.
203 S. Udenfriend and P. Saltzman, Anal. Biochem., 3 (1962) 49.
204 P. Greengard, Nature (London), 178 (1956) 632.
205 H. Börresen, Acta Chem. Scand., 17 (1963) 921.
206 O.H. Lowry, N. Roberts and J. Kapphahn, J. Biol. Chem., 224 (1957) 1047.
207 N. Kaplan, S. Colowick and C. Barnes, J. Biol. Chem., 191 (1951) 461.
208 J. Huff and W. Perlzweig, J. Biol. Chem., 167 (1947) 157.
209 C. Dhere, La Fluorescence en Biochimie, Presses Universitaires, Paris, 1937.
210 S. Schwartz, M. Berg, I. Bossenmauer and H. Dinsmore, Methods Biochem. Anal., 8 (1960) 221.
211 K. Solov'ev, Opt. Spectry (English), 10 (1961) 389.
212 K. Solov'ev, S. Shkirman and T. Kachura, Izv. Akad. Nauk. S.S.R., Ser. Fiz., 27 (1963) 767.
213 W. Runge, Science, 15 (1966) 1499.
214 C. Martinez and G. Mills, Clin. Chem., 17 (1971) 199.
215 G.G. Guilbault and D.N. Kramer, Anal. Chem., 35 (1963) 588.
216 G.G. Guilbault and D.N. Kramer, Anal. Chem., 37 (1965) 120.
217 G.G. Guilbault and D.N. Kramer, Anal. Chem., 37 (1965) 1675.
218 A. Seligman and R. Barnett, Science, 114 (1959) 579.
219 D.N. Kramer and C. Gelman, CRDL Rep., 1960, p. 541.
220 B. Gehauf and J. Goldenson, Anal. Chem., 29 (1957) 276.
221 G.G. Guilbault, M.H. Sadar, R. Glazer and C. Skou, Anal. Lett., 1 (1968) 365.
222 B. Rotman, Proc. Nat. Acad. Sci. U.S.A., 47 (1961) 1981.
223 B. Rotman, J. Zdevic and M. Edelstein, Proc. Nat. Acad. Sci. U.S.A., 50 (1963) 1.
224 J. Mead, J. Smith and R.T. Williams, Biochem. J., 61 (1955) 569.
225 M.A. Verity, R. Caper and W.J. Brown, Arch. Biochem. Biophys., 106 (1964) 386.
226 D.N. Kramer and G.G. Guilbault, Anal. Chem., 35 (1963) 588.
227 G.G. Guilbault and D.N. Kramer, Anal. Chem., 36 (1964) 409.
228 G.G. Guilbault and M.H. Sadar, Anal. Lett., 1 (1968) 551.
229 G.G. Guilbault and J. Hieserman, Anal. Chem., 41 (1969) 2006.
230 D.W. Moss, Clin. Chem. Acta, 5 (1960) 283.
231 D.B. Land and E. Jackim, Anal. Biochem., 16 (1966) 481.
232 G.G. Guilbault, S.H. Sadar, R. Glazer and J. Haynes, Anal. Lett., 1 (1968) 333.
233 G.G. Guilbault and A. Vaughan, Anal. Lett., 3 (1970) 1.
234 G.G. Guilbault and A. Vaughan, Anal. Chem., 43 (1971) 721.
235 S. Ochoa, A.H. Mahler and A. Kornberg, J. Biol. Chem., 174 (1948) 979.
236 P. Greengard, Nature (London), 178 (1956) 632.
237 P. Greengard, in H. Bergmeyer (Ed.), Methoden der enzymatische Analyse, Verlag Chemie, Germany, 1962, p. 551.
238 L.T. Graham and M.H. Aprison, Anal. Biochem., 15 (1966) 487.
239 T. Laursen and G. Espersen, Scand. J. Clin. Lab. Invest., 11 (1959) 61.
240 T. Laursen and P.F. Hansen, Scand. J. Clin. Lab. Invest., 10 (1958) 53.
241 G.G. Guilbault and D.N. Kramer, Anal. Chem., 36 (1964) 2497.

242 G.G. Guilbault and D.N. Kramer, Anal. Chem., 37 (1965) 1219.

243 A.S. Keston and R. Brandt, Anal. Biochem., 11 (1965) 1.

244 W.A. Andreae, Nature (London), 175 (1955) 859.

245 H. Perschke and E. Broda, Nature (London), 190 (1961) 257.

246 G.G. Guilbault, D.N. Kramer and E. Hackley, Anal. Chem., 39 (1967) 271.

247 H. Weissbach, T.E. Smith and S. Udenfriend, Biochemistry, 1 (1962) 137.

248 W. Lovenberg, R. Levine and A. Sjoerdsma, Fed. Proc., Fed. Amer. Soc. Exp. Biol., 20 (1961) 318.

249 O.H. Lowry, J. Passonneau, D. Schulz and M. Rock, J. Biol. Chem., 236 (1961) 2746.

Chapter III

Photometric titrations

M.A. LEONARD

1. Introduction

(A) DEFINITION

"Photometric titrations? Ah yes, I suppose they're quite useful if the indicator colour change is indistinct or one is trying to see an end-point in a simulation of brown-windsor soup, but otherwise there's no point in using them." This unfortunately widespread view is quite erroneous. Photometric titrations can in many instances yield results of amazing precision in otherwise hopeless situations.

Photometric titrations may be defined as those in which the absorbance of the solution, varying with the changing concentrations of absorbing solutes, is plotted against volume of titrant added. In most cases, of course, the equivalence-point of a definite chemical reaction is sought but it is arguable that dilution with solvent to an instrumental colour match may also be regarded as a photometric titration (see Sect. B). They may be easily applied to all the principal types of reactions utilized in titrimetry (acid—base, redox, precipitation and complex formation) and also they make possible the use of less well known interactions for the analysis of certain organic and organometallic materials. In addition to absorption, the fluorescent, light scattering or optical rotation properties of solutions may be used.

(B) HISTORY

Recent reviews ascribe the first spectrophotometric titration to Tingle [1] (1918) but I would rather suggest that a method for the determination of didymium in gadolinite proposed in 1866 by Bahr and Bunsen [2] should take precedence. Thus on p. 30 of their paper we read: "In order to establish, at least approximately, the ratio of didymium to lanthanum in gadolinite by another method we also carried out a spectralanalytical titration which, while not exact, yielded approximately the correct results. This method is based on the principle that one compares the absorption bands of an analysed solution of known didymium content with the absorption bands of the similarly examined solution of unknown didymium content in question. The stronger solution is diluted with a measured volume of water until the bands both show the same intensity."

The spectroscope used was the improved Kirchhoff and Bunsen instrument of 1861 fitted with a divided entrance slit, one half of which was covered with a small reflecting prism so that two absorption spectra could be juxtaposed. Presumably the sharp absorption bands of hydrated lanthanoid ions made the method feasible.

In 1875 Karl Vierordt, Professor of Physiology at the University of Tübingen and founder of quantitative absorptiometry, wrote a paper entitled, "The Application of Quantitative Spectral Analysis to the Titration Method", [3] which was not, however, as promising as it sounded. To determine the concentration of glucose in solution, a known volume of Fehlings solution of accurately determined glucose equivalence was mixed with a known volume of the sugar test solution insufficient to reduce all the Cu^{II} present. After the formation of Cu_2O was complete, the absorbance of the supernatant alkaline Cu^{II} solution was measured in a region of high absorption using Vierordt's own effective visual (divided variable slit) spectrophotometer. The sugar equivalence of the remaining Cu^{II} solution was deduced from the absorbance reading, and the weight of sugar in the test solution calculated by difference. This is hardly a conventional photometric titration, yet in effect Vierordt was assessing the reaction equivalence point by means of an absorbance reading taken before equivalence where equilibrium and kinetic factors are still favourable. This, after all, is the essence of the photometric titration.

Tingle [1] (1918) measured the acidity of deeply coloured (blue end absorption) alkaloidal extracts by adding an indicator with an alkaline-form absorption band distinctly separated from the general background absorption. Throughout the titration of the test solution with standard alkali the transmission characteristics were assessed using a pocket diffraction spectroscope. The titration was deemed complete when the width of the indicator absorption band was

208

identical with that given by one drop of titrant in a "blank" situation. Methyl orange proved most effective. The principle was also applied successfully to the determination of sulphuric acid in copper sulphate solutions using methyl orange as indicator where the background absorption lay at the red end of the spectrum. The method gave good results on solutions inaccessible to conventional titration though it can hardly be classed as photometry. It seems a throwback to the Hartley absorption spectra of the 1880's where absorption bandwidths were studied in relation to concentration and cell thickness. In addition, remarks made towards the end of the paper suggest that Tingle had no notion of the concept of indicator or titrand pK values and their influence on correct indicator choice.

The first true photometric titrator, exactly as we appreciate the term today, was built by Field and Baas-Becking (1926) to measure the rates of formation and destruction of biological pigments [4]. The instrument is illustrated in Fig. 1. An incandescent platinum ribbon lamp was used as light source with "colour screens of known spectral transmissivity" interposed to give wavelength selection. After partial absorption in the test solution the intensity of the beam was measured by means of a silver—bismuth thermocouple radiomicrometer connected to a torsion fibre mirror galvanometer. The starch—iodine reaction was studied in some detail and the iron(III)—thiocyanate complex formation briefly examined with no conclusive result in the latter case. Unfortunately percent transmission was plotted against titrant volume.

In 1928 Müller and Partridge produced a vertical light path photometric titration apparatus based on the type of photoemissive tube in use today [5]. The photocurrent, modified by changes in the absorbance of the titrated solution, was used to change the grid potential of a simple triode whose plate current then operated a relay which itself controlled the current supply to a solenoid burette release. Hence by suitable choice of indicator (yellow to blue transitions were especially effective) and initial light intensity settings an automatic titration could be performed. The circuit diagram is shown in Fig. 2 and a plot of plate current against acid/base ratio, the first photometric titration curve to be published, in Fig. 3. Acid—base titrations and indicator—pH transitions were studied in detail. The potential of the apparatus for executing permanganate, dichromate (*m*-phenylene diamine indicator) and iodine redox titrations was assessed, precision in all cases being greater than with visual titrations.

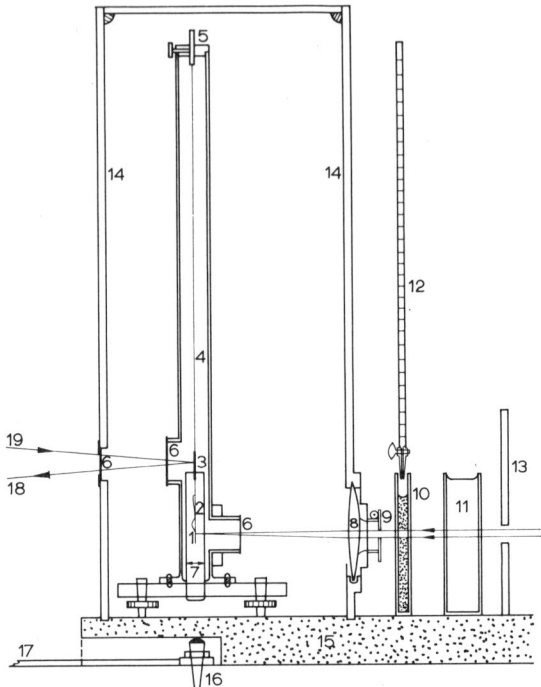

Fig. 1. The first true photometric titrator. Instrument of Field and Baas-Becking (1926). Reprinted from ref. 4 by kind permission of the Rockefeller University Press.

Turbidimetric and a form of nephelometric (absorption of scattered blue—green light by Ag_2CrO_4) silver nitrate titration were mentioned but not pursued in detail. The authors well appreciated too the value of filtering the light beam so as to use only the wavelength region of maximum absorption for some critical solution species. Undoubtedly it is to these authors that credit for the introduction of modern photometric titrimetry must go. It is, in addition, most strange that this first effective phototitrator paralleled the first good commercially available photoelectric spectrophotometer designed by Hardy in that it too was an "automatic" instrument.

The 1930's saw the development of numerous photoelectric titrators and application of these to the commoner types of titration. Notable workers of this period were Somiya and Hirano in Japan and

210

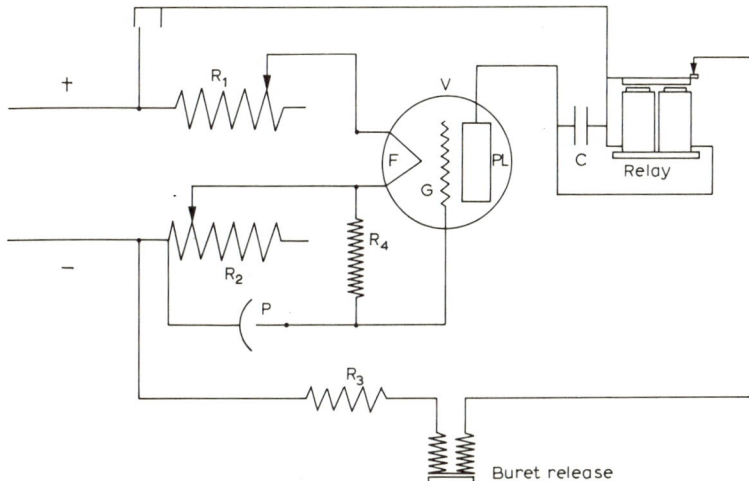

Fig. 2. Circuit diagram of the Müller and Partridge photometric titrator (1928). The photocell is component P.

Partridge and Müller in America. Somiya and Shiraishi, as early as 1930, applied photometric end-points (1000 W tungsten filament source, neon "glim" lamp detector) to neutralisation, redox and precipitation reactions [340]. Hirano, in a long series of papers, investigated Ag^+ vs. Cl^- (starch as protective colloid) [216]; $AuCl_3$ vs. KI [195] or $SnCl_2$ [242]; I^- vs. IO_3^- in strong acid [243]; Hg^{2+} or Pb^{2+} vs. Na_2S (gum arabic as protective colloid) [215] and Mn^{2+} vs.

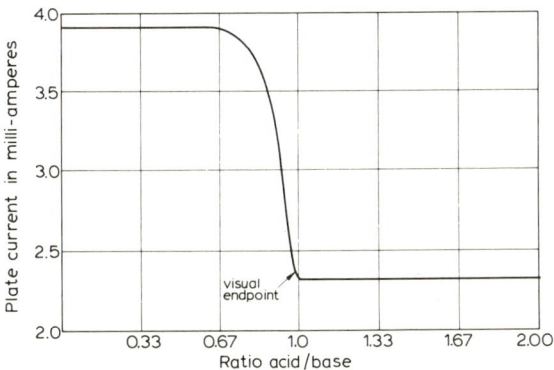

Fig. 3. Relation between plate current and reagent added for a phenolphthalein-indicated strong acid—strong base titration (Müller and Partridge 1928).

BrO_3^- titrations with photometric end-point detection [246]. Somiya's trace cobalt determination (Co^{II} oxidised to Co^{III} by H_2O_2 titrant in the presence of $P_2O_7^{4-}$ and NH_3) is remarkable [247]. Del Campo et al. (1936) titrated SO_4^{2-} with Ba^{2+} [210] while Goodhue (1938) determined nicotine by precipitation photometric titration with silicotungstate [203]. Boyer (1938) produced the first photometric complex forming titration-nickel in steel by addition of excess CN^- followed by back titration with standard Ag^+ under Liebig-Denigés conditions (turbidimetric end-point) [217].

The first German contribution was that of Müller (1934) [6]; while the paper clearly showed the usefulness of the new barrier-layer photovoltaic cells and suggested an effective 2-photocell titrator, it broke no essentially new ground. For the Soviet Union, the pioneering work of Lur'e and Tal (1940) [249] was unoriginal but contained a useful practical analysis of nichrome alloy for chromium and iron (essentially a Fe^{2+}—MnO_4^- finish in both cases).

Twin photocell instruments allowed accurate comparison titrations to be performed and it was appreciated that $-\log T$ (absorbance) plots against titrant volume gave intersecting straight lines similar to those encountered in the previously established conductimetric titration technique. However, in most cases well defined break points were taken as being identical with equivalence-points, no deeper theoretical interpretation being attempted. Complex formation studies concerning the systems Au^{3+}—Br^-; Fe^{3+}—SCN^- and starch—iodine were carried out.

The splendid contribution of Ringbom and co-workers to the development of photometric titrations began essentially in 1939 with an acid—base study in which a solution pH could be assessed to 0.01 of a unit and acids titrated to an accuracy of 0.2% (1 N solution of acid $K_a = 10^{-10}$) by suitable comparison titrations using a 2-cell Lange colorimeter [7]. Ringbom sensed the potential of the technique in his final paragraph: "The light-electric titration is, in other words, not to be considered only as a curiosity which has mostly been the case up to now; it is rather a new example of the success with which physical methods can be called upon for the solution of different analytical chemical problems". Ringbom followed this work in 1941 with a thorough study of photoelectric precipitation titrations [200].

A further advance came in 1943 with the 1 sample cell—2 photocell photoelectric titrator of Osborn et al. used principally for the alkalimetric titration of dark coloured resins [8]. Here the light beam, after passing through the titration cell, was split by an inclined glass plate, the two resulting beams then impinging on separate photocells. Before each photocell was placed a filter, one corresponding in peak transmission to the absorption maximum of some critical solution species, the other being of such transmission that the transmitted light intensity varied negligibly, or in the opposite direction at the end-point. In this way great stability and sensitivity were achieved. Graphs of (essentially) $-\log T$ vs. titre were of high quality. More significantly perhaps, a third photocell was situated at right angles to the source-titration cell axis to detect effectively fluorescent or nephelometrically scattered radiation though unfortunately no specific applications were detailed.

Determination of the formula of complexes in solution by photometric titration, though touched on earlier, was first investigated thoroughly by Yoe and Jones (1944) in their study of the iron[III]-1,2-dihydroxybenzene-3,5-disulphonate complexes [9]. While theoretically the method is little different from the continuous variation approach of Job [10], and Vosburgh and Cooper [11], Fig. 1 of the Yoe and Jones paper probably represents the first photometric chelometric titration (Fig. 4). The slope ratio method of Harvey and Manning (1950) further pursued these ideas [12].

In 1947, Lambert applied the turbidimetric titration technique to the mutual interaction of anionic and cationic surface active agents and their analysis by this means [13]. Nichols and Kindt (1950) further investigated reactions of this type by determining the critical concentration of soap solutions towards micelle formation using a photometric titration [14]. In addition they included in their paper what was probably the first fluoride—thorium nitrate photometric (indicator) titration.

Kenny and Kurtz (1952) proposed luminol as a chemiluminescent indicator [15] for photometric acid—base titrations but the suggestion never seems to have been taken up generally.

The 1950's were the halcyon days of photometric titrimetry with good papers on basic theory and useful well verified new applications. Bricker and Sweetser (1952) proposed UV photometric titrations (As[III]—Ce[IV], 320 nm) [253] and the following year applied the technique to (visible region) EDTA titrations [134], though the first conventional indicated complexometric titration had already been published by Kibrick et al. (1952) [137]. Underwood (1953) titrated

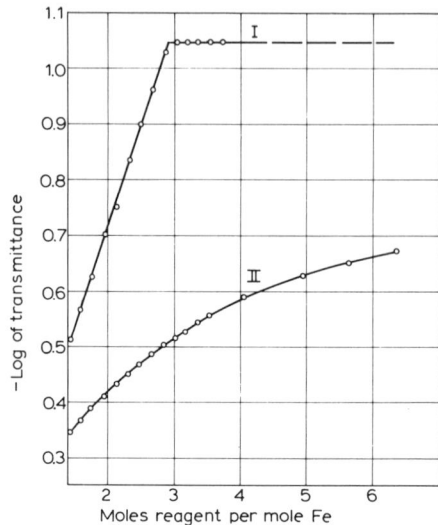

Fig. 4. The first chelometric photometric titration — that of iron(III) with 1,2-dihydroxybenzene-3,5-disulphonate. I, formation of a red complex, $\lambda = 480$ nm, pH = 9.5; II, formation of a blue complex, $\lambda = 620$ nm, pH = 4.0. Reprinted from ref. 9 by permission of the American Chemical Society.

mixtures of metal ions [132]. In 1954, perhaps the most fruitful year of all, Goddu and Hume wrote an excellent paper on the (self-indicating) photometric titration of weak acids [75] while Reilly and Schweizer extended this type of analysis by the use of non-aqueous solvents [85]. Ringbom and Vanninen [16], and Fortuin et al. [17] thoroughly examined the theory of photometric chelometric titri-metry.

Bobtelsky, whose turbidimetric titration-like technique of hetero-metry has carved itself a sizeable niche in the literature of photo-metric titrimetry reviewed progress in this specialized field (1955) [18]. In 1956 Grunwald showed how end-points could be calculated from poor photometric titration curves [19]. Furman and Fenton applied photometric end-point detection to coulometric titration (As^{III} with generated Ce^{IV}) [261] while Higuchi began his series of papers on the very accurate determination of weak acids using in-dicators — an approach later to become known as the "Higuchi plots".

Marple et al. (1958) [175] proposed the first non-aqueous photo-

214

metric reagent—metal titration in which the zinc content of oils was determined by titration in benzene—methanol solution with dithizone. Such an approach was later investigated in detail by le Goff and Trémillon (1964) [176], while Galik (1966) extended the idea to 2-phase extractive titrations [187].

Photometric titrations concerning organic molecules had previously appeared sporadically but in 1961 Schenk and Ozolins developed a particularly interesting technique utilizing π acid—base interactions for the determination of polynuclear aromatic hydrocarbons [300].

In 1962 Bruckenstein and Gracias applied a combined differential high precision photometric titration to the analysis of dilute weak acids [91]. Flaschka and Sawyer studied the titration of around 0.1 mg of calcium—magnesium mixtures using three different approaches [146].

The first photometric titration involving free radicals appeared in 1964 when Paris et al. used 2,4,6-tri-t-butylphenoxy radicals to determine oxygen and antioxidants [306].

Ironically the method of Bahr and Bunsen mentioned at the beginning of this historical survey was resurrected exactly a century later (1966) by Wallin [20] though this time the basic instrument was a double beam photoelectric spectrophotometer.

In recent years further good theoretical studies have been carried out by Groeneveld and den Boef; one on the accuracy of substitution photometric titrations (1966) [21] and the other on limits of detection (1969) [22]. Ringbom et al. have described photometric titrations using dichromatic light and two photocells [23]; Schenk and Bazzelle utilised kinetic differences and catalysis in the resolution of metal mixtures [24] while Leonard and Murray applied the technique to 2-phase diazo coupling titrations [25]. Pearson has developed the potentially valuable application of optical rotation titrimetry [26].

The subject of photometric titration has been reviewed by: F. Müller (1934) [6]; R.H. Müller (1939) [27]; Osborn et al.* (1943) [28]; Mika (1948) [29]; Underwood* (1954) [30]; Goddu and Hume* (1954) [31]; Hirt (1956) [32]; Headridge* (1958) [33], book* (1961) [34]; Connors and Higuchi (1961) [35]; Reynolds* (1963) [36]; Headridge* (1963) [37]; Eriksen and Connors* (1964) [38]; Underwood* (1964) [39]; Tanaka and Nakagawa (1965)

---

* Proved to be particularly helpful.

[40]; Flaschka* (1965) [41]; Greuter (1966) [42]; L'Her* (1967) [43].

2. Theory

(A) PHYSICO-CHEMICAL BACKGROUND

(1) Absorption of electromagnetic radiation by solutions

The sinusoidally fluctuating electric component of a monochromatic beam of electromagnetic radiation will interact with the outer-electron clouds or dipole of a molecule in solution to produce electronic or vibrational excitation of the molecule and concomitant absorption of energy from the beam in the UV, visible and IR spectral regions.

The intensity of a monochromatic beam of radiation passing through a solution containing absorbing solute decreases exponentially with the number of solute molecules intercepted. Thus for the situation illustrated in Fig. 5 where a solution is contained in a transparent cell

$$I = I_0 e^{-kcl} \tag{1}$$

This fractional geometric progression relation between intensity and number of absorbing particles was essentially first realised by the French physicist Bouguer in 1729.

Changing eqn. (1) to its base 10 log form we obtain

$$\log_{10} I_0/I = \epsilon cl \qquad \text{where } \epsilon = 0.434\ k \tag{2}$$

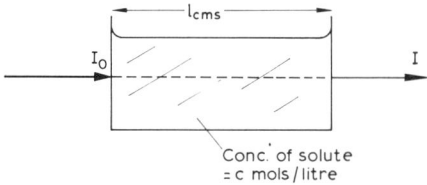

Fig. 5. Absorption of monochromatic light by a solution.

216

$\log_{10} I_0/I = A$ is the absorbance (optical density, extinction) of the solution. If c is quoted in mols per litre ϵ is the molar absorptivity (molar extinction coefficient, molar absorbancy index) of the solution which varies with wavelength, temperature and solvent. If, because the solute molecular weight is unknown, c is in gms/litre

$$A = E_s cl \qquad\qquad (3)$$

where E_s is the absorptivity (specific extinction coefficient, absorbancy index) of the solute. $\epsilon = ME_s$, where M = molecular weight of solute.

The transmission of the solution $T = I/I_0$

$$\therefore A = \log_{10} I_0/I = \log_{10}(1/T) = -\log_{10} T \qquad\qquad (4)$$

$$\% T = T \times 100$$

$$\therefore 1/T = 100/\% T$$

$$\therefore A = 2 - \log_{10} \% T \qquad\qquad (5)$$

In practice the light beam traversing the cell also suffers reflection and solvent absorption losses and while in conventional absorption spectroscopy these are conveniently eliminated using a blank absorption cell, in photometric titrations this device is scarcely necessary as only absorbance changes are of interest.

A plot of absorbance against wavelength of the irradiating light yields the absorption spectrum of the solute (Fig. 6). In photometry

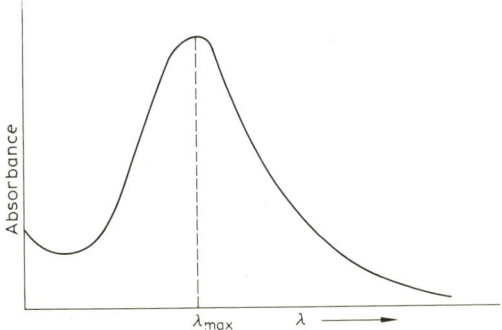

Fig. 6. A typical solute absorption spectrum.

and photometric titrations greatest sensitivity is naturally obtained by working at the wavelength of maximum absorption (λ_{max}) but often this is not necessary and indeed in some instances it may be desirable to operate elsewhere.

Pure monochromatic radiation may be obtained by isolating a particular line in an arc discharge spectrum (e.g. that of a low-pressure mercury vapour lamp) but more usually monochromation of a poly-chromatic source is used. Photometric titrators using simple optical filters are often quite adequate if the transmission characteristics of these match well the absorption spectra of solutes.

A graph of absorbance against concentration of absorbing solute is usually linear provided that the nature of the solute remains un-changed. Association or dissociation produces deviation from linear-ity if the analytical concentration of some fundamental entity is used as independent variable. Photometric titrations are confined to the UV and visible spectral regions. Infrared spectroscopy could con-ceivably be applied but problems of solvent transparency at the thickness required in a feasible titration cell would be severe.

(2) Solution fluorescence

Fluorescence is the general name given to the phenomenon in which certain atoms and molecules, when electronically excited by the absorption of electromagnetic radiation of suitable frequency, subsequently lose this energy by emitting radiation of the same or, more usually, lower frequency (i.e. longer wavelength). This effect may most commonly be observed in the UV-visible and X-ray spec-tral regions. For molecular electronic excitation in the UV-visible with which we will be exclusively concerned in this chapter, $\pi \rightarrow \pi^*$, $n \rightarrow \pi^*$ and $n \rightarrow \sigma^*$ transitions are usually utilised. High probability transitions give rise to strong absorption bands. The energy level diagram shown in Fig. 7 indicates the energy transitions involved and the spectra which result.

Excitation, which occurs in about 10^{-15} s uses the same energy levels as the normal absorption process and since this can occur to a number of vibrational sub-levels denoted by $V = 0,1,2 \ldots$ etc. from (for fairly simple molecules) the zero vibrational level of the ground state, there results the usual broad absorption or excitation band with possible vibrational fine structure. The excited molecule, if present in a low-pressure gas may well immediately lose its electronic

218

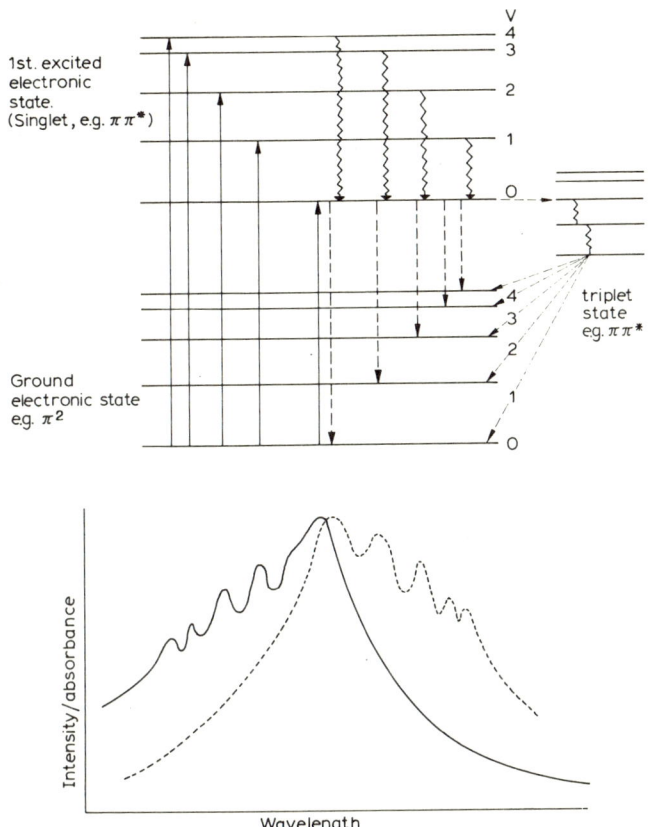

Fig. 7. Fluorescence excitation and emission spectra. ———— absorption transitions and absorption/excitation spectrum; - - - - - - emission transitions and emission/fluorescence spectrum; ⌒⌒⌒→ radiationless energy transfer processes.

energy by radiation of a photon of the same frequency as that absorbed to give resonance fluorescence. In solution, however, the excited molecule will undergo vibrational relaxation in which the excited electronic state is robbed of its excess vibrational energy by collisional processes. If the molecule, now in the zero vibrational level of the first excited electronic state, loses no further energy by such radiationless energy transfer or internal conversion processes, it emits a fluorescence photon to finish up in some vibrational level of the ground electronic state. Just as ↑↑ represent in terms of frequency or energy the most intense transitions on absorption, so ↓↓ represent the most intense transitions associated with fluorescence. Since ↑↑ are almost always longer than ↓↓ we have a physical basis for Stokes' empirical law that fluorescence bands are centred at longer wavelengths (or lower frequencies) than the associated absorption bands.

In normal fluorescence spectroscopy singlet states, in which electron spins remain paired, are involved but occasionally an excited molecule will intersystem cross to a triplet state in which the spin of the promoted electron becomes parallel to that of its former partner. An illustration of this process, which is facilitated by matched vibrational energy levels can be seen in Fig. 7. The triplet state molecule may lose its energy by a formally forbidden radiative transition (phosphorescence) or by vibrational coupling interaction with the ground state.

Organic molecules which fluoresce strongly usually contain large aromatic or conjugated systems having extensive delocalised π bonding. This gives a high value of ϵ and fluorescence photons in or near the visible (anthracene fluoresces blue; pentacene red). If the molecule can be held in a planar configuration by bridging or metal chelate formation fluorescence is enhanced because the rigid structure limits the number of possible vibrations and rotations and hence the ability of the molecule to transfer energy to its neighbours by vibrational relaxation type internal conversion. The fluorescent aluminium chelate of Alizarin Garnet R illustrates this point.

Alizarin Garnet R   Al complex

a. Spectra. Fluorescent molecules are characterised by excitation and fluorescence (emission) spectra. The fluorescence spectrum is emitted when the molecule absorbs anywhere over the excitation waveband. The excitation spectrum, when corrected for variation of lamp output, monochromator transmission and photomultiplier response with wavelength, should be identical with the usual absorption spectrum, or at least the longer wavelength portion of it.

In fluorimetric titrations the test solution, contained in a clear

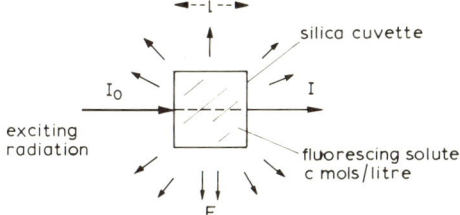

Fig. 8. Emission of fluorescence by a solution.

silica cuvette, is irradiated with more or less monochromatic light of wavelength near the centre of the excitation band of the relevant solute. For a cell of path length l cm in the exciting light direction and fluorescing solute concentration of c mols per litre (Fig. 8), the sampled fluorescent intensity F is given by

$$F = k\phi I_0 \cdot 2.3\epsilon cl$$

where k accounts for the fact that only a small solid angle of the total emitted fluorescence reaches the detector, ϕ is the fluorescence quantum yield, I_0 is the incident light intensity and ϵ the molar absorptivity of the fluorescing solute. Essentially, for fixed instrumental conditions

$$F \propto c$$

The concentration must be very small otherwise the $F = f(c)$ expression becomes

$$F = k\phi I_0 (1 - e^{-2.3\epsilon cl})$$

Also self-absorption and quenching effects become troublesome.

(3) Scattering of radiation by suspensions

The phenomenon of light scattering by finely suspended solid particles is used in analytical chemistry in two distinct ways: turbidimetry, closely akin to absorptiometry, and nephelometry, similar at least instrumentally to fluorescence.

Suspensions are very difficult to reproduce even with the use of protective colloids and great experimental care. For this reason the above techniques have never proved very popular and tend to be "last resort" methods. However, in light scattering titrations where intensity changes are of interest, absolute reproducibility is not vital and such methods form a large and useful section of the photometric titration technique.

a. Turbidimetry. In this method the decrease in intensity of a beam of monochromatic radiation transmitted directly through the sample cell is related to the "concentration" of the suspension. For a dilute suspension

$$I = I_o e^{-rl}$$

i.e. $\tau = \dfrac{2.303}{l} \log_{10} I_0/I$

where I_0 = intensity of incident light; I = intensity of transmitted light; l = cell path length (cm); τ = turbidity (cm^{-1}).

$$\tau \propto \dfrac{cM}{\lambda^4}$$

where c = concentration of suspension; M = molecular weight of particle; λ = wavelength of light used.
This shows the apparent absorbance

$$\log_{10} I_0/I \propto c$$

and also the great sensitivity towards wavelength. Lothian has proposed a similar equation for uniform opaque particles in a parallel light beam [44].

$$A = \log_{10} e \cdot nQSl$$

where $A = \log_{10} I_0/I$; n = number of particles per unit volume; S = cross-sectional area of particle; l = path length; Q = total scattering coefficient. Q takes values from 0 to 4 and depends on the absorbing properties of the material, refractive index relative to the solvent, particle size relative to wavelength used and particle shape. Again the linear relation between apparent absorbance and concentration (for constant particle size) is shown.

b. Nephelometry. A beam of electromagnetic radiation passing through a suspension of particles of principal dimension less than about 0.1 λ suffers Rayleigh scattering. The electric component of the radiation induces forced dipolar vibration in each particle at the frequency of the radiation. Particles then re-emit this same radiation in all directions. Since colloidal particles range in size from about 0.001 μm to 2 μm, the lower end of this range will exhibit such a Tyndall effect.

The intensity I_s of light distance d from a compact mass of N scattering particles each of volume V is given by Rayleigh's equation.

$$I_s = I_0 \left[\frac{(n_p^2 - n_m^2)NV^2}{n_m^2 \lambda^4 d^2} (1 + \cos^2 \theta) \right]$$

where I_0 = incident light intensity; n_p = refractive index of particle; n_m = refractive index of medium; λ = wavelength of light used; θ = angle between directly transmitted beam and viewing axis. A diagram illustrating the spatial distribution of I_s values is shown in Fig. 9.

In the technique of nephelometry, the intensity of light scattered away from the direction of the incident beam (usually at 90°) is

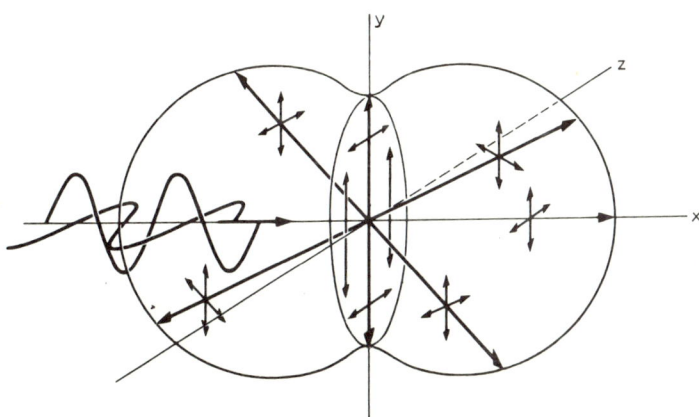

Fig. 9. Pattern of scattering of unpolarised monochromatic light by a small spherical isotropic particle. Reprinted from "Chemical Instrumentation", by H.A. Strobel, Figs. 4—13, p. 78, Addison-Wesley Publishing Company, Reading, Massachusetts, by kind permission of the publisher.

measured as a function of N, the concentration of particles.

The product NV^2 in Rayleigh's equation implies that scattered intensity increases with particle size for a constant mass of suspended solid. It was found for $BaSO_4$ suspensions, however, that I_s reached a maximum in the region 0.2—0.8 μm.

As particle size increases towards the value of the wavelength of light used, deviation from Rayleigh's equation occurs, and the realm of Debye scattering is reached in which a large proportion of the scattered light is thrown forward in the direction of the incident light. When particle size is greater than 0.5 λ, random reflection and refraction begin to take over and the relation between scattered intensity, direction and particle size etc. becomes highly complex. The elucidation of the size and shape of soluble polymer molecules by studying I_s vs. θ relations is an extensive and highly technical field of its own [45].

The turbidance and nephelos equations quoted are valid only for dilute suspensions where multiple scattering may largely be avoided. Plots of τ or I_s against suspension concentration are often curved and analyses must be performed on a strictly comparative basis. As mentioned previously though, this is no great drawback where titrations are concerned.

Nephelometry is used for very dilute suspensions as intensities are measured against a dark background. Turbidimetry is reserved for denser suspensions. This parallels the complementary nature of fluorescence and absorption spectrophotometry.

(B) GENERAL THEORY OF PHOTOMETRIC TITRATIONS

Photometric titration allows us to follow the course of a reaction such as A + B $\rightleftharpoons$ C forming the basis of a titration where A = titrand, B = titrant and C = product, provided that either A, B or C absorb or fluoresce at the wavelength used. Reserving considerations of fluorescence for the section on fluorimetric titrations, it is apparent that self-indicating photometric absorbance titration curves take the following forms depending upon the relative absorbing properties of A, B and C (Fig. 10).

The equation of a type "a" curve has been deduced by L'Her [43], thus: for the reaction A + B $\rightleftharpoons$ C let A have an initial concentration C_0 in volume V and let B be added to an analytical concentration of xC_0 where $x = 1$ at the equivalence-point. Assume C is the only ab-

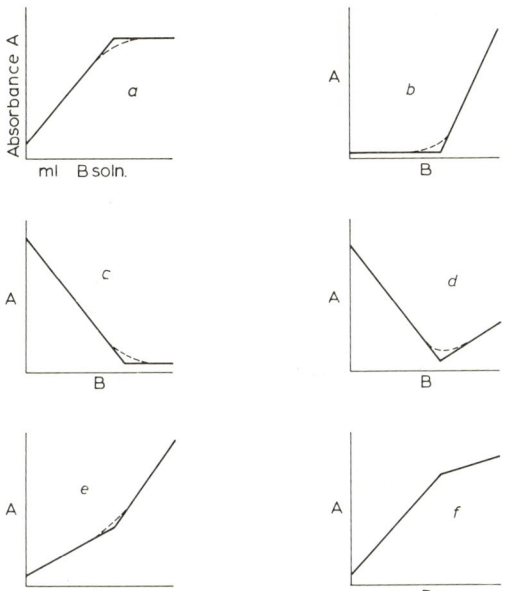

Fig. 10. Possible titration curves for the self-indicating reaction $A + B \rightleftharpoons C$ where A = titrand, B = titrant, and C = product, ϵ = molar absorptivity of designated entity. a, $\epsilon_C > \epsilon_A$, $\epsilon_B = 0$; b, $\epsilon_B > \epsilon_A = \epsilon_C$ (usually $\epsilon_A = \epsilon_C = 0$); c, $\epsilon_A > \epsilon_C$, $\epsilon_B = 0$; d, $\epsilon_A > \epsilon_C < \epsilon_B$; e, $\epsilon_C > \epsilon_A$, $\epsilon_B > \epsilon_C$; f, $\epsilon_C > \epsilon_A$, $\epsilon_B < \epsilon_C$.

sorbing species at the wavelength used and that the absorbance is corrected for volume change using the factor $(V + v)/V$ where v = volume of titrant B added. At any stage, absorbance $A = \epsilon \cdot l[C] = k[C]$ where ϵ = molar absorptivity of C at the wavelength used.

For a "complete" reaction, before the equivalence-point, $[B] = 0$; $[A] = C_0(1-x)$; $[C] = xC_0$.

$$\therefore A = kC_0 x$$

After the equivalence-point, $[B] = C_0(x-1)$; $[A] = 0$; $[C] = C_0$.

$$\therefore A = kC_0$$

Figure 11 illustrates these relations.
For a finite product dissociation constant

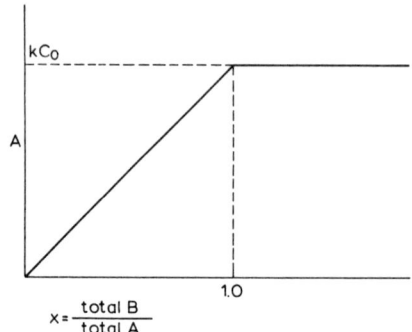

Fig. 11. Absorbance as a function of x for complete reaction $A + B \rightarrow C$.

$$K = \frac{[A][B]}{[C]}$$

at any stage

$$[A] + [C] = C_0$$

$$[B] + [C] = C_0 \cdot x$$

$$\therefore K = \frac{(C_0 - [C])(C_0 \cdot x - [C])}{[C]}$$

Since $[C] = \dfrac{A}{k}$

it follows that

$$A^2 - (kK + kC_0)A - kC_0 xA + k^2 C_0^2 x = 0 \tag{6}$$

Dividing through by x,

$$\frac{A^2}{x} - (kK + kC_0)\frac{A}{x} - kC_0 A + k^2 C_0^2 = 0$$

when $x \rightarrow \infty$

$$A = kC_0 = \epsilon \cdot l \cdot C_0 \text{ as for } K = 0 \text{ above.}$$

Again, from eqn. (6),

$$x = \frac{-A^2 + A(kK + kC_0)}{k^2 C_0^2 - kC_0 A} \tag{7}$$

226

$$\therefore \frac{dx}{dA} = \frac{(k^2C_0^2 - kC_0A)(-2A + kK + kC_0 - [-A^2 + A(kK + kC_0)](-kC_0)}{(k^2C_0^2 - kC_0A)^2}$$

$$\therefore \left(\frac{dx}{dA}\right)_{A, x \to 0} = \frac{K + C_0}{kC_0^2}$$

$$\therefore \frac{dA}{dx} = \frac{kC_0^2}{K + C_0}$$

$\therefore$ equation of line near the origin is

$$A = \frac{kC_0^2 \cdot x}{K + C_0}$$

If $K \ll C_0$ as is usual,

$A = kC_0x$, as deduced above for $K = 0$.

Simplifying eqn. (7) we obtain $x = f(A)$ with no assumptions.

$$x = \frac{-A^2 + kA(K + C_0)}{kC_0(kC_0 - A)}$$

For determination of k (to obtain ϵ_c) make $C_0 \gg K$

then $x = \dfrac{-A^2 + kAC_0}{kC_0(kC_0 - A)}$

which yields a linear relation between x and A for the initial part of the graph. For determination of K, C_0 should not exceed K by too great a factor.

For maximum sensitivity in a photometric titration the wavelength of light used should be that which produces maximum absorption by the most strongly absorbing species. Frequently, however, such sensitivity is not required and other wavelengths giving lower ϵ values may be used to avoid interfering absorbers or to give a method a wide concentration range of application.

If a UV-visible region spectrophotometer is available the UV absorption of A, B or C may be monitored, but failing this, and assuming that neither A, B nor C absorb sufficiently strongly in the visible, an indicator must be added which yields a suitably absorbing product with either A or B. Our titration reaction now becomes

$$A + Ind_A + B \rightleftharpoons C + Ind_B$$

where Ind_A = the indicator form dictated by A; and Ind_B = the indicator form dictated by B. Assuming we monitor the absorbance of indicator form Ind_B, then for the "classical" visual titration case where the indicator is half changed at the equivalence point, the titration curve resembles this:

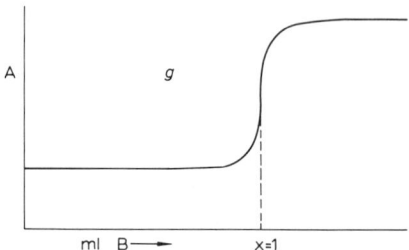

Fig. 12. Titration curve when $[Ind_A] = [Ind_B]$ at equivalence.

An indicator highly sensitive to A will yield a graph

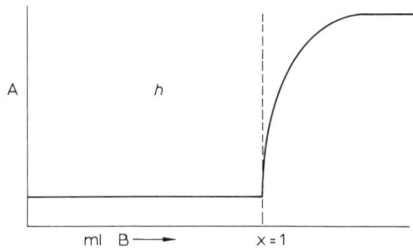

Fig. 13. Titration curve for indicator of high A sensitivity.

while one of low A sensitivity will give

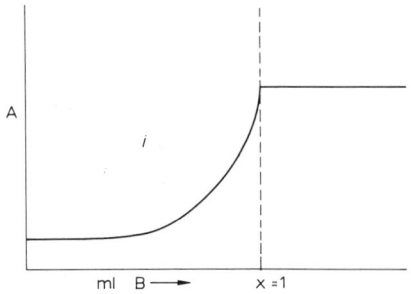

Fig. 14. Titration curve for indicator of low A sensitivity.

228

Strangely, cases (h) and (i), which would prove hopeless by classical visual titration criteria are more suited to photometric titration than case (g) which merely mirrors the appropriate sigmoidal potentiometric titration curve as $[Ind_A]/[Ind_B]$ is a logarithmic function of $[A]$.

The presence of sufficient indicator to react either with all A present or the relatively large concentration of B obtained from continuing the titration far beyond the end point yields linearly branched titration graphs of the types (a)—(f). Flaschka has named such titrations slope indicator titrations, whereas types (g), (h) and (i) he terms step indication titrations [41]. The introduction of a foreign indicator substance to give a further equilibrium system should be avoided if possible, especially where the selective titration of a multicomponent sample or work at the feasibility limit of $K \times C_A$ is being attempted. Pearson's rotation titrations score heavily here. The use of indicators will be examined in more detail under the individual titration types.

Photometric titrations owe their usefulness to the often quite staggering precision obtainable under the most adverse conditions. Potentiometric titrations in which instrument response depends upon the logarithm of some species concentration (e.g. A) via the Nernst equation i.e.

$$E = E^0 + \frac{RT}{nF} \ln C_A$$

show a sigmoidal type of titration curve in which (for a symmetrical $A + B \rightleftharpoons C$ reaction) the end-point coincides with the point of inflection. The curve must therefore be accurately defined in the region where A and B react together in almost equivalent quantities — the most unsuitable region from the point of view of completeness of reaction and kinetics. In addition, electrodes are often poorly poised and exhibit sluggish response in very dilute solutions where (if absorptivities are suitable) photometric titrations thrive. This is especially so for titrations in non-aqueous solvents. The linear branches of a photometric titration may be accurately positioned far from the equivalence-point where reaction between A and B may be considered complete. The precision with which a titration may be performed is governed by the product $C_A \times K$. Limits of feasibility for visual potentiometric and photometric titrations are shown in Table 1.

TABLE 1

Minimum values of $C_A \times K$ required to produce a usable titration

	$C_A \times K$ (K dissn. const.)	$C_A \times K$ (K stability const.)
Visual	10^{-9}	10^5
Potentiometric	10^{-10}	10^4
Photometric	10^{-12}	10^2

These values were obtained through practical experience. Until recently the only theoretical study which attempted to forecast the precision and concentration limit of a photometric titration was that carried out by Groeneveld and den Boef [46,47].

To determine the standard deviation of the equivalence-point volume for a simple self-indicating system they imagined two linear titration branches p and q, the regression lines of points x_i y_i as shown in the diagram (Fig. 15).

$$\text{Equation of line p, } A - \bar{y}_p = m_p(v - \bar{x}_p) \left.\begin{array}{c}\\ \\ \end{array}\right\}$$
$$\text{Equation of line q, } A - \bar{y}_q = m_q(v - \bar{x}_q) \qquad (8)$$

where $\bar{y}_p = \overset{i}{\Sigma} y_{ip}/N_p$; $\bar{x}_p = \overset{i}{\Sigma} x_{ip}/N_p$ etc.; $N_{p,q}$ = number of measure-

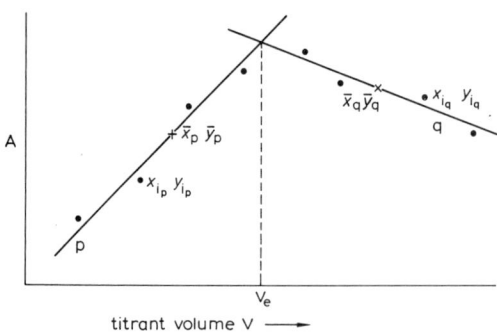

Fig. 15. Precision of a linear photometric titration. Adapted from ref. 21.

230

ments on lines p,q; $m_{p,q}$ = gradient of lines p,q.

At the equivalence-point intersection the above equations become simultaneous ($v = v_e$).

$$\therefore v_e = \frac{\bar{y}_q - \bar{y}_p + m_p \bar{x}_p - m_q \bar{x}_q}{m_p - m_q} \tag{8a}$$

From linear regression analysis, for each line

$$m = \frac{N\Sigma xy - \Sigma x \Sigma y}{N\Sigma x^2 - (\Sigma x)^2} \tag{8b}$$

For an individual point,

$$S_{A_i} = \sqrt{S_{y_i}^2 + m^2 S_{x_i}^2} \tag{9}$$

= total standard deviation of titration absorbance measurement.

$$S_y = 0.4343 \frac{S_T}{T} \tag{10}$$

where T = solution transmission and S_T (standard deviation of trans. T) depends upon the photometer stability and readout system.

S_x depends upon the burette used.

Finally,

$$S_{v_e}^2 = \frac{1}{(m_p - m_q)^2} (S_{A_p}^2 + S_{A_q}^2) \tag{11}$$

where $S_{A_{p,q}}$ = standard deviation of points on p and q of highest y value.

Equation (11) clearly illustrates the value of designing the titration so that linear branches have the greatest gradient difference. This is done by suitable control of optical path length, test solution volume and concentration, and ϵ for the absorbing constituent. Operation at high T values and with burettes of low S_x is very beneficial. The authors showed also that graphical construction of branches p and q by sight was virtually as effective as a linear regression approach. For their titration of 10 ml of 0.01 N Ce^{IV} solution with 0.2 N Fe^{II} solution, $m_p = 0.44$ ml^{-1}; $m_q = 0$; T at the end of the titration = 68%; $S_T = 0.009\%$; $S_x = 0.0001$ ml. $\therefore S_{v_e} = 0.00022$ ml theoretically, i.e. a coefficient of variation of 0.05%. Even with $\epsilon = 10$ for the absorbing species, titration of a 5×10^{-4} N solution yielded a coefficient of variation of 0.7%.

Groeneveld and den Boef have defined the limit of determination as the titrand concentration at which the coefficient of variation becomes 5%,

i.e. $S_{v_e}/V_e = 0.05$ (12)

From eqn. (11),

$$S_{v_e}^2 = \frac{1}{(\Delta m)^2} (S_{A_p}^2 + S_{A_q}^2)$$

Since close to the intersection $S_{A_p} = S_{A_q}$,

$$S_{v_e} = \frac{S_A \cdot \sqrt{2}}{\Delta m}$$ (13)

Also since $\Delta_m = m_p - m_q = \dfrac{1 \cdot t \cdot \Sigma \epsilon}{V}$ (14)

where t = titrant concentration; V = initial volume

and $C_A = \dfrac{v_e \cdot t}{V}$ (15)

C_A = initial concentration of titrand, combination of eqns. (12), (13), (14) and (15) yields concentration limit $C_A = (S_A\sqrt{2})/0.05l\Sigma\epsilon$.

Even as early as 1939, Ringbom and Sundmann assessed the precision (coefficient of variation) of a single absorbance reading to be about 1% and that of a good photometric titration to be about 0.05% [7].

Photometric titrators of the two wavelength-two photocell type are capable of yielding very precise results when used with two-colour indicators showing sharp absorption bands (triphenylmethane are preferable to azo dyes). The theory of such instruments has been examined by Ringbom et al. [23] and will be discussed in more

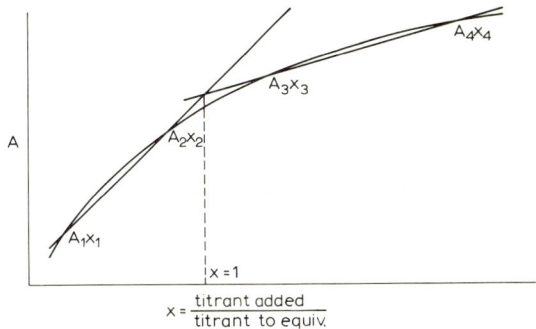

$$x = \frac{\text{titrant added}}{\text{titrant to equiv.}}$$

Fig. 16. Location of the equivalance-point from a poor titration curve. Adapted from ref. 19.

detail under complexometric titrations.

Grunwald has devised a successive approximation procedure for the accurate location of the equivalence-point from a titration curve of poor quality [48].

If the two (non-tangential) lines chosen as shown in Fig. 16 intersect at the true equivalence-point, then

$$\frac{[A_2(1-x_1)-A_1(1-x_2)]}{(x_2-x_1)} = \frac{[A_4(1-x_3)-A_3(1-x_4)]}{(x_4-x_3)}$$

For any photometric titration, at any stage

$$A = A_0 + \Sigma \epsilon_i C_i \quad \text{(1 cm cell)},$$

where A_0 = absorbance of solvent, and ϵ_i = molar absorptivities for all absorbing species i.

For a simple $A + B \rightleftharpoons C$ interaction, A can be expressed as a function of x and it can be shown that

$$\frac{x_2(1-x_1)}{(x_2-x_1)(1-x_2)} - \frac{x_1(1-x_2)}{(x_2-x_1)(1-x_1)} = \frac{x_4+x_3-2}{(x_3-1)(x_4-1)}$$

In practice, an equivalence-point would first be guessed from the titration curve, values for x_1, x_2 and x_3 chosen and x_4 calculated from the last equation. The intersection of the lines constructed as shown would yield a different, better, equivalence-point and the process would be repeated until the equivalence-point remained steady.

For more complex interactions, knowledge of the equation relating A to x leads to equations in x values only, similar in nature to the one above.

Early in the development of photometric titrations, one-wavelength two-photocell instruments of a type similar to the evergreen "Spekker" filter photometer were used to perform comparison titrations. In these, the test solution was titrated until the absorbance at some particular wavelength became identical with that of some known standard solution. This involved, at its simplest, merely dilution, but in more complex situations one chemical reaction (e.g. metal ion plus photometric reagent) or one chemical reaction interacting with one indicator system (e.g. acid—base plus indicator) were used. Titration to a previously established, or null, deflection was used. The duplication, or "Colorimetric Titration" method in which cuvette (1) contains unknown (A) plus a large excess of reagent and cuvette (2) contains the same amount of reagent together with standard A solution added as a titrant, is admirably suited to two-photocell null-reading photometers.

A further advantage of photometric titrations is that absorbance differences rather than absolute absorbance measurements are important. Thus a strongly absorbing but inert background is not harmful provided that stray light in the photometer is minimal. With appreciable stray light I_s, apparent absorbance is given by log $(I_0 + I_s)/(I + I_s)$ which tends towards a constant value of log $(I_0 + I_s)/I_s$ at high absorbance.

For titration curves of types (a), (e) and (f) (Fig. 10) where the break point appears at high absorbance, values in excess of $A = 1$ should not be used. Caution must be exercised when operating near the UV cut-off region of the titration cell. Goddu and Hume illustrate this point beautifully as shown by Fig. 17 [31]. The photometer may be tested by titrating water with, e.g. $KMnO_4$ which obeys Beer's Law.

Perhaps the most important characteristic of an efficient photometric titrator is medium-term stability since the instrument cannot be "zeroed" once the titration has begun. The main contribution to end-point uncertainty is S_y, the standard deviation of a single absorbance measurement, therefore spectrophotometers of very high stability should be ideally suited to work of this nature. It should, however, be emphasised that good instrumentation will only yield high precision of end-point detection; accuracy will still be depen-

234

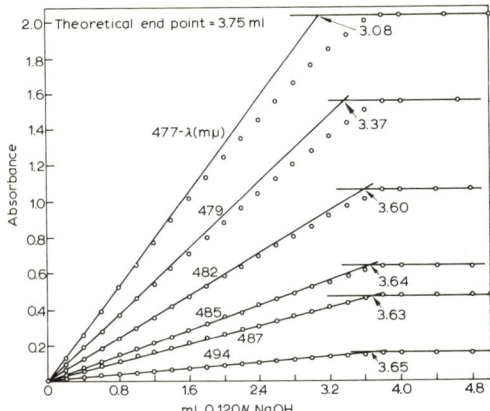

Fig. 17. The stray light problem. Titration of 10.0 ml of 0.450 N *p*-nitrophenol in 100 ml of water at various wavelengths. Reprinted from ref. 31, by kind permission of the American Chemical Society.

dent upon the balance, glassware calibration, salt and protein errors, temperature effects from stirring etc.

The graphical presentation of a photometric titration is usually of A vs. titrant volume or x since for most high equilibrium constant titrations the absorbance varies linearly with concentration of the absorbing species which in its turn varies linearly with titre. However, the reasoning of L'Her which follows [43] shows that at low absorbance values transmission varies linearly with x which may be helpful in certain instances.

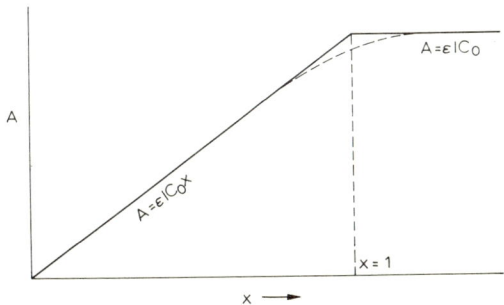

Fig. 18. Variation of absorbance with extent of titration.

Before equivalence point

$$A = -\log T$$

$$\therefore \ln T = -2.303 \, A$$

$$\therefore \quad T = e^{-2.303 \, A} = e^{-2.303 \, \epsilon l C_0 x} = e^{-kx}$$

where $k = 2.303 \epsilon l C_0$

Likewise at the equivalence-point $T = e^{-k}$ (for complete reaction). Before the equivalence-point,

$$T = e^{-kx} = e^{-k} \cdot e^{k(1-x)}$$

$$= e^{-k}\left[1 + k(1-x) + \frac{k^2(1-x)^2}{2!} + \cdots\right]$$

When $x \to 1$ and when k is small (low absorbance) the second-order term in the series expansion may be neglected whereupon $T = e^{-k} \, [\, 1 + k(1-x)]$, a linear relation between T and x, in particular, when $x = 1$, $T = e^{-k}$.

To avoid dilution-induced curvature, all appreciable absorbance values should be multiplied by $(V + v)/V$. Relatively concentrated titrant is commonly added from a small burette but care must be taken that S_x, the standard deviation of the titrant volume, does not contribute too markedly to S_A. In addition only the best quality graph paper should be used for the plot.

In the case of potentiometric titrations, 1st or 2nd derivative plots, dE/dx or d^2E/dx^2 are useful for end-point location but this does not apply to linear photometric titrations.

The precision of linearly extrapolated hyperbolic titration curves has recently been examined by Kotrly and co-workers [49,50].

3. Instrumentation

Since photometric titrations involve for the most part a plot of absorbance against titre, any arrangement of photometer and burette capable of measuring these two variables simultaneously will suffice. Instruments may be built specifically for the purpose or purchased from the few manufacturers who produce them. In addition spectrophotometers and filter photometers can quite easily be adapted for such work.

Developments concerning photometric titration instrumentation up to 1964 have been well reviewed by Headridge [34], and by Underwood [39] and will not be repeated in detail here. Significant contributions over the whole period from 1928 to 1970 have, however, been collected in Table 2. More detailed comments are offered on post 1964 designs. Some points of general importance concerning the construction of phototitrators are as follows.

The titration cell may be placed inside the photometer cell compartment whereupon provision for stirring [51] and titrant addition must be made. The compartment must remain light-tight to eliminate stray light difficulties unless the photoreceptor is very small and accepts light only over a small solid angle. Flaschka and co-workers have excelled in the production of instruments of this type. The influence of internally mounted magnetic stirrers on photometer response must be checked.

Titration may equally well be performed in an external vessel with circulation of the solution through a flow-through cuvette [74]. The centrifugal impeller of Higuchi [52] is particularly useful.

Silica spectrophotometer cells are of course required for work in the near-UV but certain glass cells (e.g. Vycor 7910, wavelength limit ca. 250 nm) transmit well into this region.

Wavelength selection has been carried out using limited wavelength response photodetectors, coloured filters, interference filters, interference wedges and prism and grating monochromators. Good quality coloured filters appear to suffice for absorbance changes due to the clear separation of broad absorption bands, a situation that applies to most practical analyses. Two-photocell optical null instruments are proposed from time to time; these offer high sensitivity and stability.

Photometric titrators appear to fall naturally into three sections namely Manual, Automatic shut-off, and Recording. Each of these can be further sub-divided into Home-made, Modification of commercially available photometer and Commercial. Table 2 is based on this classification.

The post 1964 development of photometric titrators has been dominated by Flaschka's school. In 1961 Flaschka, who was dissatisfied with available commercial titrators and attempts to modify existing spectrophotometers, clearly stated the desirable attributes of a cheap but effective phototitrator [53]. These were:

(1) ability to function in ambient light,

(2) variable path length,

(3) high medium term stability of source and photoreceptor,

(4) easy facilities for scale expansion,

(5) reasonable degree of light monochromaticity.

These requirements are to some extent mitigated by the facts that absolute values of T and A are not required and that the titration cell need not be exactly replaced for each determination. He published details of a horizontal light beam instrument based on the use of a tiny silicon diffused NPN photoduodiode as photocell. This receptor, which showed no fatigue effects, responded only to illumination received over a small solid angle and hence allowed titrations to take place in ambient light. The light source was an under-run battery powered 6 V bulb and this, together with the remarkable output stability of the solid state receiver, produced a photometer of excellent overall stability. Monochromation was produced by interference filters. The photocell was of high impedence, with conductance accurately related linearly to incident light intensity. Provision for scale expansion in both directions (zero and 100% T) was an additional valuable feature of the instrument. In 1962 the instrument was modified for micro-scale titrations (cuvettes 0.2—2 ml capacity, burette 0.1 ml capacity, vibrating rod agitation) [54].

Flaschka had long hankered after a variable light path titrator and in 1965 he, with Butcher, published details of a semi-immersion instrument produced by tipping the above titrator up on end [55]. The vertical light beam emerged from a glass tube which could be lowered into the test solution. The beam then carried on down into the lower part of the instrument through an interference wedge and was focussed onto the photoduodiode mentioned above. Spurious wedge transmission orders were removed using optical filters, special care being needed with near-IR radiation for which the photoreceptor had maximum response. Linearity between A and concentration of an absorbing species was difficult to maintain at low wavelengths (ca. 450 nm and below) due to low photocell response and consequent stray light difficulties. This instrument was an interim measure and in 1968 Flaschka and Speights proposed their full immersion spectrophotometer and titrator [56]. Here collimator and diffraction grating were obtained from the Bausch and Lomb Spectronic 20 spectrophotometer. A movable mirror was positioned below the light probe to reflect the monochromated light back at an angle of 5.5° to the incident beam, the path length thus being adjustable between 0 and 10 cm. The detection system was virtually identical with that of previous instruments. A constant flow burette and recorder could easily be fitted to produce a recording titrator.

A recently developed [57] very small photovoltaic receptor may be used in place of the photoresistive diode in all the above instruments. This permits elimination of one battery and use of a simpler circuit.

Hedrick (1965) also advocated the use of CdS or CdSe (peak response 515 nm) miniature photoconductive cells. A "photometric probe" was constructed by mounting such a cell at the bottom of a glass tube faced with a plane glass plate. Normal room light could be used as source for colour change titration with wavelength selection achieved by surrounding the cell with a suitably coloured liquid. Fluorescent or turbidimetric titrations were successful if the photocells were shielded from the incident light beam. The power supply was a 1.4 V mercury battery yielding a current of 100 μA to zero; incorporation of a log input recorder gave absorbance—volume plots directly. The photocell response time of 0.5—3 ms was satisfactory for all but the most demanding kinetic experiments.

The Radiometer Titrator TTT 1b and Titrigraph SBRI were linked to a CdS photoresistor via a simple bridge system by Skytte-Jensen (1966) to produce a recording photometric titrator [58]. Filters were used for wavelength selection. The detector showed good response in the 500—850 nm region and little temperature sensitivity. The relation between resistance and incident light intensity was given as $\log R = -k \log I$ and for the bridge circuit,

$$\Delta V = B V_0 \, \Delta A \text{ when } \Delta V \ll V$$

where ΔV = bridge unbalance potential; V_0 = voltage across bridge; ΔA = change in solution absorbance; B = apparatus constant. Thus absorbance—volume curves could be plotted directly. In addition, the apparatus could easily be adapted for kinetic investigations by functioning as a potentiostat; i.e. recording against time the volume of titrant required to maintain ΔV, and thus ΔA constant.

In 1966 Wallin produced a double beam single photomultiplier null detecting spectrophotometric titrator for the very accurate determination of solute concentrations by absorption measurement [20]. The instrument, based on the Bausch and Lomb Spectronic 505, was a modern development of the earlier ideas of Kortüm and von Halban. Figure 19 shows a diagram of the optical and electrical

240

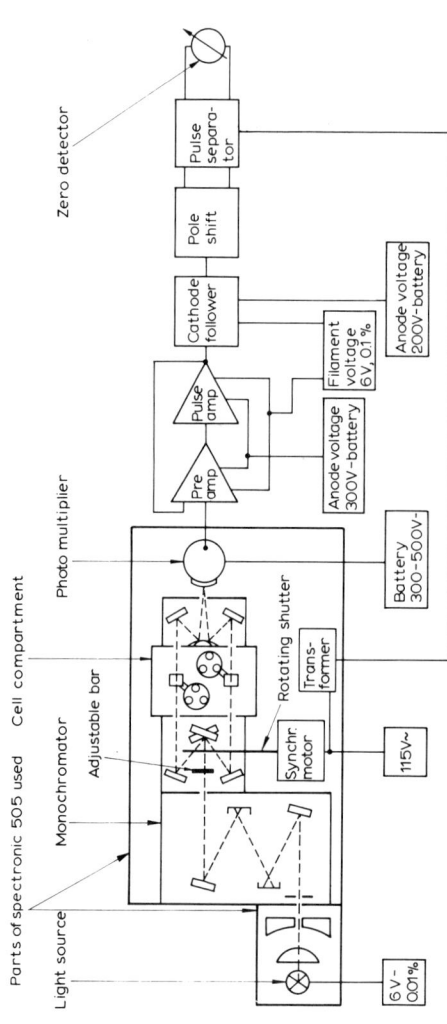

Fig. 19. The Wallin spectrophotometric titrator. Reprinted from ref. 20, by kind permission of Almquist and Wiksells Boktryckeri AB, Stockholm.

parts of the photometer. The absorbance of the sample solution was continuously compared with the absorbance of a reference solution of known composition and almost identical absorption spectrum in the wavelength range studied to eliminate stray light errors. To prepare this reference solution titration cells were used which made small stepwise changes in concentration possible (Fig. 20). When no difference in absorbance between the sample and reference solutions could be detected the concentration of the absorbing species was considered equal in the two solutions. Light pulses from the sample and reference beams were transformed into electrical pulses by the

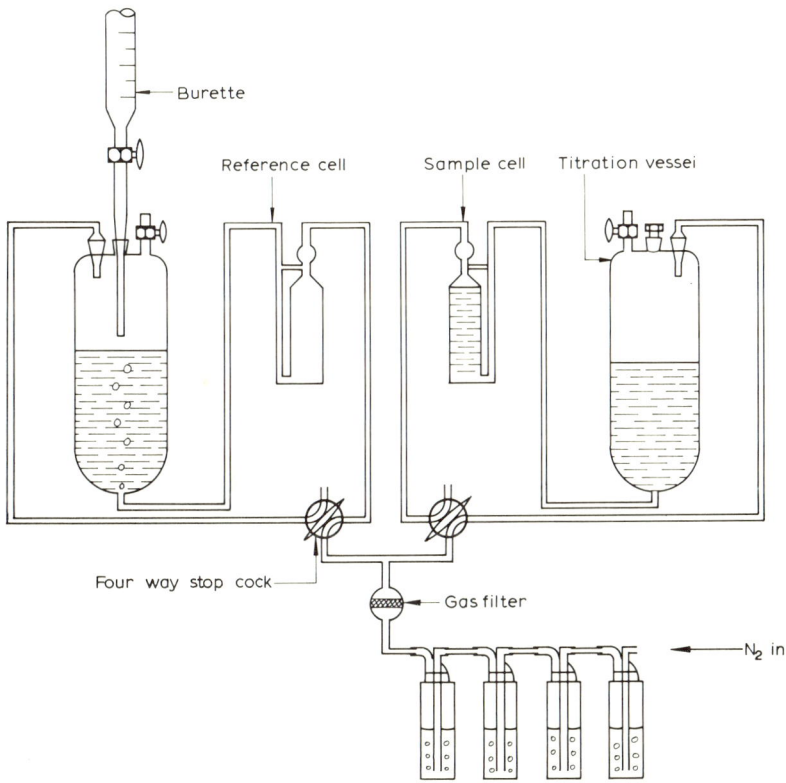

Fig. 20. The titration system used with Wallin's photometric titrator. Reprinted from ref. 20 by kind permission of Almquist and Wiksells Boktryckeri AB, Stockholm.

photomultiplier. The pulses were approximately square-wave in form with a pulse length of 5 ms and a frequency of 100 s^{-1}. After amplification the reference and sample pulses were separated by a six pole double chopper and compared by a galvanometer. This instrument was essentially a complex comparator and it is difficult to see how it might be used for conventional photometric titrations. Undoubtedly, however, it would make a superb basis for substitution titrations.

The twin photocell titrator of Osborn et al. [8], and others, was brought to an interesting stage of development by Ringbom et al. in their "Dichrotitrator" [23]. This double beam of different wavelengths, single cuvette, double photocell instrument (Fig. 21) was designed for accurate step indication titrations. The two photocells were connected in opposition over a load resistance R thus giving a total photocurrent of zero when equally illuminated. Wavelengths were chosen so that at the start of a titration one beam was strongly absorbed while after the end-point the roles were reversed with zero photocurrent at the theoretical end-point. To choose such wavelengths properly the absorption spectrum of the indicator at $[M]_{eq}$ must be known so that at the equivalence-point the solution exhibits the same absorbance to both beams. Many wavelength pairs are suitable; other factors such as photocell—wavelength response and the interrelation of free and metallised indicator absorption curves dictate the final choice. The method is concerned primarily with complexometric titration but may equally well be applied to acid—base or redox types. If operated correctly the technique is indepen-

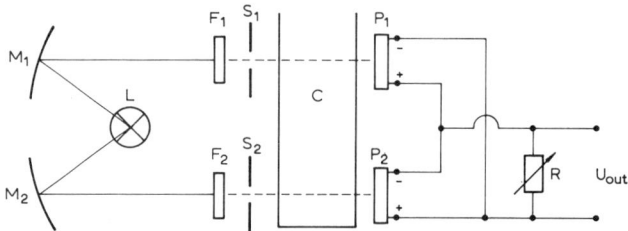

Fig. 21. The Dichrotitrator of Ringbom et al. Reprinted from ref. 23, by kind permission of the American Chemical Society. L = lamp; M_1, M_2 = mirrors; C = titration cell; F_1, F_2 = monochromators; S_1, S_2 = shutters; P_1, P_2 = photocells; R = resistor for sensitivity adjustment; U_{out} = connection to galvanometer, relay or recorder.

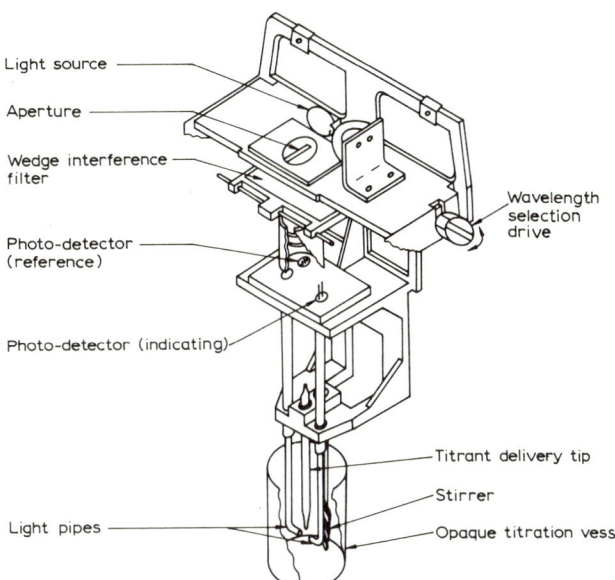

Light source

Aperture

Wedge interference filter

Wavelength selection drive

Photo-detector (reference)

Photo-detector (indicating)

Titrant delivery tip

Stirrer

Light pipes

Opaque titration vessel

Fig. 22. The light-pipe phototitrator of Young et al. Reprinted from ref. 59 by kind permission of Academic Press, New York.

dent of indicator concentration and capable of measuring metal concentrations with 1% error at $K_{MY} \cdot C_M$ values as low as 10.

It seemed inevitable that fibre optics light pipes should be introduced into photometric titrations and indeed this was done by Young et al. (1968) [59]. Their instrument, designed for automated serial titrations and shown in Fig. 22 used a tungsten lamp source and movable wedge interference filter. Monochromatic light was focussed onto a reference photoresistive cell then carried down into an opaque titration beaker by means of a transmitter coherent fibre optics light pipe. The beam traversed a 0.5 in solution gap then returned to the photometer measuring cell through an identical pipe. The reference and measuring photocells were placed in opposite arms of an electrical bridge and the bridge unbalance potential fed to an automatic titrant control device or recorder.

Adaptation of the ubiquitous Unicam SP 600 again featured in a paper by Rees and Hill (1969). Their 250 ml titration cell with ancillary fittings is shown in Fig. 23. It was supported on a platform

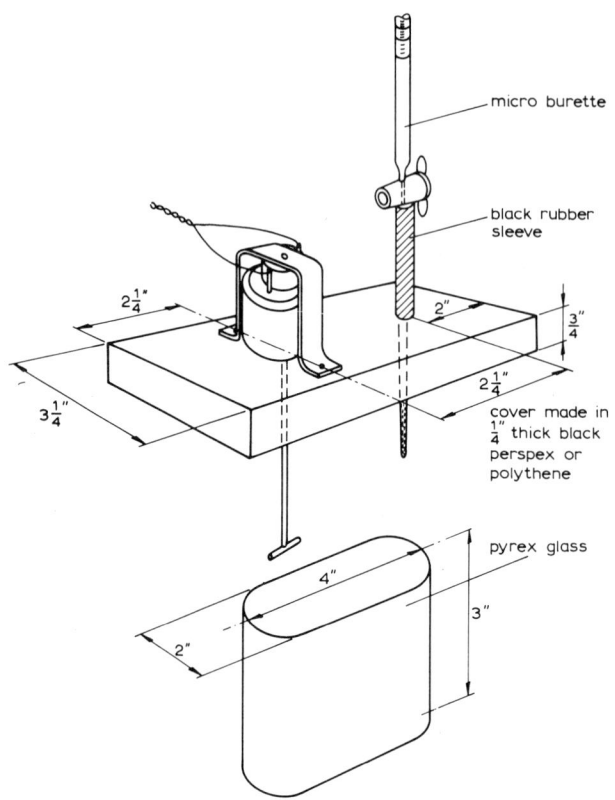

micro burette

black rubber
sleeve

$2\frac{1}{4}''$

$2''$

$\frac{3}{4}''$

$3\frac{1}{4}''$

$2\frac{1}{4}''$

cover made in
$\frac{1}{4}''$ thick black
perspex or
polythene

pyrex glass

$4''$

$3''$

$2''$

Fig. 23. The titration cell, cover and stirrer unit for spectrophotometric titrations using the Unicam SP600 as designed by Rees and Hill. Reprinted from ref. 212, by kind permission of Pye Unicam, Cambridge.

placed within the well of the instrument. A short path-length cell for high absorbance titrations was constructed by fusing a 2.5-cm diameter 4.5-cm long glass cylinder inside the cell to coincide with the light beam. Excellent titration curves were obtained for the usual systems.

Foust and co-workers (1969) produced an elegant anaerobic titration assembly, applicable to almost any spectrophotometer, in order to titrate redox active proteins and enzymes with oxygen-sensitive titrants such as sodium dithionite, but naturally such a system could command much wider use [60]. The two units consisting of burette-

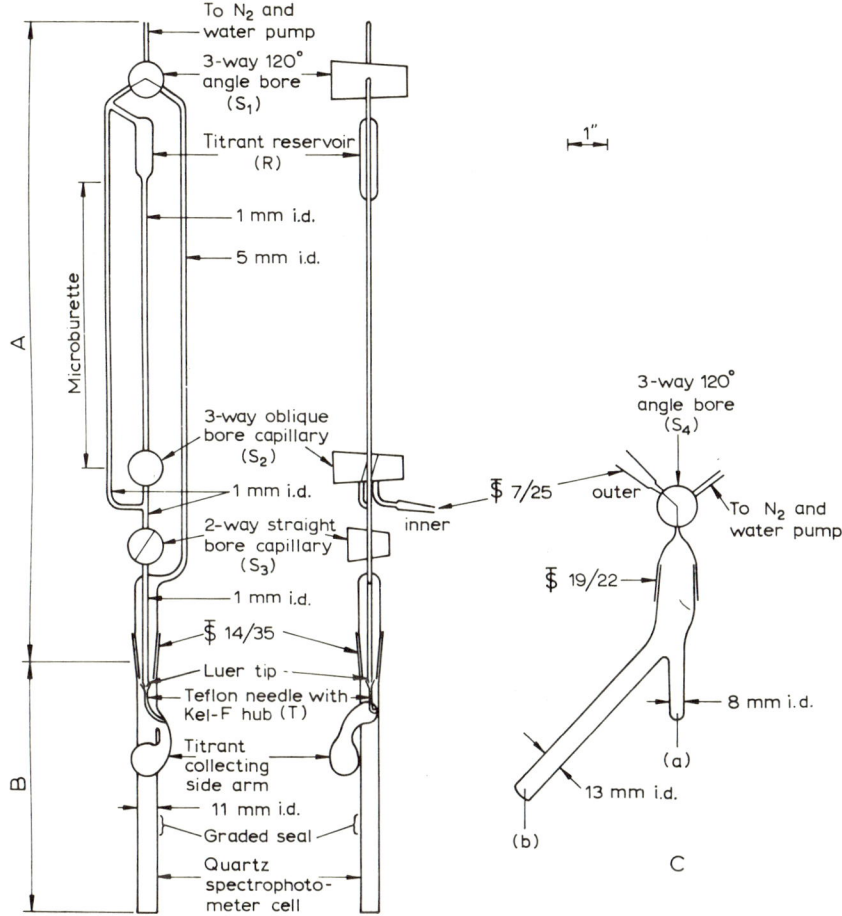

Fig. 24. The combined anaerobic titrant preparation unit-burette-cell titrator of Foust and co-workers. Reprinted from ref. 60 by kind permission of Academic Press, New York.

cell titration assembly and the titrant preparation unit are shown in Fig. 24. The quartz 2.5-ml cuvette, 0.2-ml burette and titrant preparation unit could be completely flushed with nitrogen. Titrant additions were as small as 1 μl but the apparatus had to be removed from the spectrophotometer and gently agitated to ensure mixing.

TABLE 2

Photometric titrators

Author(s) (date)	Instrument modified	Details	Ref.
MANUAL — *Photometer modification*			
Headridge and and Magee (1958)	Unicam SP 500	Large internal cell, overhead stirring. Perspex cell with quartz windows for UV	61
Goddu and Hume (1950)	Coleman Model 14	150-ml beaker with test-tube in bottom. Overhead stirrer, high stray light errors	251
Sweetser and Bricker (1953)	Beckman Model B	Large cylindrical cuvette	134
Bricker and Sweetser (1952)	Beckman Model DU	Rectangular cell sealed to bottom of 200-ml beaker. CO_2 stream stirring	253
Klingman et al. (1955)	Beckman Model DU	Magnetic stirrer	341
Herrington (1958)	Unicam SP 600	7-ml perspex cuvette, magnetic stirrer. Agla microburette	62
Fricker (1955)	Hilger Spekker	External titration cell	63
Lee et al. (1956)	Beckman Model B	External titration cell	64
Higuchi and co-workers (1959)	Beckman Models B and DU Cary recording, Bausch and Lomb Spectronic 20	External titration cell centrifugal stirrer circulation	52
Sweet and Zehner (1958)	Beckman DU	External titration cell, quartz cuvette, centrifugal stirrer circulation, 5 s. homog.	342
Goddu and Hume (1954)	Beckman Model B	Beaker, magnetic stirrer. Provision for inert gas	31, 75
Sweetser and Bricker (1954)	Beckman DU	T-shaped quartz cuvette overhead stirrer. λ down to 215 nm	135
Underwood (1953)	Beckman DU	Use of test-tube adaptor. 90-ml rect. cuvette, overhead stirrer	132

246

TABLE 2 (continued)

Author(s) (date)	Instrument modified	Details	Ref.
Underwood (1964)	Beckman DU	Magnetic stirrer, beaker, instrument insensitive to magnetic field	39
King and Hirt (1953)	Cary spectrophotometer	External titration flask, flow through cell. Absn. spectrum recorded after each titrant addition	65
Chalmers (1954)	Unicam SP 600	Square perspex cell, rod stirrer	139
Hollingworth (1957)	Unicam SP 600	Perspex cell, rod stirrer	223
Wallin (1966)	Bausch and Lomb Spectronic 505	Double beam in time, single detector, null reading. External titration vessel circulation	20
Rees and Hill (1969)	Unicam SP 600 ·	250-ml cells, overhead stirrer, short path length cell	212
Home made			
Bobtelsky (1960)		Heterometer. Vertical light path. Constant-temp. jacket. Barrier layer cell. Used for precipitation titrations	204
Flaschka and Sawyer (1961)		Interf. filters. 1-mm square photoconductive cell. "Daylight" cuvette. Scale expansion facility	53, 54
Field and Baas-Becking (1926)		Filters, Ag/Bi thermocouple. Mirror galvo. 1st photoelectric titrator	4
Osborn et al. (1943)		Split beam, one cuvette, 2 photocells. 2 filters. Provision for fluorescence. Very sensitive	8
Flaschka and Butcher (1965)		Vertical, variable length light path. Beaker, overhead stirrer, interference filters or wedge. 1 sq mm photoduodiode	55

TABLE 2 (continued)

Author(s) (date)	Instrument modified	Details	Ref.
Hedrick (1965)		Miniature CdS or CdSe photo-conductive cells in end of glass tube. Also used for turbi-dimetric and fluorimetric work	211
Ringbom et al. (1967)		2 beam, 2 photocell, 2 filter, 1 cuvette "dichrotitrator". Null indication. Very sensitive	23
Flaschka and Speights (1968)		Full immersion spectrophotom-eter. Spectronic 20 monochrom-ator. Vertical variable length light path. Photoduodiode.	56
Foust et al. (1969)		Anaerobic titrator — all-in-one titrant reservoir, burette and quartz cell. Manual agitation. Adaptable to most spectro-photometers	60
Agazzi and Bond (1961)		Light source and photocon-ductive cell placed in immers-ible tubes. Filters	66
Partridge (1932)		1st use of barrier layer photo-cell. Simple potential balancing arrangement	67
Commercially available			
(1957)	E.E.L. Titrator	Filters, external cylindrical cuvettes 4 or 50 ml cap. Barrier layer cell, separate mir-ror galvo	34
	Cambridge Sci. Corp. Spectrosym electronic colorim-eter	Lamp and photoreceptor im-mersed. Lamp modulated, tuned amplifier. Filters. Variable path length	
	Fisher Photo-titrator	Immersed light pipes, interference wedge	
AUTOMATIC SHUT OFF — *Photometer modification*			
Sunderman and Propst (1953)	Beckman K pot. titrator	Vacuum phototube and series resistance as signal source	343

TABLE 2 (continued)

Author(s) (date)	Instrument modified	Details	Ref.
Home made			
Müller and Partridge (1928)		Vertical illumination, overhead stirrer. Filters. Photocell modifies triode grid potential; plate current operates release	5
Howerton and Wasilewski (1961)		"Titrocolormat". Filters, barrier layer cell, (photomultiplier for fluorescence). Microammeter and relay operate burette. $X—Y$ recorder fitted	279
Commercially available			
	Cenco Titrator	Set value of T triggers relay. Syringe burette. Filters. CdS cell, 250 ml beaker. Needs high dA/dV	34, 39
	Stone Titrator	Filters, 2-photocells. Unbalance potential operates release. 30—600 ml beakers; delay mechanism. Needs fairly high dA/dV	34, 39
	Sargent-Malmstadt Spectro-Electro Titrator	Two units. Filters (interference). λ range 370—650 nm. 30—600 ml beakers. Barrier layer and CdS cells. d^2A/dV^2 operates solenoid	34, 39
	Leco ASD-1 Sulphur Titrator	Determines S as SO_2 via IO_3^-/I^- reaction	34
	Fisher Titrator	Wedge interference filter, 2 beam, fibre optics light pipe. Bridge unbalance burette shut-off	

RECORDING — *Photometer modification*

Malmstadt and Gohrbrandt (1954)	Cary recording	Overhead stirring. Motor driven syringe. Quartz windowed cell for UV	70
Marple and Hume (1956)	Beckman Model B	Log attenuator inserted. Constant rate delivery system	71

TABLE 2 (continued)

Author(s) (date)	Instrument modified	Details	Ref.
Menis, Manning and Ball (1958)	Warren Spectracord	Titrant added at constant rate from motor syringe	72
Mullen and Anton (1960)	Beckman Model B	Log attenuator	346
Chalmers and Walley (1957)	Unicam SP 600	dA/dV recorded	73
Underwood and Robertson (1964)	Beckman DU	Energy recording adaptor and log recorder used. 10 ml constant rate burette	39
Skytte-Jensen (1966)	Radiometer Titrator and Titrigraph	Filters. CdS photocond. cell. Wheatstone unbalance potential applied to titrigraph	58
Home made			
Shapiro and Brannock (1955)		Mariotte bottle delivery, beaker, filters, barrier layer cell, recorder	68
Brill, Holzer and Rethy (1959)		Filters, beaker, barrier layer cell. Spot galvo and photo-electric recorder. Semi-manual titrant addition	344
Barredo and Taylor (1947)		Mariotte bottle	345
Miyake and Sakamoto (1967)		Monochromator 400—800 nm. T vs. V plots	69
Commercially available			
	Quére Titrator	Filters. Absorbance vs volume plot slows with dA/dV increase	34

250

4. Acid—base photometric titrations

Acid—base titrations involving an added dyestuff indicator were naturally among the first to be examined when photoelectric titrators came upon the scene in the late 1920's. Müller and Partridge, for example, published in their 1928 paper [5] a graph of plate current against acid/base ratio for a phenolphthalein transition (Fig. 3). In general, filters were used to produce radiation coinciding with one or both of the absorption peaks involved in an indicator's transition and the change in photocurrent was plotted as such or utilised to operate a burette release. However, no serious studies of the underlying relation between absorbance and titre were undertaken until the 1950's.

Acid—base photometric titrations fall naturally into two categories,

(a) self-indicating in which the Brönsted base of the titrand has an absorption spectrum (in an accessible region) well separated from that of the acid species, and

(b) indicator titrations for which an indicator with suitable optical and chemical characteristics must be added.

The borderline between these categories, as with complexometric titrations, can be diffuse.

(A) SELF-INDICATING TITRATIONS

The theory of these was examined in the excellent paper of Goddu and Hume, 1954 [75].

For the titration of a weak acid solution, initial concentration C_0, volume V_0 with a strong base concentration C, volume added V, i.e.

$$HX + NaOH \rightarrow X^- + Na^+ + H_2O$$

$$C_0, V_0 \quad C, V$$

$$[H^+][OH^-] = K_w$$

$$\frac{[H^+][X^-]}{[HX]} = K_a$$

Mass balance on HX,

$$[HX] + [X^-] = \frac{C_0 V_0}{V + V_0}$$

Mass balance on NaOH,

$[Na^+] = CV/(V + V_0)$

Charge balance,

$[Na^+] + [H^+] = [OH^-] + [X^-]$

Combining these equations results in the cubic

$$[H^+] = \frac{K_w}{[H^+]} - \frac{CV}{V + V_0} + \frac{K_a C_0 V_0}{(V + V_0)([H^+] + K_a)}$$

Hence $[H^+]$ as $f(V)$ or $f\left(\dfrac{CV}{C_0 V_0}\right)$ can be obtained.

If X^- absorbs,

$$\alpha_{X^-} = \frac{[X^-]}{C_0} = \frac{K_a}{K_a + [H^+]} \propto \text{Absorbance}$$

via the relation $A = \epsilon_{X^-}[X^-]l$

Since $[H^+] = \text{known } f(V)$

α_{X^-} or A can be obtained as $f(V)$.

Goddu and Hume, using the above relations, produced most interesting graphs of A vs. $CV/C_0 V_0$ (the fraction of acid titrated), one of which is shown in Fig. 25. Potentiometric titrations show a point of inflection only if $C_0 K > 3 \times 10^{-13}$; also if $C_0 K < 10^{-11}$ a considerable difference exists between equivalence and inflection points, the latter being hard to locate with certainty anyway. With photometric titration, however, $C_0 K$ values as low as 10^{-12} give good end-points, though a concentration limit of 10^{-5} M is reached in practice. Similar considerations must apply to non-aqueous solvents such as glacial acetic acid, butylamine, etc. if K values are known. Indeed the authors advocate addition of ethanol or dioxan to aqueous strong acid solutions to reduce the dielectric constant and with it the dissociation constant to yield an initial linear branch of reasonable slope. Mixtures of acids may be summed or resolved; their Figure 4 (Fig. 26) is self-explanatory and suggests how a weak acid with absorbing anion can act as indicator for a stronger acid. The titration of a mixture of p-nitrophenol and m-nitrophenol where $K_1/K_2 = 20$ only (Fig. 27) is remarkable; the feasibility limit of K_1/K_2 for a potentiometric titration is about 100. If two weak acids of very similar K values are titrated together, then if only one produces an absorbing

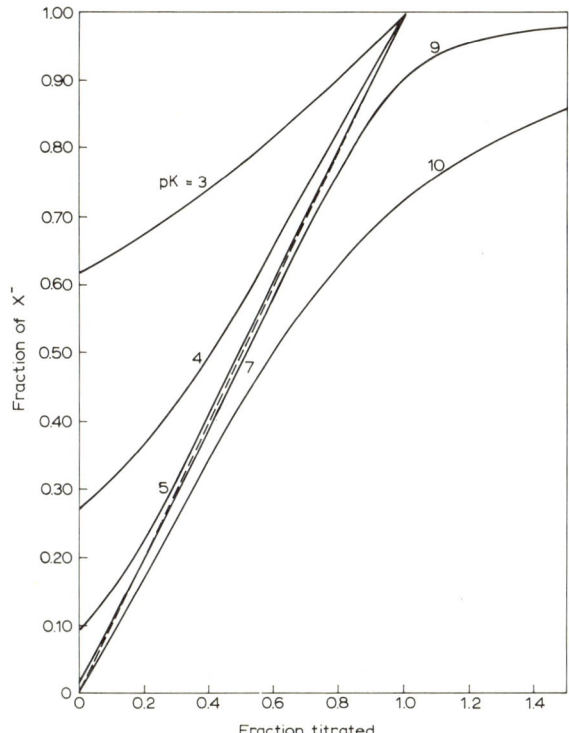

Fig. 25. Theoretical photometric titration curves for acids of various pK's at concentration = 10^{-3} N. Reprinted from ref. 75, by kind permission of the American Chemical Society.

anion the photometric titration curve is analogous to that of the absorbing acid alone but with diminished molar absorptivity. Thus, 2,4-dinitrophenol can effectively act as indicator for 10^{-4} M m-hydroxybenzoic acid. For medium strength acids of reasonable concentration, there seems little to choose between the potentiometric and photometric titration methods provided that the solvent system allows rapid electrode response. Photometric titrations come into their own in limiting situations such as the titration of 4×10^{-3} M m-nitrophenol, pK 8.3 (Fig. 28) or 10^{-3} M p-bromophenol, pK 9.2. Dissolved CO_2 or carbonate ion can readily be determined though this is a mixed blessing. Leonard and Henry minimised CO_2 problems in the analysis of telluric acid by using ammonia as titrant [76].

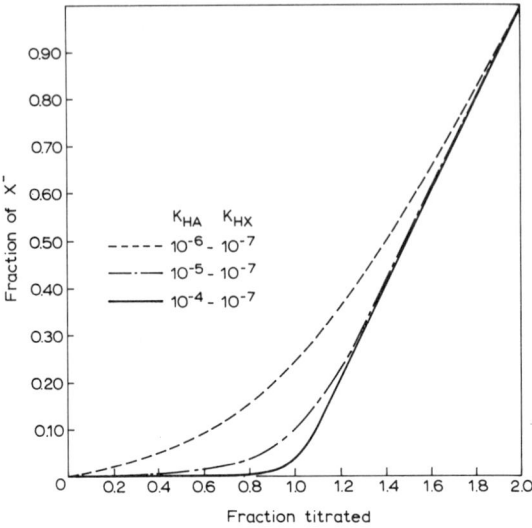

Fig. 26. Theoretical titration curve of two weak acids in the same solution where the titration is carried out at the wavelength at which the anion of the second acid absorbs all at a concentration of 10^{-3} M. Reprinted from ref. 75, by kind permission of the American Chemical Society.

Demonstration of the method's utility with regard to non-aqueous work was provided by a titration of 5×10^{-4} M p-toluidine in 1-butanol with 0.1 M aqueous $HClO_4$ at 289 nm [75].

McKinney and Reynolds (1958) [77] pursued the titration of 10^{-3} M phenols pKa (water) 5—11 in n-butylamine as solvent using NaOH in methanol as titrant. Phenate ions, which almost invariably show a bathochromic absorption wavelength shift from the parent phenols, were monitored. Many two-component mixtures were successfully resolved, but the use of aqueous pK_a values to predict the outcome of such an analysis proved unreliable. Variable CO_2 absorption limited C of V's to around 1%.

The titration of weak bases (mostly 10^{-2} M) in acetic acid or acetonitrile solvent using 0.5 M $HClO_4$ in acetic acid was examined by Hummelstedt and Hume (1960) [78]. The decrease in absorbance of the base-form peak was followed where possible to take advantage of the low scattered light and dilution errors of this type of curve. The use of weaker bases as indicators was examined. For glacial

254

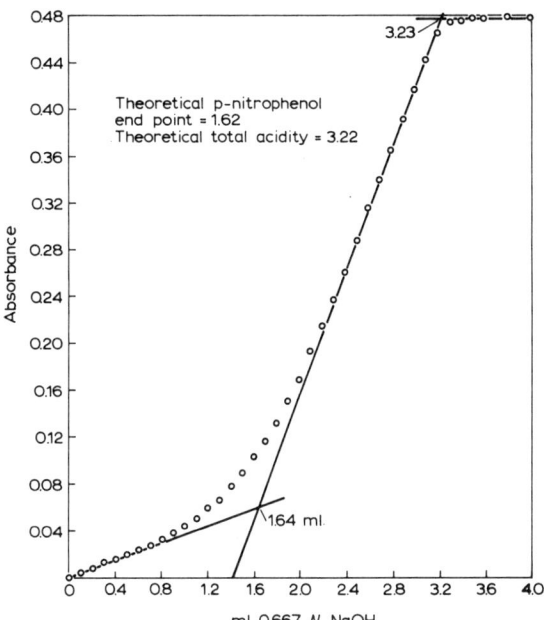

Fig. 27. Titration of a mixture of *p*- and *m*-nitrophenols. Total volume = 100 ml; $K_1/K_2 = 20$ only. Reprinted from ref. 75, by kind permission of the American Chemical Society.

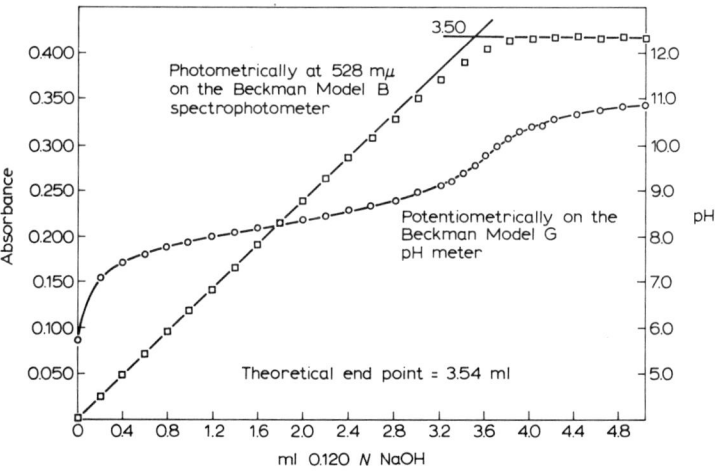

Fig. 28. Titration of 100 ml of 4.32×10^{-3} N *m*-nitrophenol under nitrogen. Comparison of potentiometric and photometric acid—base titration in a limiting situation. Reprinted from ref. 75, by kind permission of the American Chemical Society.

References pp. 380—389

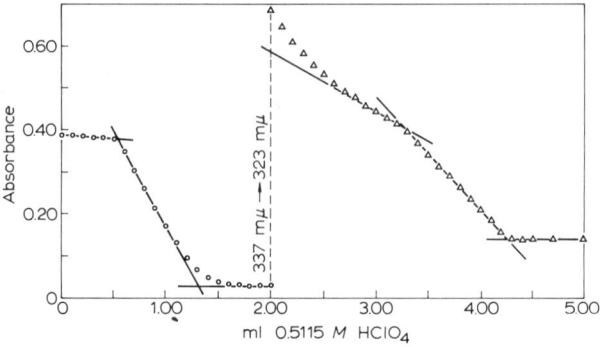

Fig. 29. Photometric titration of mixture of 0.283 meq. of di-*n*-butylamine, 0.409 meq. of *N,N*-diethylaniline, 1.014 meq. of aniline, and 0.509 meq of *o*-chloroaniline. ○, titrated at 337 mμ; Δ at 323 mμ. Reprinted from ref. 78, by kind permission of the American Chemical Society.

acetic acid as solvent, a (aqueous) pK_B limit of 13.5 was found (10^{-2} M solution) but for 5-nitro-1-naphthylamine, pK_B = 11.2, the concentration limit was 3.6×10^{-5} M. Mixtures of bases with pK_B (aq) values differing by 1.5 and sometimes only 0.8 could be resolved; Fig. 29 illustrates a four-component titration with a wavelength change half way through. Levelling solvents should naturally be avoided in such applications. In a later paper [79] these authors examined the titration of phenols in non-UV absorbing isopropyl alcohol with carbonate-free tetrabutylammonium hydroxide and showed that using photometric titrations a neutral solvent sufficed for many phenols normally requiring a strongly basic solvent for potentiometric titration. They demonstrated the formation of 1 : 1 hydrogen bonded dielectric constant-sensitive complexes between untitrated phenol and phenate ion around the mid point of the titration — a phenomenon noted also in potentiometric and conductimetric titrations, and the base-titrations mentioned above. This interaction produced curvature in the affected branch. Mixtures of phenols were differentiated; a four-component example is shown in Fig. 30 where again the mid-titration wavelength change technique was used. Recently Agarwal and Blake applied the foregoing theory most effectively to the analysis of phenobarbital-diphenylhydantoin mixtures [80].

256

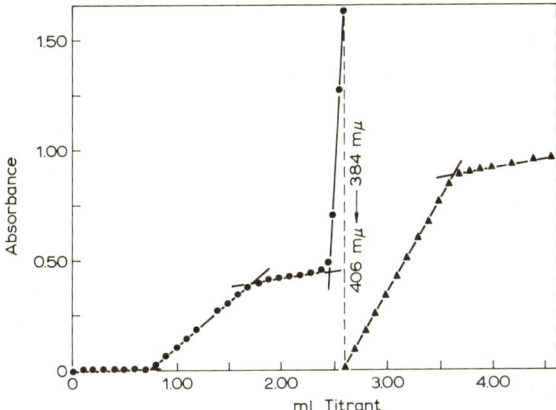

Fig. 30. Resolution of weak acids. Titration of 0.318 meq of diphenylphosphate, 0.418 meq. of pentachlorophenol, 0.291 meq. of 2,4,6-trichlorophenol and 0.462 meq. of 1-naphthol in 100 ml isopropyl alcohol with 0.4 M tetra-*n*-butylammonium hydroxide in 10% methanol in IPA. Note wavelength change following titration of the third component. Reprinted from L.E.I. Hummelstedt and D.N. Hume, Anal. Chem., 32 (1960) 1972, by kind permission of the American Chemical Society.

(B) INDICATED ACID—BASE TITRATIONS

The use made by Tingle (1918) [1] of the visual spectroscopic perception of the alkaline form absorption bands of methyl orange and phenolphthalein for the titration of acids in deeply coloured alkaloidal or copper sulphate solutions has been dealt with in the historical survey.

The first objective indicated acid—base photometric titration curve was that obtained by Müller and Partridge (1928) [5] for the titration of a strong base with strong acid using phenolphthalein as indicator. The plot of anode current against acid/base molar ratio yielded the typical titration curve shown in Fig. 3. pH values were accurately determined from acid form/base form ratios for suitable indicators.

For almost the next thirty years no real advance was made concerning the theory of these titrations. Inflection point indicators were used as in visual titrimetry, the large photocurrent change at the end-point being used either directly as an indication of equivalence or to operate a burette release electromagnetically. Comparison titrations involving the use of two-photocell instruments and comparison solutions accurately matching the equivalence-point composition of the solution did, however, yield excellent results; Ringbom and Sundman titrated carbonic acid by this means [7].

Goddu and Hume in their 1954 paper on self-indicating acid—base titrations [75] touched upon the proper use of indicators in photometric titration. They suggested that the indicator should not begin the colour change until very near the equivalence-point and that a comparatively large amount should be present to give a slope indication curve. Thus for strong acid (titrand)—strong base titrations they advocated the use of thymol blue ($pK = 9.0$) rather than bromothymol blue ($pK = 7.1$), the automatic removal of the indicator blank being useful in micro titrations.

Though for an indicator equilibrium

$$In_A \rightleftharpoons H^+ + In_B$$

$$[In_B] = \frac{K_{In}C_{In}}{[H^+] + K_{In}}$$

$$A = \epsilon_{In_B}[In_B] \cdot l$$

where K_{In} = indicator dissociation constant; C_{In} = total indicator concentration; A = absorbance if only In_B absorbs, and $[H^+]$ may be expressed as a function of the fraction of acid titrated, this approach seems never to have been pursued.

From 1956 the theory of acid—base photometric titrations has been dominated by T. Higuchi and his school. His highly successful approach has been to express the ratio of acid to base form of an indicator as a linear function of the volume of titrant added and the equilibrium constant for the indicator—titrand interaction using data obtained some distance from the equivalence-point.

(1) Type I plot [81]

Consider the titration of a weak acid HA with strong base. The indicator—titrand interaction may be written as the proton exchange process

$$A^- + IH^+ \rightleftharpoons HA + I$$

$$K_{ex} = \frac{[HA][I]}{[A^-][IH^+]} = \frac{K_I}{K_A} = K \tag{16}$$

where K_I and K_A are the indicator and titrand dissociation constants, respectively.

For $K = 0.1$ and monitoring of the absorbance of I, the following figure would be obtained

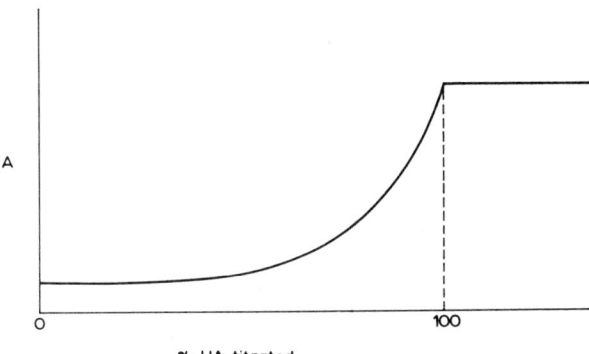

Fig. 31. Weak acid—strong base titration curve when $K = 10^{-1}$. $K = K_I/K_A$.

If the total acid present $\equiv S$ ml of standard base, and if X ml of base are added, then since $[HA] = S - X$; $[A^-] = X$, from eqn. (16)

$$K = \frac{[I]}{[IH^+]} \times \frac{S - X}{X} \tag{17}$$

This neglects hydrolysis of A^-; also K is a concentration constant which must vary somewhat with solution conditions, hence holding the ionic strength constant by introduction of a high inert electrolyte concentration is recommended.

This equation is useful when $K < 0.05$. Equation (17) may be re-arranged

$$XK \cdot \frac{[IH^+]}{[I]} = S - X \tag{18}$$

which becomes, when $X \to S$ near the equivalence-point

$$SK \cdot \frac{[IH^+]}{[I]} = S - X \tag{19}$$

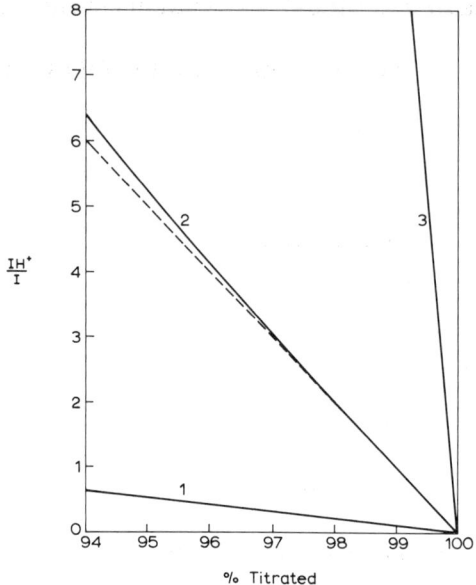

Fig. 32. Indicated acid—base titrations. The Higuchi Type 1 approach. Idealised plots of IH^+/I vs. per cent titrated. 1, $K = 0.1$; 2, $K = 0.01$; 3, $K = 0.001$. Reprinted from ref. 81, by kind permission of the American Chemical Society.

Hence a plot of $[IH^+]/[I]$ against X is linear with x-axis intercept $= S$, the equivalence titre. This approach is feasible only for small K values as indicated by Fig. 32, hence it is advantageous to select an indicator 2 or 3 pK units greater than the acid pK value. For $K = 0.001$, $[IH^+]/[I]$ may easily be determined to an accuracy of one part per 100; this corresponds to an equivalence-point accuracy of one part in 10^5.

(2) Type II plot [81]

For $K > 0.05$
From eqn. (17)

$$\frac{1}{X} = \frac{1}{S} + \frac{K}{S}\frac{[IH^+]}{[I]}$$ (20)

hence a plot of $[IH^+]/[I]$ against $1/X$ is linear with x-axis intersec-

260

tion $= S^{-1}$. K may be deduced from the slope $= S/K$

Note that $[IH^+]/[I] = (A_b - A)/(A - A_a)$ (21)

where A_b = absorbance of the indicator completely in its basic form. For the titration of weak bases with strong acid $[I]/[IH^+]$ is plotted against $1/X$. These methods probably show their greatest potential with non-aqueous titrations. Thus urea and other very weak bases whose titration is impossible by the more common procedures are easily determined by Type II plots in glacial acetic acid solution.

(3) Type III plot [82]

In yet another approach, indicators are used which are very weak in comparison with the titrand and absorbance information taken beyond the equivalence-point is utilised. Thus in the titration of a weak base the colour change occurs primarily after the addition of an equivalent of acid titrant. The indicator—titrand interaction

$$BH^+ + I \rightleftharpoons B + IH^+$$

is assumed to lie far to the left so that the concentration of BH^+ will not influence $[IH^+]/[I]$. After the equivalence-point

$$H_3O^+ + I \rightleftharpoons H_2O + IH^+$$

$$K = \frac{[I][H_3O^+]}{[IH^+]} = K_I$$ (22)

i.e. $[H_3O^+] = K_I \cdot \frac{[IH^+]}{[I]}$ (23)

If X = moles acid titrant added and S = moles of acid required to convert base to BH^+, excess acid present $= X - S$

$$\therefore X - S = K_I \cdot \frac{[IH^+]}{[I]} \cdot V$$ (24)

where V = volume of solution in vessel. Therefore a plot of $[IH^+]/[I]$ against X yields an intercept S. Only if the indicator is very weak will the colour change be governed by eqn. (23).

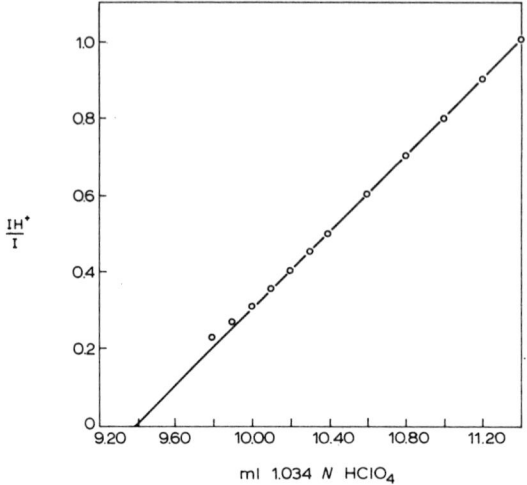

Fig. 33. The Higuchi Type III plot: $X - S = K_I \, [IH^+]/I \cdot V$. The titration of urea in acetic acid using Sudan III indicator. Reprinted from ref. 82, by kind permission of the American Chemical Society.

These titrations may be carried out equally well in glacial acetic acid solvent which need not be perfectly dry.

For the titration of very weak bases where solvolysis of the conjugate acid leads to curvature of the usual Type II plots, the modified Type II plot becomes useful [83].

Thus for

$$BHA + I \rightleftharpoons IHA + B \tag{25}$$

$$K_{ex} = \frac{C_{IHA} \cdot C_B}{C_I \cdot C_{BHA}} = \frac{K_f^{IHA}}{K_f^{BHA}} \tag{26}$$

where K_f's are the formation constants of the species involved.

If S' = total concentration of base present initially and X' = total concentration of acid added,

$$X' = C_{HA} + C_{BHA} \tag{27}$$

$$S' = C_B + C_{BHA} \tag{28}$$

Since $X' = XN/V$ and $S' = SN/V$

262

where N = normality of titrant, V = vol. of sample solution, from eqns. (26) (27) and (28).

$$X = S + \left(\frac{V}{NK_f^{IHA}}\right)\left(\frac{I_A}{I_B} - \frac{S}{(1 + (I_A/I_B)(1/K_{ex}))}\right) \qquad (29)$$

This equation cannot be used as it stands, therefore two weights of base S_1 and S_2 ($S_2 \gg S_1$) are titrated and values of X_1 and X_2 noted for similar values of $[I_B]/[I_A]$.

Then

$$\frac{1}{(X_2 - X_1)} = \frac{K_{ex}}{(S_2 - S_1)} \times \frac{[I_B]}{[I_A]} + \frac{1}{S_2 - S_1} \qquad (30)$$

since $V/(K_f^{IHA} \cdot N)$ is nearly identical for each.

Hence $S_2 - S_1$, determined from the intercept of a plot of $[In_B]/[In_A]$ against $1/X_2 - X_1$ may be related to the difference of the two samples taken. Again, contamination of glacial acetic acid with a little water or basic impurity is not important.

Since the equilateral hyperbola $x = y/(a + by)$ may also be written in the linear forms

$$1/x = a/y + b$$

$$x/y = -(bx)/a + 1/a$$

$$y/x = by + a$$

the Type II formula (determination of base B) may be used in the guises

$$\frac{1}{X} = \frac{K_{ex}}{S} \cdot \frac{[I_b]}{[I_a]} + \frac{1}{S} \qquad (31)$$

or

$$X \cdot \frac{[I_b]}{[I_a]} = \frac{S}{K_{ex}} - \frac{X}{K_{ex}} \qquad (32)$$

or

$$\frac{[I_a]}{[I_b]} = \frac{[I_a]}{[I_b]} \cdot \frac{S}{X} - K_{ex} \qquad (33) \, [84]$$

TABLE 3

Applied photometric acid—base titrations

Author(s) (date)	Compound	Titrant	Solvent	Notes: wavelength, indicator, etc.	Ref.
Goddu and Hume (1954)	m-Hydroxy benzoic acid	NaOH 0.1 N	Water	Indic. 2,4-dinitrophenol, similar pK value	75
	Bromo and nitro phenols; acetic acid—nitrophenol mixtures; HCl—nitrophenol mixtures	NaOH 0.1 N	Water	Self-indicating; λ max for anion	
	p-Nitrophenol	Sec. amyl amine 0.1 N	Water	Self-indicating	
	CrO_4^{2-}	0.1 N HCl	Water	425 nm, 370 nm; self-indicating	
	p-Toluidine	0.1 N HClO$_4$	n-Butanol	289 nm, self indicating	
Goddu and Hume (1954)	Borate	0.5 N HCl	Water	465 nm, p-nitrophenol	31
Reilley and Schweizer (1954)	o-Chloroaniline, quinoline, m-chloroaniline, acetate	0.1 N HClO$_4$ in gl. acetic acid	Glacial acetic acid	Self-indicating; 312 nm 350 nm, 314 nm, 312 nm; o-chloroaniline	85
Leonard and Henry (1956)	Telluric acid	2 M NH$_3$	Water	250—280 nm, self-indicating	76
Aronoff (1958)	Porphyrins	0.01 N HClO$_4$ in dioxan	Nitrobenzene	Absn. spec. recorded; p,p-dimethylaminoazo-benzene, p-aminonaphthol-benzene	86

Author (year)	Analyte	Titrant	Solvent	Notes	Ref.
McKinney and Reynolds (1958)	Wide range of phenols; some binary mixtures resolved	0.05 N NaOH in methanol	Butylamine	λ max of phenolate ion — self-indicating; pK_a's 5 to 11; 10^{-3} M or greater	77
Hummelstedt and Hume (1960)	Aniline derivatives	0.5 M $HClO_4$ in glacial acetic acid	Acetic acid or acetonitrile	Self-indicating, λ change during titration for resolution of 4-component mixtures	78
Hummelstedt and Hume (1960)	Various phenols	0.4 M Bu_4NOH in IPA	Isopropyl alcohol	Self-indicating; λ change during titration for resolution of 4-component mixtures	79
Agarwal and Blake (1969)	Phenobarbital	0.1 N Et_3BuNOH	Acetonitrile	Resolved from diphenylhydantoin; λ 293, 275 nm; self-indicating	80
Ringbom and Sundman (1939)	Weak acids and bases inc. CO_2	0.1 N HCl or NaOH; some titns. with weaker titrants	Water	Comparison titrations in 2-photocell Lange photometer; various indicators; acids down to pKa 10	7
Osborn et al. (1943)	Acids in dark resins	0.1 N KOH in ethanol	Water/ethanol	Thymol blue, 2-photocell instrument; sap. values determined	8
Nichols and Kindt (1950)	Weak acids and bases	0.1 to 0.01 N HCl, NaOH, NH_4OH	Water	2-photocell instrument, titrate to buffer condition; inflection indicators used	14
Higuchi et al. (1956)	Benzoic acid, triphenyl guanidine	1 N NaOH, 0.25 N $HClO_4$ in glac. acetic	Water, glac. acetic acid	Type 1 plots; bromothymol blue, quinaldine red	81
	Urea		Glacial acetic acid	Type II plot, malachite green indicator	

TABLE 3 (continued)

Author(s) (date)	Compound	Titrant	Solvent	Notes: wavelength, indicator, etc.	Ref.
Rehm and Higuchi (1957)	Acetate, benzoate, aniline, pyridine	1 N HCl	Water	Type III plots; metanil yellow	82
	Urea, antipyrine, N-Me pyrrolidone	0.1 N $HClO_4$ in glac. ac.	Glacial acetic acid	Sudan III	83
Connors and Higuchi (1960)	Very weak bases, some mixtures	0.1 N $HClO_4$ in gl. ac.	Glacial acetic acid	Modified Type II plots; various indicators	
Bruckenstein and Nelson (1961)	Weak acids and bases, mixtures	1 M NaOH, 1 M HCl	Water	If pK <9, accuracy 0.2 % at 0.1 M; Type II plots	87
Higuchi et al. (1962)	Amides and other very weak bases, e.g. ether	0.1 N $HClO_4$ in glac. acetic acid	Glacial acetic acid	Modified Type II plot; Sudan III indicator	88
Malmstadt and Vassolo (1959)	HNO_3 (or HCl) +H_2SO_4 mixtures	Bu_3MeNOH in benzene/methanol	Wet acetone	Neutral red, thymolphthalein, mixtures resolved; 575 nm	89
	Weak acids, imides	Bu_3MeNOH	Princ. acetone	Azo violet, 600 nm filters	
Tuthill et al. (1960)	Opium alkaloids	$HClO_4$ in gl. acetic acid	Glacial acetic acid	Malachite green; 620 nm	90
Karsten et al. (1960)	Benzoate, citrate phenobarbital	$HClO_4$ in gl. acetic acid	Glacial acetic acid	Methyl violet; 546 nm	347

266

Reference	Sample	Titrant	Solvent	Method/Indicator	Ref.
Ellert et al. (1960)	Weak org. bases, e.g. adrenalin, sulphaguanidine	$HClO_4$ in gl. acetic acid	Glacial acetic acid	Methyl violet; plot $\Delta A/\Delta V$ vs. V	348
Bruckenstein and Gracias (1962)	Weak acids, weak bases	1 M NaOH, 0.05 M $HClO_4$ in gl. acetic acid	Water, gl. acetic acid	Differential spectrophotometric technique; various indicators	91
Young et al. (1968)	"Acidity" of water	NaOH	Water	Phenolphthalein; fibre optics serial titration instrument	59
Schute (1970)	HCl—benzoic acid mixture	0.1 N $Ba(OH)_2$	Ethanol/water	Methyl orange; very theoretical paper concerning resolution of mixtures	92
Braid et al. (1966)	CO_2	Bu_4NOH in benzene/methanol	5 % ethanolamine in DMF	Thymolphthalein	93
Burns and Lawler (1963)	Hydrazine + 1,1-diMe hydrazine	$HClO_4$ in glacial acetic acid	Glac. acetic acid	Add salicylaldehyde—neutral hydrazine; crystal violet	94
Jasinski and Szponar (1965)	Weak bases	0.1 N p-toluene sulphonic acid in Ac_2O	Acetic anhydride	Higuchi plot, crystal violet, 625 nm; tropeolin II 496 nm	95
	Weak bases	0.1 N p-toluene sulphonic acid in Ac_2O	Propionic acid/anhydride	Malachite green	96
	Weak bases	$HClO_4$ in G—H	G—H solvent	Ethylene glycol best	97
Kelley (1965)	Fatty acids	0.01 N Bu_4NOH in methanol	Heptane/IPA	Micro scale, phenol red; acids extracted	98

TABLE 3 (continued)

Author(s) (date)	Compound	Titrant	Solvent	Notes: wavelength, indicator, etc.	Ref.
Kolling and Stevens (1962)	Weak bases	$HClO_4$ in 70 % Ac_2O in AcOH	Acetic anhydride/acetic acid	Rhodamine B; 470 nm. Hume and Higuchi plots used	99
Kreshkov and Vasil'ev (1964)	Weak bases	$HClO_4$ in glacial acetic acid	Glacial acetic acid	Self indicating; small pK difference acceptable	100
(1962)	Weak bases	$HClO_4$ in glacial acetic acid	Glacial acetic acid	Substituted anilines	101
Kreshkov (1965)	Nitro aminobenzoic acid isomers	NaOMe in acetone/methanol or $HClO_4$ in acetic acid	Gl. acetic acid or 2 : 1 acetone: methanol	Self-indicating, λ 420—600 nm	102
Kreshkov et al. (1967)	p-Subs. benzoic acid salts	HCl in IPA	Ethanol	Self-indicating	103
Paalman (1968)	Barbituric acid derivatives	0.05 N NaOH	Water or ethanol	Self-indicating; UV; automatic titration	104
Powers et al. (1958)	Mg and Al oxinates	$HClO_4$ in glacial acetic acid	Glacial acetic acid	450 nm; self-indicating best for 0.5—10 mg Al oxinate	105
Underwood and Howe (1962)	CO_2, HCO_3^- and CO_3^{2-}	1 M NaOH, 1 M HCl	Water	UV absorption of CO_3^{2-} used; mixtures resolved	106

Schuele et al. (1956)	BBr_3 and its addition compounds with organic bases; essentially titration of $HBr + H_3BO_3$	NaOH	Water	Methyl red or phenolphthalein	107
van Lingen (1969)	Carboxyl end-groups in polyethylene terephthalate	KOH in ethanol 0.03N	o-Cresol/chloroform	Bromophenol blue indicator; $\lambda = 602$ nm	108
Jagner (1970)	SO_4^{2-} in sea water	HCl 0.02 M aq.	Dimethyl sulphoxide	Bromocresol green indicator; 5-ml sample + 120 ml DMSO; $\lambda = 580$ nm	109
Merz (1970)	O in organic comp + metal via combuston and CO_2 titration	Bu_4NOH in isobutanol	DMF + monoethanolamine	Thymolphthalein indicator	110

To counteract solvolysis effects a difference titration can be performed using eqn. (32), i.e. plot $X[I_b]/[I_a]$ vs. X for two samples of widely differing size. Then plot $(X_2 - X_1)[(I_b)/(I_a)]$ vs. $X_2 - X_1$ to determine $S_2 - S_1$.

Table 3 summarises the practical use made of acid—base photometric titrations. Simple strong acid—strong base titrations used to test instruments are not included.

5. Photometric complexometric titrations

Though most photometric complex-forming titrations are based upon EDTA or similar chelating agents it must be remembered that early in the development of the method a few non-chelate forming interactions were studied by essentially photometric titration techniques, e.g. the starch—iodine [4,111], thiocyanate—iron(III) [113] and cobalt(II) [112], pararosaniline—palladium [114] and di-(p-biphenyl)-thiocarbazone—arsenical [115] reactions. In 1944 the ideas expressed in these papers were crystallised by Yoe and Jones in their study of a phenol—iron (III) complex [9]; Fig. 4, taken from their paper, is essentially representative of the first chelometric photometric titration. This mode of attack on the elucidation of complex compositions was later developed into the "slope ratio" method of Harvey and Manning (1950) [12]. Numerous complex formation studies have been carried out in later years but as these are not photometric titrations in the strict sense they will not be discussed further. However, one recent rather disquieting paper concerning complex formation between iron(III) and 7-bromo-1,3-dihydro-5-(2-pyridyl)-2H-1,4-benzodiazepin-2-one shows that the results obtained in a Yoe—Jones plot may depend on the concentrations used [116] (Fig. 34). A plot of the ratio C_{lig}/C_{Fe} at the break point against C_{Fe}^{-1} extrapolated to $C_{Fe}^{-1} = 0$ gave $C_{lig}/C_{Fe} = 2.8$.

The detailed study of photometric chelometric titrations requires some understanding of the basic theory of such methods. This is briefly summarised as follows.

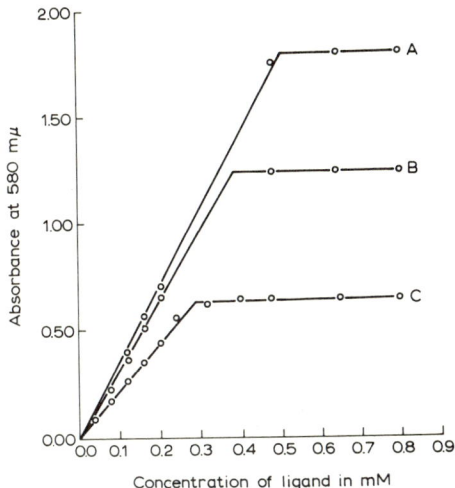

Fig. 34. Complex formation between iron(III) and 7-bromo-1,3-dihydro-5-(2-pyridyl)-2H 1,4-benzodiazepin-2-one as examined by a modified Yoe—Jones method. Effect of ligand A concentration on the complex. pH = 5. A. C_{Fe} = 10.8×10^{-5} M; $C_A : C_{Fe}$ = 4.6; B. C_{Fe} = 7.2×10^{-5} M; $C_A : C_{Fe}$ = 5.4; C. C_{Fe} = 3.6×10^{-5} M; $C_A : C_{Fe}$ = 8.1. Reprinted from ref. 116, by kind permission of the American Chemical Society.

The fundamental complex-forming reaction $M^{n+} + Y^{4-} \rightleftharpoons MY^{n-4}$ is beset by two principal competing side reactions; one in which secondary, usually monodentate, ligands attack the metal ion, and the other where protons attack the EDTA moiety, i.e.

$$M^{n+} \qquad + \qquad Y^{4-} \qquad \rightleftharpoons MY^{n-4}$$
$$\updownarrow \qquad\qquad \updownarrow$$
$$pL^{q-} \qquad\qquad mH^+$$
$$mL_p^{n-pq} \qquad\qquad H_m Y^{m-4}$$

$$K_{MY} = \frac{[MY^{n-4}]}{[M^{n+}][Y^{4-}]}$$

is the absolute stability constant.

$$K_{M'Y'}^{eff} = \frac{[MY^{n-4}]}{[M]'[Y]'}$$

where $[M]' = \sum_p [ML_p]$; $[Y]' = \sum_m [H_m Y]$.

$K_{M'Y'}^{eff}$ is the conditional or effective stability constant for the complex forming process taking account of the side reactions mentioned above.

If $\alpha_{Y(H)} = \dfrac{[Y]'}{[Y^{4-}]}$ and $\alpha_{M(L)} = \dfrac{[M]'}{[M^{n+}]}$

$$K_{M'Y'}^{eff} = \frac{K_{MY}}{(\alpha_{Y(H)} \cdot \alpha_{M(L)})}$$

where the α terms are referred to as side reaction coefficients and can be found from tables [119].

Ringbom has succinctly summarised the advantages offered by photometric detection of the end-point thus

(1) Photometric complexometric titrations can be carried out with a very high degree of accuracy; the relative error can often be kept considerably below 0.1%.

(2) Photometric titrations also can be performed with satisfactory accuracy in extremely dilute solution (concs. $< 10^{-5}$ M).

(3) Complexes of low stability giving a titration curve with a slope that is too small for visual titration can be successfully titrated photometrically.

(4) Indicators unsuitable for visual titrations can be successfully used for photometric titrations.

(5) Even strongly coloured and turbid solutions can be titrated.

(6) Light outside the visible range can be used.

The subdivision of chelometric photometric titrations, necessary for an ordered examination of the topic, is a difficult problem but one which has probably been most logically tackled by Flaschka [41].

Self-indicating titrations

(a) The simplest form of this sub-division is found when either the complex formed (MY) or the titrant (Y) absorbs strongly under the conditions used. Graphs of types a and b, respectively (Fig. 10) are produced. Only the equilibrium $M + Y \rightleftharpoons MY$ need be considered which is a great advantage when metal mixtures are analysed. Use of the near-UV soectral region can greatly increase the scope of this type of approach.

(b) For the situation in which no fundamental solution species absorbs, a reagent R, which can form an absorbing complex with the metal ion, may be added in large excess. Titration graphs then take the form shown in Fig. 10(c) provided that $K_{MR} \ll K_{MY}$ where K_{MR} is the formation constant of the reaction $M + R \rightleftharpoons MR$. The use of conditional constants is, of course, obligatory in all considerations of this nature.

(c) The chromogenic reagent added in excess may frequently be a conventional two-colour metallochromic indicator. In this case the destruction of the metallised form MI is followed always in the presence of a large excess of the metal-free form I. There is no theoretical difference between subgroups (b) and (c), I absorbs in the visible, R does not.

In many cases techniques (b) and (c) give excellent resolution of metal ion mixtures. Conditional K_{MY} values are influenced by the added reagent R or indicator I as well as by the more conventional secondary ligands such as NH_3 and if ϵ_{MR} values are sufficiently varied suitable resolution can often be obtained by manipulation of solution conditions. Flaschka has produced several excellent examples of this approach.

Slope indication

An alternative means of indicating the equivalence-point in a titration of ion M_1 with titrant Y, where $M_1 Y$ does not absorb, is by the addition of a further metal M_2 where $M_2 Y$ does absorb strongly. Provided that $K_{M_2 Y} \ll K_{M_1 Y}$ the satisfactory titration curve shown in Fig. 35 will result. Note that the concentration of M_2 may also be determined, hence metal ion mixtures which fortuitously contain suitable K_{MY} and ϵ_{MY} relationships may be analysed by this means. In this case, however, the category merges into division (a) for metal mixtures. Sometimes the above technique is referred to as the "Pilot ion" method.

Step indication

Here a conventional pM indicator is added at a concentration well below that of the metal being titrated. The resulting titration graph

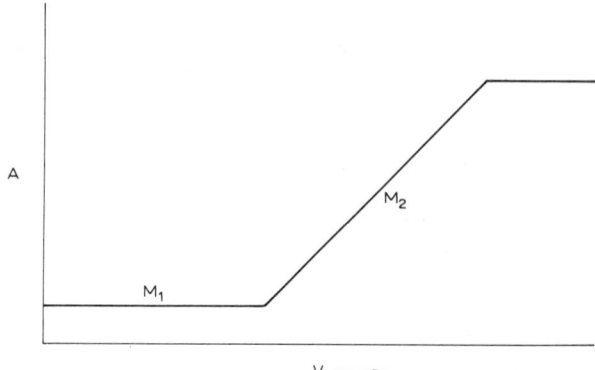

Fig. 35. Titration of metal ions M_1 and M_2 where $K^{eff}_{M_2Y} < K^{eff}_{M_1Y}$ and ϵ_{M_2Y} is high.

consists of two more or less horizontal portions connected by a steep section whose shape depends upon the relation between the pM for equivalence and indicator half-transition.

The borderlines between all these sub-divisions are diffuse. Thus step titrations merge into self-indicating type (b) titrations as the ratio of indicator to metal increases. Likewise, the addition of ion M_2 in a slope indication titration is little different from adding reagent in a type (b); it is simply a chromogen for the titrant rather than the metal.

The literature on complexometric photometric titrations largely divides itself between self-indicating and slope titrations, and "Indicator" titrations in which stress is always laid on the fraction of indicator in metallised or free form. This general division will be adhered to in the following review.

A further general tendency is to regard all titrations in which equivalence-points are located by linear intersections (not tangents) as of the slope type.

(A) SLOPE AND SELF-INDICATING EDTA TITRATIONS

The first EDTA self-indicating titrations were carried out by Sweetser and Bricker in 1953 [134]. Copper and nickel were titrated at pH 2.8, 745 nm and pH 4, 1000 nm, respectively to give graphs of the type shown in Fig. 36. Iron(III) was titrated via destruction of its salicylate complex (Fig. 37) and copper by the application of ex-

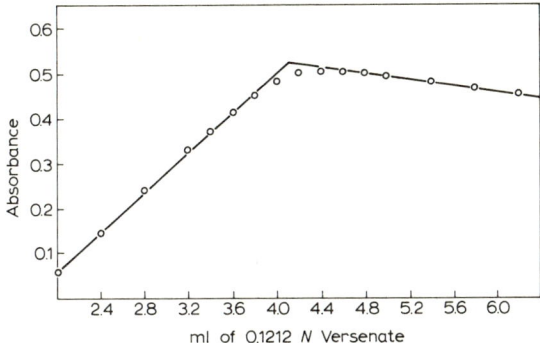

Fig. 36. The self-indicating titration of copper with EDTA at pH 2.8, λ = 745 nm. Interferences are few. Reprinted from ref. 134, by kind permission of the American Chemical Society.

cess ammonia (580 nm). The methods were applied to analysis of alloys. Note that in self-indicating systems concentrations must be compatible with the rather low ϵ values of Cu · EDTA etc. complexes. Foreign ion interference often manifests itself in poorly defined curves and unhelpful kinetics. In 1954 the above authors extended their idea to the UV region by titrating various metals at pH 10 using the selective absorption of the Y^{3-} entity in the 220 nm region. Ca/Mg and Cd/Zn mixtures were resolved. However, extraneous ab-

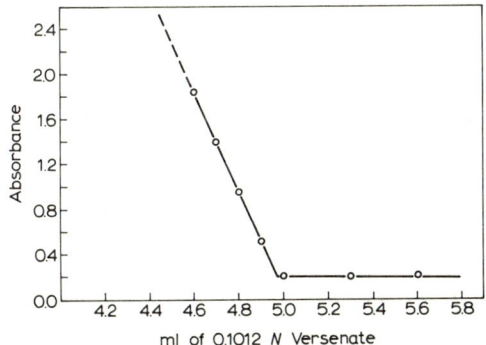

Fig. 37. Titration of iron(III) with EDTA followed via decomposition of the iron—salicylate complex. pH = 2.4; λ = 525 nm. Reprinted from ref. 134, by kind permission of the American Chemical Society.

References pp. 380—389

275

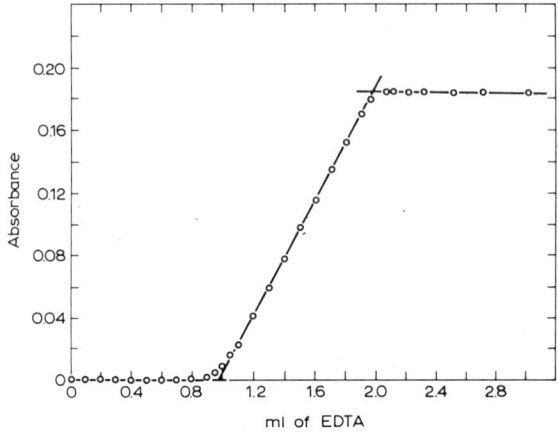

Fig. 38. Titration of an Iron(III)—Copper(II) mixture with 0.1 M ethylenediamine-tetraacetic acid (5.40 mg Fe, 6.36 mg Cu; 745 mμ). Reprinted from ref. 132, by kind permission of the American Chemical Society.

sorptions, common at this wavelength, could prove troublesome.

Underwood (1953) was first to resolve a metal ion mixture by photometric EDTA titration. Iron(III) and copper(II) were titrated sequentially at pH 2 (chloracetate buffer) and 745 nm to give Fig. 38. Conditional stability constants should be large and different (see later). Relevant spectra must be suitable with regard to λ_{max} and ϵ values. Kinetic effects may be troublesome. For the titration of iron(III) alone, copper(II) could be added as a "pilot ion" indicator.

Malmstadt and Gohrbandt (1954) utilised the high UV absorption of the Cu · EDTA complex at 290 nm to titrate thorium—copper mixtures at pH 3.1, either sequentially or by the addition of excess EDTA and back titration with copper. Fig. 39 shows the final destruction of the Cu · EDTA complex. Again, Wilhite and Underwood analysed mixtures of lead and bismuth using the facts that at pH 2 $K_{BiY}^{eff} \gg K_{PbY}^{eff}$ and at 240 nm $\epsilon_{PbY} > \epsilon_{BiY}$. Relevant absorption spectra and the titration graph are shown in Figs. 40 and 41. G. den Boef et al. have recently utilised the high UV absorbance of the lead—EDTA chelate in some interesting methods [120].

Metal ion mixtures can be resolved by using masking agents just as in visual titrations. A clear example lies in the titration of nickel in the presence of cobalt using nitroso R salt as mask [117].

276

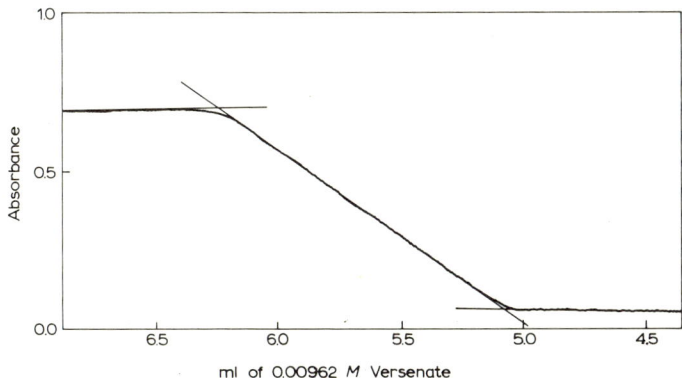

Fig. 39. Titration of thorium using decomposition of the Cu · EDTA complex for location of equivalence. pH = 3.1, λ = 290 nm. Reprinted from ref. 70, by kind permission of the American Chemical Society.

Probably the most lucid theory of complexometric self-indicating titrations was that produced by Flaschka in 1961 [118]. For the simple complexing reaction M + Y ⇌ MY

$$A = \sum_i \epsilon_i [i] \, l \tag{34}$$

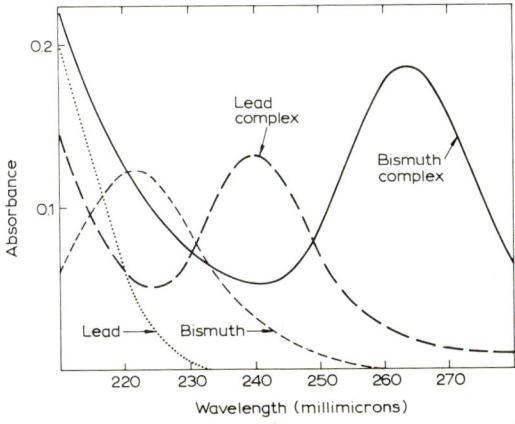

Fig. 40. Absorption spectra of bismuth and lead perchlorates and their ethylene-diaminetetraacetic acid complexes (2×10^{-5} M Bi and Pb in 10^{-2} M HClO$_4$ complexes at same concentration). Reprinted from ref. 136, by kind permission of the American Chemical Society.

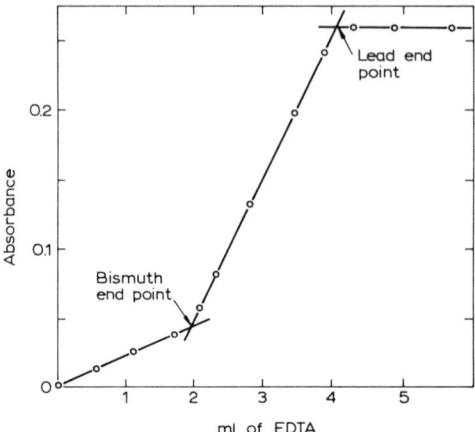

Fig. 41. Simultaneous titration of bismuth and lead (0.5 mg of Bi and 0.5 mg of Pb titrated at 240 nm). Reprinted from ref. 136, by kind permission of the American Chemical Society.

where i is a species present in the reaction.

Let $\epsilon_i \cdot l = k_i$

then $A = \Sigma k_i[\text{i}]$ (35)

Neglecting dilution effects,

$C_M = [\text{M}] + [\text{MY}]$ (36)

$C_Y = [\text{Y}] + [\text{MY}]$ (37)

$K = \dfrac{[\text{MY}]}{[\text{M}][\text{Y}]}$ (38)

where K is the conditional stability constant relevant to the existing conditions. From eqn. (35)

$A = [\text{M}]k_M + [\text{MY}]k_{MY} + [\text{Y}]k_Y$ (39)

Let $a = \dfrac{C_Y}{C_M}$, the titration factor (40)

From eqns. (36), (37), (39) and (40)

$A = C_M[k_{MY} + k_Y(a-1)] + (k_M + k_Y - k_{MY})[\text{M}]$ (41)

278

From eqns. (36)—(38) and (40)

$$[M] = \frac{[KC_M(1-a)-1] + \sqrt{[KC_M(1-a)-1]^2 + 4KC_M}}{2K} \qquad (42)$$

Since we now have $A = f([M])$ and $[M] = f(a)$, the titration curve $A = f(a)$ is available.

If $k_M = k_Y = 0$ which is common, eqn. (41) reduces to

$$A = k_{MY}(C_M - [M]) \qquad (43)$$

Figure 42 shows titration and pM curves for a likely situation where $K = 10^5$, $C_M = 10^{-2}$, $k_{MY} = 100$. It can be shown that the feasibility limit for a good titration is $KC_M = 50$, an extraordinarily low value, though to be fully effective k_{MY} must have a suitably high value. If dilution is taken into consideration

$$C_M = (C_M^0 V)/(V + b)$$

where b = mls of titrant of molarity m added; C_M^0 = initial metal ion

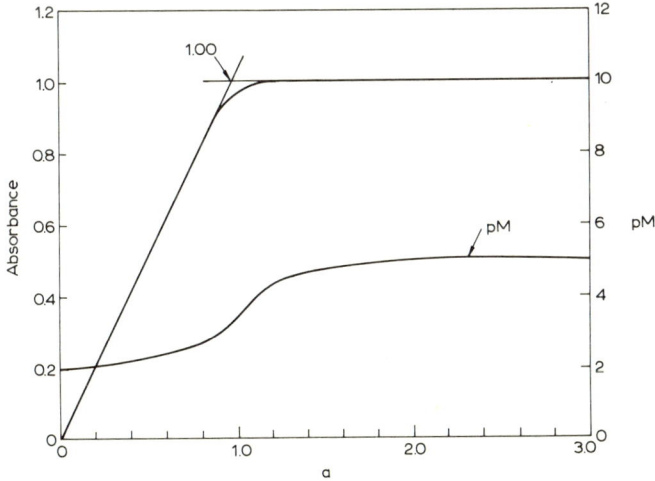

Fig. 42. Titration curve and pM plot for the determination of a single metal. $a = C_Y/C_M$, $K = 10^5$, $C_M = 10^{-2}$, $k_{MY} = 100$. Reprinted from ref. 118, by kind permission of the American Chemical Society.

concentration; C_M = "analytical" metal concentration at any stage; and V = initial solution volume.

Also $C_Y = \dfrac{m \cdot b}{V + b}$

$\therefore a = \dfrac{mb}{C_M^0 V}$

These modifications are independent of the influence of dilution upon A which should also be considered. For a relatively concentrated titrant, however, these dilution corrections may often be ignored, especially when it is born in mind that $K_{M'Y'}^{eff}$ can seldom be calculated exactly. Skytte-Jensen [58] evolved the relation

$$A = \frac{C_Y}{2}(\epsilon_{MY} - \epsilon_M) + \frac{C_M}{2}(\epsilon_M + \epsilon_{MY}) + C_M(\epsilon_M - \epsilon_{MY})\frac{\sqrt{B} - 1}{2K_Y C_M}$$

where $B = [1 + K_Y(C_Y - C_M)]^2 + 4K_Y C_M$.

At the equivalence-point where $C_Y = C_M$,

$$A = C_M(\epsilon_M - \epsilon_{MY})\frac{\sqrt{1 + 4K_Y C_M} - 1}{2K_Y C_M} + C_M \epsilon_{MY}$$

Also for $K_Y C_M \gg 1$,

$$A_{C_Y \to 0} = C_Y(\epsilon_{MY} - \epsilon_M) + C_M \epsilon_M$$

$$A_{C_Y \to \infty} = C_M \epsilon_{MY}$$

Therefore at the equivalence-point the absorbance axis deviation of the titration curve from the point of intersection of the branch projections

$$\Delta A = C_M(\epsilon_{MY} - \epsilon_M)\frac{\sqrt{1 + 4K_Y C_M} - 1}{2K_Y C_M}$$

With regard to the precision likely to result from such titrations, Ringbom [119] deduced the equation

$$\Delta a = \frac{1}{(1 - a)C_M K_{MY}}$$

280

This referred to the intersection of a line drawn through A at $a = 0$ and A at $a = 0.5$ with the A_{max} horizontal for an MY absorbing system. The minimum value of $C_M K_{MY}$ to yield a 1% error on this basis is 200.

Sequential titration of two metals [118] (Flaschka)

For a mixture of metal ions M and N, let

$K_M = [MY]/[M][Y]$, $K_N = [NY]/[N][Y]$

$C_M = [M] + [MY]$

$C_N = [N] + [NY]$

$C_Y = [Y] + [MY] + [NY]$

Let $Q = K_M/K_N$

It can be shown that

$$a = \frac{C_Y}{C_M} = \frac{1}{C_M} \cdot \frac{C_M - [M]}{[M]} \left(\frac{C_N}{Q + \dfrac{C_M - [M]}{[M]}} + [M] + \frac{1}{K_M} \right)$$

and if only NY absorbs,

$$A = k_{NY}[NY] = k_{NY} \left[QC_M - (C_M - [M]) - \frac{C_M - [M]}{[M]K_M} \right] \ {}^*$$

Figure 43 illustrates a typical limiting situation. Values of $Q \geqslant 100$ usually give acceptable resolution though this factor is sensitive to K^{eff} values and, as was mentioned above, side reaction coefficients are not accurately known in many cases. Ionic strength variation and the presence of acid and hydroxy complexes of reactants and product for which constants are not available can cause additional trouble.

Reilley and co-workers in 1961 produced an interesting paper

---

* It is difficult to see how this equation, quoted in Flaschka's text, is derived. The Author and Editor would suggest:

$$A = K_{NY}[NY] = k_{NY} \cdot \frac{C_N(C_M - [M])}{Q[M] + (C_M - [M])}$$

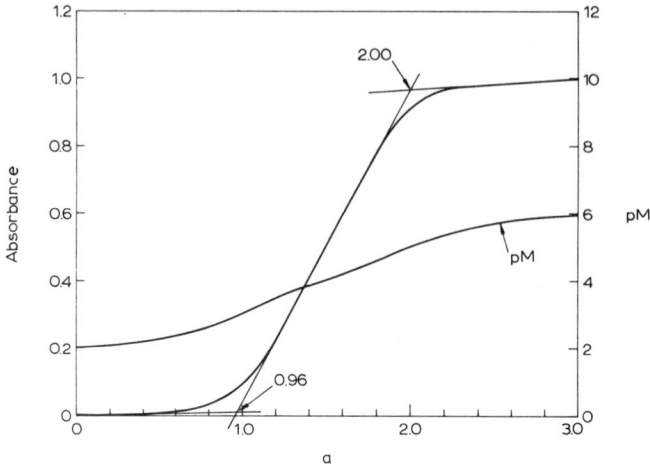

Fig. 43. Titration curve for the determination of two metals. $K_M = 10^6$, $K_N = 10^4$, $C_M = C_N = 10^{-2}$ M. $k_M = k_N = k_{MY} = 0$, $k_{NY} = 100$. Reprinted from ref. 118, by kind permission of the editor of Talanta.

describing how manipulation of $K_{Cu'Y'}^{eff}$ could cause Cu^{II} to function as a slope indicator for the resolution of binary mixtures, in particular Ca/Mg and Zn/Cd [121]. In particular they pointed out how $\Delta \log K^{eff}$ for two metals requires to be 5—6 for a successful visual titration, 4 for a potentiometric titration but only 2 for a good photometric titration.

Den Boef et al. have utilised the high UV absorbance of the lead—EDTA complex (ϵ = 10000 at 244 nm) for the titration of Ca and Ba at pH 12.8 and 13.7, respectively [120]. Generally speaking, for two complexing reactions

$$M + L \rightleftharpoons ML \text{ and } P + L \rightleftharpoons PL$$

where $K_{ML} > K_{PL}$, PL and possibly P absorb strongly,

$$K_{ML} = [ML]/[M][L] \quad K_{PL} = [PL]/[P][L]$$

$$[M] + [ML] = C_M$$

$$[L] + [ML] + [PL] = C_L = fC_M \quad [P] + [PL] = C_P$$

282

These combine to give

$$f = \frac{[\mathrm{PL}]}{C_{\mathrm{M}}} + \frac{[\mathrm{PL}]}{K_{\mathrm{PL}}C_{\mathrm{M}}(C_{\mathrm{P}} - [\mathrm{PL}])} + \frac{K_{\mathrm{ML}}[\mathrm{PL}]}{K_{\mathrm{PL}}C_{\mathrm{P}} + (K_{\mathrm{ML}} - K_{\mathrm{PL}})[\mathrm{PL}]}$$

and

$$f = \frac{C_{\mathrm{P}} - [\mathrm{P}]}{C_{\mathrm{M}}} + \frac{C_{\mathrm{P}} - [\mathrm{P}]}{K_{\mathrm{PL}}C_{\mathrm{M}}[\mathrm{P}]} + \frac{K_{\mathrm{ML}}(C_{\mathrm{P}} - [\mathrm{P}])}{K_{\mathrm{PL}}C_{\mathrm{P}} + (K_{\mathrm{ML}} - K_{\mathrm{PL}})(C_{\mathrm{P}} - [\mathrm{P}])}$$

In the titration of calcium, for example, $\mathrm{ML} = \mathrm{Ca} \cdot \mathrm{EDTA}$, $\mathrm{PL} = \mathrm{pB} \cdot \mathrm{EDTA}$. From these relations the photometric titration curve A vs. f can be calculated provided that ϵ values of P and PL are known. It can be shown that suitable titration curves are obtained when $K_{\mathrm{ML}}/K_{\mathrm{PL}} \geqslant 10^2$ and $K_{\mathrm{PL}}C_{\mathrm{M}} \geqslant 10^2$, irrespective of the value of C_{p}. The same conditions are valid for the titration of M with PL. The titration curve for calcium is shown in Fig. 44.

Ringbom's approach to the problem of two-metal titrations is one of elegant simplicity.

$$\alpha_{\mathrm{Y(N)}} = 1 + [\mathrm{N}]K_{\mathrm{NY}} \backsimeq [\mathrm{N}]K_{\mathrm{NY}}$$

where $\alpha_{\mathrm{Y(N)}}$ is the side reaction coefficient for the attack of N upon the titrant.

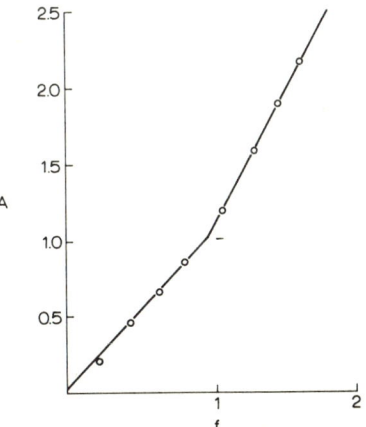

Fig. 44. Titration curve of $9 \cdot 10^{-5}$ M Ca with 10^{-2} M PbEDTA; log $K_{\mathrm{CaEDTA}} = 10.4$; log $K_{\mathrm{PbEDTA}} = 7.9$. Reprinted from ref. 120, by kind permission of Springer—Verlag, Heidelberg.

$$K_{MY'} = \frac{K_{MY}}{\alpha_{Y(N)}}$$

$$\therefore \log K_{MY'} = \Delta \log K + pN$$

(Note that $K_{MY'}$ increases with dilution.)

N only interferes in the pH range where $\alpha_{Y(N)} > \alpha_{Y(H)}$.

Theory given by Dean and Harris for metal ion indicator ampero-metric EDTA titrations where $K_{MY}/K_{NY} < 10$ is relevant also to photometric titrations [122].

G. den Boef and co-workers have utilised the effect of hydrogen ions released during chelate formation on acid—base indicators. Typical slope indication curves can be obtained [123].

The theory of self-indicating complexometric titrations utilising complexes of low stability has been investigated in detail by Fischer et al. [124].

(B) STEP (INDICATOR) EDTA TITRATIONS

The theory of photometric chelometric titrations of this type in which essentially the ratio of unmetallised to metallised indicator form at low C_I/C_M values is plotted as a function of titrant added has been approached in two distinct ways by Fortuin, Karsten and Kies, and Flaschka; and by Ringbom's school. The former group used as a basis the titration curve $a = f(\alpha)$ while the latter anticipated A_{eq} from calculated values of $[M]_{eq}$.

Remarkably early in the history of these titrations Fortuin et al. (1954) [17] developed a formula for the titration curve in the form $a = C_Y/C_M = f(\alpha)$ where α = fraction of free indicator,

$$\text{i.e. } a = \frac{C_Y}{C_M} = 1 - \frac{(1-\alpha)}{C_M}\left(C_I + \frac{1}{\alpha K_{MI}}\right) + \frac{\alpha K_{MI}}{K_{MY}}\left\{\frac{1}{1-\alpha} - \frac{1}{C_M}\left(C_I + \frac{1}{\alpha K_{MI}}\right)\right.$$

$$\alpha = \frac{[I]}{[MI]+[I]} = \frac{[I]}{C_I}; \quad A = (\epsilon_I - \epsilon_{MI})l\alpha C_I$$

For any titration α lies between the values

$$1/(1 + K_{MI}(C_M - C_I)) < \alpha < 1$$

For the usual case where $C_M \gg C_I$ and $K_{MY} \gg K_{MI}$, the equation reduces to

284

$$\frac{C_Y}{C_M} = 1 - \frac{(1-\alpha)}{C_M}\left(C_I + \frac{1}{\alpha K_{MI}}\right)$$

There followed a series of graphs showing α vs. a for situations likely to be met in titrations of this type. Note that

$$\frac{d\alpha}{da} = \frac{C_M}{C_I + \dfrac{1}{\alpha^2 K_{MI}} + \dfrac{K_{MI}}{K_{MY}}\left[\dfrac{C_M}{(1-\alpha)^2} - C_I\right]}$$

Figure 45 shows a graph of the locus of the maxima of $d\alpha/da$ vs. a (i.e. points of inflection of α vs. a) for various ratios of K_{MY} to K_{MI}. To plot A vs. V substitute $a = (MV/V_0)/(1/C_M)$. Fortuin and co-workers summarised their paper by suggesting that for a very accurate photometric chelometric titration

$$\log K_{M'Y'}^{eff}/K_{M'I'}^{eff} > 4$$

$$\log K_{M'I'}^{eff} > 4$$

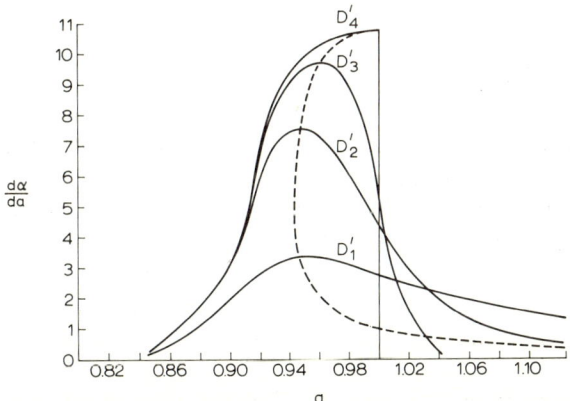

Fig. 45. Plot of $d\alpha/da$ against a for an indicated complexometric titration. Variation with K_{MY}, $\alpha = [I]/C_I$, $a = C_Y/C_M$. - - - - - - is the locus of the maxima. $C_M = 3.3 \times 10^{-3}$; $C_I = 3 \times 10^{-4}$; $K_{MI} = 1.25 \times 10^5$; $K_{MY} = $ (for D_1') 1.25×10^6; (for D_2') 1.25×10^7; (for D_3') 1.25×10^8; (for D_4') 1.25×10^{12}. Reprinted from ref. 17.

C_I should be as low and C_M as high as possible. The log K limits now seem rather pessimistic; they are suitable for a visual titration (see ref. 125).

Flaschka and Khalafallah in 1957 produced a paper [125] on the theory of visual end-point determination in complexometric titration in the course of which this equation relating a and $\phi = [MI]/C_I$ was derived

$$a = \frac{C_Y}{C_M} = \frac{1}{C_M}\left(C_M - \phi C_I - \frac{\phi}{1-\phi}\cdot\frac{1}{K_{MI}}\right)\left(\frac{1-\phi}{\phi}\cdot\frac{K_{MI}}{K_{MY}} + 1\right)$$

or

$$a = \frac{1}{C_M}\left(C_M - (1-\alpha)C_I - \frac{(1-\alpha)}{\alpha}\cdot\frac{1}{K_{MI}}\right)\left(\frac{\alpha}{1-\alpha}\cdot\frac{K_{MI}}{K_{MY}} + 1\right)$$

A series of figures showing a vs. ϕ for various C_M values at K_{MY}, K_{MI} and C_I values fixed for one figure but varied among figures followed.

Skytte-Jensen (1966) [58] has extended Flaschka's reasoning specifically to the photometric titration situation. His equation

$$\delta = \frac{C_Y - C_M}{C_M} = \frac{\alpha}{1-\alpha}\cdot\frac{K_{MI}}{K_{MY}} - \frac{1-\alpha}{\alpha}\cdot\frac{1}{C_M K_{MI}} - \frac{1}{C_M K_{MY}}$$

$$+ \frac{C_I}{C_M}\left\{\alpha\left(1 - \frac{K_{MI}}{K_{MY}}\right) - 1\right\}$$

where δ = relative titration error leads to the representative titration curves shown in Figs. 46, 47 and 48.

Skytte-Jensen further shows that satisfactory break points will appear at $\alpha \simeq 0$ if $C_M K_{MY} > 10 C_M K_{MI} > 10^4 C_M > 10^2 C_I$ and at $\alpha \simeq 1$ if $C_M K_{MY} > 10^3 C_M K_{MI} > 10^4 C_M K_{MY}/C_I K_{MI} > 10^2$. The last requirement shows that in some cases useful titration curves will be produced for the slope indication system where $C_I > C_M$ as shown in Fig. 49.

Ringbom [119] begins from the premise that for any actual titration curve of A vs. V,

$$A_{eq} = \frac{(A_I^{max} + A_{MI}^{max}[M]_{eq}K_{MI})}{(1 + [M]_{eq}K_{MI})}$$

286

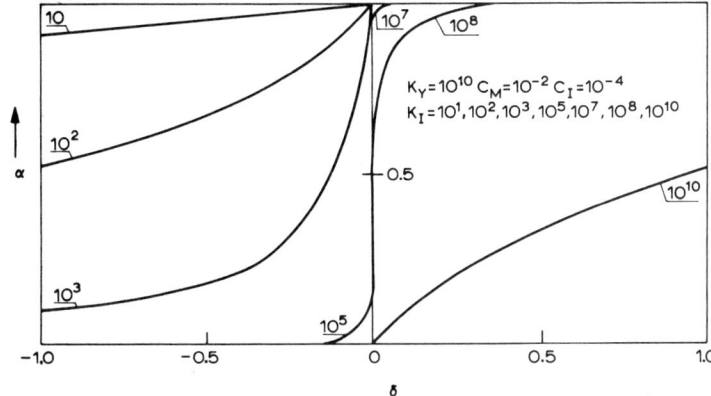

Fig. 46. Plot of $\alpha = [I]/C_I$ against $\delta = (C_Y - C_M)/C_M$ for various values of K_{MI}. Reprinted from ref. 58, by kind permission of the Radiometer Company, Copenhagen.

where $[M]_{eq}$ is calculated from $[M]_{eq} = (C_M/K_{MY})^{1/2}$.

The A^{max} values are apparent from the following titration curves which also illustrate clearly Ringbom's classification of indicators into low sensitivity ($pM_{tr} < pM_{eq}$, $pM_{infl} < pM_{eq}$), medium sensitivity ($pM_{tr} = pM_{eq} = pM_{infl}$) and high sensitivity ($pM_{tr} > pM_{eq}$, $pM_{infl} > pM_{eq}$) types. See Fig. 50. For a low sensitivity indicator no indicator

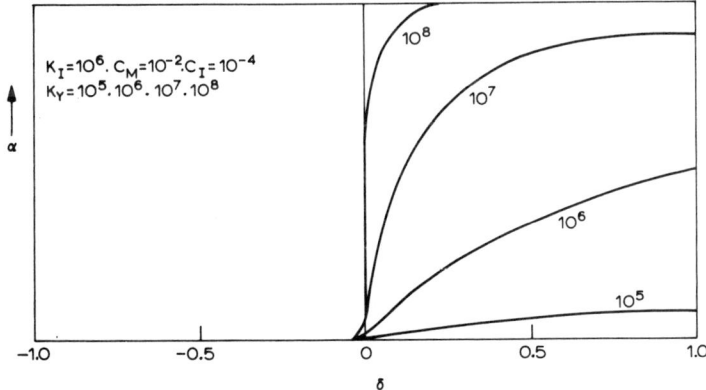

Fig. 47. Plot of α against δ for various values of K_{MY}. Reprinted from ref. 58, by kind permission of the Radiometer Company, Copenhagen.

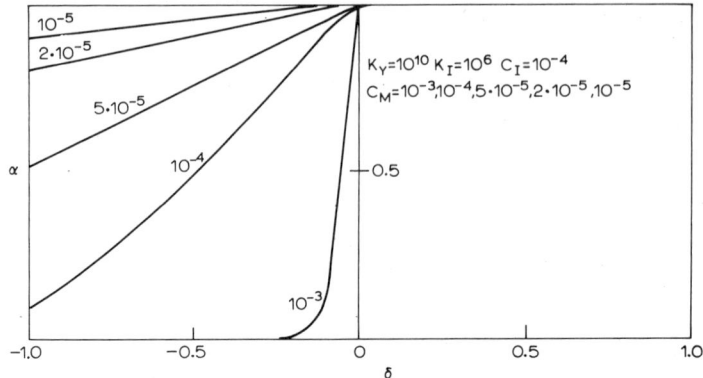

Fig. 48. Plot of α against δ for various values of C_M. Reprinted from ref. 58 by kind permission of the Radiometer Company, Copenhagen.

correction is necessary as little MI complex remains at the break point. A medium sensitivity indicator requires a correction of $0.5\,C_I$ and a high sensitivity indicator one of C_I to allow for MI remaining at the equivalence-point. For intermediate situations the equation $[MI]_{eq} = C_I((A_{eq} - A_I^{max})/(A_{MI}^{max} - A_I^{max}))$ can be used. Ringbom was first to study the likely precision of indicated photometric titrations [119]. For a pM determination by absorbance measurements

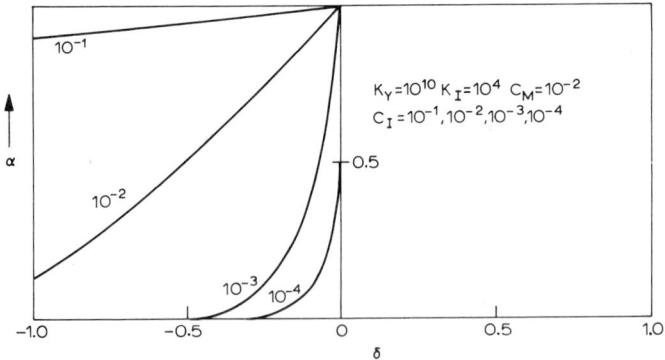

Fig. 49. Plot of α against δ for various values of C_I. Reprinted from ref. 58 by kind permission of the Radiometer Company, Copenhagen.

288

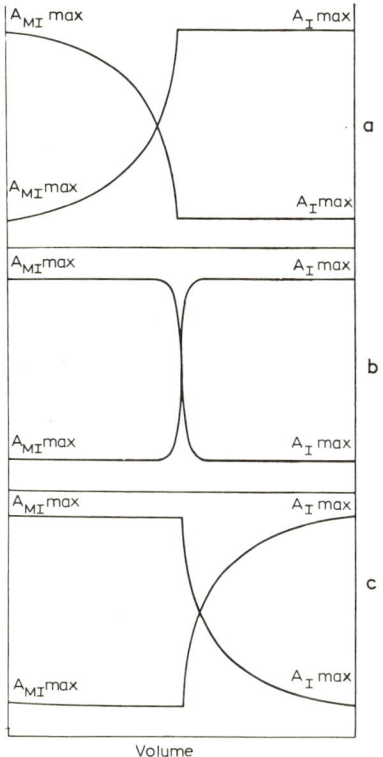

Fig. 50. Ringbom's classification of metallochromic indicators into (a) low sensitivity; (b) medium sensitivity and (c) high sensitivity types. Reprinted from ref. 119.

on a metallochromic indicator,

$$\Delta pM = -\frac{0.188(1 + [M]K_{MI})}{(A - A_{MI}^{max}) \cdot T \cdot [M]K_{MI}} \cdot \Delta T$$

where ΔT is the smallest change in transmittance that can be measured. Application of the calculated ΔpM for $[M]_{eq}$ to Ringbom's log $C_M K_{MY}$ error curve yields the likely error of the titration.

If $[M]_{eq}$ is uncalculable because $K_{M'Y'}^{eff}$ is not known, then for an asymmetric titration curve of type intermediate between those shown in the above figure the correct course is to take the inflection point as the equivalence-point and apply an indicator correction given by

$$[MI]_{infl} = C_I \frac{(A_{infl} - A_I^{max})}{(A_{MI}^{max} - A_I^{max})}$$

The graphical methods suggested by Higuchi's school for acid—base titrations may be applied to the complexometric case as long as $C_M \gg C_I$. If only the uncomplexed indicator absorbs light

$$(A_I^{max} - A)/A = [MI]/[I]$$

and this ratio is plotted against titrant volume. If K_{MI}/K_{MY} is large (>0.05) the plot is against V^{-1}.

Still and Ringbom (1965) extended the above theory [126] by noting that since

$$[M]_{infl} = (C_M/K_{MY}K_{MI})^{1/3},$$

$$[M]_{tr} = \frac{1}{K_{MI}} \quad \text{and} \quad [M]_{eq} = (C_M/K_{MY})^{1/2}$$

$$[M]_{infl}^3 = [M]_{tr}[M]_{eq}^2$$

or

$$3pM_{infl} = pM_{tr} + 2pM_{eq}$$

If in a titration the inflection point is taken as the equivalence-point the error

$$\Delta pM = pM_{infl} - pM_{eq} = \tfrac{1}{3}(pM_{tr} - pM_{eq})$$

Since $[M]_{eq}/[M]_{infl} = (C_M K_{MI}^2/K_{MY})^{1/6}$

$$\Delta pM = 1/6 \log (C_M K_{MI}^2/K_{MY})$$

The percentage titration error can then be read from a Ringbom error plot. A symmetric titration curve results when $\Delta pM = 0$, i.e. $K_{MI} = (K_{MY}/C_M)^{1/2}$. A most useful graph showing the percentage colour change of the indicator at the inflection and equivalence points together with ΔpM as a function of $\log C_M K_{MI}^2/K_{MY}$ is shown in Fig. 51. Note that ΔpM may be determined even if K_{MI} is un-

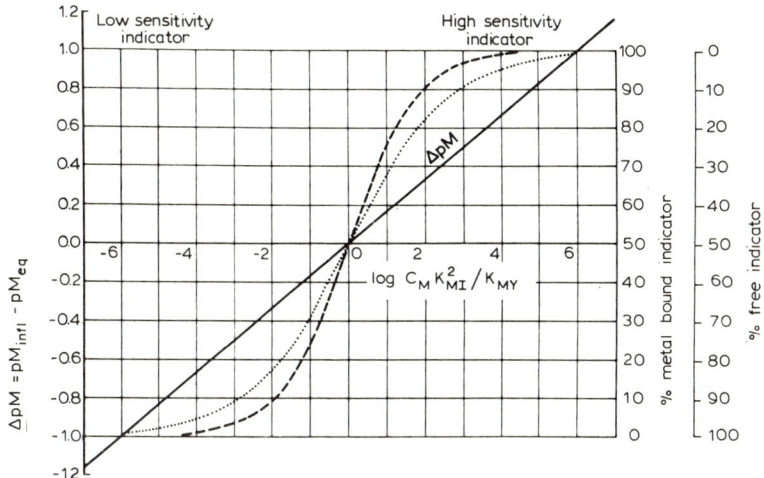

Fig. 51. Reprinted from ref. 126.

known if the percentage indicator colour change at the point of inflection of the titration curve is found.

Still and Ringbom also point out that the inflection point of a transmittance vs. volume plot lies at the same volume as the inflection of an absorbance plot if the absorbances involved are low.

Reilley and Schmid in their theory of visual complexometric titrations [127] used the concepts Δ_1 and Δ_2 defined by $\Delta_1 = \log K_{MI} + \log C_M$ (a measure of the indicator's complex forming ability) and $\Delta_2 = \log K_{MY} - \log K_{MI}$ (a measure of the titrant's ability to strip metal from the indicator).

$$\Delta_1 - \Delta_2 = \log (C_M K_{MI}^2/K_{MY}) = 6\Delta pM$$

and is therefore the abscissa of Fig. 51.
Reilley's titration equation is

$$a = \frac{C_Y}{C_M} = \left(\frac{1-\alpha}{\alpha}\right) \cdot \frac{1}{10^{\Delta_1}} + \left(\frac{\alpha}{1-\alpha}\right) \frac{1}{10^{\Delta_2}} - \frac{C_I}{10^{\Delta_2}C_M} [1 + 10^{\Delta_2}(1-\alpha)]$$

and a series of titration curves showing $100\,\alpha$ vs. a for various combinations of Δ_1 and Δ_2 is shown in Fig. 52.

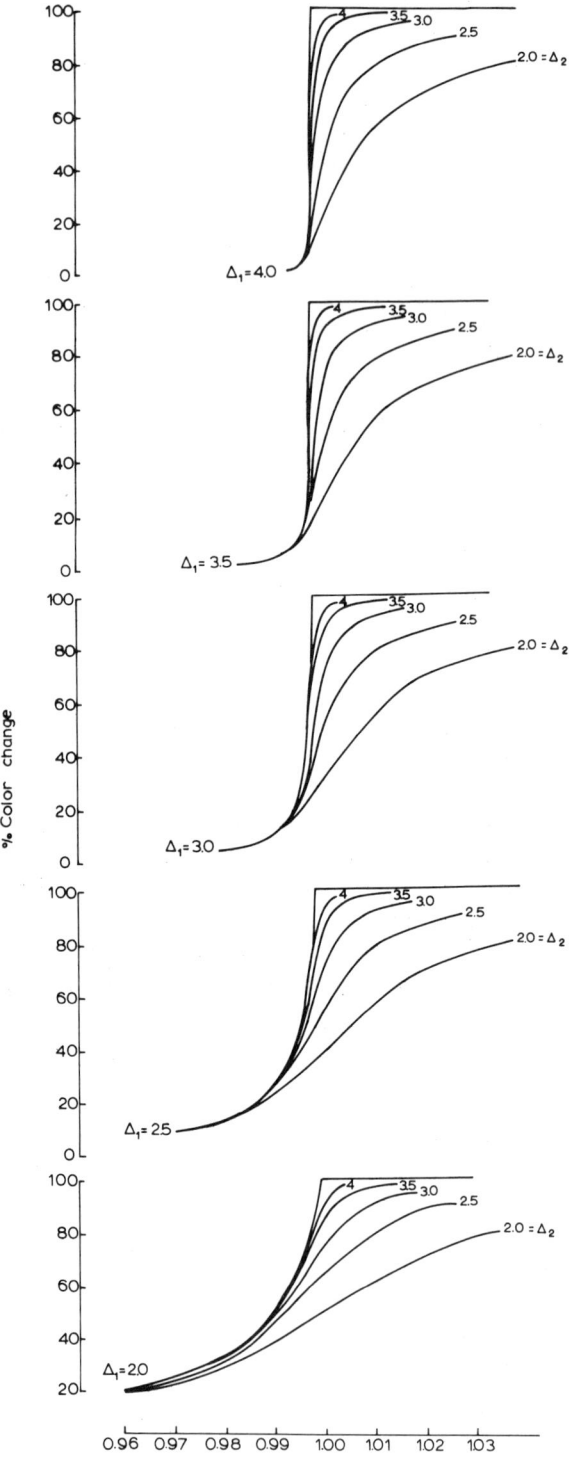

Fig. 52. Plot of α against a for various Δ_1 and Δ_2 values. $\Delta_1 = \log K_{\mathrm{MI}} + \log C_{\mathrm{M}}$; $\Delta_2 = \log K_{\mathrm{MY}} - \log K_{\mathrm{MI}}$. Reprinted from ref. 127 by kind permission of the American Chemical Society.

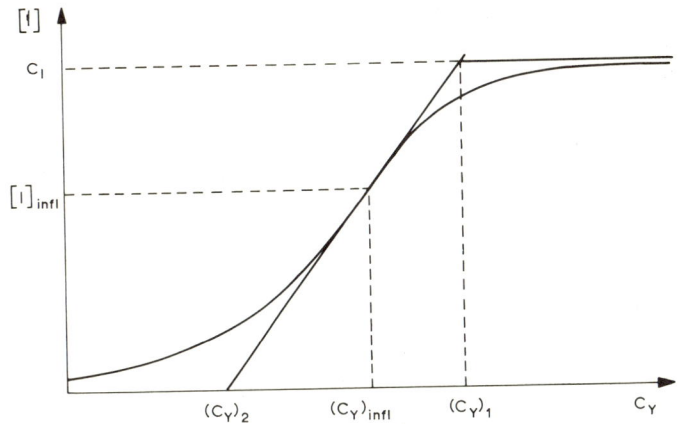

Fig. 53. Selection of the equivalence-point. Plot of [I] against C_Y. Reprinted from ref. 128 by kind permission of the Finnish Chemical Society.

Still (1968) [128], in a paper discussing the errors involved in taking as the equivalence-point the intersection of the tangent to the [I] vs. C_Y curve at the point of inflection with the [I] = 0 and [I] = C_I axes (Fig. 53), deduces the titration equation

$$C_Y = C_M - C_I + \frac{1}{K_{MI}} - \frac{1}{K_{MY}} + [I]\left(1 - \frac{K_{MI}}{K_{MY}}\right) - \frac{C_I}{K_{MI}[I]} + \frac{C_M K_{MI}[I]}{K_{MY}(C_I - [I])}$$

The relative errors are

$$\frac{(C_Y)_1 - C_M}{C_M} = \frac{1}{\sqrt{C_M K_{MY}}}\left(10^{\Delta pM} + 3 \cdot 10^{\frac{\Delta pM}{3}}\right) - \frac{1}{C_M K_{MY}} - \frac{C_I K_{MI}}{C_M K_{MY}}$$

$$\frac{(C_Y)_2 - C_M}{C_M} = -\frac{1}{\sqrt{C_M K_{MY}}}\left(10^{-\Delta pM} + 3 \cdot 10^{\frac{-\Delta pM}{3}}\right) - \frac{1}{C_M K_{MY}} - \frac{C_I}{C_M}$$

where $\Delta pM = pM_{tr} - pM_{eq} = \log K_{MI} - \frac{1}{2}(\log K_{MY} - \log C_M)$
The titration error is smaller when the end-point is determined by

extrapolating the transmittance curve than by extrapolating the corresponding absorbance curve provided that the titration leads to a solution of higher absorbance.

Kotrly, in a rather complex paper [129], has examined the course of a photometric titration curve when the indicator forms a stepwise system of complexes with the metal ion MI, MI_2, MI_3 ... . For an MI, MI_2, MI_3 system, for example,

$$A = \epsilon_{11}[MI] + \epsilon_{12}[ML_2] + \epsilon_{13}[MI_3] + \epsilon_{01}[I]$$

where ϵ values refer to the appropriate indicator species which are assumed to be the only absorbers in the system.

In terms of [I] only,

$$A = \epsilon_{01}[I] + (C_I - [I]) \cdot \frac{\epsilon_{11}K_1 + \epsilon_{12}K_2[I] + \epsilon_{13}K_3[I]^2}{K_1 + 2K_2[I] + 3K_3[I]^2}$$

where K values are overall formation constants for the indicator—metal complexes,

e.g. $K_3 = \dfrac{[MI_3]}{[M][I]^3}$

Introduction of $\alpha = [I]/C_I$ yields a relation $A = f(\alpha)$.

Kotrly's equation for $a = f(\alpha)$ is

$$a = 1 - \left\{ \frac{C_I}{nC_M} \cdot (1-\alpha) \left[1 + \frac{3(C_I\alpha)^{-1} + K_2 C_I \alpha + 2K_1}{3K_3 C_I^2 \alpha^2 + 2K_2 C_I \alpha + K_1} \right] + \frac{1}{C_M K_{MY}} \right\}$$

$$+ \frac{K_1}{K_{MY}} \cdot C_I \alpha \left[\frac{1}{C_I(1-\alpha)} - \frac{1}{C_M} \right] + \frac{K_2}{K_{MY}} (C_I\alpha)^2 \left[\frac{2}{C_I(1-\alpha)} - \frac{1}{C_M} \right]$$

$$+ \frac{K_3}{K_{MY}} (C_I\alpha)^3 \left[\frac{3}{C_I(1-\alpha)} - \frac{1}{C_M} \right]$$

Hence graphs of $A = f(a)$ can be constructed. Particularly interesting are plots of species percentage vs. a for various situations derived from the equation

$$[MI_{(3-i)}] = \frac{K_{3-i}(C_I\alpha)^{(2-i)}}{3K_3 C_I^2 \alpha^2 + 2K_2 C_I \alpha + K_1} \cdot (1-\alpha)C_I$$

294

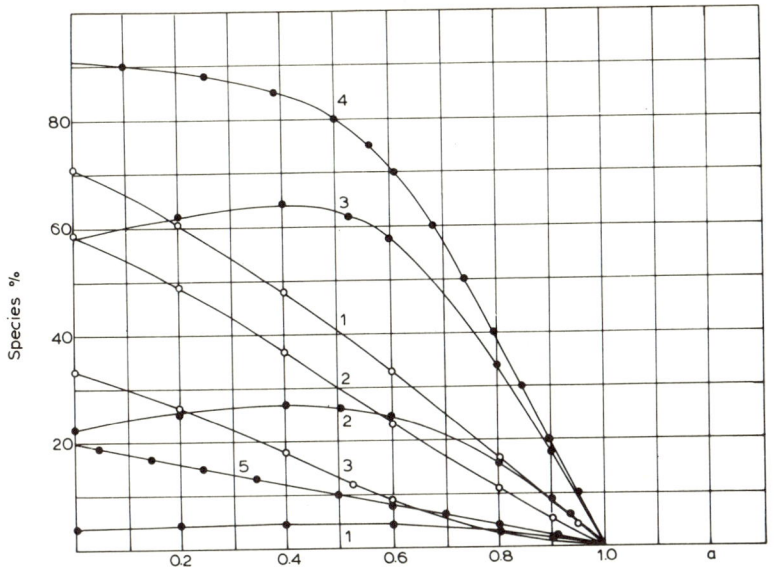

Fig. 54. Effect of the stability of the complex MI_2 on curves for $MI = f(a)$ and $MI_2 = f(a)$ for $C_I = 10^{-5}$, $K_1 = 10^6$, $C_M = 10^{-5}$. $\bigcirc$, $[MI]/C_I = f(a)$; $\bullet$, $2[MI_2]/C_I = f(a)$. Values of K_2: (1) 10^{10}; (2) 10^{11}; (3) 10^{12}; (4) $K_1 = K_2 = 10^{12}$; (5) $K_1 = 10^6$; $K_2 = 10^{12}$, $C_I = 10^{-4}$. Reprinted from ref. 129.

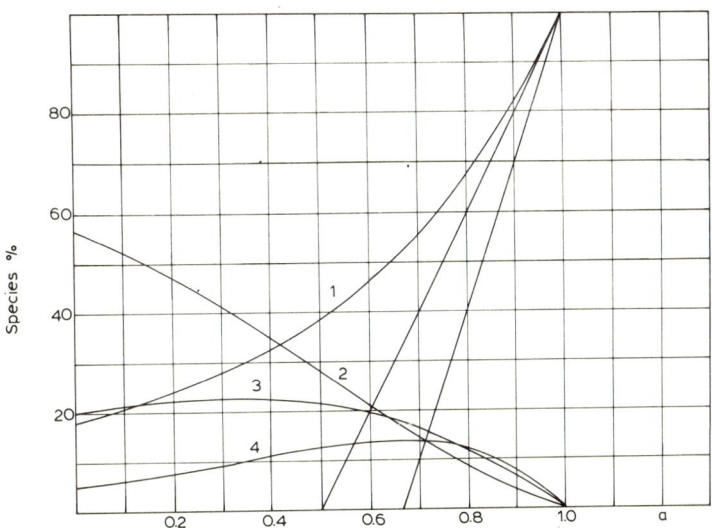

Fig. 55. Distribution of the indicator into its ionic species during titration for $C_I = 10^{-5}$; $C_M = 10^{-5}$; $K_1 = 10^6$; $K_2 = 10^{11}$; $K_3 = 10^{16}$. (1) $\alpha = f(a)$; (2) $[MI]/C_I = f(a)$; (3) $2[MI_2]/C_I = f(a)$. (4) $3[MI_3]/C_I = f(a)$. Reprinted from ref. 129.

TABLE 4

Bibliography of photometric complexometric titrations using EDTA or related compounds

Metals determined	Author(s) (date)	Titra-tion type [a]	Indicator	General conditions	Ref.
Fe^{III}, Cu^{II}	Underwood (1953)	2	Cu^{II}	pH 2 (chloracetate); 745 nm; applied to Al alloys	132
Bi, (Cu^{II})	Underwood (1954)	1b 2	Cu^{II}	Thiourea complex; pH 1.5—2.4; heat solution to 70° for 10 min; 400 nm; pH 1.5—2.4; 745 nm; use thiourea for low Bi concentrations	133
Fe^{III}	Sweetser and Bricker (1953)	1b 1a 1a	—	Salicyclic acid; 525 nm; pH 1.7—2.3 745 nm; pH 2.4—2.8 (acetate) 1000 nm; pH 4.0 (acetate)	134
Cu^{II} Ni					
Mg, Ca, Zn, Cd	Sweetser and Bricker (1954)	1a 1a 1a 1b	—	Absorbance of HY^{3-} at 222—228 nm; pH 10 (NH_3/NH_4^+) NaOH/citrate; 234 nm NaOH/KCN; 236 nm Fe^{III}—salicylate back titration; pH 4; 525 nm	135
Ca/Mg Cd/Zn Zr					
Bi, Pb	Wilhite and Underwood (1955)	1a	—	Bi only — pH ≃ 2, 265 nm. Bi + Pb (resolved); pH 2, 240 nm	136
Ni/Co	Brake et al. (1957)	1a	—	Nitroso R salt mask; pH 5—7; heat, add HNO_3, titrate at pH 3.5—4, 1000 nm	117
Ca/Mg	Reilley et al. (1961)	2	Cu^{II}	pH 10; EGTA; 550 nm	121

Analyte	Author	Type	Indicator	Conditions	Ref.
Zn, Cd		2	Cu^{II}	pH 10; DTPA; 700 nm	120
Ca, Ba	Den Boef et al. (1970)	2	Pb^{II}	242 nm; pH(Ca) 12.8, pH(Ba) 13.7; (NaOH)	
Ca	Kibrick et al. (1952)	1c	Murexide	In blood, pH 11.6 (borate—NaOH); 580 nm	137
Ca/Mg	Fales (1953)	1c	Murexide	In serum, 620 nm; dilute NaOH	138
Ca	Chalmers (1954)	3	Murexide	pH 10 (dil NaOH); 610 nm	139
Cu	Ringbom and Vanninen (1954)	3	Murexide	pH 5.5 (HCO_3^-/CO_3^{2-}); blue filter	140
Mg		3	Erio. bl. T.	pH 10; red filter	
Cu	Lane and Fritz (1957)	3	Naphthyl azoxine	pH 6; 452, 440, 500 nm	141
Rare earths/ UO_2^{2+}		3	Arsenazo	pH 6 (pyridine); pH 7—8 (triethanolamine); 570 nm, expanded scale technique, for very dilute solution	
Ag	Gedansky and Gordon (1957)	1c or 3	Murexide	435 nm; dil NH_3 via $2\,Ag^+ + Ni(CN)_4^{2-} \rightarrow 2\,Ag(CN)_2^- + Ni^{2+}$	142
Fe^{III}	Hedrick (1965)	1b	SCN^-	pH 2 (trichloroacetate)	211
Mg	Bauditz (1967)	1c	Pyrocatechol violet	pH 10 ($NH_3—NH_4^+$); 644 nm Cu, Fe masked with CN^-	143
Ca		1c	Murexide	Dil. NaOH; 509 nm (in biological fluids)	
Bi	Ramaiah et al. (1968)	1b	Fe^{III} salicylate	pH 0.5; 520 nm	144
Al	Belcher et al. (1969)	3	Xylenol org.	4.20 μg of metal; pH $\sim$ 6 (hexamine); Zn^{2+} back titration of excess EDTA; 550 nm	145
Ba		3	Methylthymol blue	Dil. NaOH soln; 570 nm	
Bi		3	PAN	pH $\simeq$ 2; 550 nm	

TABLE 4 (continued)

Metals determined	Author(s) (date)	Titration type [a]	Indicator	General conditions	Ref.
Cd		3	Xylenol org.	pH ≅ 6 (hexamine); 570 nm	
Ca		3	Methylthymol blue	Dil. NaOH; 570 nm	
Co		3	Xylenol org.	pH ≅ 6 (hexamine); 570 nm	
Cu^{II}		3	Pyrocat. violet	pH ≅ 6 (hexamine); 570 nm	
Mg		3	Erio. bl. T.	pH 10; >650 nm	
Mn		3	Erio. bl. T.	Replacement titration using Mg—EDTA; pH 10; >650 nm	
Hg		3	Erio. bl. T.	Replacement using Mg—EDTA; pH 10; >650 nm	
Ni		3	Murexide	Dil NH$_3$; 440 nm	
Zn		3	Xylenol org.	pH ≅ 6 (hexamine); 570 nm	
Ca, Mg	Rees and Hill (1969)	1c	Sol. bl. WDFA	pH 10; 650 nm; mixture resolved	212
Ca, Mg	Flaschka and Sawyer (1962)	1c	Calmagite	pH 10, 622 nm, mixture resolved at 1 μg level or below	146
Cd, Zn	Flaschka and Carley (1964)	1c	Murexide	pH 10 (NH$_3$— NH$_4^+$); 450 nm; EGTA titrant; mixture resolved	147
Ca	Skytte-Jensen (1966)	3	Calmagite	pH 11	58
Mn, Ca	Skrifvars and Ringbom (1966)	3	Metalphthalein	DTPA titrant; pH (Mn) 9.3, Ca 10; green filter; mixture resolved	148
Mn, Mg		3	Metalphthalein	pH (Mn) = 9; pH (Mg) = 10; DTPA titrant; green filter; mixture resolved	

					Ref.
Ca, Mg		3	Metalphthalein	EGTA titrant; pH 10; green filter; mixture resolved	
Mn, Ca, Mg		3	Metalphthalein	DTPA, EGTA used in conjunction; green filter; mixture resolved; pH $\simeq$ 10	
Zn, Mg		3	Erio. bl. T.	520 nm; pH 9.7; mixture resolved	
Ba	Cohen and Gordon (1956)	3	Metalphthalein	pH 11; 570 nm	149
Ca, Mg	Conradi (1967)	3	Calcein erio. bl. T.	pH 12; 506 nm [$Mg(OH)_2$] (Ca + Mg) pH 10 (PO_4^{3-}, Fe^{III} etc., int. removed)	150
Th	Datta and Saha (1962)	3	β-SNADNS	pH 2.5; 590 nm	151
Ca	Fedorov et al. (1967)	3	Murexide	Dilute NaOH	152
Ca/Mg Mg	Flaschka and Ganchoff (1961)	3	Murexide Erio. bl. T.	EGTA titrant; pH 10; 490 nm EDTA; pH 10; 622 nm (consecutive titrations)	153
Cu(Ni)	Flaschka and Soliman (1957)	1a	—	Triethylene tetramine titrant; pH 12; highly selective	154
Co	Flaschka and Ganchoff (1961)	3	Catechol violet	Co^{III}–EDTA–H_2O_2 complex formed in alkali; back titration with Bi; pH 0.5–2	155
Cd, Zn	Flaschka and Butcher (1964)	3	Zincon	pH 9.5; 622 nm; c_{NH_3} 0.01–0.03 M; EGTA titrant; mixture resolved; Cd, Ca, Zn resolution possible	156
Cd/Zn	Flaschka and Ganchoff (1962)	2	$Cu(NH_3)_4^{2+}$	EGTA titrant; pH 10; 742 nm	157

TABLE 4 (continued)

Metals determined	Author(s) (date)	Titra- tion type [a]	Indicator	General conditions	Ref.
Mg, Ca, Sr, Ba	Headridge and Magee (1958)	3 1a	Erio. bl. T. —	pH 10; 630 nm pH 10; 225 nm	61
Zn	Hunter and Miller (1956)	3	Erio. bl. T.	pH 9—10; 665 nm (good in presence of KCN + chloral)	158
Zn/Cd	Ideno and Hozumi (1968)	3	Xylenol org.	pH 4.4—5.6 (acetate); 562 nm (Cd masked with I^-)	159
Ca, Mg Ca Mg	Karsten et al. (1955)	3	Metalphthalein murexide Erio bl. T.	pH 10; 570—575 nm pH 12; 580 nm pH 9—10; 546 nm; mixture resolved	160
Pb	Kotrly (1964)	1c—3	Dithizone	pH 5; 50 % EtOH; 590 nm	161
Ca, Mg	Lacy (1963)	1c	Erio. bl. T.	pH 10; red filter; mixture resolved	162
Zn/Mn	Levine and Golden (1967)	3	Erio. bl. T.	pH 8—10; 660 nm; mask Zn with KCN, demask with HCHO	163
Th	Malmstadt and Gohrbandt (1954)	2	Cu^{II}	pH 3.1 (acetate); 290 nm	164
Ca Mg	Malmstadt and Hadjiioannou (1958)	3	Calcon Erio. bl. T.	pH 12.5; 650 nm pH 10 (Cu + Mg); 650 nm (applied to dolomite, etc.)	165
Th	Menis et al. (1958)	3	Quercetin	pH 3.0; 422 nm (automatic titration)	166
Zr	Milner and Edwards (1955)	1b	Salicylate	pH 3—7 (back titration excess EDTA with Fe^{III}); 480 nm	167
Zr		1b	Benzohydroxamate	pH 1.8—3.3 (Fe^{III} back titration) 520 nm	

Element	Type	Author (year)	Indicator	Conditions	Ref.
Cd Cd	1c	Nechiporenko et al. (1969)	PAR Xylenol org.	pH 10.2; 495 nm pH $\simeq$ 6 (hexamine); 575 nm	168
Ca	2	Ramaiah and Vishnu (1957)	$Cu(NH_3)_4^{2+}$	630 nm; 2 M NH_3	169
Ca Mg	3	Rauterberg and Ossenberg-Neuhaus (1968)	Murexide Erio. bl. T.	pH 12; 578 nm pH 10; 546 nm; Fe^{3+}, Al^{3+}, Mn^{2+} PO_4^{3-} int. elim.	170
Ba	3	Gordon et al. (1956)	Erio. bl. T.	pH 11; 650 nm 15 % methanol	349
Ca	3	Shapiro and Brannock (1955)	Murexide	pH 12 (dilute NaOH) green and orange filters	171
Mg	3		Erio. bl. T.	pH 10 (precipitate Cu^{2+} with tungstate); orange filter (analysis of carbonate rocks)	
Ca Mg	3	Wharton and Chapman (1964)	Erio blue SE Calmagite	pH 12; 610 nm pH 10 (sum); 610 nm	172
Ca Mg	3	Zak et al. (1956)	Erio. bl. T.	pH 10; 660 nm pH 10; 660 nm (Ca separated as Ca Ox)	173
Bi	3	Suk and Miketukova (1958)	Pyrocatechol-violet	pH 2–3, yellow filter; Fe int. elim.	350
Mg	3	Robinson and Rathbun (1959)	Erio. bl. T.	pH 10.4; 660 nm (precipitate Ca as oxalate)	351
Ca			Erio. bl. T.	pH 10.4; 660 nm (sum)	
Zn	3 mod.	Still (1965)	Dithizone	pH = 6.7–7.0; 50 ml $CHCl_3$, 10 ml H_2O	174
Pb			Dithizone	pH = 7.0, 50 ml $CHCl_3$, 10 ml H_2O	
Cd			Dithizone	pH = 6.5, 50 ml $CHCl_3$, 10 ml H_2O	

[a] 1a, self-indication (slope), no reagent; 1b, slope, excess reagent; 1c, slope, excess indicator; 2, slope, metal ion indicator; 3, step indication, C_I/C_M low (see introduction). M_1/M_2 signifies M_1 determined in presence of M_2.

The change in shape of the titration curve with variation in wavelength of the light used if the various metallised indicator species have different absorption characteristics is well explored. In particular, if

$$\epsilon_{13} = \epsilon_{01} + \epsilon_{12}$$

the final portion of the titration curve leading into the equivalence-point will be linear provided that only species MI_2 and MI_3 are involved. Increase in indicator concentration helps to achieve this condition. The MI_n formation problem has also been investigated by Kragten [131].

Two-phase extractive titrations are sometimes used for the determination of metal ion solutions of low concentration with EDTA. The only application of a photometric end-point to such systems has been by Still (1965) [130] in which Zn, Pb and Cd were titrated with EDTA using dithizone as indicator in water/$CHCl_3$ or water/CCl_4 mixtures. pM_{trans} values at which the indicator is distributed equally between the phases can be very high in extraction systems, hence the feasibility of titrating very low concentration solutions, $pM_{eq} = \frac{1}{2}(\log K_{MY} - \log [MY])$.

(C) COMPLEXOMETRIC TITRATIONS INVOLVING A PHOTOMETRIC REAGENT AS TITRANT (ONE PHASE)

In view of the success of Sweetser and Bricker's self-indicating titration of metal ions using the UV absorption of the HY^{3-} species (1954) [135] it seems odd that the first photometric titration using one of the many available organic reagents was not accomplished until 1958 (Marple et al. [175]). The basic theory of these titrations is the same as that of self-indicating titrations already discussed; the difference is that in these titrations the (usually rather high) absorbance of the complex is due to excitation of reagent molecular orbitals rather than excitation of inner atomic orbitals of the metal ion, which had largely been the case hitherto.

The only paper to deal thoroughly with the theory of these titrations was that of Le Goff and Trémillon (1964) [176] which described the titration of suitable transition metal ions with dithizone in a 2 : 1 acetone : water solvent. This reagent

has two replaceable protons and forms complexes with metal ions
thus

$H_2D_z + M^+ \rightleftharpoons H^+ + MHD_z$ (primary)

$MHD_z + M^+ \rightleftharpoons H^+ + M_2D_z$ (secondary)

$2H_2D_z + M^{2+} \rightleftharpoons 2H^+ + M(HD_z)_2$ (primary)

$M(HD_z)_2 + M^{2+} \rightleftharpoons 2H^+ + 2MD_z$ (secondary)

Primary and secondary dithizonates are formed, the latter only with
difficulty. HD_z^- shows one absorption peak at 475 nm, H_2D_z shows
almost equal peaks at 450 and 610 nm. Most dithizone–metal complexes peak in the 490–530 region; those of Cu^{2+}, Co^{2+} and Zn^{2+} retain considerable absorption at 610 nm. For the case where only
H_2D_z absorbs appreciably, titration curves were of the form shown in
Fig. 10(b), though for ions which form complexes absorbing appreciably at 610 nm an initial positive slope is evident. Le Goff and
Trémillon's equation in A and x where x = total concentration of
dithizone added was

$$A^2\left[1 + \frac{K_A}{a_{H^+}}\right] - l \cdot \epsilon_{H_2D_z} \cdot x \cdot A + l\epsilon_{H_2D_z}\left[K_c + C_0 + \frac{K_A}{K_c} \cdot a_{H^+} + A\right]$$

$$- l^2(\epsilon_{H_2D_z})^2 \frac{K_c}{K_A} \cdot a_{H^+} \cdot x = 0$$

where $K_A = \dfrac{[H^+] \cdot C_{HD_z^-}}{C_{H_2D_z}}$, $K_c = \dfrac{C_{M^+} \cdot C_{HD_z^-}}{C_{MHD_z}}$

and C_0 = concentration of total metal ion present.

The equation is that of a hyperbola in which the directions of the
asymptotes are given by the solutions of the equation

$$A^2 \left[1 + \frac{K_A}{a_{H^+}} \right] = l \cdot \epsilon_{H_2D_z} \cdot x \cdot A$$

i.e. $\dfrac{dA}{dx} = 0$ and $\dfrac{dA}{dx} = \dfrac{l\epsilon_{H_2D_z}}{1 + \dfrac{K_A}{a_{H^+}}}$.

The equation of the rising asymptote is

$$A = \frac{l\epsilon_{H_2D_z}}{1 + \dfrac{K_A}{a_{H^+}}} (x - C_0)$$

The gradient is thus controlled by a_{H^+}. With decreasing a_{H^+} the gradient decreases but so does the degree of curvature at the equivalence-point. The absorbance at the equivalence-point is given by

$$A_{eq} = \frac{1}{2} l\epsilon_{H_2D_z} \cdot \frac{K_c}{K_A} \cdot a_{H^+} \left[\left(1 + \frac{4C_0}{1 + \dfrac{K_A}{a_{H^+}}} \right)^{1/2} - 1 \right]$$

The authors showed titration curves obtained with all the metals examined using as titrant 4×10^{-4} M dithizone in $2 : 1$ acetone : water.
Feasible lower pH limits were

Ag^+, 0; Hg_2^{2+}, 0; Hg^{2+}, 0; Cu^{2+}, 0; Cd^{2+}, 3; Co^{2+}, 3; Zn^{2+}, 4.2.

Co and Zn (and presumably Cu^{2+}) may be titrated with advantage at 550 nm using NH_4^+ or Na^+ HD_z^- where formation of the complex is monitored. The following mixtures were resolvable by careful control of pH and wavelength: Hg^{2+}, Cd^{2+}; Ag^+, Zn^{2+}; Hg^{2+}, Cu^{2+}; and Cu^{2+}, Zn^{2+}. A lower concentration limit of 1—2 μg of metal in 50 ml was achieved. At this sensitivity reagent blank titrations were imperative. Solution pH values <3 were fixed by addition of dilute $HClO_4$; those in the range 3—4.7 by the use of tartrate, benzoate and acetate buffers. pH measurements were made before the addition of acetone.

This thorough investigation of Le Goff and Trémillon was anti-

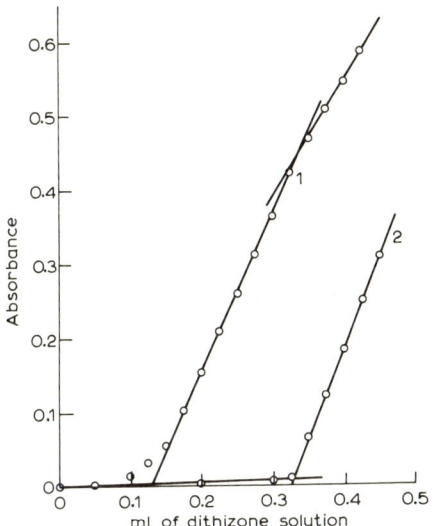

Fig. 56. Titration of mercury(II) and zinc with dithizone. 1, wavelength = 550 nm; 2, wavelength = 600 nm. Reprinted from ref. 175, by kind permission of the American Chemical Society.

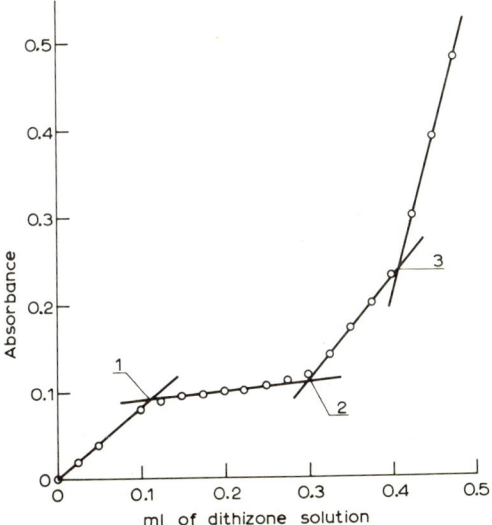

Fig. 57. Titration of copper(II) and zinc with dithizone (wavelength, 600 nm). Branch to point 1 corresponds to CuDz formation; branch to point 2 corresponds to ZnDz$_2$ formation; branch to point 3 corresponds to CuDz$_2$ formation; branch after point 3 corresponds to Dz alone. Reprinted from ref. 175, by kind permission of the American Chemical Society.

TABLE 5

Further analyses based on the principle discussed

Metal determined	Author(s) (date)	Titrant	General conditions	Ref.
Cu	Takahashi and Robinson (1960)	1-nitroso-2-naphthol, 0.005 M in DMF	600 nm; solvent 40 ml DMF 0.0015 M in acetic acid	177
Ni			525 nm; solvent 5 ml butylamine + 35 ml DMF	
Ni	Behm and Robinson (1963)	Dimethylglyoxime, 0.005 M in 1 : 1 $CHCl_3$: IPA	370 nm; solvent 2.5% n-butylamine in 1 : 1 $CHCl_3$: IPA	178
Co			370 nm; solvent 2.5 % cyclohexyl-amine in 1 : 1 $CHCl_3$: IPA	
Cu	Poppe (1969)	PAN, 10^{-4} M in 1 : 1 methanol : water	pH 4.0 aqueous (acetate); 550 nm; 25 ml, 1 μg Cu; very selective	179
Cu	Boyle and Robinson (1958)	Oxine, 0.015 M in DMF or 1 : 1 dioxan : n-propanol	470 nm, 40 ml; solvent 12 % n-butyl-amine in DMF, or 470 nm in 35 ml dioxan-n-PrOH + 1 ml conc. NH_3 + 2 ml triethylamine; 500 nm, 40 ml, 12 % butylamine in DMF also	180
Zn			478 nm; 30 ml 1 : 1 dioxan : n-PrOH + 1 ml NH_3 + 2 ml triethylamine	
Cu, Mo VI, UO_2^{2+}	Silver and Bowman (1967)	Eriochrome blue-black R 2×10^{-3} M in water	635 nm; pH 1.5—3.0 (acetate), 510 nm; pH 4 (acetate) 500 nm; other ions examined	181

Arsenicals, e.g. $EtAsCl_2$	Tarbell and Bunnett (1947)	Di-(p-biphenyl) thiocarbazone	Solvent benzene + 1 drop pyridine; 655 or 510 nm; 1 : 1 ratio	182
$(C_2H_5)_2PbCl_2$	Pilloni and Plazzogna (1966)	4-(2-pyridylazo) resorcinol (PAR) 2×10^{-3} M in water	pH 9, 512 nm; 1 : 1 complex; also $PbCl_2/Et_2PbCl_2$ by addition of excess PAR and EDTA titration	183
CH_3Hg^+	Ingman (1971)	Dithizone 50 μM in $CHCl_3$	To 1 μmol CH_3Hg^+ in 10 ml water add 1 ml buffer (formate pH 2.5—3.0) and 40 ml ethanol; titrate at 590 nm	184

cipated by the interesting determination of zinc in lubricating oil additives proposed by Marple et al. (1958) [175]. The oil sample was diluted with 8% methanol in benzene saturated with ammonium acetate and titrated with 0.05% dithizone in benzene at 640 nm. Titration graphs similar to the H_2D_z absorption curves of Le Goff were obtained. Mixtures of Zn^{2+} and Hg^{2+} were easily resolved by wavelength control (see Fig. 56). Copper—zinc mixtures gave a complex but quite usable graph (Fig. 57). The advantage in time saved over conventional ashing procedures can be imagined.

(D) SPECTROPHOTOMETRIC EXTRACTIVE TITRATIONS USING HIGH ABSORBANCE REAGENTS AS TITRANTS

Just as an extraction process can be used to indicate equivalence in very dilute metal solution—EDTA titrations, so such a two phase system can be used to indicate the equivalence-point in "reagent"—metal titrations. This approach to trace metal analysis has been pursued almost exclusively by the Czech chemist Galík over the period 1966 to date. His theoretical paper (No. V in the series) [185] produced the parametric equations $A = f(E,a)$ and $a = f(E,a)$ where parameter E is the extent of the overall reaction defined by

$$[MA_M]_{org} = E \cdot C_{HA}/M$$

The extraction constant K_M is the equilibrium constant for the reaction

$$M^{M+}_{aq} + MHA_{org} \rightleftharpoons MA_{M(org)} + MH^+_{aq}$$

i.e. $K_M = \dfrac{[MA_M]_{org}[H]^M}{[M][HA]^M_{org}}$

The influence of a second, less easily complexed, metal N is considered. Titration curves for mono-, bi- (Fig. 58) and tervalent cations at various values of B_M are shown where

$$B_M = \frac{[H]^M \left(\dfrac{V_{org}}{V}\right)^{N-1}}{K_M \cdot C^M_M \cdot M^M}$$

308

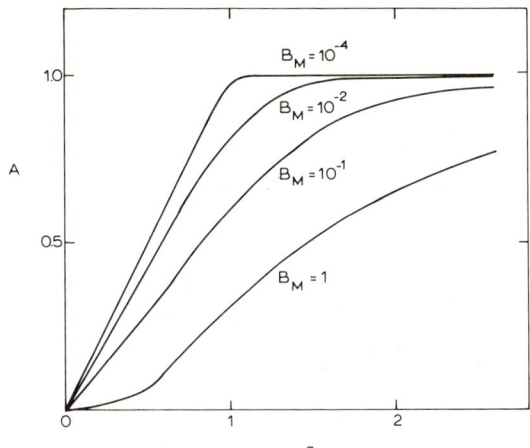

Fig. 58. Titration of bivalent cation. Plot of absorbance against $a = C_{HA}/C_M$ for a typical extractive titration ($\epsilon_{MA_M}lC_M = 1$; $V_{org}/V = 1$; $C_N = 0$; $\epsilon_{HA} = 0$). B_M is defined in the text. Reprinted from ref. 185, by kind permission of the editor.

Threshold (minimum) pH values to give a 1% error for lines drawn through $a = 0$, $A = 0$, $a = 0.5$, $A = A_{0.5}$ and $A = l\epsilon_{MA_M} \cdot C_M \cdot (V/V_{org})$ ($a \to \infty$) are quoted for ions of charge +1 to +4. See Table 6. Alternatively, the substoichiometry equation of Starý and Růžička may be used

TABLE 6

Critical values in the titration of metal ion M

Charge on the metal ion	Critical B_M value	Threshold pH [a]
1	2.338×10^{-3}	$2.63 - \log K_M - \log c_M$
2	1.321×10^{-5}	$2.14 - \dfrac{\log K_M}{2} - \log c_M$
3	4.723×10^{-8}	$1.97 - \dfrac{\log K_M}{3} - \log c_M$
4	2.125×10^{-10}	$1.82 - \dfrac{\log K_M}{4} - \log c_M$

[a] Supposing that $V = V_{org}$.

References pp. 380—389

TABLE 7

Metals determined	Author(s) (date)	Reagent	General Conditions	Ref.
Hg(II) Cu Cd Zn Co Ag Bi	Galík (1966)	Dithizone, 100 μM in CCl_4	485 nm; 20 ml; pH 8.5—10 (tartrate) 550 nm; 20 ml; pH 8.5—10 (tartrate) 520 nm; 20 ml; pH 8.5—10 (tartrate) 532 nm; 20 ml; pH 8.5—10 (tartrate) 542 nm; 20 ml; pH 8.5—10 (tartrate) Cu—Bi; Ag—Co; Ag—Zn; Ag—Cu mixtures resolved; 1—4 μg titrated 620 nm; pH 4.5—6.0	187
Total "heavy" metals	Galík and Knizek (1966)	Mercury(II) 5.00 μg ml^{-1}, (dithizone) 10^{-4} M in CCl_4	Impurities in GaAs. Extract metals with excess H_2Dz (pH 9.4—9.9 (tartrate)). Remove excess H_2Dz into 0.1 M NH_3 solution. Transfer to fresh CCl_4 (add H_2SO_4) and titrate with Hg^{II}; pH 3; 620 nm	188
Ag, Cu	Galík and Knizek (1966)	Dithizone 10^{-4} M in CCl_4	In Pb. pH 4.3—5.5 (tartrate); 550 nm; mixture resolved	189
Cd Zn Cu	Galík (1968)	PAN $1-3 \times 10^{-4}$ M in ethanol	1 M NaOH; 550 nm; pH 10.5 (borate NaOH); 555 nm; 0.1 M SCN^-; pH 4.5—7; 565 nm	191
Zn	Galík (1967)	Dithizone 10^{-4} M in CCl_4	In $GeCl_4$ or GeO_2; 532 nm; pH 8—9 (tartrate). Dithiocarbamate removal of interferences	190
Ni	Nasouri et al. (1967)	DMG 0.1 M in dilute NaOH	pH 8—9 (NH_3); $CHCl_3$ solvent; violet filter	192
V^V	Hartkamp (1959)	Pyridine-2,6-dicarboxylic acid (0.02 M)	432 nm, H_2O_2, dilute acid, H_3PO_4; 1:1:1 complex formed with H_2O_2	193

| Ag, Hg/Zn | Galík (1970) | Di-2-naphthyl-thio-carbazone 10^{-5} M in $CHCl_3$ or CCl_4 | 10 ml $CHCl_3$ + 50 ml solution containing 10 μg Ag + 1 ml 0.1 % EDTA, pH 3—11 with NH_3; λ = 650 nm. For Hg/Zn use NaAc solution, λ = 580—590 nm | 194 |

$$\text{pH} \geqslant \frac{1}{N} \cdot \log \frac{C_{HA}}{N} - \frac{1}{N} \log \left[C_M - \frac{C_{HA}}{N} \cdot \frac{V_{org}}{V} \right] - \frac{1}{N} \log K - \log 10^{-3} \cdot C_{HA}$$

Maximum pH values are dictated by the ease of formation of hydroxo complexes or selectivity requirements. Further graphs showing the titration of metal ion M in the presence of a ten-fold excess of N (M and N both univalent or both bivalent) for various values of $Q = K_M/K_N$ are produced (Fig. 59). Note that the chelates MA_M and NA_N should not have identical absorption spectra and a wavelength must exist where ϵ values differ sufficiently for distinct breaks on the titration curve to be obtained. Reagent titrations of this nature are apparently less sensitive but more selective than straightforward photometric procedures because of their substoichiometric nature. Real systems studied by Galík and co-workers, and others are summarised in Table 7. Special note should be taken of Galík's paper VI [186] in which association constants for the ion pairs $Cu \cdot PAN^+ Br^-$ and $CuPAN^+ NCS^-$ are determined. Extraction of such species under the usual conditions is thoroughly studied on a quantitative basis.

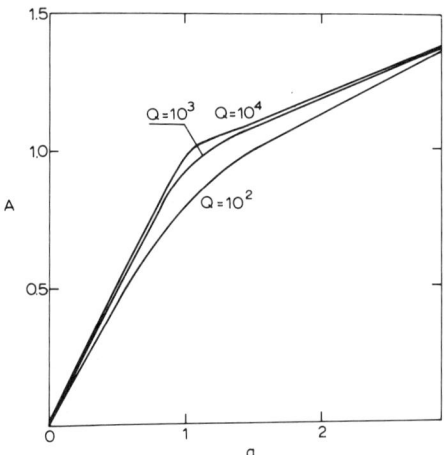

Fig. 59. Titration of bivalent cation M in the presence of ten-fold excess of bivalent cation N. $\epsilon_{MA_M}lC_M = 1$; $\epsilon_{MA_M} = 5\epsilon_{NA_N}$; $B_M = 10^{-4}$; $V_{org}/V = 1$; $\epsilon_{HA} = 0$; $K_M = 10^6$. Reprinted from ref. 185 by kind permission of the editor.

TABLE 8

Monodentate ligand photometric titrations

Species determined	Author(s) (date)	Titrant	General Conditions	Ref.
Au^{III}	Hirano (1934)	KI (0.01 M)	$AuCl_3$ in acid solution < 0.05N. Excess Cl_2—water. Chem. obscure; end point either $AuI_3\downarrow$ (dark green) or AuI_4^-	195
NCO^-	Trusell et al. (1967)	$Co(ClO_4)_2$ (0.025 M in appropriate solvent)	645 nm; solvent DMF (best), DMSO or EtOH. End-point $Co(NCO)_4^{2-}$	196
CN^-, SCN^-	Nomura (1968)	$Hg(NO_3)_2$ (0.5 mM)	240 nm; pH 7 (phosphate); 0.01 M EDTA. Only Hg—SCN—EDTA absorbs; Hg—CN—EDTA forms first. Species resolved	197
Li^+	Specker et al. (1968)	$Cu(ClO_4)_2$ (0.01 M in acetone)	366 nm; solvent acetone or cyclo-hexanone. 3 $LiCl + Cu(ClO_4)_2 \rightarrow$ $Li(CuCl_3) \cdot xR_2CO + 2\ LiClO_4$	198
Cl^-	Young et al. (1968)	$Hg(NO_3)_2$ (ca. 0.01 M)	Indicator diphenylcarbazone screened with bromophenol blue; 510 nm; pH 2.5	59
Hg in organic	Yeh (1969)	KSCN (0.01 M)	Fe^{3+} indicator; Initial wet oxidation	199

313

(E) PHOTOMETRIC MONODENTATE LIGAND COMPLEX-FORMING TITRATIONS

Monodentate ligands such as ammonia, halide and pseudo-halide ions usually form complexes with metal ions in a series of steps in which the stepwise formation constants gradually tend towards zero as the ligands are added. Such a process invalidates the use of the reaction for visual titration purposes as the overall stoichiometry is so uncertain. However, in certain instances stepwise formation constants are grouped in such a way that stoichiometry is exact over a wide range of conditions, the best known examples being the formation of the complex species $Ag(CN)_2^-$ and $HgCl_2$. The few photometric titrations involving this kind of complex formation mentioned in the literature are for the most part adaptations of existing visual methods but occasionally the objective photometric method is able to detect definite combining ratios quite imperceptible to the eye. The $Co^{2+} - SCN^-$ system is, perhaps, the outstanding example of this [112].

Some form of photometric titration has been used regularly over the years to investigate the nature of absorbing complexes, e.g. starch—iodine; $Fe^{3+} - SCN^-$; $Co^{2+} - SCN^-$; $PdCl_2$ — pararosaniline, but where such an investigation is not the basis of an analytical method it will not be mentioned further. Purely analytical applications of these reactions are summarised in Table 8.

6. Precipitation titrations

(A) GENERAL CONSIDERATIONS

Consider a general precipitation titration taking place in one litre of solvent [200]:

$$A^+ \quad + \quad B^- \quad \rightleftharpoons AB \quad \rightleftharpoons AB \downarrow$$

at start: a mols   —   —   —

before E.P.: x mols z mols y mols

where A^+ = titrand; B^- = titrant; y = mols of precipitated salt; x = mols precipitant added; a = initial conc. titrand; z = solubility of AB, and L = solubility product of AB.

314

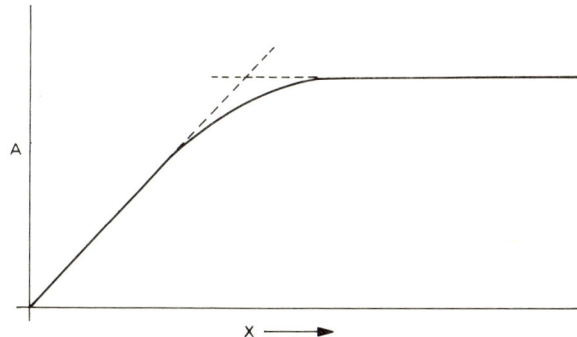

Fig. 60. Titration curve of a self-indicating precipitation reaction.

Then,

$$y = \frac{a + x}{2} - \sqrt{\frac{(a - x)^2}{4} + L}$$

Well before the equivalence-point L is negligible with respect to $(a - x)^2/4$

$$\therefore y = \frac{a + x}{2} - \frac{a - x}{2} = x$$

and if absorbance $\propto y$ we get the titration curve shown in Fig. 60. When $x = a$, $y = a - \sqrt{L}$ i.e. $a - y = \sqrt{L} \equiv A^+$ remaining in solution. Hence if L is large gross curvature results. This illustrates the compromise which often needs to be reached in precipitation titrations. If the temperature is low L is small but rate of crystal growth is low so that absorbance readings become steady only slowly; at higher temperatures crystal growth and ordering is facilitated but L increases.

When titrant B is added to A two extreme cases can arise. Either (1) constant fresh nucleation occurs with a small amount of uniform growth on each particle or (2) there is initial nucleation, then growth only on these [201]. Turbidimetry depends upon the surface area of the particles and the relation between total surface area and mass precipitated for cases 1 and 2 is as shown in Fig. 61. Most titrations are intermediate in nature; in fact they unfortunately favour (2) as

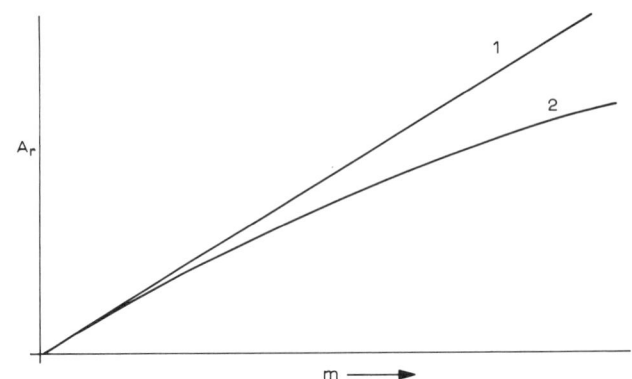

Fig. 61. Variation of total particle surface area with mass precipitated. A_r = total surface area ($\propto$ turbidance) and m = mass precipitated. For curve 2, A_r = $(36\pi/\rho)^{1/2} m^{2/3}$ where ρ = particle density. Significance of 1 and 2 is described in the text.

the concentration of titrand diminishes. Turbidimetric and nephelo-metric titrations may be self-indicating or indicated.

The first precipitation titration mentioned in the literature was by Müller and Partridge (1928) and was, in fact, an indicated nephelo-metric type based on Mohr's method for chloride [5].

$$\xrightarrow[\text{green}]{I_0} \boxed{}$$

$$\downarrow I_s \text{ (green)}$$

$$Cl^- + Ag^+ \rightleftharpoons AgCl \downarrow$$

$$2Ag^+ + CrO_4^{2-} \rightleftharpoons Ag_2CrO_4 \downarrow$$

(red-absorbed green)

Titration was continued to a sudden drop in I_s intensity. During the 1930's Hirano (in Japan) investigated silver halide titrations using starch as protective colloid and titration of Hg^{2+} and Pb^{2+} with Na_2S solution using gum arabic for protection. Del Campo and co-workers (Spanish) were the first of many to investigate the barium—sulphate interaction.

As with other types of photometric titration, background theory papers were sparse until in 1941 A. Ringbom produced an excellent

316

theoretical discussion which has probably not been surpassed [200]. He pointed out that heterogeneous light effects are dependent on such factors as rate of precipitation, temperature, aging of precipitate, hydrogen ion concentration, neutral salt concentration, presence of uncharged colloidal material etc., and therefore no easily established relation between absorbance and concentration really exists. Note, however, that the long established Ag^+—Cl^- visual self-indicating precipitation titration of Gay Lussac was one of the most accurate ever produced. Ringbom suggested the addition of ethanol or acetone to reduce precipitate solubility and use of a protective "colloid" (sugar or gum arabic) to maintain a small particle size and discourage flocculation. His findings for scattered light intensity and light loss as a function of particle radius for barium sulphate are shown in Fig. 62. He found that flocculation gave an absorbance increase, but the opposite has been found by later workers. He cited the best conditions as

(1) The test solution should be of such a concentration that one drop of titrant gives an instantaneous precipitate, though excessive concentration should be avoided so that the precipitate may remain colloidal throughout the titration. Add an organic solvent and/or a colloid stabiliser to achieve this.

(2) Choose titrant concentration so that one drop corresponds to the accuracy of measurement of the instrument.

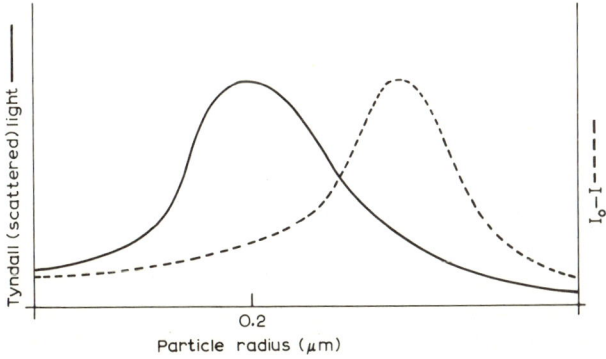

Fig. 62. Variation of scattered light intensity and light loss with barium sulphate particle size. Adapted from ref. 200, by kind permission of Springer—Verlag, Heidelberg.

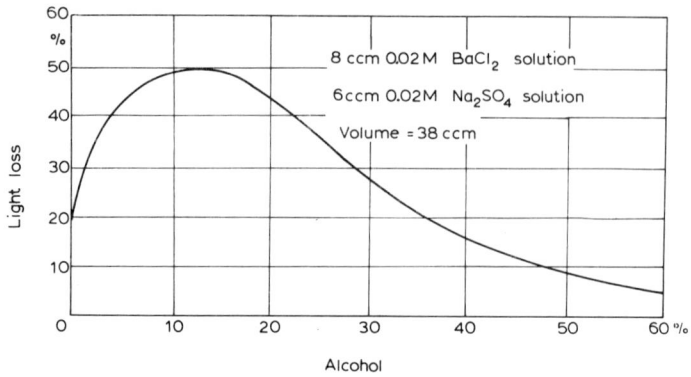

Fig. 63. Influence of ethanol addition on the percentage light loss produced by a barium sulphate suspension. No light filter used. Reprinted from ref. 200, by kind permission of Springer—Verlag, Heidelberg.

(3) Choose cell path length so that percent transmission falls in the 20—60% region at the equivalence-point.

He applied his ideas to the valuable Ba^{2+}—SO_4^{2-} interaction, and produced the interesting relationship for alcohol addition (Fig. 63). Ringbom studied the charge (zeta potential) on the precipitate as a function of free barium ion concentration (Fig. 64). At high Ba^{2+} concentration strong $BaSO_4 + Ba^{2+} \rightarrow [BaSO_4 \cdot Ba]^{2+}$ adsorption takes place giving the desired peptising effect. This is less evident with sulphate excess. For the above reason Ringbom found titration of Ba^{2+} by SO_4^{2-} best, but later workers found no difference due to direction of titration. Quality of the titration curve suffered with addition of HCl or highly charged cations, e.g. Mg^{2+}, Fe^{3+} as coarser precipitates were produced. The optical effects of streaming birefringence could be awkward but were minimised at small crystal size.

As in gravimetry, nitrate ions proved very troublesome, hence if these were present sulphate was titrated with barium to avoid a positively charged precipitate.

A polarographic recording galvanometer was used in this work.

The titration curves, excellent even by modern standards, produced by Ringbom in this work are illustrated in Fig. 65.

In 1943, Osborn et al. produced the first effective fluorimetric and nephelometric titrator [8]; this will be described in the section on fluorimetric titrations.

318

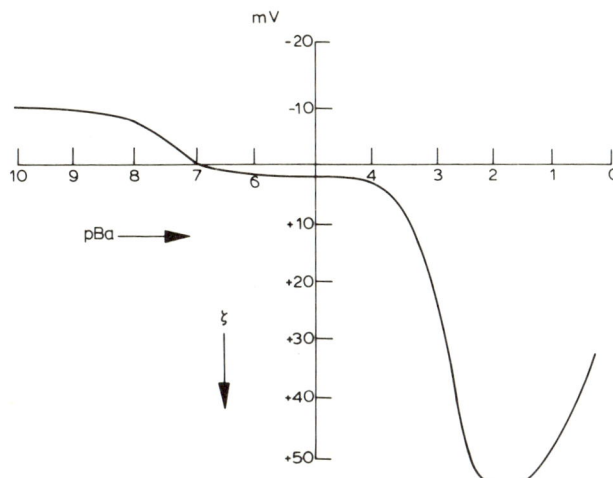

Fig. 64. The charge (in terms of zeta potential) acquired by a suspension of barium sulphate as a function of free barium ion concentration. Reprinted from ref. 200, by kind permission of Springer—Verlag, Heidelberg.

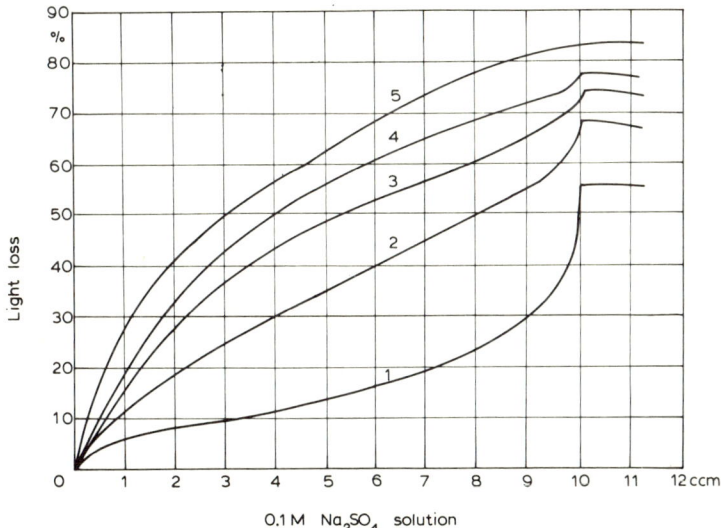

0.1 M Na$_2$SO$_4$ solution

Fig. 65. The photometric precipitation titration of 10.0 ml of 0.1 M BaCl$_2$ solution mixed with 25 ml of ethanol and 5 ml of glycerine together with these additional solutions: (1) nil; (2) 2 ml 1 M NaCl; (3) 0.25 ml 1 M HCl; (4) 1 ml 1 M MgCl$_2$; (5) 1 ml 1 M FeCl$_3$ + 0.25 ml 1 M HCl. Reprinted from ref. 200, by kind permission of Springer—Verlag, Heidelberg.

(1) Heterometry

In 1953 Bobtelsky (Hebrew University in Jerusalem) introduced the technique of Heterometry [18] which, despite protestations to the contrary, appears to be merely a sensitive, vertical light path form of turbidimetric titration in the absence of protective colloid [202]. Production of crystalline intermediates is shown by kinks (critical points) and titration curves of all possible shapes corresponding to the formation and dissolution of precipitates have been produced. Over the years Bobtelsky must have examined a large proportion of all the aqueous precipitation reactions ever quoted in the literature both in and out of analytical chemistry. Metal reagent, e.g. nickel—DMG and metal—anion, e.g. barium—sulphate reactions were particularly well represented.

Heterometric—potentiometric curves are useful for showing optimum regions of precipitation [18], especially for reagent reactions of the type $M^{2+} + 2HR \rightleftharpoons MR_2 \downarrow + 2H^+$ (neutral salts are often needed to give reasonable graphs). Another useful interaction is that of alkaloids, etc. $- R_4N^+$ with complex anions, e.g. BiI_4^-, silicotungstate or other heteropoly acid anions. However, an interaction of this type was used in 1938 by Goodhue who titrated nicotine with silicotungstate [203]. Quite complex metal ion mixtures may be titrated using selective titrants and perhaps suitable masking agents. The usual concentration range was $10^{-2} - 10^{-4}$ M. Some other workers have difficulty in reproducing Bobtelsky's graphs and of inducing kinks in mixed systems which ought to show them [202]. pH monitoring of ion—reagent titrations gives the metal—reagent ratio and number of protons produced.

Due to the vast range and complexity of Bobtelsky's work on these precipitation systems no attempt has been made to catalogue them here. The reader is referred to his many papers (usually in Anal. Chim. Acta) and in particular to his book [204].

Of anion titrations, those of sulphate and fluoride have been most frequently studied because of their importance and the dearth of good alternative methods. They are discussed exclusively in a following section.

(2) Surface active agents

Anionic surfactants may be titrated with cationic agents to produce a colloidal precipitate near the 1 : 1 equivalence-point which redissolves or coagulates in excess titrant. The end-point is that of maximum turbidity.

The Igepon, $C_{17}H_{33}\overset{\overset{\displaystyle O}{\|}}{C}-\underset{\underset{\displaystyle CH_3}{|}}{N}-C_2H_4SO_3^- Na^+$—cetylpyridinium chloride,

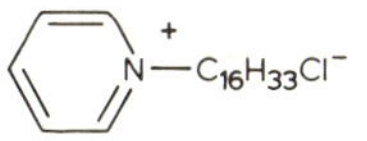

titrations of Lambert are a good example [13].

The titration must feature interaction of micelles where low solubility is due to proximity to the isoelectric point, i.e.

$$A^-c^+ + C^+a^- \rightarrow AC\downarrow + c^+a^-$$

Addition of a non-ionic detergent often sharpens up a broad peak. Large concentration ranges are coped with by changing filters (Rayleigh).

(3) Polymer titrations

Light scattering titrations are used to determine molecular weight distributions but this is a vast and specialised subject [239].

(4) Indicated titrations

Any precipitation titration process $A + B \rightleftharpoons C\downarrow$ may be indicated if an indicator exists for A or B. If a titrant-sensitive indicator is used sufficient is present to give a well defined linear portion to the graph (Fig. 66). λ corresponds to the absorption maximum of BIn. This seems to have been applied almost exclusively to $SO_4{}^{2-} - Ba^{2+}$ (Thorin), and $F^- - Th^{4+}$ (alizarin S, etc.) titrations, probably because adsorption is so often a problem.

In the above, precipitate absorbance change and indicator absorbance change compete; it would therefore be more logical if free In λ_{max} were monitored, but this seems to have been tried in only very few cases.

(B) THE PHOTOMETRIC TITRATION OF FLUORIDE AND SULPHATE

The titrimetric determinations of fluoride and sulphate have, one might well argue, been the twin Holy Grails most ardently pursued

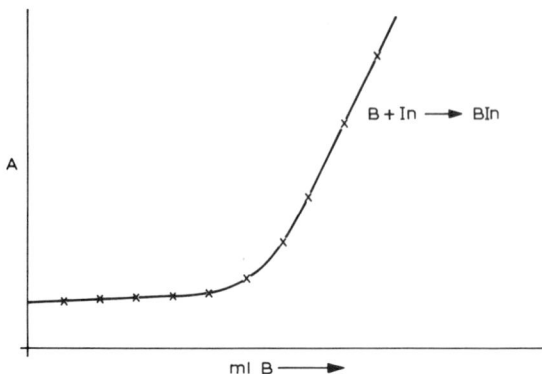

Fig. 66. Titration curve of an indicated precipitation titration.

by classical analytical Sir Galahads over many years. The first reasonably good visual fluoride titration was that proposed by Willard and Winter in the early 1930's involving the use of thorium nitrate titrant and initially the ZrO^{2+}—alizarin S complex but soon after simply the Th^{IV} — alizarin S complex as indicator. This reaction is unsatisfactory because the stoichiometry varies widely around the expected ThF_4 and the gelatinous precipitate obscures the indicator colour change. The titration was thus an early customer to be fitted with a photometric end-point and this was achieved by Nichols and Kindt in 1950 [14]. A glance at the bibliography shows that the titration was subsequently modified by numerous workers but until the work of Harzdorf none improved upon Nichols and Kindt's precision of 0.01 ml in 5—10 ml. Only two attempts were made to determine fluoride by phototurbidimetric titration without indicator. Brandt and Duswalt used thorium nitrate and calcium chloride titrants achieving coefficients of variation of 0.8—4% for both. The finely-divided CaF_2 precipitate compared favourably with gelatinous "ThF_4" and was favoured for low fluoride weights. Cerous was also satisfactory as a turbidimetric titrant but lead chloride (leading to PbClF precipitation) was not. Roberts had better success with the more stoichiometric neodymium—fluoride reaction; his best error was 0.09 in 20. Initial seeding with NdF_3 was necessary but the method was very insensitive to pH or alcohol content.

In recent years, however, Claus Harzdorf of Bayer has produced some excellent and well researched methods concerning the deter-

322

mination of fluoride with metal titrants and this problem which has bedevilled analysts for years may well now be said to have been beaten. He initially examined the thorium titration achieving a precision of 0.6% but soon moved on to an examination of aluminium nitrate as titrant. He valued the high inherent precision of the photometric titration as a means for determining exactly the combining ratio of fluoride with metal under various conditions and in the case of aluminium showed the stoichiometric range to be Al $F_{5.86-5.94}$. Under his final conditions end-point breaks were very sharp and he achieved a precision of 0.034 in 120. He then turned his attention to the zirconyl—fluoride reaction. Though this is still plagued by formation of a variety of complexes and hence uncertain stoichiometry — his combining ratios centered around $ZrF_{5.5}$ — it does have the great advantage that no precipitate is produced, thus circumventing the usual adsorption difficulties. He showed that many earlier visual zirconyl—fluoride titrations were carried out at too low a pH and that instead of titrating in 1 M HCl, 0.1 M was far better. Here, though, a compromise is struck between quality of titration graph and extent of hydrolysis of the titrant solution. He showed that extent of hydrolysis and with it the combining ratio, depended upon room temperature during the titration. Organic solvent present in the titration solution reduced hydrolysis problems. Standard deviation in this method was 33 μg in 25 mg. In a comparison with the thorium titration he concluded that thorium nitrate was satisfactory for low fluoride weights — say up to 4 mg — but for greater amounts zirconyl chloride was more satisfactory. Finally Harzdorf examined the lanthanum—fluoride titration. The great advantage here was that though the reaction was again heterogeneous the stoichiometry was far more certain than in any of the previous methods. The range was La $F_{2.96-3.05}$ over a wide variety of conditions, thus conferring an agreeable lack of susceptibility to environment. Even the presence of alcohol — a subject of much controversy in early work — was unimportant; variation of content from zero to 50% mattered little. Cerium(III) titrant proved similar but yttrium nitrate, while not being quite so good in that adsorption of fluoride gave a combining ratio somewhat over 3 : 1 proved useful since the YF_3 precipitate was far less opaque than that of LaF_3. This allowed titration of up to 35 mg of fluoride. Precision of lanthanum titrations was characterised by a standard deviation of 5 μg F in 9 mg. In all Harzdorf's papers he examined a wide variety of indicators and conditions and it

TABLE 9

Bibliography of self-indicating and indicated photometric precipitation titrations

Species determined	Author(s) (date)	Titrant	General conditions	Ref.
Self-indicating				
F^-	Brandt and Duswalt (1958)	$Th(NO_3)_4$ (0.1 M) $CaCl_2$ (0.2 M)	F^- solution 0.04–1.0 M: pH 1–7 F^- solution 0.01–0.09 M; pH 1–7; tungsten lamp, no wavelength selection	205
F^-	Roberts (1967)	$Nd(NO_3)_3$ (0.2 M)	1–20 mg F; pH 1–2. Add 2–3 drops 50 % polyethylene glycol 400 and NF_3 suspension seeder; $\lambda = 525$ nm	206
SO_4^{2-} Ba^{2+}	Ringbom (1941)	Na_2SO_4 (0.02 M)	30–50 ml 0.005 M in Ba^{2+}; 35 % ethanol. Add 1 ml 1 % gum arabic; No filter.	200
SO_4^{2-} (in vulcanised rubber)	Frey (1952)	$BaCl_2$ (0.02 M)	$HClO_4 + HNO_3$ digestion. 2 mg S 40 % ethanol; pH 5.0–7.5. 50°C; Blue–green filter; Total volume 100 ml.	207
SO_4^{2-}	Wickbold (1953)	$BaCl_2$ (0.0015 M)	90 % methanol; blue filter. 100 μg S titrated; Total volume = 33 ml, Include 3 ml electrolyte solution (20 g $MgCl_2 \cdot 6H_2O$ + 3.5 g $NH_4Cl\ l^{-1}$); pH = 8–8.2.	208
SO_4^{2-}	van Nieuwenburg and Englebert van Bevervoorde (1958)	$Ba(NO_3)_2$ (0.02 M)	Solvent water : ethanol : butanol 2 : 1 : 1. 2–20 mg SO_4^{2-}; pH 1.5 (dilute HCl); volume 24 ml; 50 % alcohols at end-point. Ethanol alone and protective colloids deleterious; Alkali salts interfere.	209

324

Species	Reagent	Author	Conditions	Ref.
SO_4^{2-}	$BaCl_2$ (0.03 M)	Nichols and Kindt (1950)	5 ml 0.03 M SO_4^{2-} + 35 ml water + 25 ml ethanol + 1 ml 1 % gum arabic; $\lambda = 490$ nm.	14
SO_4^{2-}	$BaCl_2$ (0.05 M)	del Campo et al. (1936)	20—70 ml 0.05 M H_2SO_4; Agar concentration. 0.05—0.12 %.	210
SO_4^{2-}	$BaCl_2$ (0.1 M)	Blakeley and Ryan (1964)	10 ml 0.05 M H_2SO_4 + 2 ml 4N HCl + 8 ml 95 % ethanol	202
SO_4^{2-}	$BaCl_2$ (0.1 M)	Hedrick (1965)	40 % isopropanol; 0.5 mmol SO_4^{2-} as $(NH_4)_2$ SO_4	211
SO_4^{2-}	$BaCl_2$ (0.002 M)	Rees and Hill (1969)	$\lambda = 400$ nm; Solvent, aqueous methanol; 0.4—0.8 mg SO_4^{2-}; Volume $\sim$ 200 ml	212
Various metal ions	Various precipitating anions and reagents	Blakeley and Ryan (1964)	General study of shape of absorbance — titrant precipitation graphs	202
Sodium poly-vinyl sulphonate	$Ba(OH)_2$ (0.02 M)	Staszewska and Ulinska (1968)	Nephelometric titration; Solute 2—12.5 mmolar; N_2 atmosphere	213
Anionic surfactants	Cetyl pyridinium chloride (0.01 M)	Lambert (1947)	Sample 40 ml 10^{-3} M; blue filter	13
Cationic surfactants	Igepon T.D. (sodium N-acyl-N-methyl taurate). (0.01 M)		Similar	
Ca in limestone	Na oxalate (0.1 M)	Blakeley and Ryan (1964)	pH 4.5—5.0 (acetate); 0.5 g sample in 100 ml 0.2 N HCl; Titrate 10 ml	202
Cd in alloys	NaBr (0.01 M) in glacial acetic acid	Lebedeva and Isaeva (1971)	1—1.5 mg Cd in 7 ml acetic acid; Red filter; Zn, Bi, Mg, Cu, Ca, Ba do not interfere	214
Hg^{2+}, Pb^{2+}	Na_2S (0.01 M or 0.001 M)	Hirano (1935)	Gum arabic as protective colloid	215

TABLE 9 (continued)

Species determined	Author(s) (date)	Titrant	General conditions	Ref.
Cl^-, Br^-, I^-	Hirano (1935)	$AgNO_3$ (0.1 N)		216
Ni in steel	Boyer (1938)	$AgNO_3$ (5.789 g l^{-1})	$Ni^{2+} + 4\ CN^- \rightarrow Ni(CN)_4^{2-}$ Excess CN^- determined by Liebig–Denigés titration. 0.3 g Ni in 20 ml 1 : 1 HNO_3, dilute to 300 ml + 60 ml 20 % citric acid + NH_3 till alkaline, then 5 ml excess. Add 2 ml titrant + 10 ml 10 % NaI. Dilute to 500 ml, temperature 30°C. Add standard NaCN until solution clears then 1 ml excess. Adjust to 100 % T, titrate with $AgNO_3$ to pre-determined galvo deflection	217
Nicotine	Goodhue (1938)	Silicotungstic acid (5g/litre)-back titn. with nicotine	To 0.05–0.75 mg nicotine add 2 ml silicotungstic acid solution, 4 drops 1.5 M formic acid, 4 drops Irish moss extract, 10 ml water. Titrate with nicotine formate (0.5 g nicotine l^{-1})	203
Ca^{2+} (in rubber)	Frey (1951)	Ammonium oxalate (0.0125 M)	2–4 mg Ca, 50 % low alcohol, dilute NH_3 (1 ml conc/100 ml); temperature 50°C blue–green filter; multiply titre by factor 1.13; Mg > Ca	218
Zn^{2+} (in rubber)	Frey (1951)	$K_4Fe(CN)_6$ (0.04 M)	10–35 mg ZnO equiv. + 30 mls EtOH + 3 ml 1 : 1 H_2SO_4 to 100 ml with water; temperature 40°C; blue–green filter; Iron interferes	219

Ni^{2+} (in steel)	Nasouri et al. (1967)	Dimethyl glyoxime (0.01 M) in EtOH	0.06—0.7 mg Ni + 2 ml 2 M NH$_4$Cl + 1 ml 2 M NH$_3$; pH 8—9. Add 10 ml protective colloid, e.g. 2% gum arabic, dilute to 30 ml; red filter	192
Oxalate	Curran and Fletcher (1969)	La^{3+} elec. gen. from LaB$_6$	0.5—2.6 mg; volume = 20 ml. pH 4.6 (2 M acetate). Vary current to suit solute concentration La$_2$(C$_2$O$_4$)$_3$ forms stoich. but beware of acetate ternaries and La$_2$(C$_2$O$_4$)$_3$ · Na$_2$C$_2$O$_4$ double salts	220

Indicated

F$^-$	Nichols and Kindt (1950)	Th(NO$_3$)$_4$ (0.0125 M)	2—10 mg F$^-$ in 50 ml water + 20 ml ethanol + 15 drops 0.1% alizarin S + 3.5 ml 0.2 M chloracetate buffer + 10 ml 1% starch solution; pH 3.15; λ = 515 nm	14
F$^-$ (in radioactive samples)	Lee et al. (1956)	Th(NO$_3$)$_4$ (0.0125 M)	Exactly as for Nichols and Kindt.	221
F$^-$ (in organic compounds)	Ma and Gwirtsman (1957)	Th(NO$_3$)$_4$ (0.0025 M) (0.00025 M)	Approximately 1 mg F$^-$ in 250 ml water pH 3.0. Add 2 ml 0.01 or 0.035% sodium alizarin sulphonate. Titrate to fixed galvo deflection; λ = 515 nm	222
F$^-$ (in rocks)	Hollingworth (1957)	Th(NO$_3$)$_4$ (10^{-3} M)	70 μg F in 25 ml water; pH = 3.1. Add 0.5 ml 0.02% SPADNS indicator; λ = 580 nm. Use of buffer deleterious	223
F$^-$ (in plants air and water)	Gwirtsman et al. (1957)	Th(NO$_3$)$_4$	As for Ma and Gwirtsman	224
F$^-$	Dean et al. (1957)	Th(NO$_3$)$_4$ (0.005 N)	0—200 μg F$^-$ in 100 ml; λ = 540 nm. Add 1.5 ml 0.1% sodium alizarin sulphonate; pH = 3.0	225

TABLE 9 (continued)

Species determined	Author(s) (date)	Titrant	General conditions	Ref.
F^-	Menis et al. (1958)	$Th(NO_3)_4$ (5×10^{-4} M)	$5–50$ μg F^-, adj pH to 3, add 5 ml buffer (30 g chloracetic acid in 700 ml to pH 3.0 with NaOH) + 2 ml 0.01% alizarin sulphonic acid: Fix volume to 35 ml; $\lambda = 520$ nm. Keep salt concentration to minimum	72
F^- (in organic compounds)	Steyermark et al. (1959)	$Th(NO_3)_4$	Schöniger combustion then titration under Ma and Gwirtsman conditions	226
F^-	Pickhardt (1962)	$Th(NO_3)_4$ + indicator + ethanol	200 μg F^- in 250 ml water + 5 ml 0.006% sodium alizarin sulphonate. Fix to pH 3.00. Titrant: 80 mg indicator + 250 ml water + 25 ml 0.025 M $Th(NO_3)_4$ + 200 ml ethanol to 500 ml with water; $\lambda = 525$ nm	227
F^-	Harzdorf and Steinhauser (1965)	$Th(NO_3)_4$ (0.05 M)	1 mg F^- in 25 ml, fix pH = 3.2. Add 5 ml 1 M chloracetate buffer pH 3.2 + 2 mls 0.07% morin in alcohol. Adjust volume to 35 ml; $\lambda = 445$ nm. Many indicators examined	228
F^- (in silicates)	Lorec (1967)	$Th(NO_3)_4$ (0.0025 M) to (0.0125 M)	To ~ 0.4 mg F^- in 100 ml add 1 ml 0.1% alizarin sulphonic acid then 0.2% HCl dropwise just to yellow colour. Add 5 ml pH 3.9 chloracetate buffer (0.12 M); $\lambda = 520$ nm	229

Determinand	Author (year)	Titrant	Details	Ref.
F^- (general and in silicates)	Harzdorf (1967)	$Al(NO_3)_3$ (0.007; 0.033; 0.1 M)	1.5–120 mg F^- in 25 ml water–methanol; 10 ml 1 M pH 6.0 acetate buffer + 2 ml 0.01% chrome azurol S. Volume = 40 ml, 50% methanol; λ = 570 nm	230
F^-	Harzdorf (1967)	$ZrO\ Cl_2$ (0.012 M; 0.05 M). 0.1 N in HCl	0.5–25 mg F^- in 80 ml. Up to 6 mg F^- with more dilute titrant. 10 ml pH 3.2 (1 M chloracetate) buffer. Indicator 0.01% aqueous purpurin-3-sulphonic acid (4 ml); λ = 550 nm; For 6–25 mg F^- fix pH using 2,4-dinitrophenol as glass elec. unreliable. Keep salt concentration low and fixed	231
F^-	Harzdorf (1968)	$La(NO_3)_3$ (0.017 M and 0.003 M; 0.01 N in HNO_3) also $Y(NO_3)_3$ (0.066 M)	0.2 to 9 mg F^- in 25 ml, pH 6.5. Add 5 ml buffer (1 M pyridine/HCl pH 6.5) 1 ml indic. (0.1% aq PAR) and ethanol to 1 : 1 with water. λ = 510 nm; titration rate 0.4 ml min^{-1}. For 0.003 M La^{3+} range = 0.1–2 mg. For 0.066 MY^{3+} range limit = 35 mg F	232
F (in organic)	Cheng (1970)	$Th(NO_3)_4$ (0.004 M)	1 mg F^- in 15 ml 0.2% sodium formate + 60 ml isopropyl alcohol: water 5 : 1, fix pH to 3.3 with buffer solution (0.5N $HClO_4$, 7% glycine, 10% sodium perchlorate). Add 1 drop 0.2% arsenazo 1 indicator; λ = 580 nm	233
SO_4^{2-} (organic sulphur)	Walter (1950)	$BaCl_2$ (0.02 M)	1–3 mg S in 15 ml solution. Make just acid with 0.01 N HNO_3. Add 15 ml alcohol + 0.15 g Na rhodizonate (1 : 300 with sucrose); λ = 537 nm. Alternative indicator tetrahydroxyquinone. Absorbance decrease used	234

TABLE 9 (continued)

Species determined	Author(s) (date)	Titrant	General conditions	Ref.
SO_4^{2-}	Menis et al. (1958)	$Ba(ClO_4)_2$ (0.00125 M)	6—50 μg SO_4^{2-} in 20 ml isoamyl alcohol, 10 ml methanol, 5 ml water + Thoron 0.01%; $\lambda = 520$ nm	72
SO_4^{2-} (SO_3 in flue gases)	Fielder and Morgan (1960)	$Ba(ClO_4)_2$ (5×10^{-4} M)	Equiv. 8 μg SO_3 in 10—50 mls 4 : 1 isopropyl alcohol : water + 5 drops 0.04% Thoron. Apparent pH 2.5—4.0; $\lambda = 550$ nm	235
SO_4^{2-}	Niwa and Parry (1960)	$Pb(NO_3)_2$ (0.005 M)	25 ml 20—50 p.p.m. solution. Adj. to pale green of bromophenol blue, add 1 ml 20% acetic acid, 100 ml acetone then 0.1% dithizone in acetone to $A = 0.2$—0.3; $\lambda = 620$ nm (free dithizone)	236
Ba^{2+} Sr^{2+} together	Kreshkov and Kuznetsov (1970)	Na_2SO_4 (50 to 500 μM)	μg amounts in 65% acetonitrile in water. Indicator nitchromazo (3,6-bis-(4-nitro-2-sulphophenylazo) chromotopic acid). 6 : 1 to 1 : 3 mixtures resolved	237
SO_4^{2-} from organic sulphur	Yih and Mowery (1971)	$BaCl_2$ (0.005 M)	SO_4^{2-} equiv. to 4 mg S in 15 ml water (neutral) + 15 ml methanol. Indicator 130 mg tetrahydroxyquinone (nephelometric titration)	238
SO_4^{2-}	Rees and Hill (1969)	$BaCl_2$ (0.001 M)	pH 2—4, aqueous-alcoholic medium. Thorin (thoron) indicator. 1 ml 10^{-3} M SO_4^{2-} titrated. Initial seeding carried out; $\lambda = 520$ nm	212

would seem that these investigations must represent the ultimate in fluoride titration. Even silicate species, present following separation of fluoride by Willard—Winter distillation, cause very little trouble.

Early investigators into the turbidimetric titration of sulphate with barium salts appear to have obtained a high degree of precision not equalled by later workers. Thus del Campo, Burriel and Garcia Escolar (1936) obtained a mean error of 0.4% in the titration of 2.5 mmols of H_2SO_4 and Ringbom (1941) an error of 0.2—0.4%.

All workers advocated addition of miscible organic solvent — usually ethanol — but were not in agreement about the addition of colloid stabiliser. Thus van Nieuwenburg and van Bevervoorde in a particularly thorough study recommended a water—ethanol—butanol solvent and concluded that with good magnetic stirring a protective colloid was unnecessary — even harmful. The general impression gained from these papers is that 20 μg—100 mg of sulphate may be titrated under good conditions with a coefficient of variation of about 1%. Again, all indicated sulphate titrations used organic/aqueous mixtures as solvent. Menis et al. introduced isoamyl alcohol and Kreshkov and Kuznetsov used acetonitrile which were a little unusual. The first to put forward this type of analysis was Walter (1950); his paper is interesting in that he monitors the absorbance due to free indicator (rhodizonate). Niwa and Parry were the only analysts to use a titrant other than barium; this was a lead nitrate solution and the production of free dithizone indicator allowed location of the end-point. Rees and Hill found that an initial seeding process greatly improved titration curve quality. The most recently noted titration, that of Yih and Mowery, is interesting in that it is an indicated nephelometric titration. The only other example found by the author was the Mohr titration of Müller and Partridge in 1928.

7. Redox titrations

(A) GENERAL SURVEY

As might be anticipated redox quantitative interactions were among the first to be subjected to the photometric end-point technique. Thus Tingle's idea of detecting the equivalence-point of an acid—base titration in a deeply coloured solution by noting with a visual spectroscope the characteristic absorption band of some suitable indicator was anticipated for many years by Brücke (1877) [240]. He noted that a very dilute $KMnO_4$ solution showed five separate rather sharp absorption bands and used this fact to determine Fe^{II} in Fe^{III} salts and I^- in I_2 by $KMnO_4$ titration — both analyses involving high background absorption. The pioneering photometric titrator of Müller and Partridge (1928) [5] was applied to iodine titrations and, interestingly, to $KMnO_4$ and $K_2Cr_2O_7$ titrations using m-phenylenediamine as redox indicator. As noted earlier, Japanese workers in the 1930's exploited photometric titrations for difficult applied analyses and often used rather subtle redox processes for this purpose. These are summarised at the end of the section.

The value of photometric end-point detection in situations of high background absorbance was well appreciated by early workers. Examples are the determination of manganese in chrome steel by Hirano and Nakamura, of cobalt in nickel sulphate by Somiya and Yasuda, and the work of Lur'e and Tal, whose investigation of the titration of iron(II) with $KMnO_4$ against nickel sulphate and chrome alum backgrounds gave an elegant method for the determination of iron and chromium in nichrome alloy.

Later analysts exploited the singularly effective application of photometric titrimetry to selective and stepwise redox processes. Thus Goddu and Hume (1950) determined vanadium in steel by modification of a popular but complex visual titrimetric method in which the sample was heated to fuming with $HClO_4$ (Fe^{III}, V^V, Cr^{VI}, Mn^{VII}) then cooled, diluted and treated with H_3PO_4 and excess Fe^{II} (Fe^{III}, Fe^{II}, V^{IV}, Cr^{III}, Mn^{II}). Addition of persulphate selectively oxidised excess Fe^{II}, then the selective V^{IV} to V^V titration was carried out using standard $KMnO_4$ photometrically at 525 nm. Two years later Bricker and Sweetser demonstrated the stepwise photometric titration of As^{III} and Sb^{III} in 7 N HCl with standard BrO_3^-/Br^- at 325 nm. The $As^{III} \rightarrow As^V$ oxidation occurs first but no absorbance change is evident as the As^{III} and As^V species present do not absorb. Because $Sb^{III}Cl_6^{3-}$ absorbs at the wavelength used far more strongly than $Sb^VCl_6^-$ a steep absorbance decrease occurs during the antimony oxidation. Following this the absorbance again rises rapidly due to the accumulation of Br_3^- species. Their earlier $As^{III} - Ce^{IV}$ titration at 320 nm was the first example of a UV photometric titration.

Some redox titrations employing coulometric titrant generation have been modified by introduction of photometric end-point detection but bubble formation sometimes proved troublesome.

The fact that redox reactions vary enormously in rate has been nicely utilised by Schenk and Bazzelle (1968) [241]. Symmetrical redox reactions, e.g. $Ce^{IV} + Fe^{II} \rightarrow Ce^{III} + Fe^{III}$ involving exchange of equal numbers of electrons are usually fast but reactions involving the exchange of unequal numbers of electrons (and little structural change) are usually slow. They thus selectively titrated a thallium(I)—iron(II) mixture at 425 nm with cerium(IV) solution. Though both the Fe^{II} and Tl^I oxidations are thermodynamically feasible, Fe^{II} is completely oxidised first due to kinetic differences. Following this, Mn^{III} which acts as a catalyst for the $Tl^I \rightarrow Tl^{III}$ oxidation is added and the thallium determination completed as for iron. In the presence of Mn^{III}, the rate of oxidation of reduced species by Ce^{IV} is again very variable and mixtures such as $Tl^I + Hg^I$ and $Tl^I + Cr^{III}$ may be resolved because oxidation of Tl^I is completed before any appreciable transformation of the other species present occurs. The authors determined rate constants and mechanisms for many of the reactions studied and used these to forecast the feasibility of mixture analyses.

Some further subtle applications of photometric redox titrations were as follows.

Grand and Tamres analysed metal diethyldithiocarbamate chelates by the reaction

$$M\left(\begin{array}{c} S \\ S \end{array}C-N\begin{array}{c} Et \\ Et \end{array}\right) + I_2 \longrightarrow MI_2 + \begin{array}{c} Et \\ Et \end{array}N-\overset{\overset{S}{\|}}{C}-S-S-\overset{\overset{S}{\|}}{C}-N\begin{array}{c} Et \\ Et \end{array}$$

carried out in chloroform. The chelate solution was used as titrant and the disappearance of the I_2 absorption at 510 nm monitored. Complexes of Pb^{II}, Hg^{II}, Zn^{II} and Cd^{II} were successfully studied.

An all-inclusive small scale (2.5 ml titrand solution) anaerobic titration assembly was described by Foust and co-workers in 1969. It was tested by the titration of lumiflavin-3-acetic acid with sodium dithionite and subsequently used for the anaerobic redox titration of enzyme solutions and in studying the thermodynamics and kinetics of protein reactions in oxygen-sensitive redox states.

The Amsterdam analytical school of G. den Boef has been active in the field of redox photometric titrimetry as well as in complexometry. Analyses studied were the use of $KMnO_4$ under strongly alkaline conditions for the direct determination of As^{III} and indirect determination of Cr^{III}, $S_2O_3^{2-}$ and HCOOH. Precision was assessed at 0.3% coefficient of variation for the determination of 1 mequiv. Again on the subject of precision the $Ce^{IV}-Fe^{II}$ reaction was studied and the coefficient of variation to be expected at various dilutions carefully determined theoretically and practically. Den Boef has exploited the performance of inorganic redox photometric titrations in the presence of excess chelating agent. Thus Ti^{III} has been titrated with Fe^{III} in 1 N H_2SO_4 in the presence of 0.1—0.5 M acetylacetone which forms a strongly absorbing complex at 490 nm with both Ti^{III} and Fe^{III}. Analysis of a Ti^{IV} solution proceeds, at 50°C, in these steps where species are present as their acetylacetone complexes

$$Ti^{IV} + Cr^{II}(\text{excess}) \rightarrow Ti^{III} + Cr^{III}$$

$$Cr^{II} + Fe^{III} \rightarrow Cr^{III} + Fe^{II}$$

$$Ti^{III} + Fe^{III} \rightarrow Ti^{IV} + Fe^{II}$$

(Excess Fe^{III} acetylacetonate absorbs strongly and indicates the equivalence-point.) The titration curve appears thus (Fig. 67). The initial reduction of Ti^{IV} may also be performed using V^{II} though the resulting titration curve then shows four branches. Oxygen present is converted to water by excess reductant and does not interfere. Finally the den Boef school has comprehensively examined the determination of both inorganic and organic oxidised species by chromium(II) in alkaline or neutral solution in the presence of excess

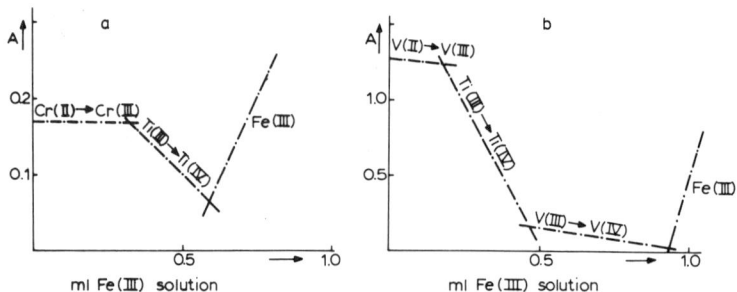

Fig. 67. Photometric titration of (a) 0.142 mg of titanium with 10^{-2} M iron(III) solution after prereduction with chromium(II); (b) 1.42 mg of titanium with 10^{-1} M iron(III) solution after prereduction with vanadium(II) (λ = 490 nm; temp. = 50°; Zeiss PMQ II). Reprinted from ref. 265.

EDTA. The chelating agent enhances the reducing power of the Cr^{II} and eliminates problems due to $Cr(OH)_2$ and $Cr(OH)_3$ precipitation. Also $Cr^{II} \cdot EDTA$ absorbs quite strongly at 850 nm where few other species — including $Cr^{III} \cdot EDTA$ — can interfere. Excess $Cr^{II} \cdot EDTA$ is titrated finally with standard $K_3Fe(CN)_6$ or KIO_3 solution, the decrease of Cr^{II} absorbance being followed. A separate paper deals with the useful determination of nitrate using this principle though here other reducible substances are initially selectively reduced with hydrazine.

(B) THEORY OF PHOTOMETRIC REDOX TITRATIONS

No paper concerning the theory of these titrations or producing an $A = f(\text{mls titrant})$ relation has been found but presumably for the half-equations

$$Ox_1 + e \rightleftharpoons Red_1$$

$$Ox_2 + e \rightleftharpoons Red_2$$

leading to the overall equation

$$Ox_1 + Red_2 \rightleftharpoons Red_1 + Ox_2 \tag{44}$$

where $K = [Red_1][Ox_2]/[Ox_1][Red_2]$

the approach used, by e.g. Flaschka, for metal complex formation may be used. Thus from the photometric viewpoint, eqn. (44) is essentially the same as, e.g.

$$M^{n+} + HY^{3-} \rightleftharpoons MY^{n-4} + H^+$$

the fundamental chemical equation of a self-indicating complexometric system. However, in the redox system four potentially absorbing species are involved, the concentration changes of which may be monitored. For the simple redox system shown

$$\log K = (E_1^0 - E_2^0)/0.0591$$

$$\text{and } E = E_1^0 + \frac{RT}{F} \ln \frac{[Ox_1]}{[Red_1]} = E_2^0 + \frac{RT}{F} \ln \frac{[Ox_2]}{[Red_2]} \tag{45}$$

If a redox indicator is involved,

TABLE 10

Bibliography of photometric redox titrations

Species determined	Author(s) (date)	Titrant	General conditions	Ref.
Fe^{II} in Fe^{III}, I^- in I_2	Brücke (1877)	$KMnO_4$	Dilute H_2SO_4	240
Not specified	Müller and Partridge (1928)	$KMnO_4$ $K_2Cr_2O_7$ I_2	Dilute H_2SO_4; m-phenylenediamine indicator	5
$AuCl_3$	Hirano (1934)	KI	Acidity <0.05 N; use excess Cl_2 water to maintain Au^{III}. No interference from Pb, Cu, Fe.	195
$AuCl_3$	Hirano (1934)	$SnCl_2$	Acidity <0.05 N; use excess Cl_2 water. Cu, Pb and small amounts Fe do not interfere	242
I^-	Hirano (1934)	KIO_3	Acid solution; low I^- concentrations only	243
Mn in chrome steel	Hirano and Nakamura (1934)		(Not mentioned in abstract)	244
V in steel	Somiya and Nakamura (1934)		(Not mentioned in abstract)	245
Mn in steel	Hirano (1937)	$KBrO_3$	Dissolve in 7N H_2SO_4, boil with H_2O_2, add $(NH_4)_2S_2O_8$ and boil 15 min. Add H_2SO_4 to 9 M and H_3PO_4 to 20%. Co and Cr do not interfere	246
Co in $NiSO_4$	Somiya and Yasuda (1938)	H_2O_2	Dissolve 10 g $NiSO_4$ in water, add 2 g $Na_4P_2O_7$ and purge with N_2. Add 5 ml. concentrated NH_3, 10 g $(NH_4)_2CO_3$. Red $\rightarrow$ violet transition. Best if Mn absent	247
Fe		$KMnO_4$ 0.01 N	Fe^{II} produced by Zn/Hg reduction; $\lambda = 525$ nm.	

Element	Author (year)	Reagent	Remarks	Ref.
Cu		I_2	Cu precipitated using zinc powder. Dissolve Cu + Zn in dilute HNO_3, treat with Br_2 water, make just ammoniacal, then acid with acetic. (Cu_2I_2 precipitate appears not to interfere)	
MnO_4^-	Rowland (1939)	H_2O_2	Dilute H_2SO_4; $\lambda = 525$ nm	248
Fe Cr (in nichrome)	Lur'e and Tal (1940)	$KMnO_4$	For Cr, dissolve 0.5 g in 1 : 4 H_2SO_4, dilute to 80 ml, add 3 drops H_3PO_4, crystal of $MnSO_4$ and 2 ml 0.1 N $AgNO_3$, heat, add 3 g $(NH_4)_2S_2O_8$, boil 8 min to red colour, add 20 drops 4N HCl, boil 5 min. Cool, filter, dilute to 250 ml, take 50 ml aliquot, add 20 ml 0.1 N $FeSO_4$. $(NH_4)_2SO_4$ and titrate at 520 nm. For Fe, titrate Fe^{II} directly	249
Fe	Weber (1942)	$KMnO_4$, $K_2Cr_2O_7$	Lecture report only, no details given	250
Fe	Osborn et al. (1943)		"Instrument can be applied to redox systems"	8
V (in steel)	Goddu and Hume (1950)	$KMnO_4$	Sample (5–15 mg V) heated with 10 ml water + 12 ml 60% $HClO_4$ to fuming. Dilute to 80 ml, add 3 ml 85% H_3PO_4 then 0.1 N $FeSO_4$ in 0.5% $HClO_4$ until 5 ml in excess. Add 2.5 ml 15% $(NH_4)_2S_2O_8$. Titrate after 1 min; $\lambda = 525$ nm	251
Np	Hindeman et al. (1949)	Sn^{II}, Fe^{II}, Ce^{IV}	Dilute acid. Investigation of spectra of various oxidation states	252
Fe	Juliard, van Cakenberghe and Heitner (1952)	$K_2Cr_2O_7$	2 M H_2SO_4, $SnCl_2/HgCl_2$ prior reduction. Müller–Partridge type titrator. Indicator Ba diphenylamine sulphonate. C of V 0.2%.	352
Mn		Fe^{II}	Follows $S_2O_8^{2-}$ oxidation of Mn^{II}; $\lambda = 525$ nm	

TABLE 10 (continued)

Species determined	Author(s) (date)	Titrant	General conditions	Ref.
As^{III}	Bricker and Sweetser (1952)	Ce^{IV}	At 320 nm, $\epsilon_{Ce} = 5.6 \times 10^3$ in 1 N H_2SO_4. 4 drops 0.01 M OsO_4 catalyst. Titrant concentration range high, 4×10^{-4} N minimum	253
Organic addition and substitution of Br; oxidn. of As^{III} Sb^{III}	Sweetser and Bricker (1952)	BrO_3^-/Br^-	Br_3^- absorbs 270–360 nm — choose λ (i.e. ϵ) to suit titrant strength 0.25 N–0.001 N max $\epsilon = 104$. Used for olefins, phenols, amines, etc. If reaction slow use larger Br_3^- excess. Hg^{II}, Zn^{II} possible catalysts. As/Sb mixture titrated — 325 nm — see earlier. If sample sensitive to Br_2 use low excess and high ϵ wavelength	254
As^{III}	Wise et al. (1953)	I_2 by coulometry	Titrand 10 ml 0.1 N As^{III}. Electrolyte 0.5 M KI, 0.25 M NaH_2PO_4 adjusted to pH 7; Current 175 mA	255
U, Fe, singly and mixture	Bricker and Sweetser (1953)	Ce^{IV} (0.01 N)	2N H_2SO_4, Cd/Hg reduction to U^{IV} and Fe^{II}; $\lambda = 360$ nm. For mixture H_2SO_4 concentration = 0.2–0.4 N; U^{IV} titrates 1st at 340 nm; Fe^{II} 2nd at 360 nm	256
Fe^{II}, $Fe(CN)_6^{4-}$; Ce^{III}	Bricker and Loeffler (1955)	Co^{III} (prep. by electrolytic. oxidn.)	Dilute H_2SO_4; $\lambda = 610$ nm (Co^{III}) or to suit titrand. $Co_2(SO_4)_3$ solution fairly stable in cold	257
Cr, V	Miles and Englis (1955)	Fe^{II} (0.1 N) As^{III} (0.1 N)	For Cr^{VI} alone, $\lambda = 350$ nm, 5N H_3PO_4, 2N H_2SO_4. For $Cr^{VI} + V^V$. Sample + 6 ml 85% H_3PO_4 + 2 ml concentrated $H_2SO_4 \to 50$. 1st break corresponds to Cr^{VI}. Continue titration with Fe^{II} for V^V. Determination of Cr and V in steel described	258

Fe(in Ti sponge etc.)	Malmstadt and Roberts (1956)	Ti^{III} by coulometry	Ti^{III} produced from Ti^{IV} of sample. Leuco methylene blue indicator C of V = 0.1%. Dissolve 1 g sample in 30 ml 5M H_2SO_4 + 10 ml 45% fluoboric acid. Add 2 ml concentrated HCl, $KMnO_4$ to pink — remove excess with Na azide. Dilute with 4M H_2SO_4 to fill cell, boil 5 min, cool to 75°C, then titrate using λ = 665 nm	259
Ce^{III}	Marple et al. (1956)	$KMnO_4$	Neutral pyrophosphate medium; λ = 525 nm. Reducing subs, F$^-$, Cr^{III} interfere	260
As^{III}	Furman and Fenton (1956)	Ce^{IV} by coulometry	Generating solution saturated $Ce_2(SO_4)_3$ in 1 N H_2SO_4; OsO_4 catalyst; λ = 320 nm. Oxidisable impurities titrated first	261
Thiourea + Me derivs.	Deshmukh and Bapat (1957)	KIO_3 (0.025 M)	0.5N HCl, 20% w/v KBr; λ = 450—480 nm. Complex A vs. titre behaviour	262
Ce(in Bi alloys)	Edwards and Milner (1957)	$KMnO_4$	Pyrophosphate solution; pH 5.5—7.0; λ = 525 nm.	263
U^{IV}	Menis et al. (1958)	Ce^{IV} (0.1 N)	340 nm; 0.5M H_2SO_4 10—200 μg U	72
As, Cr $S_2O_3^{2-}$ H·COOH	Polak et al. (1963)	$KMnO_4$ in alkali. As^{III} back titrant	2M KOH, As^{III} in burette. As^{III} by direct titration; Cr^{III}, $S_2O_3^{2-}$, HCOOH by addition of excess $KMnO_4$ in 2 M KOH then back titration with saturated As^{III}; λ = 525 nm.	264
Ti^{III}	van der Linden and den Boef (1967)	Fe^{III}	1N H_2SO_4; 0.1—0.5 M acetylacetone. Reduce Ti^{IV} with small excess of Cr^{II} or V^{II}. 50°C; λ = 490 nm; 60—500 μg Ti (for details see text)	265
I$^-$	Tanase and Shimomura (1960)	$KMnO_4$ KIO_3	λ = 290 nm. Break point equiv. varies with acid used, e.g. H_2SO_4 or HCl. KIO_3 in H_2SO_4 good for I$^-$ in presence of Cl$^-$, Br$^-$, I_2 and EtOH	302
Fe^{II}	Miyake and Sakamoto (1967)	$KMnO_4$	Dilute H_2SO_4. Test of a recording phototitrator	266

TABLE 10 (continued)

Species determined	Author(s) (date)	Titrant	General conditions	Ref.
Ce^{IV}	Groeneveld and den Boef (1966)	Fe^{II}	1M H_2SO_4; ΔA only 0.01 when titre 0.4460 ml 0.2 N Fe^{II} C of V = 0.141%; when titre 0.514 mls 0.02 N Fe^{II} C of V = 0.234%. Precision study	21
IO_3^-, IO_4^- $Fe(CN)_6^{3-}$ CrO_4^{2-} $Cr_2O_7^{2-}$ BrO_3^- MnO_4^- $S_2O_8^{2-}$ VO_3^- VO^{2+} $UO_2(Ac)_2$ BO_3^- ClO^- NH_2OH NO_2^- methyl yellow, p-amino azo benzene, p-hydroxy azo benzene, picric acid, o-nitro phenol, 2,4-dinitro-phenyl hydrazine, 1-nitro-naphthalene, p-nitroben-zoic acid,	Groeneveld and den Boef (1968)	Cr^{II} in alkali. $K_3Fe(CN)_6$ or KIO_3 back titrants (0.2 N)	In excess EDTA; λ = 850 nm. pH 7–10 (NH_4^+/NH_3 buffer) + tartrate. Solution covered with paraffin oil prior to Cr^{II} addn. Details in script	267

Compound	Titrant	Reference	Conditions	Ref.
maleic acid, fumaric acid, NO_3^-	Cr^{II}, KIO_3 as back titrant	Groeneveld and den Boef (1968)	pH 7.0—7.5; excess EDTA. Add large excess Cr^{II}. λ = 850 nm. 0.4—1.0 mg NO_3^-, C of V 0.5%. Ti^{III} catalyst. For further details see script	268
Metal diethyldithio carbamates	I_2	Grand and Tamres (1968)	In $CHCl_3$; λ = 520 nm. See text for details	269
Tl^I Fe^{II}	Ce^{IV}	Schenk and Bazzelle (1968)	0.1M H_2SO_4; λ = 425 nm. Catalysed by Mn^{III}. Selectivity by kinetic differences — see text	241
V^{IV}	$KBrO_3$ (0.1 M)	Fuller and Ottaway (1969)	H_2SO_4 > 0.06 M: λ = 660 nm (V^{IV}). Mechanism study. V concentration > 2×10^{-3} M	270
Lumiflavin-3-acetic acid, enzymes, proteins etc.	$Na_2S_2O_4$ or general	Foust et al. (1969)	Anaerobic titration unit. Tested with lumiflavin-3-acetic acid; λ = 445 nm, buffer 0.04 M pyrophosphate pH 8.3	60
Te^{IV}	$KMnO_4$	Naidu and Rao (1970)	In "condensed" H_3PO_4 2 Te^{IV} + MnO_4^- + 8 H^+ → 2 Te^{VI} + Mn^{III} + 4 H_2O; λ = 530 nm. Pause 1 min following additions	271
As^{III}	Os^{VIII} (3×10^{-4} M)	Naidu and Rao (1970)	In dilute H_2SO_4 + pyrocatechol — gives blue complex with Os^{VIII} in presence of trace As^{III}; λ = 660 nm	272
As^{III}	$K_3Mo(CN)_8$ (0.005 M)	Duque-Macias and Lucena-Conde (1970)	λ = 400 nm; 3—4 M KOH. 5—50 μeq As per 50 ml. Titrant reduced to $Mo(CN)_8^{4-}$	273
V^{IV} Cr^{VI} V^V	Fe^{II} (0.1 N)	Dikshitulu (1971)	9M H_3PO_4. V^{IV} at 430 nm; V^V at 680 nm (2 breaks). Sequential V^V, Cr^{VI} at 640 nm (3 breaks). V^V—U^{VI}; V^V—Mo^{VI} also	274
Phenothiazine derivatives	$Ce(SO_4)_2$ (0.005 M)	Chatten et al. (1971)	Good for thiethylperazine and thioridazine. Titrate 0.25 mg of drug in 1M H_2SO_4 at 275 nm	275

$$E = E_{In}^0 + \frac{RT}{nF} \ln \frac{[In_{ox}]}{C_{In} - [In_{ox}]} \tag{46}$$

If E can be determined from eqn. (45) then $[In_{ox}]$ can be calculated from eqn. (46) whereupon $A = \epsilon_{In_{ox}}[In_{ox}]\,l$ assuming it is the oxidised form of the indicator which absorbs.

8. Spectrofluorimetric phototitrations

Establishment of the value of photometric absorption titrations made the utilisation of photometrically measured fluorescent indicator changes inevitable. Relatively few papers have appeared on this topic but all have made a definite contribution. The first instrument designed to use fluorescent light was that of Osborn et al. (1943) [8]. In this, a third photocell lying at right angles to the main light beam intercepted fluorescent radiation and was connected in a bridge circuit with the photocell receiving directly transmitted light. An enhanced effect was thus obtained since a quantity proportional to the ratio of the two responses was actually measured. No specific fluorimetric titrations were detailed but the instrument was obviously applicable to all situations where a fluorimetric indicator could be used.

The first specific spectrofluorimetric phototitration was that of Willard and Horton (1950) which used the fluorescence of the thorium—quercetin complex at 575 nm for the determination of fluoride.

$$4F^- + Th^{IV} + nQuerc. \rightarrow ThF_4 \downarrow + Th(Querc)_n$$

(titrant) slightly   fluorescent
        fluorescent

Because this was a heterogeneous reaction, photocell response was due to both fluorescent light from the metal complex formed and Tyndall scattering from ThF_4 particles. Curvature of the fluorescence intensity vs. mls $Th(NO_3)_4$ solution graph in the region of the endpoint, which was lessened by allowing the precipitate to settle, was rather marked but intersection points could still be located with good precision. Thus 95% confidence level deviation for 1 mg F was 1.3% and for 15 mg F 0.6%. Interferences were those usually encountered in metal ion—fluoride titrations but PO_4^{3-} added on exactly and large masses of suspended ThF_4 were well tolerated.

Luminol,

in the presence of H_2O_2 as oxidant and haemoglobin as catalyst emits light by a chemiluminescent process. The emission is sensitive to hydroxide ion concentration and Kenny and Kurtz (1952) used the process for the titration of 0.1 N NaOH with 0.1 N HCl in the presence of very high background colour. A photomultiplier tube was used to monitor the light produced at pH >6.6. Light levels undetectable by eye were easily measured; this allowed a great reduction in luminol, H_2O_2 and haemoglobin concentrations to be made, which greatly reduced the buffer capacity of solutions. This high capacity had been an unfortunate feature of earlier visual methods. The precision of the instrumental method was 0.7 p.p.t.

Howerton and Wasilewski (1961) designed an automatic colorimetric and fluorimetric (photomultiplier) titrator and examined in detail the titration of calcium with EDTA using calcein as indicator. They commented clearly on the somewhat critical indicator concentration required; too little gave a weak signal and poor slope difference while too much produced concentration quenching effects. Borle and Briggs (1968) took this calcium—EDTA—calcein—pH 13 determination to very low levels of calcium in a remarkable and significant piece of work. Their automatic titrator built around a Turner fluorometer could determine Ca specifically over a range 0.04 μg—1μg with coefficient of variation $<2\%$. No interference was encountered from biologically reasonable concentrations of phosphate, magnesium, A.T.P. or protein. The need to use calcium-free plastic ware wherever possible was stressed.

One of the most consistent investigators of photofluorimetric titrations has been J.A. Bishop of Newark Engineering College. 8-Quinolinol-5-sulphonic acid forms fluorescent complexes with "complete shell" ions Zn^{2+}, Cd^{2+}, Mg^{2+} and non-fluorescent complexes with incomplete shell transition metal ions. Thus Cu^{2+} ions may be titrated with a standard solution of the magnesium complex of the reagent $Mg(HQSO_3)_2$

$$Mg(HQSO_3)_2 + Cu^{2+} \rightarrow Cu(HQSO_3)_2 + Mg^{2+}$$

titrant (fluorescent)   non-fluorescent

The equivalence-point is indicated by the appearance of fluorescence. The quality of titration curves is related to relative stability constant values. Mixtures of metal ions with suitably related formation constants e.g. Cu^{2+}, Zn^{2+}, Mn^{2+} may be resolved. Titration of Cu + Cd or Cu + Zn mixtures with non-fluorescent $Hg(HQSO_3)_2$ yielded stepwise plots by a related process. Bishop has extended his work by combining the photodecomposition of the $Co^{III}(oxalate)_3{}^{3-}$ ion with the fluorescence of $HQSO_3$ complexes. Thus a fluorescent complex $M(HQSO_3)_2$, where M = Zn, Cd or Mg, may be determined by the reactions

$$Co^{III}(ox)_3^{3-} \overset{h\nu}{\rightarrow} Co^{2+}$$

$$M(HQSO_3)_2 + Co^{2+} \rightarrow Co(HQSO_3)_2 + M^{2+}$$

fluorescent   non-fluorescent

Here fluorescence diminishes with time to a sharp cut-off point. The same mercury discharge lamp is used to produce the photodecomposition and provide the fluorescence excitation energy. Ions producing non-fluorescent $HQSO_3$ complexes may be determined by the reactions

$$Cu^{2+} + Cd(HQSO_3)_2 \rightarrow Cu(HQSO_3)_2 + Cd^{2+}$$

     (excess)   non-fluorescent

$$Cd(HQSO_3)_2 + Co^{2+} \qquad \rightarrow Co(HQSO_3) \qquad + Cd^{2+}$$

(remainder)   (from   non-fluorescent
                photodecomp.)

Wiersma and Lott (1968) have studied the titration of metal ions in low concentration (total ~ 30 μg) with EDTA using morin as indicator. Direct titration suffices for metals forming fluorescent morin complexes but gallium(III) whose morin complex is particularly strongly fluorescent is used for back titration of excess EDTA in the determination of ions which do not fluoresce with morin. Mixtures of Ga^{III} and In^{III} were resolved using an adaptation of the EDTA—

TTHA process of Pribil and Vesely [276]. They stress that concentration quenching is not important and that contamination from other fluorescing compounds only increases the residual fluorescence level resulting in a slight loss of sensitivity whereas in ordinary fluorescence methods such a contamination results in extremely high errors.

A recent example of a simple fluorescence titrator is that described by Clements and Sergeant [290]. It has a capacity of 100 ml and light selection by means of filters. The instrument was tested using the Ca^{2+}—EDTA—calcein reaction and also the interesting Cu^{2+}—EDTA—o-dianisidine tetraacetic acid system of Belcher et al. [277].

Finally Wronski, pioneer of mercury—sulphur titrations, has produced some remarkable trace analyses of sulphur-containing species by photofluorimetric titration using tetramercurated fluorescein (tetramercuryacetate fluorescein — TMF).

In alkaline solution TMF fluoresces strongly at 520 nm whereas complexes formed with sulphur species such as sulphide, mercaptans or thiophenols are far less fluorescent. Thus titration of, for example, H_2S with a standard 5×10^{-6} N solution of TMF yielded an initial almost horizontal line of low fluorescence intersected by a line of considerable slope as equivalence was passed. An H_2S solution of concentration 0.8 μg H_2S l^{-1} was satisfactorily analysed. Thiophenols were equally well determined but mercaptans alone yielded less suitable titration curves. Strangely H_2S + R—SH mixtures showed curves with well defined breaks for both constituents. Sulphide and mercaptan solutions were stabilised with formaldehyde which gave R S CH_2OH or HO CH_2S CH_2OH species; these reacted with TMF in an identical manner to RSH and H_2S. For the determination of sulphur species in organic solvents in which TMF is unstable, weak complexes of TMF were used as titrants. The TMF—oxine complex is

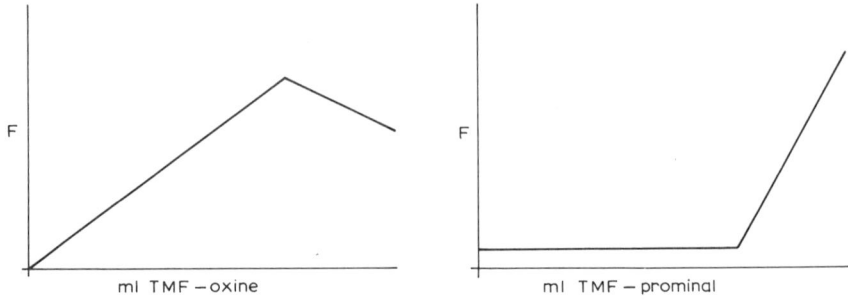

Fig. 68. Titration of sulphur compound — product fluorescent.

Fig. 69. Titration of sulphur compound — product non-fluorescent.

non-fluorescent while the TMF—methylphenobarbitone (prominal) complex is highly fluorescent. Oxine or prominal is displaced from its TMF complex by the sulphur species being determined. If the titrand produces a complex with a strong fluorescence the determination is carried out in the presence of oxine to give a titration curve shaped thus (Fig. 68). The sudden decrease is due to light absorption by the orange TMF—oxine complex which builds up after equivalence. On the other hand, substances strongly quenching the fluorescence of TMF are titrated in the presence of prominal to give the type of graph shown in Fig. 69. TMF—oxine was used for n-alkyl thiols, dithiocarbamates (from CS_2 + sec. amine) and HCN; TMF—prominal was used for thioacids and H_2S.

Wronski applied the method to the selective determination of H_2S, CS_2, COS, RSH and HCN in air. Thiocarbamates and thiols were stabilised using triethyl lead chloride. An elegant procedure for the determination of traces of H_2S in water was recently proposed in which H_2S was extracted as $Et_3PbSPbEt_3$ into n-hexanol using a solution of Et_3PbCl in hexanol. The extract was titrated with 2×10^{-5} N TMF—prominal in 80% ethanol. Fluorescence is minimal at first but increases sharply on completion of the reaction between TMF and triethyl lead sulphide.

The remarkable sensitivity and specificity shown by many fluorimetric analyses during recent years would seem to indicate that photofluorimetric titrations are a fruitful field for research; indeed the school of den Boef has recently embarked on such a program [353]. The work of Borle and Briggs, and Wronski in particular demonstrates the potential of the technique.

TABLE 11

Bibliography of photofluorimetric titrations

Species determined	Author(s) (date)	λ excit.	λ emiss.	Titrant	Other conditions	Ref.
F^-	Willard and Horton (1950)		575 nm	Th^{IV} 0.015 M 0.0032 M	pH 3.0, 50% ethanol. Ranges 0.1–2.0 mg F, 1–40 mg F. Indicator 0.02% quercetin in ethanol	278
OH^-	Kenny and Kurtz (1952)			HCl (0.1N) + 10 mg/ litre of luminol	30 ml NaOH + 0.5 ml 3% H_2O_2 + 0.1 ml 5% haemoglobin. Chemiluminescent reaction.	15
Ca^{2+}	Howerton and Wasilewski (1961)	470 nm	524 nm	EDTA, Ca^{2+} back titn.	20 ml 2×10^{-4} M EDTA titrated with 2.5×10^{-3} M Ca^{2+}. 40 μg calcein as indicator; 0.2 N in NaOH.	279
Ca^{2+}	Borle and Briggs (1968)			EDTA (0.25 mM)	0.2–1 μg Ca in 3.25 ml calcein–KOH solution, KOH = 5 N.	280
Cu, Ni, Co, Fe^{II}, Mn^{II}, Hg^{II}	Bishop (1966)	365 nm	495 nm	Mg (HQ SO_3)$_2$ 10^{-2} M	pH 7–8, titrands 10^{-5} M Cu + Zn + Mn; Cu + Zn + Hg mixtures resolvable. Cu + Cd and Cu + Zn solutions resolvable with Hg(HQSO$_3$)$_2$ titrant	281
Pd^{II}	Bishop (1967)	365 nm	510–525	Mg (HQSO$_3$)$_2$, Cd (HQSO$_3$)$_2$, Zn (HQSO$_3$)$_2$ 10^{-3} M	pH 7. Titrand concentration 10^{-5} M. Titration done in external vessel. Exchange very slow; boil solution 10 min after each addition	282

Species determined	Author(s) (date)	λ excit.	λ emiss.	Titrant	Other conditions	Ref.
Zn, Cd, Mg, Cu	Beck et al. (1970)	Hg lamp. prob. 365 nm	500 nm	$Co(Ox)_3^{3-}$ photo-decomp. to given Co^{2+}	$1-15$ μmols of $M(HQSO_3)_2$ complex + 20 ml 0.5 M PO$_4^{3-}$ buffer; pH 9 + 4 ml 5×10^{-2} M $Co(Ox)_3^{3-}$, dilute to 200 ml. Also Cu by initial exchange — see text	283
Ca, Mg, Fe together. In limestone, cement, serum	Escarrilla (1966)			EGTA (0.005 M) EDTA (0.005 M) Cu. EDTA (0.005 M)	Calcein blue indicator. Metal complexes fluoresce. For Ca take 1–6 ml of sample solution. Mask Fe, Al, Mn with 0.5 ml 20% triethanolamine. Add 1 ml 0.02% indicator. Adjust to pH 13 with 2M KOH. Titrate with EGTA to 100% past equivalence point. For Mg, add 2 ml more of 0.005 M EGTA and 0.1 ml 1 M HCl to above solution. Add 1 ml indicator solution, 2 ml ammonia buffer (67.5 g NH_4Cl + 570 ml concentrated NH_3 in 1 litre) and 1 ml 0.1M KCl. Fix pH to 11, titrate with Cu–EDTA complex. For Fe, to above solution add 0.1 ml 37% H.CHO, 0.5 ml 30% H_2O_2, 2 ml indicator and 1 ml EtOH.Check pH is 10.5. Titrate with EDTA of suitable concentration. Note: Fe is demasked from its TEA complex by H.CHO. Then {Fe^{III}. C.B} + EDTA + $H_2O_2 \to$ [Fe.EDTA.H_2O_2] + CB	284

Substance	Reagent		Author	Ref.	Procedure
Fe^{III} Sm^{III}, Hg^{II} Al^{III}, Ga^{III}, In^{III} in mixture — see text					metal — final volume 90 ml. To suitable volume of water add 2 ml buffer (pH 2.0–3.0 — 2 M $ClCH_2COOH$ + 1 M NaAc) 0.5 ml indicator and a few drops of Ga^{III} solution. Titrate with Ga^{III} solution. Add test solution, excess EDTA and titrate with Ga^{III}
H_2S. >0.01 $\mu g\,l^{-1}$	Tetra mercurated fluorescein (TMF) 2×10^{-5} M in EtOH	520	Wronski (1971)	286	Shake 500 ml of water sample containing 0.5 ml solution A (2.5 N NaOH, 0.5 N in EDTA) with 10 ml 0.01% Et_3PbCl in hexanol. To 3.5 ml of extract add 0.5 ml solution B (20 ml saturated prominal in EtOH + 1 ml triethylamine diluted to 100 ml with 96% etOH) then titrate Multiply result by 1.8 to allow for solubility of hexanol in water
Sulphide, mercaptans, thiophenols	TMF 5×10^{-6} N — 5×10^{-5} N		Wronski (1961)	287	Dilute NaOH 0–0.4 μg H_2S in 25 ml
Thiols, thiophenols, thioacids, thiocarbamates, xanthates, subs. thiourea, H_2S, HCN	TMF—oxine 2×10^{-5} N; TMF—prominal 2×10^{-5} N in triethylamine—ethanol-isopropanol	520	Wronski (1970)	288	Take 3 ml sample in organic solvent + 0.5 ml EtOH + 0.2 ml 1% triethylamine in IPA + 0.2 ml 0.5% oxine in IPA. Exclude oxine for TMF—prominal titrations

TABLE 11 (continued)

Species determined	Author(s) (date)	λ excit.	λ emiss.	Titrant	Other conditions	Ref.
H_2S, CS_2, COS, R—SH, HCN in air	Wronski (1971)		505 520	TMF 2×10^{-5} N in water or alcohol	H_2S, range 3–200 ng. Mix 2 ml of test sample with 0.4 ml 0.16 M EDTA and titrate; λ_{em} = 505 nm..For CS_2 and COS, titrate 4 ml of absorption solution with alcoholic 2×10^{-5} N TMF—oxine — likewise for thiols	289

TABLE 12

Bibliography of organic functional group photometric titrations

Species determined	Author(s) (date)	Titrant	Principle and general conditions	Ref.
Procaine, propoxycaine, tetracaine, general aromatic amines	Pratt (1957)	HNO_2 (0.1M)	$R-NH_2 + HNO_2 \rightarrow R-N\equiv N^+$ $R_2NH + HNO_2 \rightarrow R_2N-NO$ 6 N HCl; λ = 385 nm (HNO_2). Can use product absorbances. Sample wt $\sim$ 1.6 mg. Vol. 150 ml	293
Olefins	Miller and de Ford (1957)	Br_2 (coulometric generation)	$\Large \diagdown \!\! C \!\! = \!\! C \!\! \diagup \, + \, Br_2 \longrightarrow$	294

Substance	Reference	Reagent	Remarks	Ref.
Water in ketones, esters and ethers	Jackwerth and Specker (1959)	$Cu(ClO_4)_2$ 10^{-3} M in acetone	Monitor Br_3^- absorbance at 360 nm. Solvent glacial acetic–methanol $3:1$ + KBr + HCl. $HgCl_2$ catalyst. Vary λ to change sensitivity. Sample 3–12 mg in $\sim$ 100 ml. $3\,LiCl + Cu(ClO_4)_2 \rightarrow LiCuCl_3 + 2\,LiClO_4$. Absorbance of $LiCuCl_3$ depends on water content. $\lambda = 366$ nm. 25 ml titrand + 3.00 ml 0.01 M LiCl in acetone	295
Sulphanilamides	Tanase and Shimomura (1960)	HNO_2	$\lambda = 355$ nm. Effective for sulphanilamide and sulphacetamide. $R-NH_2 + HNO_2 \rightarrow R-N{\equiv}N^+$	302
Aromatic amines	Reynolds et al. (1960)	Acetic anhydride 10^{-3} M in pyridine	$Ph-NH_2 + 2(CH_3CO)_2O \rightarrow Ph-N(COCH_3)_2 + 2\,CH_3COOH$. Monitor dec. in abs. of amine. Rather specific. Solvent, dry pyridine saturated with HCl (0.23 M). Vol., 100 ml, wt. 1 mg minimum. Selectivity by rate difference	296
Sulfhydryl groups	Klotz and Carver (1961)	Salyrganic acid 10^{-3} M	$R\,Hg^+ + R^1SH \rightarrow R\,Hg\,S\,R^1 + H^+$. pH 5.5–6.5 (acetate). Indicator, pyridine-2-azo-p-dimethylaniline. Sample, 1 ml 10^{-3} M thiol etc., $\lambda = 550$ nm. Works with suitable proteins	297
Mercaptans	Fritz and Palmer (1961)	$Hg(ClO_4)_2$ 0.05 M aq.	$2\,R\,SH + Hg^{2+} \rightarrow (RS)_2Hg + 2\,H^+$. Sample 0.3–1 mmol. Indicator thio-Michler's ketone. Solvent water or acetone; $\lambda = 580$ nm	298
1,3-dienes	Schenck and Ozolins (1961)	Tetracyano ethylene; back titn. cyclopentadiene 0.05 M in ethanol	$CH_2 = CH-CH=CH_2 + (NC)_2-C=C-C(CN)_2 \rightarrow$ (cyclohexene ring bearing two CN groups and CH_2 groups) Diels–Alder reaction. Solvent, CH_2Cl_2. Indicators, phenanthrene in benzene then pentamethyl benzene in CH_2Cl_2. Diene added to excess TCNE. Indicator functions by π interaction.	299

TABLE 12 (continued)

Species determined	Author(s) (date)	Titrant	Principle and general conditions	Ref.
Polynuclear aromatic hydrocarbons	Schenk and Ozolins (1961)	Tetracyano ethylene (0.1 M) in CH_2Cl_2	Hydrocarbons are π bases; TCNE is a strong π acid. Sample, minimum 0.3 mmol; solvent $CHCl_3$, volume 60 ml. π complex formed absorbs strongly.	300
Prim. aliph. amines	Liu and Reynolds (1962)	2-ethyl hexanal. (0.1 M) in dioxan	$R-NH_2 + R^1CH=O \xrightarrow{H^+} R^1CH=NR + H_2O$. Solvent, dioxan; glacial acetic acid catalyst (0.005 M). Vol. $\sim$ 35 ml; λ = 305 nm (aldehyde)	301
Amides	Reynolds (1963)	OBr^-, actually $Ca(OCl)_2 + Br^-$.	$RCONH_2 + OBr^- \rightleftharpoons RCONHBr + OH^-$, aqueous solution 0.2 M in KBr, pH 9.9 (borax); λ = 350 nm (O Br^-). Sample 0.3 meq.	291
Aldehydes	Cochran and Reynolds (1961)	$NaBH_4$. (1.6 N) in DMF	$R\,CHO \xrightarrow{BH_4^-} R.CH_2OH$. Solvent, IPA—water. Use λ max. of CH=O chromophore. Selectivity by rate variation. Vol. = 40 ml. Use 1 mmol of solute. Heat if necessary	303
Primary aliphatic amides	Post and Reynolds (1964)	OBr^-	Reynolds (1963) in greater detail. Sample 0.08–0.34 mmol. Can be rather specific	304
R_3Al R_2AlH	Wadelin (1963)	Isoquinoline (0.25 M) in toluene	These complexes form: R_3Al. IQ, R_2AlH. IQ, R_2AlH $(IQ)_2$ (red). Solvent, dry toluene; λ = 460 nm; Sample 1–2 mmol active Al. Add some R_2AlH if not present. Perform under dry N_2	305
O_2 in solvents and antioxi-	Paris et al. (1964)	2,4,6-tri-t-butyl phenoxy free radicals.	2 TTBP· + O_2 → TTBP–O–O–TTBP. TTBP· + RH → TTBP + P·.	306

352

Substance	Reference	Reagent	Remarks	Ref.
dants		2×10^{-3} M in various solvents	Prepare titrant by passing solution of 2,4,6-tri-t-butyl phenol over PbO_2—celite column; $\lambda = 625$ nm (radicals). Perform under N_2	307
Thiols	Haglund and Lindgren (1965)	$PdCl_2$ (0.05 M) in 0.3 M HCl	$RSH + PdCl_2 \rightarrow RSPdCl + HCl$. $\lambda = 490$ nm; indicator p-nitrosodimethylaniline. Solvent 1 ml 0.3 M HCl + 15 ml ethanol. Sample $\sim 10^{-5}$ mol. Selective. Perform under N_2	308
Starch polysaccharides	Richter and Szejtli (1966)	I_2 (KI + KIO_3) various concentrations	$\lambda = 650$ nm. Complex study. Clear breaks obtained in many cases	309
Protein—SH groups, and thiol comps.	Boyer (1954)	p-mercuribenzoate (7×10^{-5} M) neut. aq.	$2\,RSH + HO.Hg.Ph.COOH \rightarrow Hg(SR)_2 + PhCOOH + H_2O$. pH 4.5—7.5. Measure inc. in abs. at $\lambda = 250$ nm; absorbance increases due to mercaptide formation	
Primary and tertiary amines together	Citron and Dolan (1965)	Cu.EDTA	Total amine: $2\,(amine) + CuH_2Y \rightarrow CuY\,(amine)_2 + 2\,H^+$. Measure dec. in abs. of CuH_2Y at 720 nm. For constituents: Treat prim. and tert. amine mixture of total concentration 0.01 M (preparable from determination of total concentration) with CuY^{2-}; $\lambda = 720$ nm. Complex analysis, see original	310
Phenols, olefins, general unsaturates	Williams et al. (1964)	Pyridine bromide, perbromide (10^{-3} N min) in acetic acid + + 10% CH_3OH	$C_5H_5NH^+\ Br_3^-$ functions as Br_2 for bromine addition. Phenols: 0.1 to 0.6 mmol. in 25 mls acetic acid + 1 ml tetramethylguanidine. $\lambda = 400$—450 nm (reagent). Aromatic ethers and unsaturates: solvent, 10 ml 0.05 M $HgCl_2$ in methanol + 40 ml methanol	311
Olefins	Fritz and Wood (1968)	Bromine, (0.12 M) in glacial acetic acid	0.3—1.0 mmol of sample in 5 ml CCl_4 + 85 mls glacial acetic acid + 10 mls water; $\lambda = 400$ nm	312

TABLE 12 (continued)

Species determined	Author(s) (date)	Titrant	Principle and general conditions	Ref.
Primary nitro-paraffins	Reynolds and Underwood (1968)	NaOH (0.05 M) aqueous	$R-CH_2NO_2 + OH^- \rightarrow R-CH = NO_2^- + H_2O.$ $R-CH=NO_2^- + HNO_2 \xrightarrow{H^+} R-C-NO_2$ "aci" form $\underset{\text{nitrolic acid}}{\overset{\parallel}{N}-OH}$ Nitrolic acid anion absorbs at 420 nm, follow absorbance change during NaOH titration	313
1—10, phen-anthroline; 2,4,6-tri pyridyl-s-triazine	Fitzgerald and Beck (1970)	Fe²⁺ from photo-chemical decomposition of ferric citrate	Formation of pH 3.5—5.5; $\lambda = 546$ nm (Hg line). 1—40 μmols. titrated. Vol. = 140 ml	314
Aromatic amines	Williams and Wakeham (1970)	Pyridine bromide, per-bromide (0.05 M) in 10% MeOH in acetic acid	60 ml glacial acetic acid + 15 ml 0.05 M HgCl₂ in methanol + 5 mls 0.05 M amine solution; $\lambda = 425$ nm	315
Ethane-1-hydroxy-1,1-diphosphonic acid (EHDP)	Liggett and Libby (1970)	Th-diamino cyclohexane tetra acetic acid (0.001 M) aqueous	EHDP + 2 Th.DCTA → (Th.DCTA)₂.EHDP End point: 2 Th.DCTA + 2 XO → Th₂(DCTA)₂ (XO)₂ (XO = xylenol orange). Sample 100—150 μg EHDP. Vol. = 100 ml; pH = 6.5; $\lambda = 575$ nm	316

| Phenols, aromatic amines, active $-CH_2-$ group comps. | Leonard and Murray (1970) | Various diazonium salts, e.g. p-nitrobenzene diazonium chloride 10^{-1}–10^{-3} M | $Ph-OH + Ph^1N{\equiv}N^+ \rightarrow HO-PhN=N-Ph^1$. Azo dye production by coupling reaction. To 15 ml of test solution, pH ~ 6, add 1.5 ml 1 M NaCl, 2.5 ml ethanol, 20 ml $CHCl_3$ and 2.5 ml iso-butanol. λ to suit azo dye produced. 1 phase titration suitable for soluble dyes | 25 |

9. Organic functional group photometric titrations

The value of applying a photometric end-point technique to organic functional group titrations has been nicely summarised by Reynolds, one of the principal exponents of the idea [291]. Reactions involved in functional group analysis are often slow and of low equilibrium constant. Linear titration techniques which can to a large extent overcome these drawbacks by taking measurements far from the equivalence-point are of the amperometric, conductimetric, thermo-metric and photometric types. Amperometric titration requires that at least one component undergo a reproducible electrode reaction in the solvent used. For conductimetric titrations to succeed, a change in the type of freely mobile ionic species must occur which is dif-ficult in the non-aqueous solvents often used. Again, for successful thermometric titration a large enthalpy change must be evident which is rare for the types of organic reactions involved. For a photo-metric titration to be applicable only one reacting or product species must absorb in the near-UV—visible region of the spectrum and this easily satisfied criterion is the basis of the very wide applicability of photometric titrations to organic analysis.

Because such a great variety of chemical reactions has been used in organic photometric titration the logical sub-division of this section is difficult though certain trends have become evident. Analyses in which both titrand and titrant are organic are fairly uncommon but a good example is the determination of primary aliphatic amines using 2-ethylhexanal proposed by Liu and Reynolds.

Here the amine sample is dissolved in dioxan containing acetic acid as catalyst and titrated directly with a standard solution of the alde-hyde in dioxan. The first excess of aldehyde is detected at 305 nm where the —CH=O chromophore absorbs strongly. Imine absorbance seldom exceeds 270 nm and therefore does not interfere. The reac-tion is specific for primary amines with respect to secondary or tertiary types but aromatic primary amines interfere badly — not be-cause they react at an appreciable rate but because of their large UV absorbance in the 300 nm region.

356

Again, Reynolds has determined aromatic amines by acetylation using acetic anhydride

$$PhNH_2 + 2(CH_3CO)_2O \rightarrow PhN(COCH_3)_2 + 2CH_3COOH$$

The amine, dissolved in pyridine saturated with HCl, is titrated with acetic anhydride dissolved in pyridine. The decrease in absorbance of the unreacted amine is followed at an appropriate wavelength. Amines are preferentially acetylated with respect to alcohols etc. and different aromatic amines may be distinguished by varying rates of acetylation, e.g. aniline and 2-naphthylamine may both be determined in admixture. Anilines substituted with electron donor groups are especially amenable to acetylation.

π acid—base interactions have been used by Schenck and Ozolins in the determination of Diels—Alder active dienes and aromatic hydrocarbons. In the former case

Here the diene, e.g. butadiene, is reacted with excess tetracyano-ethylene (TCNE) in dichloromethane to form the Diels—Alder adduct shown. After reaction excess TCNE is titrated with standard cyclopentadiene in alcohol. The disappearance of the absorbance of the strongly coloured π complexes formed between TCNE (a very strong π acid) and phenanthrene — then pentamethylbenzene — yields the equivalence-point. In the latter type of analysis numerous aromatic hydrocarbons (π-bases) have been titrated with TCNE in dichloromethane. Such titrations are self-indicating due to the high absorbance of the (usually 1 : 1) π complexes formed but this is just as well since equilibrium constants are very low. Thus only relatively concentrated solutions can be titrated together and slope differences at graph intersections are small. (High-precision or "expanded scale" photometry can be useful here.) However, strong π bases can often be determined in the presence of weak ones and stereochemical factors (e.g. in phenols) sometimes allow differentiation. A mixture of anthracene and fluoranthene may be resolved by determining anthracene via the Diels—Alder reaction described above and the sum of both polynuclear hydrocarbons by the TCNE titration.

The remaining case of a truly organic—organic system is the determination of phenols, aromatic amines and active methylene group compounds by a diazo coupling reaction with various standard diazonium salt solutions proposed by Leonard and Murray, e.g.

$$Ph\text{—}OH + Ph^1N\equiv N^+ \rightarrow HO\text{—}PhN\text{=}N\text{—}Ph^1$$

azo dye

Reactions producing soluble dyes were carried out in neutral or weakly alkaline one-phase aqueous systems but those giving water-insoluble dyes were done in a two-phase water—chloroform system. The absorbance of dye extracted into the chloroform layer was monitored. The most generally useful titrant was p-nitrobenzene-diazonium chloride but others of a similar nature were sometimes used. Titrands, such as resorcinol, capable of coupling in more than one position yielded titration curves showing two breaks (Fig. 70). The accuracy of this method proved disappointing in many cases but apparently has great value in studying the course of a diazotisation reaction.

Organic materials have been analysed using a wide variety of inorganic titrants though bromination and heavy metal—sulphur reactions predominate. Olefins were determined by coulometric bromine generation in glacial acetic acid—methanol solvent by Miller and de Ford.

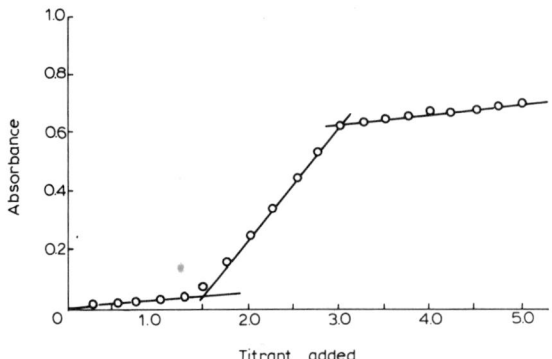

Fig. 70. Photometric coupling titrations. The titration of resorcinol (10^{-3} M) with p-toluene diazonium chloride (10^{-2} M). $\lambda = 580$ nm; initial pH = 6. Reprinted from ref. 25 by permission of Marcel Dekker, New York.

358

The equivalence-point was detected through the rise in absorbance at 360 nm of the Br_3^- ion produced from KBr present in the reaction mixture.

$$Br_2 + Br^- \to Br_3^-$$

This followed the initial idea of Sweetser and Bricker mentioned in the Redox section. Mercuric chloride was used as catalyst and sensitivity was controlled by varying the wavelength used about the Br_3^- absorption band. A great feature of this analysis is that excess bromine is present for only a short period at very low concentration at the end of the titration. This eliminates many interfering oxidation or substitution reactions. However, rate of bromination and ratio of bromine substitution to addition vary widely with titrand structure. The problems of peroxide formation were well brought out. Fritz and Wood later used bromine in glacial acetic acid as titrant and an acetic acid—water—carbon tetrachloride mixture as solvent. Bromine absorption at 400 nm was used for end-point location and a high degree of selectivity introduced by adding or withholding the mercuric chloride catalyst. Thus simple olefinic unsaturation could be distinguished from that due to $C=C$ groups conjugated with $C=O$, $C\equiv N$ or other electron-withdrawing groups. The danger of preparing solutions of unsaturates in active solvents was stressed. Williams and co-workers have advocated standard solutions of pyridine bromide perbromide

— a stable solid commonly used in preparative brominations — as brominating titrant. The titrant solvent was 10% methanol in glacial acetic acid and the reaction solvent generally acetic acid. Mercuric chloride was used as catalyst for substituted olefins and amines, and 1,1,3,3-tetramethylguanidine as catalyst for phenols. The reagent shows λ_{max} at 278 and 380 nm but various wavelengths in the range 325—525 nm were used according to sensitivity required to monitor the first excess of reagent. Compounds analysed were phenols, general unsaturates, aromatic ethers, and aromatic amines. Phenols sensitive to oxidation such as m-aminophenol were not oxidised; there was no precipitation of products and no replacement of functional groups. However, in common with the other bromination procedures reaction rate was very variable. Finally Post and Reynolds utilised the first stage of the Hofmann rearrangement to determine amides.

N-bromoamide

The titrant was standard calcium hypochlorite, and hypobromite was generated in situ from bromide ions present in the aqueous pH 10 titration solvent

$$Ca(OCl)_2 + 2Br^- \rightarrow 2OBr^- + CaCl_2$$

The end-point was determined from the absorbance at 350 nm of excess hypobromite; the usual N-bromoamide products absorbed very little here. Some N-bromoamides rearranged at an appreciable rate to give products which themselves consumed hypobromite. To combat this the first leg of the titration curve was drawn hypothetically using the previously determined extinction coefficient of the particular N-bromoamide. The analysis could be made reasonably specific.

Determination of a wide variety of thiols by utilising sulphur — B type metal ion interactions has proved to be a fruitful field. The first analysis of this type, though in fact a "standard flask" titration, was that of Boyer who determined SH groups in sulphur amino acids, peptides and proteins by means of "p-mercuribenzoate".

$$2 \ R \ SH + HO \cdot Hg \cdot Ph \cdot COOH \rightarrow Hg(SR)_2 + Ph \ COOH + H_2O$$

mercaptide

The reaction was carried out in aqueous solution pH 4.5—7.5 and the increase in absorbance at 250 nm due to mercaptide formation noted. Titration of simple molecules such as cysteine $HS-CH_2-CH-COOH$

$\qquad\qquad\qquad\qquad\qquad\qquad\qquad\qquad\qquad NH_2$

or mercaptosuccinic acid gave sharp breaks at the 1 : 1 R—SH—mercuribenzoate ratio. The number of SH groups present in protein molecules and indeed their environment could be deduced from titration curves, reaction rates and product absorption spectra. A somewhat similar type of analysis was later proposed by Klotz and Carver who determined sulfhydryl groups using salyrganic acid (the anhydride of O-{[3-(hydroxymercuri)-2-methoxypropyl] carbamyl} -phenoxy acetic acid) — R Hg$^+$.

$$R\ Hg^+ + R'SH \rightarrow R\ Hg\ S\ R' + H^+$$

$$R\ Hg^+ \text{ (excess)} + \text{Indic.} \rightarrow R\ Hg \cdot \text{indic.}$$

The titration was carried out under nitrogen in aqueous medium of pH 5.5—6.5 using pyridine-2-azo-p-dimethylaniline as indicator. This compound works well because mercury—nitrogen interactions, though quite strong, are not as powerful as those between mercury and sulphur. Thiols, sulphur-containing proteins and thiolated dextrans were successfully titrated. Masked sulfhydryl groups in proteins were unmasked using urea. Fritz and Palmer determined mercaptans by mercuric perchlorate titration in neutral aqueous or acetone medium.

$$2\ R\ SH + Hg^{2+} \rightarrow (RS)_2Hg + 2H^+$$

The indicator used was thio-Michler's ketone [4,4'-bis(dimethylamino)thiobenzophenone] which reacted with excess mercury(II) to give a complex absorbing at 580 nm. The method is remarkably specific. As part of a paper on the analysis of aminoalkylthiophosphorus compounds Haglund and Lindgren describe the titration under nitrogen of thiols with palladium chloride solution in dilute HCl—ethanol.

$$R\ SH + Pd\ Cl_2 \rightarrow R\ S\ Pd\ Cl + HCl$$

The indicator is p-nitrosodimethylaniline which reacts with excess Pd^{II} to give a bright red complex. Because palladium ions react instantaneously with thiols but not with most other sulphur compounds the method is highly selective.

The determination of primary and secondary aromatic amines using nitrous acid as titrant has been a feature of several analyses

$$R\ NH_2 + HNO_2 \rightarrow RN\equiv N^+OH^- + H_2O$$

$$R_2NH + HNO_2 \rightarrow R_2N-NO + H_2O$$

The appearance of absorption due to free HNO_2 at 385 nm is usually used to locate equivalence though often the completion of product absorbance increase is equally feasible. Pratt determined the analgesics procaine, propoxycaine and tetracaine. He pointed out that a primary/ secondary amine mixture could often be resolved as diazotisation of a primary amine was much faster than nitrosation of a secondary. Various sulphanilamides were determined by Tanase and Shimomura in a similar manner, though these authors pointed out that some diazonium salts interfered because of their high absorbance.

Jackwerth and Specker have determined water in ketones, esters and ethers using a cupric perchlorate in acetone solution by a form of photometric titration.

$$3 \text{ Li Cl} + Cu(ClO_4)_2 \rightarrow \text{Li Cu Cl}_3 + 2 \text{ Li ClO}_4$$

Lithium chloride is added to the test solution. The absorbance at 366 nm of the lithium trichlorocuprate which peaks at the stoichiometry shown depends upon the water content of the solvent thus the real working graph is that of absorbance at 366 nm vs. percent water. Lower alcohols interfere.

Certain aromatic and aliphatic aldehydes have been determined by Cochran and Reynolds using a sodium borohydride in DMF solution.

$$R \cdot CHO \xrightarrow{BH_4^-} R \cdot CH_2OH$$

The solvent was 1 : 1 isopropyl alcohol—water and the disappearance of absorption due to the $-CH=O$ chromophore noted at an appropriate wavelength. Selectivity could be achieved, as in bromination analyses, by rate variation and it was sometimes advantageous to heat the titration cell by a simple electrical means. Easily reduced ketones interfered but otherwise the method was very selective.

The determination of organometallic compounds has featured very little in photometric titrimetry. Obviously many mercury compounds could be analysed using standard thiol solutions etc. but this never seems to have been done specifically. A good example of the possibilities, however, is Wadelin's photometric adaptation of the determination of trialkyl aluminium and dialkyl aluminium hydride using as titrant a solution of isoquinoline in toluene designed by Bonitz. These organoaluminium compounds catalyse the polymerisation of olefins; their hydrolysis products such as dialkylaluminium hydroxide do not. The active compounds form complexes with iso-quinoline thus

$$R_3Al + C_9H_7N \rightarrow R_3Al \cdot C_9H_7N \text{ (yellow)}$$

$$R_2AlH + C_9H_7N \rightarrow R_2AlH \cdot C_9H_7N \text{ (colourless)}$$

$$R_2AlH \cdot C_9H_7N + C_9H_7N \rightarrow R_2AlH(C_9H_7N)_2 \text{ (red)}$$

The first two complexes form together, the last one forms only when the first two are complete. The sum of the two active species is thus determined from the rise in absorbance of the red complex at 460 nm in dry toluene. R ranges from CH_3 to C_6H_{13} and reaction rate is inversely proportional to alkyl group size. Oxygen and moisture must be rigorously excluded. R_2AlH is purposely added if none is present in the titrand.

Bulky aromatic free radicals absorb strongly in the visible region of the spectrum, yet have only once been used in photometric titration. Paris, Gorsuch and Hercules determined oxygen and antioxidants in various organic solvents using a solution of the 2,4,6-tri-t-butylphenoxy free radical produced by treatment of a solution of tri-t-butyl phenol with lead dioxide.

The first is a radical—radical coupling reaction, the latter hydrogen abstraction. The end-point is shown by the appearance of the free radical which absorbs at 625 nm. The table of the oxygen content of common solvents is of interest.

Citron and co-workers have determined primary and tertiary aromatic amines by their influence on the copper(II) complex of EDTA

$$CuH_2Y + 2\,R\,NH_2 \rightarrow Cu\,Y\,(RNH_2)_2^{2-} + 2H^+$$

References pp. 380—389

Titration of the amine solution with CuH_2Y using light of 720 nm yields a break at the given stoichiometry where formation of the ternary complex is complete. If tertiary amines are present the break point gives the sum of primary + tertiary. The mixture may then be resolved by means of a varying ratio photometric calibration plot in which primary + tertiary amine concentration equals the value found earlier.

The starch—iodine interaction has received attention from time to time as might be expected. Such investigations culminated in the paper of Richter and Szetjli in which the authors titrated a variety of starch polysaccharides with iodine (generated by the KIO_3—KI—acid reaction) under various conditions. Clear combining ratios were evident in many cases; however, this long paper is not reviewed intensively here as the work appears more of a theoretical study than an analytical method.

Reynolds and Underwood have used rather a complex series of reactions to produce a very selective determination of primary nitroparaffins

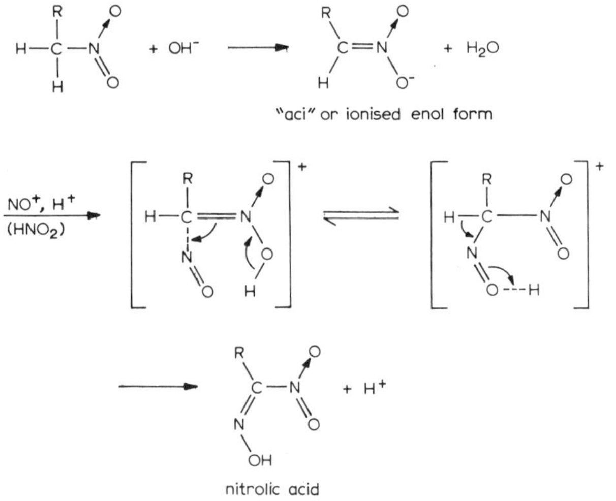

"aci" or ionised enol form

nitrolic acid

The nitrolic acid finally produced is titrated photometrically with standard NaOH by following the absorbance at 420 nm of the yellow salt produced. Excess HNO_3 used in the nitrous acid treatment is first neutralised. Interfering weak acids etc. are determined by a blank titration following destruction of the aci nitroparaffin with hypochlorite. Nitroparaffins present in non-aqueous solvents are treated in a dioxan—methanol—water solvent. Although nitrolic acid conjugate bases show absorbance maxima around 330 nm the titration is carried out at 420 nm to keep absorbance values below 1 and to minimise background absorbance.

Fitzgerald has applied his photodecomposition approach to the determination of the heterocyclic reagents 1,10-phenanthroline and 2,4,6-tripyridyl-*s*-triazine by generation of Fe^{2+} from ferric nitrate.

Finally, ternary complex formation has been used in an interesting determination of ethane-1-hydroxy-1,1-diphosphonic acid (EHDP) by Liggett and Libby based on a visual method of Pribil and Vesely [292]. The sample was titrated with a solution of the thorium complex of the diaminocyclohexanetetraacetic acid (DCTA) with xylenol orange (XO) as indicator.

$$EHDP + 2\,Th \cdot DCTA \rightarrow (Th \cdot DCTA)_2 \cdot EHDP$$

After equivalence

$$2\,Th \cdot DCTA + 2\,XO \rightarrow Th_2(DCTA)_2(XO)_2$$

Absorption of the indicator complex at 575 nm is followed; the colour change is apparently due to deprotonation of the XO upon its incorporation into the complex.

10. Miscellaneous photometric titration methods

Certain titrations have appeared in the literature which, while not wholly based on solution absorbance change, do rely on photoelectric evaluation of light intensity.

(A) SPECTROPOLARIMETRY

In this technique the course of an acid—base or metal complexation titration is followed by noting the change in optical rotation of the solution [317]. The titration is self-indicating if titrant, titrand or product is optically active, and has all the advantages inherent in such a system. Thus only one equilibrium system per titrand entity need be considered giving good likelihood of sequential titration of mixtures, and since light in or very near the visible region is used, interference from UV absorbers is minimal.

Acid—base and complexometric titrations may be followed by adding a suitable optically active indicator but the system then has no advantage over more conventional types. Spectropolarimetric titrations were introduced by Kirschner and Bhatnager of Wayne State University, Detroit, in 1963 [318] and then pursued with vigour by Kirschner's student Pearson to date. They are feasible because of the availability of rapid-reading, accurate UV—visible photoelectric spectropolarimeters.

The determination of an optically active acid, whose conjugate base has a much different molecular rotation, by plotting optical rotation as a function of volume of standard strong base is self evident [318]. A polyprotic weak acid can give a series of breaks. An inactive acid of pK_a two units lower may be determined by using the active acid as indicator. The spectropolarimetric titration of a hydrochloric and d-tartaric acid mixture with NaOH is illustrated in Fig. 71; both acids can obviously be determined with considerable accuracy. Active bases may be similarly determined.

Metal ions may be determined by stereospecific titration with optically active chelating agents. L-(+)-histidine [318], D-(—)-1,2-

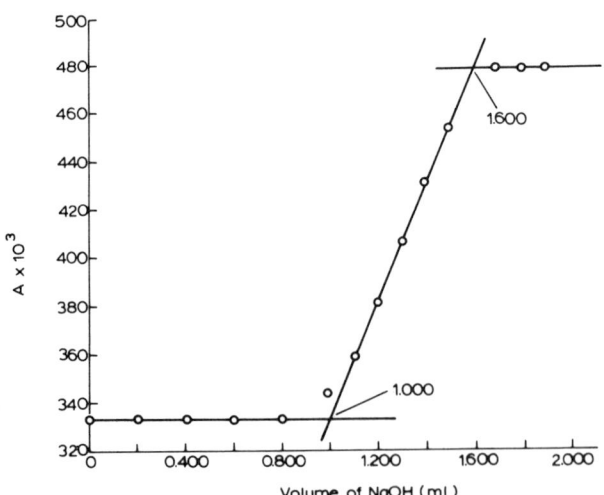

Fig. 71. Spectropolarimetric titration of HCl and d-tartaric acid with standard NaOH (λ = 4000 A). Reprinted from ref. 318, by kind permission of the American Chemical Society.

366

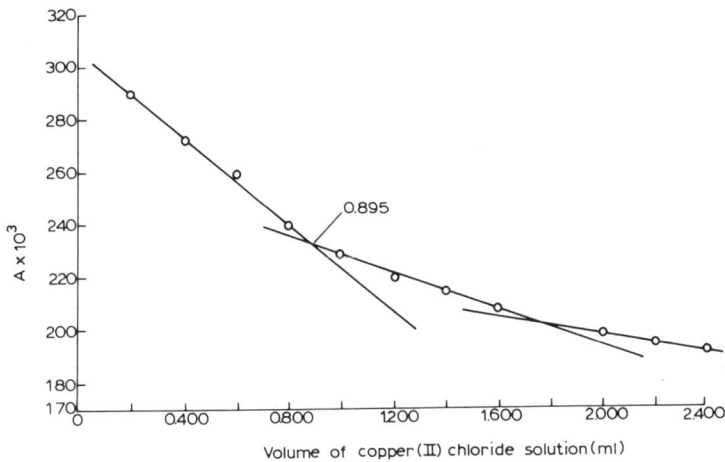

Fig. 72. Spectropolarimetric titration of l-(+)-histidine monohydrochloride with copper(II) chloride (λ = 4000 A). Reprinted from ref. 318 by kind permission of the American Chemical Society.

propylenediaminetetraacetic acid (D(−)PDTA) [319,320] and D-(−)-*trans*-1,2-cyclohexanediaminetetraacetic acid (D(−)CDTA) [321] are very stable in solution and have been used so far with great success. Insight into the nature of the complex-forming reaction may be obtained; Fig. 72 shows the formation of 2 : 1 and 1 : 1 histidine–copper complexes. Studies of this type using visual polarimeters have been fairly common. Optical rotatory dispersion (ORD) curves (molecular rotation vs. wavelength) for titrant and complex(es) indicate the wavelength to use to give best precision to the graphical break. This is usually close to the Cotton region. Figure 73 shows ORD curves for (D(−)PDTA) and its complexes with thorium and zirconium [322]. The observed rotations of metal chelates and ligands are a simple linear function of their concentration over a very wide range, e.g. lead is titratable over 10^{-1}—10^{-5} M. D(−)PDTA shows marked optical rotation changes with pH due to the different rotations of its different ionised forms, therefore titration solutions where this reagent is involved must be thoroughly buffered. In the important pH 10 region D(−)CDTA is much superior to D(−)PDTA in this respect. Mixtures of metal ions may be resolved if conditional log K values differ by more than 2.5 due to differences in log K_{abs} values or $\alpha_{M(L)}$ or a combination of the two. The rather spectacular

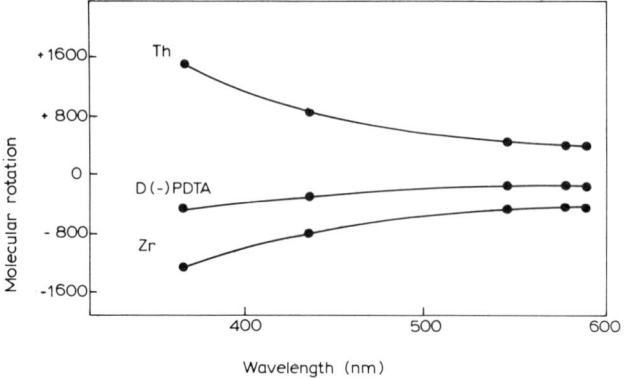

Wavelength (nm)

Fig. 73. ORD-spectra of zirconium and thorium complexes of D(—)PDTA. Reprinted from D.L. Caldwell, P.E. Reinbold and K.H. Pearson, Anal. Lett., 3 (1970) 93 by permission of Marcel Dekker, New York.

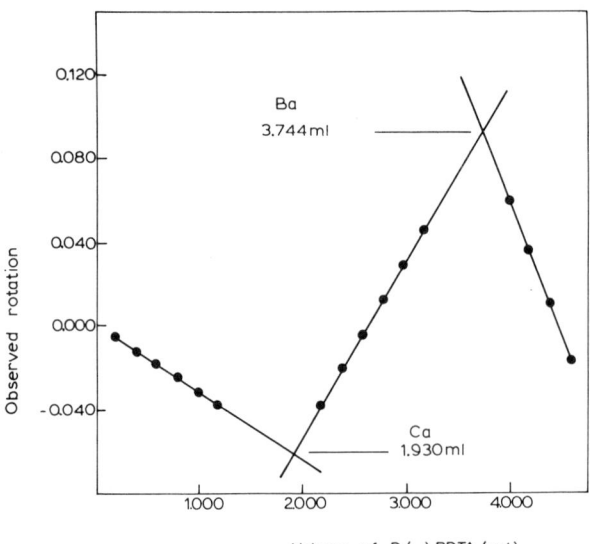

Volume of D(-)PDTA (ml)

Fig. 74. Sequential spectropolarimetric titration of 128.8 mg of barium and 40.08 mg of calcium in 100 ml with 0.5 M D(—) PDTA. pH = 10, λ = 365 nm. Reprinted from ref. 323.

graph for the titration of calcium + barium with D(−)PDTA [323] is shown in Fig. 74.

When a metal—EDTA titration is carried out using an optically active indicator such as histidine the usual $K^{eff}_{M'Y} / K^{eff}_{M'In'} > 10^4$; $K^{eff}_{M'In'} > 10^4$ relations must be observed. Precision of end-point location (coefficient of variation around 0.15% in ideal cases) is not adversely affected by high neutral salt concentrations.

Change of secondary monodentate ligand concentration can have an enormous influence on chelate rotation, e.g. the effect of NH_3 on Cu-D(−)PDTA [320]. So too can temperature change.

Pearson has pointed out the very real advantages of his self-indicating spectropolarimetric titrations over indicated complexometric titrations but perhaps a fairer comparison would have been one against self-indicating absorbance titrations in the UV or visible region. After all, UV—visible spectrophotometers are a little more common than photoelectric spectropolarimeters.

(B) ATOMIC ABSORPTION INHIBITION TITRATION

In this technique, proposed by Huber in 1971, solutions of anions which form refractory compounds with certain metals in the flame are titrated continuously, during aspiration, with a standard solution of a suitable metal. In the first example given [328], silicate over the concentration range 1 p.p.m.—20 p.p.m. was titrated with magnesium chloride solution (100 or 200 p.p.m.) using efficient magnetic stirring to the breakthrough of the magnesium atomic absorbance (Fig. 75). The ratio of magnesium to silica was 2 : 1 indicating formation of Mg_2SiO_4 in the clotlets. A calibration graph was plotted of silica concentration (in 50 ml) against intercept on volume axis. Loss of analyte due to aspiration was small — approximately 3% per min for 50 ml of solution. Errors due to this loss were compensated in the calibration plot. A relatively cool hydrogen-air flame was found most suitable. The proposed method thus exploits the high degree of inhibition by silicate of the very sensitive magnesium atomic absorption signal. Metal ions interfering by compound formation with silicate were removed by cation exchange. Coefficient of variation for five 5 p.p.m. silica solutions was 0.45%. Phosphate [328] and sulphate [329] have been determined by similar titrations with magnesium chloride, though strangely phosphate interferes little

TABLE 13

Bibliography of spectropolarimetric titrations

Species Determined	Author(s) (date)	Titrant	General conditions	Ref.
HCl + d-tartaric acid together	Kirschner and Bhatnagar (1963)	NaOH (5 N)	50 ml 0.1 N HCl + 1 ml 2 M d-tartaric acid; $\lambda = 400$ nm	318
l-(+)-histidine mono-hydrochloride	Kirschner and Bhatnagar (1963)	$CuCl_2$ (1.5 M)	50 ml 0.06 M histidine	318
Trifluoroacetic, trichloroacetic, dichloroacetic, l-mandelic acids; piperidine, triethylamine d-tartrate bases	Pearson and Kirschner (1969)	NaOH (5 N)	50 ml 0.1 N acid; $\lambda = 400$ nm. Indicator 10 ml 1 N l-mandelic acid. 50 ml 0.1 N base; $\lambda = 400$ nm. Indicator 10 ml 1 N sodium d-tartrate	317
		$HClO_4$		
EDTA	Pearson and Kirschner (1969)	$Zn(NO_3)_2$ (1.5 M)	0.5 g EDTA in 40 ml containing 10 ml strong pH 10 buffer (NH_3/NH_4Cl). Indicator l-histidine; $\lambda = 350$ nm.	317
l-propylene diamine diamine		$Ni(ClO_4)_2$ (1.5 M)	15 ml 0.23 M l-propylenediamine + 20 ml water; $\lambda = 546$ nm or 436 nm. 3 : 1 and 2 :1 ligand–metal complexes formed	
Al^{3+}, Tl^{3+}	Caldwell et al. (1970)	D(−) PDTA (0.25 M)	1 mmol in 100 ml. Indicator pH 5; Al, pH 3.5; 50°C; Tl, pH 3.5 (all acetate buffers)	324
Mg, Ca, Sr, Ba	Baker and Pearson (1970)	D(−) CDTA (0.15 M)	1 mmol in 100 ml inc. 10 ml; pH 10 NH_3 buffer (280 g NH_4Cl + 568 ml concentrated NH_3 to 1 litre); $\lambda = 365$ nm	321
Sc^{III}, Ti^{IV}, VO^{2+}, Cr^{3+}, Mn^{2+}, Fe^{3+}, Co^{2+}, Ni^{2+}, Cu^{2+}, Zn^{2+}	Palma et al. (1970)	D(−) PDTA (0.25 M)	1 mmol of metal in 40 ml. Mn, pH 5; Sc, pH 5; Cr, pH 5, back titration with Zn; Ti, pH 1.4; VO, pH 5; Fe, pH 1.3; Co, pH 4.5; Ni, pH 4.3; Cu^{2+}, pH 9.7 or 4.3; Zn, pH 9.6; $\lambda = 365$ nm	320

Cd^{2+}, Hg^{2+}, Pb^{2+}, Bi^{3+}	Palma and Pearson (1970)	D(−) PDTA (0.25 M)	1 mmol of metal in 40 ml. Cd, pH 10; Hg^{2+}, pH 5; Pb, pH 5; Bi, pH 1; $\lambda = 365$ nm	325
Zr^{IV}, Th^{IV}	Caldwell et al. (1970)	D(−) PDTA (0.25 M)	1 mmol in 100 ml. Zr in 1 N HNO_3, Th, pH 5; $\lambda = 365$ nm	322
Lanthanoids	Caldwell et al. (1970)	D(−) PDTA (0.3 M)	1 mmol in 100 ml; pH = 5; $\lambda = 365$ nm	326
Pb^{2+}	Reinbold and Pearson (1970)	D(−) CDTA	2.0 mg Pb in 100 ml; pH = 4.7; $\lambda = 365$ nm	327
Ba-Ca, Cd-Cu, Cd-Zn mixtures	Palma et al. (1970)	D(−) PDTA	1 mmol of each in 100 ml, varying amounts of pH 10 NH_3 buffer; $\lambda = 365$ nm	323

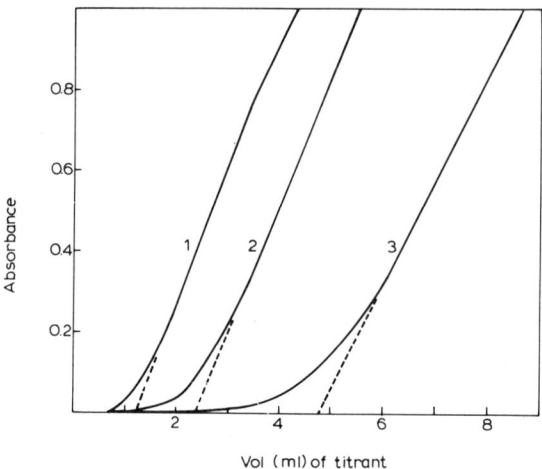

Fig. 75. Atomic absorption inhibition titration. Signals obtained for semiauto-
matic titration of standard silicate solutions with 100 μg ml^{-1} magnesium standard. (1) 2.5 μg ml^{-1} SiO$_2$; (2) 5.0 μg ml^{-1} SiO$_2$; (3) 10.0 μg ml^{-1} SiO$_2$. End-
points designated by intercept on abscissa. Titrant delivery rate 1.50 ml min^{-1}.
Reprinted from ref. 328, by kind permission of the American Chemical Society.

with the silicate analysis mentioned above. These titrations are
reminiscent of the lanthanum-phosphate flame emission titrations of
Slevin and Svehla [330].

(C) CHELATE EXCHANGE TITRIMETRY

The concentration of a metal chelate dissolved in a non-aqueous
solvent may be determined by titration with a standard solution of
some other chelating agent again in a suitable solvent provided of
course that the stability constant of the titrant complex is considerably greater than that of the complex to be determined. The theoretical basis of such determinations was laid in 1967 by Růžička and
Starý [331] who gave as an illustrative example the reaction

$$(AgHDz)_{CCl_4} + (HDDC)_{CCl_4} \rightleftharpoons (AgDDC)_{CCl_4} + (H_2Dz)_{CCl_4}$$

golden yellow   colourless   colourless   green

where H$_2$Dz = dithizone, HDDC = diethyldithiocarbamic acid. The
equilibrium constant for the exchange reaction is

$$E_{\text{AgHDz}-\text{HDDC}} = \frac{[\text{AgDDC}][\text{H}_2\text{Dz}]}{[\text{AgHDz}][\text{HDDC}]} = \frac{K_{\text{AgDDC}}}{K_{\text{AgHDz}}}$$

where $K_{\text{AgDDC}} = \dfrac{[\text{AgDDC}]_{\text{org}}[\text{H}^+]_{\text{aq}}}{[\text{Ag}^+]_{\text{aq}}[\text{HDDC}]_{\text{org}}}$

and $K_{\text{AgHDz}} = \dfrac{[\text{AgHDz}]_{\text{org}}[\text{H}^+]_{\text{aq}}}{[\text{Ag}^+]_{\text{aq}}[\text{H}_2\text{Dz}]_{\text{org}}}$

are the extraction constants of the respective complexes. Such reactions have been applied to photometric titrimetry by Grey and Cave. In their first investigation [332] they titrated microgram amounts of the dithizonates of Ag, Pb, Cu(II), Cd, Zn and In in chloroform with sodium diethyldithiocarbamate in ethanol. Titrations were successful in the presence of free dithizone which would result from a solvent extraction procedure and in addition good procedures were advanced for the one-phase titration of dithizone and diethyldithiocarbamate themselves.

(1) Titration of dithizone in benzene

Titrate approximately 0.1 μeq of dithizone in 50 ml of benzene with standard 4×10^{-4} M AgNO$_3$ in 1 : 20 methanol—benzene at 618 nm using a 1 ml microburette. (To prepare titrant dilute an 8×10^{-3} M AgNO$_3$ in methanol solution 20-fold with benzene.)

(2) Titration of diethyldithiocarbamate in ethanol

Titrate 0.7 μeq Cu(NO$_3$)$_2$ in 50 ml of ethanol with 2×10^{-3} M sodium diethyldithiocarbamate in ethanol at 435 nm.

(3) Titration of metal dithizonates

Titrate 0.3 μeq of complex in 50 ml of chloroform with 2×10^{-3} M sodium diethyldithiocarbamate in ethanol at the wavelength of maximum absorption of the dithizone complex or 615 nm, the wavelength of maximum absorption of dithizone itself.

The binary dithizonate mixtures Cu + Pb (560 nm); Cu + Cd (518 nm); Cu + Zn (530 nm); In + Pb (510 nm); Pb + Ag (510 nm); Cu +

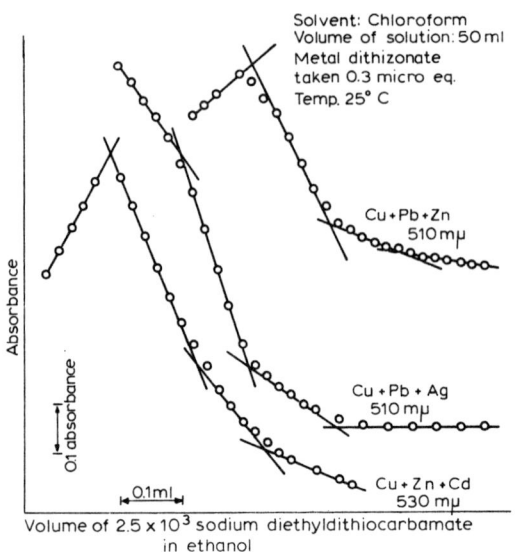

Fig. 76. The chelate exchange titration of some ternary mixtures of metal dithizonates with sodium diethyldithiocarbamate. Reprinted from ref. 332, by kind permission of the National Research Council of Canada.

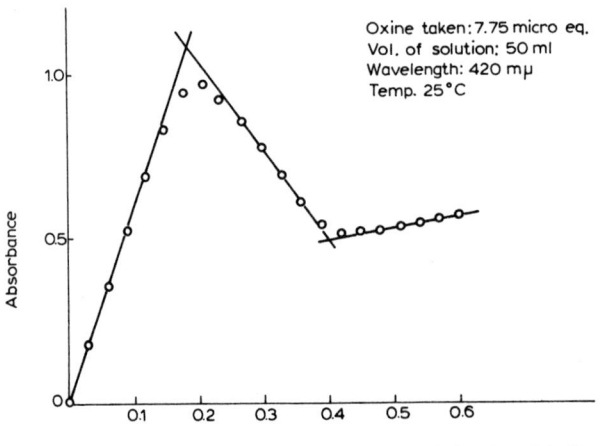

Fig. 77. Titration of 8-hydroxyquinoline with copper (II) in solution illustrating stepwise formation. Reprinted from ref. 333, by kind permission of the National Research Council of Canada.

374

Ag (465 nm) and ternary mixtures Cu + Pb + Zn (510 nm); Cu + Pb + Ag (510 nm); Cu + Zn + Cd (530 nm) were successfully resolved by a similar titration. Figure 76 shows the titration curves for the ternary mixtures. In a further paper [333] Grey and Cave titrated microgram amounts of metal 8-hydroxyquinoline chelates in benzene with standard dithizone in benzene. Chelates satisfactorily titrated were those of Pb, Cu(II), Cd, Zn and In; 8-hydroxyquinoline in benzene was well determined using a standard 2×10^{-3} M solution of $Cu(NO_3)_2$ in ethanol at 420 nm. Titrations were performed in a subdued red light. The stepwise nature of this complex formation process is beautifully illustrated in Fig. 77. Mixed complex formation, particularly with indium, sometimes becomes evident in these systems and kinetic factors may be very unfavourable.

(D) DETERMINATION OF CRITICAL MICELLE CONCENTRATION

Aqueous solutions of most soaps exhibit a more or less abrupt change in physical properties over a relatively short concentration range [14]. This phenomenon has been attributed to the formation of oriented soap aggregates and the concentration at which it occurs has been termed, "The critical concentration for the formation of micelles (CMC)". Corrin et al. showed long ago [334] that the absorption spectrum of a cationic dye placed in a solution of an anionic surface active agent changed drastically as the concentration of the anionic surfactant passed through its CMC. Likewise the absorption spectrum of an anionic dye changes as the solution of a cationic surfactant in which it is dissolved passes through its CMC. Corrin et al. devised visual titration procedures to determine the CMCs of various surfactants. Such titrations were put on an instrumental basis by Nichols and Kindt in 1950 [14]. The CMC of sodium dodecyl sulphate was determined by titrating an aliquot of a solution containing 1.900 g of the soap in 1 litre of 10^{-5} M pinacyanol chloride with 10^{-5} M pinacyanol chloride at 610 nm. The CMCs of various alkyl trimethyl ammonium bromides were determined similarly using 10^{-5} M Sky Blue FF; 625 nm would be a suitable wavelength. It is of interest to compare these titrations with the photometric titration procedure of Lambert (1947) [13] in which an anionic agent was titrated against a cationic agent (or vice versa) at concentrations above the CMCs of both reactants. A colloidal precipitate was produced near the equivalence-point and solubilised or coagulated by a

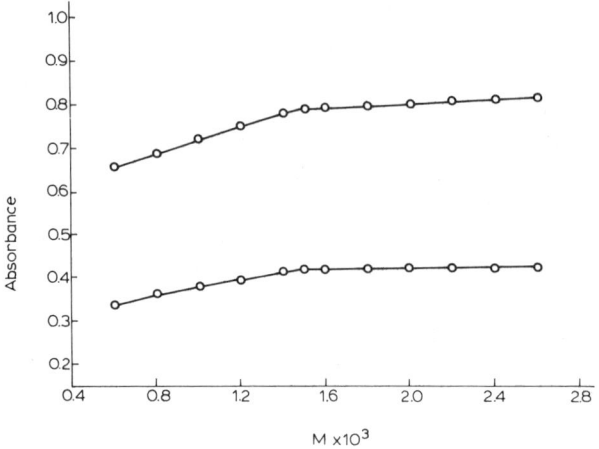

Fig. 78. Determination of CMC for myristyldimethylbenzyl-ammonium chloride. 1.00×10^{-5} M and 5.00×10^{-6} M in eosin. Reprinted from ref. 335, by kind permission of the American Chemical Society.

small excess of reagent. Recently Ledbetter and Bowen [335] have determined the CMCs of alkyldimethylbenzylammonium chlorides by concentration variation in solutions 1×10^{-5} M and 5×10^{-6} M in eosin at 528 nm. Figure 78 shows their results for the myristyl derivative. The emission characteristics of fluorescent dyes such as dichlorofluorescein and eosin are also markedly altered by complex formation with micelles.

(E) PHOSPHATE DETERMINATION

The titrimetric determination of low phosphate concentrations has been investigated by Alt et al. [336]. Phosphomolybdic acid is stable and stoichiometric (P : Mo = 1 : 12) in 40% acetone at pH 1—3 and absorbs preferentially over molybdic acid in the wavelength range 420—480 nm. Arsenic(V) may be determined similarly as the kinetics of phosphorus and arsenic heteropolymolybdate formation are reasonable under the conditions given.

Treat a neutral solution containing 0.01—5 mg P in 5 ml with 1 ml of 2 N H_2SO_4 or HCl or $HClO_4$ and 4 ml of acetone. Titrate slowly with 3×10^{-2} M or 3×10^{-3} M sodium molybdate solution in a 2 cm titration cell using light of wavelength 420—480 nm. As(V),

Si(IV), Ge(IV), Fe(III), V(V) cause interference though that of Si and Ge is not serious as the rate of formation of their polymolybdates is very small. Coefficient of variation is 1.6%.

(F) THE KARL FISCHER DETERMINATION

The Karl Fischer titration of low concentrations of water in, for example, methanol has a poorly defined visual end-point for all but the smallest amounts of water. A reliable dead stop amperometric end-point is usually used to indicate equivalence in this titration but Connors and Higuchi have shown that a photometric end-point is equally feasible [337] and after mentioning some earlier attempts propose the following method.

Pipette 30—50 ml of methanol containing 1—6 mg of water into a dry titration assembly and titrate with fresh commercial Fischer solution at a wavelength of 525 nm. A sharp __/ shaped graph results. The coefficient of variation is 3% on a 0.01% water in methanol sample.

(G) ION-PAIR PARTITION TITRATIONS

The titration against each other of bulky anions and cations with transport of the resulting ion-pair into an immiscible organic phase has been used for many years, especially in the detergent industry. Equivalence is indicated by the transport of a suitable indicator across the phases involved or by indicator—micelle interaction of the type mentioned previously. Such titrations naturally lend themselves to photometric treatment provided emulsion difficulties can be overcome but the theory of these titrations has only recently been put on a sound footing by Behrends [338].

In the case of an anion A such as MnO_4^- titrated with a cation (onium ion) O in the absence of other species the titration equation is

$$T = \lambda \left[\frac{\lambda}{(1 - \lambda) V_A} + 1 \right]$$

where T = extent of titration = $\dfrac{\text{total mols O}}{\text{total mols A}}$

$\lambda = \dfrac{A_0}{A}$ = fraction of A in organic phase

$$V_A = K_A \left(\frac{V_0}{V_w}\right)^2$$

$$K_A = \frac{[O_0][A_0]}{[O_w][A_w]}$$

V_0, V_w = volumes of organic and aqueous phases.

A graph showing λ as a function of T is shown in Fig. 79.

In the presence of an indicator B,

$$T = \left[\frac{1}{\dfrac{1-\mu}{\beta \cdot \mu} + 1} + C\mu\right]\left[1 + \frac{\mu}{(1-\mu)V_B}\right]$$

where $\mu = B_0/B$; $\beta = V_A/V_B$, the separation factor.

When B/A is very small ($<10^{-3}$) and V_B is very large,

$$T = 1 - \frac{1-\mu}{1-\mu+\beta \cdot \mu}$$

Figure 80 shows a series of plots of μ as a function of T for various separation factors when V_B is very high. The equivalence-point is

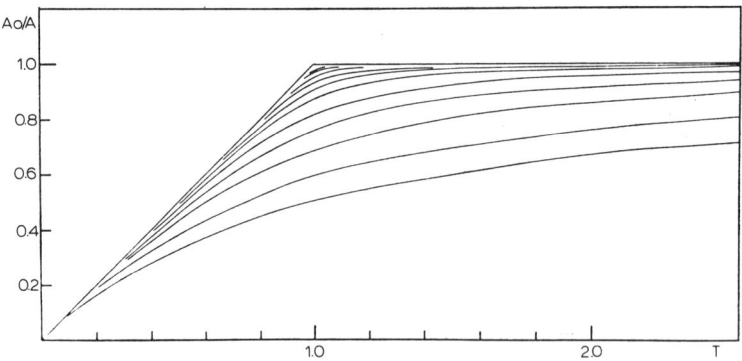

Fig. 79. Photometric ion pair partition titrations. Graph of fraction of anion A in the organic phase as a function of extent of titration T = total mols onium species/total mols anion. From bottom to top; V_A = 1, 2, 5, 10, 20, 50, 100, 200, 500, 1000, 2000. Reprinted from ref. 338, by kind permission of Springer—Verlag, Heidelberg.

378

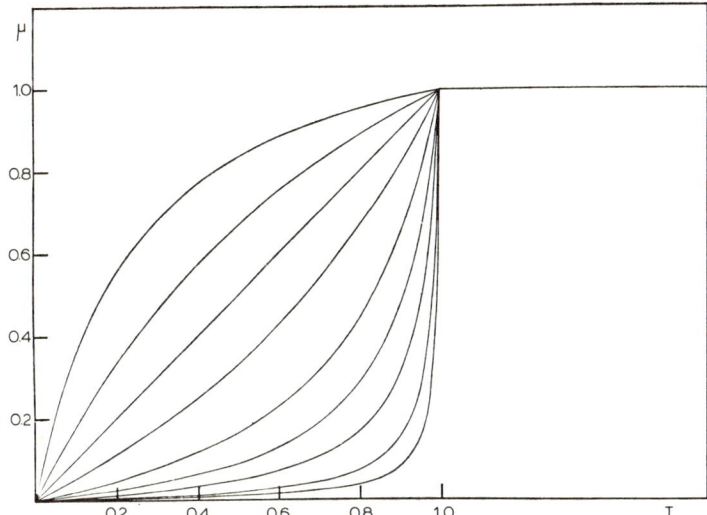

Fig. 80. Fraction of the indicator in a partition titration in the organic phase as a function of the extent of titration when the partition constant of the indicator is very high. The separation factor β between the ion to be determined and the indicator ion varies from 0.2, 0.5, 1.0, 2, 5, 10, 20, 50 to 100 from top left to bottom right. Reprinted from ref. 338, by kind permission of Springer—Verlag, Heidelberg.

reached when all the indicator is transferred to the organic layer. If the indicator concentration is appreciable a correction should be applied for the volume of titrant needed to transfer the indicator. Alternatively the indicator—titrant ion-pair itself may be used as indicator.

If β is very large and V_B very small then for $\mu > 0.5$

$$T = 1 + \frac{\mu}{(1-\mu)V_B} + C \cdot \mu$$

In this case the equivalence-point is reached when the first trace of indicator enters the organic layer.

If B/A is again very small, but neither β nor V_B very large, then

$$T = 1 + \frac{\mu}{(1-\mu)V_B} - \frac{1-\mu}{1-\mu+\beta\cdot\mu} - \frac{\mu}{(1-\mu+\beta\cdot\mu)V_B}$$

When $T = 1$ (equivalence point)

$$\frac{\mu}{1-\mu} = \sqrt{\frac{V_B}{\beta}} = \frac{V_B}{\sqrt{V_A}}$$

For a particular indicator—titrand combination V_B and β should be predetermined in order to anticipate the indicator distribution at equivalence.

To summarise: when the titrand as well as the indicator is easily extracted, the end-point is reached when the indicator is completely extracted. If, on the other hand, the indicator is only poorly extracted then the end-point corresponds to the first extraction of the indicator. When the titrand and indicator are only moderately extracted, the end-point can be located by graphical or calculated extrapolation of the indicator distribution during the titration.

As an example Behrends gives the titration of hexafluorophosphate and hexafluoroarsenate with tetraphenylarsonium hydroxide in 1,2-dichloroethane—water using permanganate as indicator [339]. At equivalence the indicator is completely extracted.

Experimental

Place 2 ml of 10^{-3}—10^{-2} M hexafluorophosphate or hexafluoroarsenate in a centrifuge tube with 2 ml of water and 4 ml 1,2-dichloroethane. Add sufficient indicator (tetraphenylarsonium permanganate in dichloroethane) to give a reasonable colour to the organic phase. Titrate with 0.05 N Ph_4AsOH solution until MnO_4^- begins to transfer to the organic phase, then monitor the absorbance of the aqueous phase at 525 nm as the titration proceeds. The equivalence-point is easily determined graphically. The coefficient of variation obtained was 0.3%.

References

1 A. Tingle, J. Amer. Chem. Soc., 40 (1918) 873.
2 J. Bahr and R. Bunsen, Ann. Chem. Pharm., 137 (1866) 1.
3 K. Vierordt, Annal. Chem., 177 (1875) 31.
4 J. Field and L.G.M. Baas-Becking, J. Gen. Physiol., 9 (1926) 445.
5 R.H. Müller and H.M. Partridge, Ind. Eng. Chem., 20 (1928) 423.
6 F. Müller, Z. Electrochem., 40 (1934) 46.

7 A. Ringbom and F. Sundman, Z. Anal. Chem., 116 (1939) 104.
8 R.H. Osborn, J.H. Elliott and A.F. Martin, Ind. Eng. Chem., Anal. Ed., 15 (1943) 642.
9 J.H. Yoe and A.L. Jones, Ind. Eng. Chem. Anal. Ed., 16 (1944) 111.
10 P. Job, Ann. Chim., 9 (1928) 113.
11 W.C. Vosburgh and G.R. Cooper, J. Amer. Chem. Soc., 63 (1941) 437.
12 A.E. Harvey and D.L. Manning, J. Amer. Chem. Soc., 72 (1950) 4488.
13 J.M. Lambert, J. Colloid Sci., 2 (1947) 479.
14 M.L. Nichols and B.H. Kindt, Anal. Chem., 22 (1950) 785.
15 F. Kenny and R.B. Kurtz, Anal. Chem., 24 (1952) 1218.
16 A. Ringbom and E. Vanninen, Anal. Chim. Acta, 11 (1954) 153.
17 J.M.H. Fortuin, P. Karsten and H.L. Kies, Anal. Chim. Acta, 10 (1954) 356.
18 M. Bobtelsky, Anal. Chim. Acta, 13 (1955) 172.
19 E. Grunwald, Anal. Chem., 28 (1956) 1112.
20 T. Wallin, Ark. Kemi, 26(2) (1966) 13.
21 E.R. Groeneveld and G. den Boef, Z. Anal. Chem., 219 (1966) 328.
22 E.R. Groeneveld and G. den Boef, Analyst (London), 94 (1969) 860.
23 A. Ringbom, B. Skrifvars and E. Still, Anal. Chem., 39 (1967) 1217.
24 G.H. Schenk and W.E. Bazzelle, Anal. Chem., 40 (1968) 162.
25 M.A. Leonard and M.P. Murray, Anal. Lett., 3 (1970) 67.
26 K.H. Pearson and S. Kirschner, Anal. Chim. Acta, 48 (1969) 339.
27 R.H. Müller, Ind. Eng. Chem. Anal. Ed., 11 (1939) 1.
28 R.H. Osborn, J.H. Elliott and A.F. Martin, Ind. Eng. Chem. Anal. Ed., 15 (1943) 642.
29 J. Mika, Z. Anal. Chem., 128 (1948) 159.
30 A.L. Underwood, J. Chem. Educ., 31 (1954) 394.
31 R.F. Goddu and D.N. Hume, Anal. Chem., 26 (1954) 1740.
32 R.C. Hirt, Anal. Chem., 28 (1956) 579.
33 J.B. Headridge, Talanta, 1 (1958) 293.
34 J.B. Headridge, Photometric Titrations, Pergamon, Oxford, 1961.
35 K.A. Connors and T. Higuchi, Anal. Chim. Acta, 25 (1961) 509.
36 C.A. Reynolds, Record of Chem. Progr., 24(3) (1963) 157.
37 J.B. Headridge, Ind. Chem., (1963) 44, 105.
38 S.P. Eriksen and K.A. Connors, J. Pharm. Sci., 53 (1964) 465.
39 A.L. Underwood, Advan. Anal. Chem. Instr., 3 (1964) 31.
40 M. Tanaka and G. Nakagawa, Anal. Chim. Acta, 32 (1965) 123.
41 H. Flaschka, Pure Appl. Chem., 10 (1965) 165.
42 E. Greuter, Chem. Rundsch. Solothurn, 19(2) (1966) 25.
43 M. L'Her, Les Titrages Spectrophotometriques, Dépt. de Chim., Centre d'Études Nucl. Fontenay-aux-Roses, France, Rapp. CEA R No. 3140 1967.
44 G.F. Lothian, Absorption Spectrophotometry, 3rd edn., Hilger, London, 1969.
45 C.C. Gravatt, Appl. Spectrosc., 25 (1971) 509.
46 E.R. Groeneveld and G. den Boef, Z. Anal. Chem., 219 (1966) 328.
47 E.R. Groeneveld and G. den Boef, Analyst (London), 94 (1969) 860.
48 E. Grunwald, Anal. Chem., 28 (1956) 1112.
49 J. Vrestal and S. Kotrly, Talanta, 17 (1970) 151.
50 P. Jandera, S. Kolda and S. Kotrly, Talanta, 17 (1970) 443.

51 R.H. Conrad, Anal. Chem., 39 (1967) 1039.
52 C. Rehm, J.I. Bodin, K.A. Connors and T. Higuchi, Anal. Chem., 31 (1959) 483.
53 H. Flaschka and P. Sawyer, Talanta, 8 (1961) 521.
54 P.O. Sawyer and H. Flaschka, Microchem. J. Symp. Ser., 2 (1962) 825.
55 H. Flaschka and J. Butcher, Talanta, 12 (1965) 913.
56 H. Flaschka and R. Speights, Talanta, 15 (1968) 1467.
57 H. Flaschka and J. Garrett, Anal. Lett., 1 (1967) 185.
58 B. Skytte-Jensen, Radiom. News, No. 7 (1966) 5.
59 M.G. Young, T.H. Clarke and R.T. Schlick, Microchem. J., 13 (1968) 712.
60 G.P. Foust, B.D. Burleigh, S.G. Mayhew, C.H. Williams and V. Massey, Anal. Biochem., 27 (1969) 530.
61 J.B. Headridge and R.J. Magee, Talanta, 1 (1958) 117, 416.
62 J. Herrington, Unicam Spectrovision, No. 6 (1958) 5.
63 D.J. Fricker, Chem. Ind., 16 (1955) 426.
64 J.E. Lee, J.H. Edgerton and M.T. Kelley, Anal. Chem., 28 (1956) 1441.
65 F.T. King and R.C. Hirt, Appl. Spectrosc., 7 (1953) 164.
66 E. Agazzi and G.W. Bond, Anal. Chem., 33 (1961) 972.
67 H.M. Partridge, Ind. Eng. Chem. Anal. Ed., 4 (1932) 315.
68 L. Shapiro and W.W. Brannock, Anal. Chem., 27 (1955) 725.
69 S. Miyake and I. Sakamoto, Japan Anal., 16 (1967) 238.
70 H.V. Malmstadt and E.C. Gohrbandt, Anal. Chem., 26 (1954) 442.
71 T.L. Marple and D.N. Hume, Anal. Chem., 28 (1956) 1116.
72 O. Menis, D.L. Manning and R.G. Ball, Anal. Chem., 30 (1958) 1772.
73 R.A. Chalmers and C.A. Walley, Analyst (London), 82 (1957) 329.
74 J.T. Stock and W.C. Purdy, Lab. Pract., 11 (1962) 116.
75 R.F. Goddu and D.N. Hume, Anal. Chem., 26 (1954) 1679.
76 G.W. Leonard and R.W. Henry, Anal. Chem., 28 (1956) 1079.
77 R.W. McKinney and C.A. Reynolds, Talanta, 1 (1958) 46.
78 L.E.I. Hummelstedt and D.N. Hume, Anal. Chem., 32 (1960) 576.
79 L.E.I. Hummelstedt and D.N. Hume, Anal. Chem., 32 (1960) 1792.
80 S.P. Agarwal and M.I. Blake, Anal. Chem., 41 (1969) 1104.
81 T. Higuchi, C. Rehm and C. Barnstein, Anal. Chem., 28 (1956) 1506.
82 C. Rehm and T. Higuchi, Anal. Chem., 29 (1957) 367.
83 K.A. Connors and T. Higuchi, Anal. Chem., 32 (1960) 93.
84 K.A. Connors and T. Higuchi, Anal. Chim. Acta, 25 (1961) 509.
85 C.N. Reilley and B. Schweizer, Anal. Chem., 26 (1954) 1124.
86 S. Aronoff, J. Phys. Chem., 62 (1958) 428.
87 S. Bruckenstein and D.C. Nelson, Anal. Chem., 33 (1961) 438.
88 T. Higuchi, C.H. Barnstein, H. Ghassemi and W.E. Perez, Anal. Chem., 34 (1962) 400.
89 H.V. Malmstadt and D.A. Vassolo, Anal. Chem., 31 (1959) 862.
90 S.M. Tuthill, O.W. Kolling and K.H. Roberts, Anal. Chem., 32 (1960) 1678.
91 S. Bruckenstein and M.M.T.K. Gracias, Anal. Chem., 34 (1962) 975.
92 J.B. Schute, Pharm. Weekbl., 105 (1970) 1.
93 P. Braid, J.A. Hunter, W.H.S. Massie, J.D. Nicholson and B.E. Pearce, Analyst (London), 91 (1966) 439.

94 E.A. Burns and E.A. Lawler, Anal. Chem., 35 (1963) 802.
95 T. Jasinski and Z. Szponar, Chem. Anal. (Warsaw), 10 (1965) 619.
96 T. Jasinski and Z. Szponar, Chem. Anal. (Warsaw), 10 (1965) 665.
97 T. Jasinski and Z. Szponar, Zesz. Nauk. Mat., Fiz. Chem. Wyzsza Szk. Pedagog. Gdansku, 6 (1966) 69.
98 T.F. Keiley, Anal. Chem., 37 (1965) 1078.
99 O.W. Kolling and T.L. Stevens, Anal. Chem., 34 (1962) 1653.
100 A.P. Kreshkov and V.I. Vasil'ev, Anal. Abstr., 12 (1965) 756.
101 A.P. Kreshkov and V.I. Vasil'ev, Zhur. Anal. Khim., 17 (1962) 908.
102 A.P. Kreshkov, V.I. Vasil'ev and L.A. Tumovskii, Tr. Mosk. Khim-Tekhnol. Inst., 48 (1965) 39, Chem. Abstr. 65, 16039 h.
103 A.P. Kreshkov, L.P. Senetskaya and T.A. Malikova, Mosk. Khim-Tekhnol. Inst., 54 (1967) 122. Anal. Abstr., 15 (1968) 5402.
104 A.C.A. Paalman, Pharm. Weekbl. Ned., 103 (1968) 961.
105 R.M. Powers, R.A. Day and A.L. Underwood, Anal. Chem., 30 (1958) 254.
106 A.L. Underwood and L.H. Howe, Anal. Chem., 34 (1962) 692.
107 W.J. Schuele, J.F. Hazel and W.M. McNabb, Anal. Chem., 28 (1956) 505.
108 R.L.M. van Lingen, Z. Anal. Chem., 247 (1969) 232.
109 D. Jagner, Anal. Chim. Acta, 52 (1970) 483.
110 W. Merz, Anal. Chim. Acta, 50 (1970) 305.
111 R.H. Müller and M.H. McKenna, J. Amer. Chem. Soc., 58 (1936) 1017.
112 P.W. West and C.G. de Vries, Anal. Chem., 23 (1951) 334.
113 H.E. Bent and C.L. French, J. Amer. Chem. Soc., 63 (1941) 568.
114 P.W. West and E.S. Amis, Ind. Eng. Chem. Anal. Ed., 18 (1946) 400.
115 D.S. Tarbell and J.F. Bunnett, J. Amer. Chem. Soc., 69 (1947) 263.
116 J.D. Sabatino, O.W. Weber, G.R. Padmanabhan and B.Z. Senkowski, Anal. Chem., 41 (1969) 905.
117 L.D. Brake, W.M. McNabb and J.F. Hazel, Anal. Chim. Acta, 17 (1957) 314.
118 H. Flaschka, Talanta, 8 (1961) 381.
119 A. Ringbom, Complexation in Analytical Chemistry, Wiley, New York, 1963, p. 103.
120 G. den Boef, T. Out and H. Poppe, Mikrochim. Acta, (1971) 366.
121 D.A. Aikens, G. Schmuckler, F.S. Sadek and C.N. Reilley, Anal. Chem., 33 (1961) 1664.
122 J.R. Dean and W.E. Harris, Anal. Chem., 40 (1968) 1213.
123 H. Poppe, G. den Boef and F. Freese, Anal. Chim. Acta, 51 (1970) 199.
124 J. Fischer, Z. Slovak and J. Borak, Talanta, 18 (1971) 615.
125 H. Flaschka and S. Khalafallah, Z. Anal. Chem., 156 (1957) 401.
126 E. Still and A. Ringbom, Anal. Chim. Acta, 33 (1965) 50.
127 C.N. Reilley and R.W. Schmid, Anal. Chem., 31 (1959) 887.
128 E. Still, Suomen Kemistilehti B, 41 (1968) 33.
129 S. Kotrly, Anal. Chim. Acta, 29 (1963) 552.
130 E. Still, Talanta, 12 (1965) 817.
131 J. Kragten, Analyst (London), 96 (1971) 106.
132 A.L. Underwood, Anal. Chem., 25 (1953) 1910.
133 A.L. Underwood, Anal. Chem., 26 (1954) 1322.
134 P.B. Sweetser and C.E. Bricker, Anal. Chem., 25 (1953) 253.
135 P.B. Sweetser and C.E. Bricker, Anal. Chem., 26 (1954) 195.

136 R.N. Wilhite and A.L. Underwood, Anal. Chem., 27 (1955) 1334.
137 A.C. Kibrick, M. Ross and H.E. Rogers, Proc. Soc. Exp. Biol. Med., 81 (1952) 353.
138 F.W. Fales, J. Biol. Chem., 204 (1953) 577.
139 R.A. Chalmers, Analyst (London), 79 (1954) 519.
140 A. Ringbom and E. Vanninen, Anal. Chim. Acta, 11 (1954) 153.
141 W.J. Lane and J.S. Fritz, U.S. Atomic Energy Commission Report ISC-945, 1957.
142 S.J. Gedansky and L. Gordon, Anal. Chem., 29 (1957) 566.
143 W. Bauditz, Clin. Chim. Acta, 17 (1967) 207.
144 N.A. Ramaiah, G.D. Tewari, S.R. Trivedi and S.S. Katiyar, Talanta, 15 (1968) 352.
145 R. Belcher, B. Crossland and T.R.F.W. Fennell, Talanta, 16 (1969) 1335.
146 H. Flaschka and P. Sawyer, Talanta, 9 (1962) 249.
147 H. Flaschka and F.B. Carley, Talanta, 11 (1964) 423.
148 B. Skrifvars and A. Ringbom, Anal. Chim. Acta, 36 (1966) 105.
149 A.J. Cohen and L. Gordon, Anal. Chem., 28 (1956) 1445.
150 G. Conradi, Chemist—Analyst, 56 (1967) 87.
151 S.K. Datta and S.N. Saha, Chemist—Analyst, 51 (1962) 49.
152 A.A. Fedorov, F.A. Ozerskaya and E.N. Strebulaeva, Zav. Lab., 33 (1967) 1502; Anal. Abstr., 16 (1969) 1239.
153 H. Flaschka and J. Ganchoff, Talanta, 8 (1961) 720.
154 H. Flaschka and A. Soliman, Z. Anal. Chem., 159 (1957) 30.
155 H. Flaschka and J. Ganchoff, Talanta, 8 (1961) 885.
156 H. Flaschka and J. Butcher, Mikrochim. Acta, (1964) 401.
157 H. Flaschka and J. Ganchoff, Talanta, 9 (1962) 76.
158 J.A. Hunter and C.C. Miller, Analyst (London), 81 (1956) 79.
159 E. Ideno and K. Hozumi, Jap. Anal., 17 (1968) 727. Anal. Abstr., 18 (1970) 1521.
160 P. Karsten, H.L. Kies, H.Th.J. van Engelen and P. de Hoog, Anal. Chim. Acta, 12 (1955) 64.
161 S. Kotrly, Mikrochim. Ichnoanal. Acta, (1964) 407.
162 J. Lacy, Talanta, 10 (1963) 1031.
163 S.L. Levine and H.J. Golden, Anal. Lett., 1 (1967) 39.
164 H.V. Malmstadt and E.C. Gohrbandt, Anal. Chem., 26 (1954) 442.
165 H.V. Malmstadt and T.P. Hadjiioannou, Anal. Chim. Acta, 19 (1958) 563.
166 O. Menis, D.L. Manning and R.G. Ball, Anal. Chem., 30 (1958) 1772.
167 G.W.C. Milner and J.W. Edwards, Analyst (London), 80 (1955) 879.
168 A.P. Nechiporenko, I.P. Kalinkin, N.G. Ventov and V.B. Aleskovskii, Zavod. Lab., 35 (1969) 432; Anal. Abstr., 19 (1970) 92.
169 N.A. Ramaiah and Vishnu, Anal. Chim. Acta, 16 (1957) 569.
170 E. Rauterberg and H. Ossenberg-Neuhaus, Z. Pfl. Ernähr. Düng. Bodenk, 121 (1968) 193; Anal. Abstr., 17 (1969) 3333.
171 L. Shapiro and W.W. Brannock, Anal. Chem., 27 (1955) 725.
172 H.W. Wharton and L.R. Chapman, Anal. Chem., 36 (1964) 1679.
173 B. Zak, W.M. Hindman and E.S. Baginski, Anal. Chem., 28 (1956) 1661.
174 E. Still, Talanta, 12 (1965) 817.
175 T.L. Marple, G. Matsuyama and L.W. Burdett, Anal. Chem., 30 (1958) 937.

176 P. le Goff and B. Trémillon, Bull. Soc. Chim. Fr., (1964) 350.
177 I.T. Takahashi and R.J. Robinson, Anal. Chem., 32 (1960) 1350.
178 R.K. Behm and R.J. Robinson, Anal. Chem., 35 (1963) 1010.
179 H. Poppe, Talanta, 16 (1969) 1519.
180 W.G. Boyle and R.J. Robinson, Anal. Chem., 30 (1958) 958.
181 G.L. Silver and R.C. Bowman, Talanta, 14 (1967) 893.
182 D.S. Tarbell and J.F. Bunnett, J. Amer. Chem. Soc., 69 (1947) 263.
183 G. Pilloni and G. Plazzogna, Anal. Chim. Acta, 35 (1966) 325.
184 F. Ingman, Talanta, 18 (1971) 744.
185 A. Galík, Talanta, 15 (1968) 771.
186 A. Galík, Talanta, 16 (1969) 201.
187 A. Galík, Talanta, 13 (1966) 109.
188 A. Galík and M. Knizek, Talanta, 13 (1966) 589.
189 A. Galík and M. Knizek, Talanta, 13 (1966) 1169.
190 A. Galík, Talanta, 14 (1967) 731.
191 A. Galík, Talanta, 16 (1969) 201.
192 F.G. Nasouri, S.A.F. Shahine and R.J. Magee, Microchem. J., 12 (1967) 26.
193 H. Hartkamp, Z. Anal. Chem., 171 (1959) 262.
194 A. Galík, Talanta, 17 (1970) 115.
195 S. Hirano, J. Soc. Chem. Ind. Jap., 37 (1934) 561.
196 F. Trusell, P.A. Argabright and W.F. McKenzie, Anal. Chem., 39 (1967) 1025.
197 T. Nomura, J. Chem. Soc. Jap., Pure Chem. Sect., 89 (1968) 580.
198 H. Specker, H. Hartkamp and E. Jackwerth, Z. Anal. Chem., 163 (1958) 111.
199 C.S. Yeh, Microchem. J., 14 (1969) 279.
200 A. Ringbom, Z. Anal. Chem., 122 (1941) 263.
201 R.B. Fischer, M.L. Yates and M.M. Batts, Anal. Chim. Acta, 20 (1959) 501.
202 St. J.H. Blakeley and D.E. Ryan, Anal. Chim. Acta, 30 (1964) 346.
203 L.D. Goodhue, Ind. Eng. Chem. Anal. Ed., 10 (1938) 52.
204 M. Bobtelsky, Heterometry, Elsevier, Amsterdam, 1960.
205 W.W. Brandt and A.A. Duswalt, Anal. Chem., 30 (1958) 1120.
206 J.E. Roberts, Anal. Chem., 39 (1967) 1884.
207 H. Frey, Anal. Chim. Acta, 6 (1952) 28.
208 R. Wickbold, Angew. Chem., 65 (1953) 159.
209 C.J. van Nieuwenburg and B.F. Engelbert van Bevervoorde, Anal. Chim. Acta, 19 (1958) 32.
210 A. del Campo, F. Burriel and L. Garcia Escolar, Bol. Acad. Cienc. (Madrid), 2 (1936) No. 7, 10. Chem. Abstr., 30 (1936) 5523.
211 C.E. Hedrick, J. Chem. Educ., 12 (1965) 660.
212 T.D. Rees and S.R. Hill, Spectrovision, No. 21 (1969) 13.
213 D. Staszewska and A. Ulinska, Chemia Anal. (Warsaw), 13 (1968) 1207.
214 M.I. Lebedeva and B.I. Isaeva, Zavod. Lab., 37 (1971) 410; Anal. Abstr., 22 (1972) 647.
215 S. Hirano, J. Soc. Chem. Ind. Jap., Suppl. Bind., 38 (1935) 646.
216 S. Hirano, J. Soc. Chem. Ind. Jap., Suppl. Bind., 38 (1935) 175.
217 W.J. Boyer, Ind. Eng. Chem. Anal. Ed., 10 (1938) 175.
218 H. Frey, Z. Anal. Chem., 133 (1951) 328.
219 H. Frey, Z. Anal. Chem., 132 (1951) 276.

220 D.J. Curran and K.S. Fletcher, Anal. Chem., 41 (1969) 267.

221 J.E. Lee, J.H. Edgerton and M.T. Kelley, Anal. Chem., 28 (1956) 1441.

222 T.S. Ma and J. Gwirtsman, Anal. Chem., 29 (1957) 140.

223 R.P. Hollingworth, Anal. Chem., 29 (1957) 1130.

224 J. Gwirtsman, R. Mavrodineanu and R.R. Coe, Anal. Chem., 29 (1957) 887.

225 J.A. Dean, M.H. Buehler and L.J. Hardin, J. Ass. Offic. Agr. Chem., 40 (1957) 949.

226 A. Steyermark, R.R. Kaup, D.A. Petras and E.A. Bass, Microchem. J., 3 (1959) 523.

227 W.P. Pickhardt, Anal. Chem., 34 (1962) 863.

228 C. Harzdorf and O. Steinhauser, Z. Anal. Chem., 210 (1965) 106.

229 S. Lorec, Chim. Anal. (Paris), 49 (1967) 557.

230 C. Harzdorf, Z. Anal. Chem., 227 (1967) 161.

231 C. Harzdorf, Z. Anal. Chem., 232 (1967) 172.

232 C. Harzdorf, Z. Anal. Chem., 233 (1968) 348.

233 F.W. Cheng, Mikrochim. Acta, (1970) 841.

234 R.N. Walter, Anal. Chem., 22 (1950) 1332.

235 R.S. Fielder and C.H. Morgan, Anal. Chim. Acta, 23 (1960) 538.

236 U. Niwa and E.P. Parry, Chemist—Analyst, 49 (1960) 102.

237 A.P. Kreshkov and V.V. Kuznetsov, Z. Anal. Khim, 25 (1970) 49.

238 C.M. Yih and D.F. Mowery, Microchem. J., 16 (1971) 194.

239 F.W. Peaker, Analyst (London), 85 (1960) 235.

240 E. Brücke, Chem. Centr., (1877) 139; J. Chem. Soc. Abstr., 34 (1878) 242.

241 G.H. Schenk and W.E. Bazzelle, Anal. Chem., 40 (1968) 162.

242 S. Hirano, J. Soc. Chem. Ind. Jap., Suppl. Bind., 37 (1934) 178.

243 S. Hirano, J. Soc. Chem. Ind. Jap., Suppl. Bind., 37 (1934) 177.

244 S. Hirano and Y. Nakamura, J. Soc. Chem. Ind. Jap., Suppl. Bind., 37 (1934) 147.

245 T. Somiya and Y. Nakamura, J. Soc. Chem. Ind. Japan, Suppl. Bind., 38 (1935) 262.

246 S. Hirano, J. Soc. Chem. Ind. Jap., Suppl. Bind., 40 (1937) 412.

247 T. Somiya and Y. Yasuda, J. Soc. Chem. Ind. Japan, Suppl. Bind., 41 (1938) 314.

248 G.P. Rowland, Ind. Eng. Chem. Anal. Ed., 11 (1939) 442.

249 Y.Y. Lur'e and E.M. Tal, Zavod. Lab., 9 (1940) 702.

250 O.H. Weber, Die Chem., 55 (1942) 364.

251 R.F. Goddu and D.N. Hume, Anal. Chem., 22 (1950) 1314.

252 J.C. Hindeman, L.B. Magnusson and T.J. La Chapelle, J. Amer. Chem. Soc., 71 (1949) 687.

253 C.E. Bricker and P.B. Sweetser, Anal. Chem., 24 (1952) 409.

254 P.B. Sweetser and C.E. Bricker, Anal. Chem., 24 (1952) 1107.

255 E.N. Wise, P.W. Gilles and C.A. Reynolds, Anal. Chem., 25 (1953) 1344.

256 C.E. Bricker and P.B. Sweetser, Anal. Chem., 25 (1953) 764.

257 C.E. Bricker and L.J. Loeffler, Anal. Chem., 27 (1955) 1419.

258 J.W. Miles and D.T. Englis, Anal. Chem., 27 (1955) 1996.

259 H.V. Malmstadt and C.B. Roberts, Anal. Chem., 28 (1956) 1412.

260 T.L. Marple, E.P. Przybylowicz and D.N. Hume, Anal. Chem., 28 (1956) 1892.

261 H.N. Furman and J.A. Fenton, Anal. Chem., 28 (1956) 515.
262 G.S. Deshmukh and M.G. Bapat, Z. Anal. Chem., 156 (1957) 276.
263 J.W. Edwards and G.W.C. Milner, Analyst (London), 82 (1957) 593.
264 H.L. Polak, G. den Boef and L. de Galan, Z. Anal. Chem., 198 (1963) 321.
265 W.E. van der Linden and G. den Boef, Anal. Chim. Acta, 38 (1967) 517.
266 S. Miyake and I. Sakamoto, Japan Anal., 16 (1967) 238.
267 E.R. Groeneveld and G. den Boef, Z. Anal. Chem., 237 (1968) 85.
268 E.R. Groeneveld and G. den Boef, Z. Anal. Chem., 238 (1968) 19.
269 A.F. Grand and M. Tamres, Anal. Chem., 40 (1968) 1904.
270 C.W. Fuller and J.M. Ottaway, Analyst (London), 94 (1969) 32.
271 P.P. Naidu and G.G. Rao, Z. Anal. Chem., 251 (1970) 301.
272 P.P. Naidu and G.G. Rao, Z. Anal. Chem., 251 (1970) 302.
273 F. Duque-Macias and F. Lucena-Conde, An. Real Soc. Espanol. Fis. Quim., Ser. B, 66 (1970) 79. Anal. Abstr., 20 (1971) 964.
274 L.S.A. Dikshitulu, Ind. J. Chem., 9 (1971) 872.
275 L.G. Chatten, R.A. Locock and R.D. Krause, J. Pharm. Sci., 60 (1971) 588.
276 R. Pribil and V. Vesely, Talanta, 9 (1962) 939.
277 R. Belcher, D.I. Rees and W.I. Stephen, Talanta, 4 (1960) 78.
278 H.H. Willard and C.A. Horton, Anal. Chem., 22 (1950) 1194.
279 H.K. Howerton and J.C. Wasilewski, in D.S. Jackson (Ed.), Titrimetric Methods, Plenum Press, New York, 1961, p. 51.
280 A.B. Borle and F.N. Briggs, Anal. Chem., 40 (1968) 339.
281 J.A. Bishop, Anal. Chim. Acta, 35 (1966) 244.
282 J.A. Bishop, Anal. Chim. Acta, 39 (1967) 189.
283 J.L. Beck, J.M. Fitzgerald and J.A. Bishop, Anal. Chim. Acta, 51 (1970) 191.
284 A.M. Escarrilla, Talanta, 13 (1966) 363.
285 J.H. Wiersma and P.F. Lott, Anal. Lett., 1 (1968) 603.
286 M. Wronski, Anal. Chem., 43 (1971) 606.
287 M. Wronski, Z. Anal. Chem., 180 (1961) 185.
288 M. Wronski, Mikrochim. Acta, (1970) 955.
289 M. Wronski, Z. Anal. Chem., 253 (1971) 24.
290 R.L. Clements and G.A. Sergeant, Lab. Pract., 19 (1970) 813, 816.
291 C.A. Reynolds, Record of Chem. Progress, 24(3) (1963) 157.
292 R. Pribil and V. Vesely, Talanta, 14 (1967) 591.
293 E.L. Pratt, J. Amer. Pharm. Ass., Sci. Ed., 46 (1957) 724.
294 J.W. Miller and D.D. de Ford, Anal. Chem., 29 (1957) 475.
295 E. Jackwerth and H. Specker, Z. Anal. Chem., 171 (1959) 270.
296 C.A. Reynolds, F.H. Walker and E. Cochran, Anal. Chem., 32 (1960) 983.
297 I.M. Klotz and B.R. Carver, Arch. Biochem. Biophys., 95 (1961) 540.
298 J.S. Fritz and T.A. Palmer, Anal. Chem., 33 (1961) 98.
299 G.H. Schenck and M. Ozolins, Anal. Chem., 33 (1961) 1035.
300 G.H. Schenck and M. Ozolins, Anal. Chem., 33 (1961) 1562.
301 Y.G. Liu and C.A. Reynolds, Anal. Chem., 34 (1962) 542.
302 Y. Tanase and S. Shimomura, Yakugaku Zasshi, 80 (1960) 516. Chem. Abstr., 54 (1960) 17796.
303 E. Cochran and C.A. Reynolds, Anal. Chem., 33 (1961) 1893.
304 W.R. Post and C.A. Reynolds, Anal. Chem., 36 (1964) 781.

387

305 C.W. Wadelin, Talanta, 10 (1963) 97.
306 J.P. Paris, J.D. Gorsuch and D.M. Hercules, Anal. Chem., 36 (1964) 1332.
307 H. Haglund and I. Lindgren, Talanta, 12 (1965) 499.
308 M. Richter and J. Szejtli, Die Stärke, 18 (1966) 95.
309 P.D. Boyer, J. Amer. Chem. Soc., 76 (1954) 4331.
310 I.M. Citron and D. Dolan, Anal. Chim. Acta, 33 (1965) 612.
311 T. Williams, J. Krudener and J. McFarland, Anal. Chim. Acta, 30 (1964) 155.
312 J.S. Fritz and G.E. Wood, Anal. Chem., 40 (1968) 134.
313 C.A. Reynolds and D.C. Underwood, Anal. Chem., 40 (1968) 1983.
314 J.M. Fitzgerald and J.L. Beck, Anal. Lett., 3 (1970) 531.
315 T.R. Williams and S. Wakeham, Anal. Chim. Acta, 52 (1970) 152.
316 S.J. Liggett and R.A. Libby, Talanta, 17 (1970) 1135.
317 K.H. Pearson and S. Kirschner, Anal. Chim. Acta, 48 (1969) 339.
318 S. Kirschner and D.C. Bhatnagar, Anal. Chem., 35 (1963) 1069.
319 R.J. Palma, P.E. Reinbold and K.H. Pearson, Chem. Commun., (1969) 254.
320 R.J. Palma, P.E. Reinbold and K.H. Pearson, Anal. Chem., 42 (1970) 47.
321 K.H. Pearson and J.R. Baker, Anal. Chim. Acta, 50 (1970) 255.
322 D.L. Caldwell, P.E. Reinbold and K.H. Pearson, Anal. Lett., 3 (1970) 93.
323 R.J. Palma, P.E. Reinbold and K.H. Pearson, Anal. Chim. Acta, 51 (1970) 329.
324 D.L. Caldwell, P.E. Reinbold and K.H. Pearson, Anal. Chim. Acta, 49 (1970) 505.
325 R.J. Palma and K.H. Pearson, Anal. Chim. Acta, 49 (1970) 497.
326 D.L. Caldwell, P.E. Reinbold and K.H. Pearson, Anal. Chem., 42 (1970) 416.
327 P.E. Reinbold and K.H. Pearson, Talanta, 17 (1970) 391.
328 R.W. Looyenga and C.O. Huber, Anal. Chem., 43 (1971) 498.
329 R.W. Looyenga and C.O. Huber, Anal. Chim. Acta, 55 (1971) 179.
330 P.J. Slevin and G. Svehla, Z. Anal. Chem., 246 (1969) 5.
331 J. Růžička and J. Starý, Talanta, 14 (1967) 909.
332 P. Grey and G.C.B. Cave, Can. J. Chem., 47 (1969) 4543.
333 P. Grey and G.C.B. Cave, Can. J. Chem., 47 (1969) 4555.
334 M.L. Corrin, H.B. Klevens and W.D. Harkins, J. Chem. Phys., 14 (1946) 480.
335 J.W. Ledbetter and J.R. Bowen, Anal. Chem., 43 (1971) 773.
336 F. Alt, F. Umland and G. Wuensch, Z. Anal. Chem., 251 (1970) 95.
337 K.A. Connors and T. Higuchi, Chemist—Analyst, 48 (1959) 91.
338 K. Behrends, Z. Anal. Chem., 250 (1970) 241.
339 K. Behrends, Z. Anal. Chem., 250 (1970) 246.
340 T. Somiya and S. Shiraishi, J. Soc. Chem. Ind. Jap., Suppl. Bind., 33 (1930) 300.
341 D.W. Klingman, D.T. Hooker and C.V. Banks, Anal. Chem., 27 (1955) 572.
342 T.R. Sweet and J. Zehner, Anal. Chem., 30 (1958) 1713.
343 D.N. Sunderman and R.C. Propst, U.S. Atomic Energy Comm., DP-SPU 53 (1953).
344 K.Y. Brill, S. Holzer and B. Rethy, Anal. Chem., 31 (1959) 1353.
345 J.M.G. Barredo and J.K. Taylor, Trans. Electrochem. Soc., 92 (1947) 437.

388

346 P.W. Mullen and A. Anton, Anal. Chem., 32 (1960) 103.

347 P. Karsten, H.L. Kies and P. de Hoog, Rec. Trav. Chim. Pays—Bas, 79 (1960) 610.

348 H. Ellert, T. Jasinski and K. Marcinkowska, Acta Pol. Pharm., 17 (1960) 29. Chem. Abstr., 54, 11806.

349 K. Rowley, R.W. Stoenner and L. Gordon, Anal. Chem., 28 (1956) 136.

350 V. Suk and V. Miketukova, Chem. Listy, 52 (1958) 2408.

351 H.M.C. Robinson and J.C. Rathbun, Can. J. Biochem. Physiol., 37 (1959) 225.

352 A. Juliard, J. van Cakenberghe and C. Heitner, Ind. Chim. Belge, 17 (1952) 25.

353 R. van Slageren, Doctoral thesis, Fluorimetric Indication of Complexometric Titrations of Trace Elements. Determination of Zinc, University of Amsterdam, 1972.

354 J. Starý and J. Růžička, in G. Svehla (Ed.), Wilson and Wilson's Comprehensive Analytical Chemistry, Vol. VII, Elsevier, Amsterdam, 1976, p. 212.

Chapter IV

Analytical applications of interferometry *

W. NEBE

1. Historical background of analytical interferometry

The discovery of the interference of light is due to the English scholar Thomas Young (1773—1829). In his discourse on the 12th November, 1801, Young formulated the interference principle by analogy with water waves as follows [1]: "I maintain that similar effects take place whenever two portions of light are thus mixed; and this I call the general law of the *interference of light*". He described his famous experiment with two slits in lecture 39 of his "Lectures on Natural Philosophy" [2], adding numerous sketches of his observations. Young's ideas, however, found no sympathy among his contemporaries. First, in 1815 the French engineer Augustin-Jean Fresnel (1788—1827) quite independently published a series of valuable experiments in which, without knowing it, he confirmed Young's theories [3]. The first practical application of the interference of light in the measurement of the refractive index of air was reported by Arago and Biot in France. François Arago [4] (1786—1853) had constructed the first apparatus to measure interference which, however, on account of the state of technology at the time, showed serious shortcomings.

As with Young's interference at the double slit, so also the interference phenomenon already observed in 1817 by the Scottish physicist Brewster [5] (1781—1868) between the four surfaces of two glass plates slightly inclined to each other, remained unregarded for many years. Jules Celestin Jamin (1818—1886) was the first to make amplitude division at a surface the basis of an efficient interference set-up. With the interferometer which bears his name Jamin investigated the refractive index of water, aqueous solutions and of air as

---

* The manuscript for this chapter was completed in 1972.

References pp. 536—546

well as the effect of pressure and temperature on these [6,7]. This formed the basis of the application of interference phenomena to chemical analysis, though this at first remained purely a research problem. The first comprehensive work by Mascart [8] appeared in 1871 on the theoretical basis of the standard interferometer. The subjects covered included a description of an interesting variation of the Jamin interferometer which in recent times has been used for the construction of very small interferometers. At the end of last century further interference set-ups were developed, which have remained important to the present day. Gouy (1854—1926) [9] described in 1880 a simple interferometer for diffusion investigations. In 1882 Michelson (1852—1931) [10] designed an interferometer which still bears his name, and which is today an even more versatile instrument. Mach [11] and Zehnder [12] independently of each other created an interference set-up for testing homogeneity.

Lord Rayleigh [13] (1842—1919) constructed in 1896, in connection with analytical determinations of noble gases, a relatively simple interferometer on the Young—Arago double slit principle, which was in the future to prove very useful for gas and liquid analyses. Because the apparatus was rather unwieldy, the measurement of interference was until then exclusively a method for pure research problems, and so it was not until the beginning of this century that its real advance as a routine analytical method took place through the joint initiative of Fritz Haber (1868—1934) and Fritz Löwe (1874—1955). In 1910 Löwe developed from the Rayleigh type the first interferometer to be manufactured commercially on a large scale [14]. In addition to a highly sensitive instrument for laboratory analysis, he constructed, using the autocollimation principle, further portable interferometers for liquid analysis [15,16], and for gas analysis in underground mining. These instrumental developments as well as Löwe's methodical work and publishing activity pioneered interferometry and established it as an independent analytical method for gases and liquids.

In more recent times, in addition to the Löwe interferometer, the Jamin type has also been mass produced in a much reduced size. This design was re-introduced by the Japanese Uzumi Doi [17] (born 1895). Meanwhile several special interferometer set-ups for concentration determinations of liquid and gaseous substances have been described, but as yet they have found no wide application.

In contrast to these instruments and methods for the measurement of refractive index of homogeneous substances, the techniques for the investigation of stratified liquids are relatively young. The method of electrophoresis, in its present form, is due to A. Tiselius [18] who succeeded in 1937 in determining the migration velocity of high molecular weight material in an electric field by using a suitable cell and observing it with a schlieren optical method. The optical equipment was further developed by Svensson [19] in 1949 through the introduction of a modified Rayleigh arrangement, so that it showed both the refractive index and also the gradient curve with high accuracy.

Interferometry has reached a state of perfection for the investigation of diffusion only in the last three decades. Interesting methods were found for it even earlier, not only by Gouy [9] but, for example, by Kroepelin [20] who in 1926 followed the course of the diffusion process with the Löwe interferometer, and also by Calvet and Chevalerias [21] in 1946.

The development of the ultracentrifuge goes back to Svedberg [22] who in 1925, with his co-workers, first constructed an ultracentrifuge capable of 45,000 rev min^{-1} and a centrifugal force of 100,000 g. The modern refractometric optical observation technique has been introduced by Lamm, Philpot and Svensson since 1937.

The investigation of inhomogeneous substances with interferometers of the Michelson and Mach—Zehnder type has found varied applications in the examination of solubility, miscibility and separability, thermal conductivity, etc. The quantitative, easily interpreted interference method has been supplemented from the outset by the more obvious schlieren optical method of August Toepler [23]. In the last two decades apparatuses have appeared in which both methods are combined. A further advance was brought by the discovery of the laser. The very great coherence of these light sources led to the construction of greatly simplified instruments on the basis of the Michelson and the Mach—Zehnder principles. Similarly the technique of diffraction interference resulted in simpler interference apparatus. This goes back to Zernike [24] who in 1937 developed the phase-contrast technique for microscopy which in recent years has been introduced for homogeneity investigations on the macroscale. Interferometry ranks today as a routine method for the busy analytical chemist. It makes possible quick and accurate estimations of the concentration of homogeneous mixtures and solutions with very little effort. Its application extends to all three states of matter. It is extremely valuable both for objects with a one-dimensional gradient and for those with a random gradient distribution because the phenomena observed can be evaluated comprehensively and very precisely. In this short historical sketch only the most important stages in the development of interferometry could be mentioned so far as they are directly concerned with the interference techniques used today for chemical analysis. Parallel to this run equally fertile lines of development in interference optics, for example in microscopy and spectroscopy, which represent immediate border areas to the interferometric methods of chemical analysis discussed here.

Further applications of the interference methods are:

Optical testing (homogeneity testing, thickness measurement, surface control, control of image quality, determination of the optical transfer function).

Material testing (surface control, strain testing).

Information technology (holography).

Heat technology (conduction of heat, heat convection, two-phase exchange effects).

Gas chemistry (combustion and explosion processes, pressure waves).

Aerodynamics (sonic and supersonic waves, eddy effects).

Hydrodynamics (flow streams).

Metrology (definition of length, standard testing).

Construction, mechanical engineering (direction and alignment testing, linear measurement).

Surveying (distance measurement, goniometry).

Astronomy, space travel (distance determination).

This list of applications is by no means complete; it only indicates the extensive significance of interferometric measurements in the fields of technology and research. A comprehensive description is given by Dyson [25].

A more complete historical description can be found in refs. 26–28.

2. Symbols

A synopsis is given below of the symbols used in the text and their meanings. References to the defining formula or the section in the text in which the idea is used, are added. The symbols are sometimes used in the text with indices, when they need a special distinction.

The indices are not listed in this synopsis. Also some characters, used in illustrating certain figures, are left out.

a slit width (cm)
        distance of coherent light rays (cm)
        constant, eqn. (67)
b width of a light entrance slit (cm)
        constant, eqn. (67)
c concentration
        speed of light (m s^{-1})
c_{vac} 2.9979 · 10^8 m s^{-1}, speed of light in a vacuum
d separation of two apertures (cm)
e 546.1 nm Hg spectral line
        base of natural logarithms (2.7183)
f focal distance (cm)
        coefficient of friction (g s^{-1}), eqn. (91)
g 435.8 nm Hg spectral line
        gravitational constant (9.807 m s^{-2})
        integer
h interferometric fringe number
$i =$ $\sqrt{-1}$
        interferometer value, i.e. the value read on the scale of measure-
        ment of an interferometer
j consecutive index
k constant
l geometric path length, especially path length in a cuvette (cm)
m refractive power, eqn. (28)
n refractive index, eqn. (26)
p pressure (g cm^{-1} s^{-2}) or (Torr), 760 Torr = 1013.25 mb
        power (g cm s^{-2})
q conditional value in chromatic effects, eqn. (50)
        electric charge (A s)
r radius (cm)
        specific refraction (cm^3 g^{-1}), eqn. (33)
s optical path length (cm), eqn. (36)
        sedimentation constant (s), eqn. (86)
t time (s)
        temperature (°C)
u half angle of aperture, eqn. (11)
v velocity (cm s^{-1})
w distance of a point from the axis (cm)
        fringe spacing (cm)   395

References pp. 536—546

x	height coordinate in a cell with stratified liquids (cm)
y	auxiliary line
	coordinate
z	coordinate
A	amplitude
B	2nd virial coefficient (cm^3), Sect. 6.E
C	656.3 nm H$_2$ spectral line
C'	643.8 nm Cd spectral line
D	589.3 nm Na spectral line
	relative effective dispersion, eqns. (47), (48)
	diffusion constant (cm^2 s^{-1}), eqn. (75)
E	electric field strength (V cm^{-1})
	plane
F	486.1 nm H$_2$ spectral line
	area (cm^2)
	factor, eqn. (52)
F'	480.0 nm Cd spectral line
H	achromatic length
	height of the gradient curve (cm), eqn. (83)
I	intensity, eqn. (7)
K	coherence length (cm), eqn. (5)
L	2.687 · 10^{19} Loschmidt's number, number of molecules in 1 cm^3 of perfect gas at 0°C, 760 Torr
	abbreviation, eqn. (12)
M	molar mass, molecular weight (g mol^{-1})
N	6.023 · 10^{23} Avogadro's number, number of molecules in one mole
	refractive index of glass
P	percentage
	point symbol
Q	substance flux (cm s^{-1}), eqns. (75), (93)
R	8.314 J K^{-1} mol^{-1}, gas constant
	refraction (cm^3), eqn. (33)
	intensity of a reflected light beam
T	temperature (K)
	intensity of a transmitted light beam
V	volume (cm^3)
W	Gauss probability function, eqn. (72)
α	1/273.16 K^{-1} cubic expansion coefficient of gas
	angle of a light beam to the axis
β	angle of inclination of a straight line

γ electrical polarizability of a molecule (cm^3), eqn. (32)

δ error

       angular difference

ϵ angle between two light beams

       dielectric constant

η viscosity (g cm^{-1} s^{-1})

κ constant correction value, eqn. (30) and Sect. 5C

λ wavelength (cm)

ν frequency (s^{-1})

       Abbe's number, Sect. 3.E

$\tilde{\nu}$ wave number (cm^{-1})

ξ width of a slit element (cm)

       tangential angle at a point of inflection, eqn. (82)

π 3.1416

ρ density (g cm^{-3})

σ reciprocal value of the vacuum wavelength (cm^{-1}), eqn. (31)

τ coherence time (s)

φ phase, phase angle, eqns. (2), (3)

ψ phase, phase angle, Sect. 3.D(4)

ω angular velocity (s^{-1})

Γ magnification factor

Λ abbreviation, eqn. (49)

ϕ state of phase; phase vector, eqn. (1)

Ψ state of phase; phase vector, Sect. 3.D(4)

3. Theoretical background of interference measurements

(A) THE WAVE NATURE OF LIGHT

The phenomenon of optical interference can only be explained by the wave-character of light. It is therefore necessary to have an understanding of the important properties of vibrations and waves to understand interferometric observations and measurements. Although complete descriptions of interference theory exist in the literature [31—38], the most essential facts, from the point of view of its use for analytical purposes, should be briefly mentioned.

Light is, in the wave theory interpretation, an electromagnetic vibration process connected with the transport of energy, which fills up the space in which the light propagates. The vibration occurs in a plane E perpendicular to the direction of propagation of the light. The particular state of vibration can be described mathematically as a vector, briefly called a "light vector" to which the laws of vectorial addition apply.

Each vibration in the plane E can be split up into two linear vibrations mutually perpendicular. It is, therefore, no limitation of the general case that linear vibrations are assumed in what follows. For the investigation of non-absorbing media only harmonic, i.e. undamped vibrations, are of interest.

For linear harmonic vibrations the light vector can be expressed in the plane of complex numbers by its state of phase, ϕ

$$\phi = A \exp(i\varphi) = A(\cos \varphi + i \sin \varphi) \qquad (1)$$

A is the amplitude and φ designates the phase position or, briefly, phase, starting from the origin, say the positive zero passage of ϕ (Fig. 1). If the vibration occurs at a fixed point with frequency ν, then the dependence of the state of phase ϕ on the time of vibration t (Fig. 2) is given by

$$\varphi = 2\pi\nu t \qquad (2)$$

The transference of the vibration between neighbouring points forms a three-dimensional progressive wave. The smallest distance between points of equal state of phase is the wavelength λ. In visual interferometry this is measured in nanometers ($1 \text{ nm} = 10^{-6} \text{ mm}$). At a distance s of the wave from the point of origin the state of phase is defined by the amplitude and by the phase (Fig. 3).

$$\varphi = 2\pi \frac{s}{\lambda} \qquad (3)$$

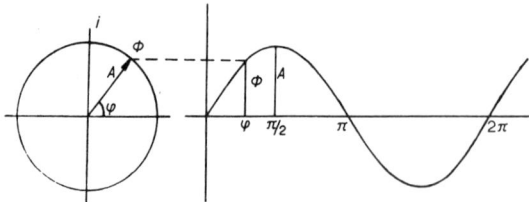

Fig. 1. Representation of the state of the phase ϕ in the plane of the complex number, and its corresponding vibrational pattern.

398

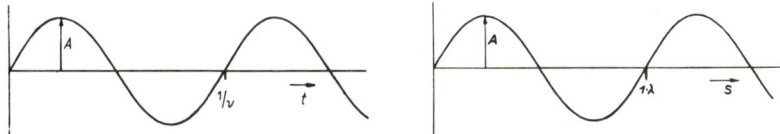

Fig. 2. The vibration as a function of time.

Fig. 3. The wave as a function of distance.

The velocity of propagation of light c is given by the product of the wavelength and the frequency

$$c = \lambda\nu \qquad (4)$$

In a vacuum the velocity of light is independent of the wavelength and has a constant value $c_{vac} = 2.9979 \cdot 10^8$ m s^{-1}.

In addition to the frequency ν and the wavelength λ one can define for the propagation of light the wave number $\tilde{\nu}$ as the number of wavelengths that fall in a distance of 1 cm. The value of $\tilde{\nu}\lambda$ is 10. In analytical interferometry the wavelength λ is preferred for the characterization of light rather than the frequency ν or the wave number $\tilde{\nu}$.

Of the whole wavelength region of the electromagnetic spectrum only visible light between 400 and 700 nm is of interest for optical interferometry. Interferometric measurements in the UV and in the IR have as yet been made only occasionally, and so the contents of this chapter have been restricted to the visible spectral region.

The spectrum of visible light is seen by the eye as a series of spectral colours, which show a well defined relationship to the wavelength. Thus the human eye has its greatest sensitivity for yellow-green light at 555 nm; for the dark-adapted eye, the maximum sensitivity is shifted to the blue. Spectrally undispersed light from the whole visible spectrum is seen as white.

In the spectrum of sunlight one can detect black lines, the Fraunhofer lines. These are caused by the passage of the light through certain substances, in this case hydrogen, sodium, etc., absorbing light of definite wavelengths. The energy exchange between light and matter, which is the cause of the absorption of definite spectral colours, can conversely give rise to the emission of light. Excited atoms or molecules emit radiation at definite wavelengths. On the basis of energy relationships derived from the quantum theory of light, the

Fig. 4. Some important spectral lines in the visible part of the spectrum.

emitted light has the same wavelength as that absorbed from white light entering the same substance. In interferometry the emission spectral lines are used as monochromate light for standardization and for special analytical procedures. Some important spectral lines are given in Fig. 4 and in Sect. 4.E(1).

(B) MONOCHROMATICITY AND COHERENCE

Light is described as monochromatic when it contains only one single spectral colour, strictly speaking when all the vibrations included in it lie in a very narrow spectral range. The magnitude of the wavelength band of monochromatic radiation is defined as its half-intensity width $\Delta\lambda$, i.e. the width of the region in which the intensity is not less than 50% of maximum intensity.

Closely associated with the monochromaticity of light is its coherence. Two light waves are said to be coherent when they vibrate with the same frequency and fixed phase relationship. As is well known, light is always emitted in quanta, i.e. an emitting atom always radiates only discrete wave-trains of finite length (Fig. 5). The length of the wave-train is constant for similar atoms under identical excitation conditions and is described as the coherence length K. Further the coherence time τ is defined as the time taken by a wave-train with coherence length K to pass through a point. There exists a simple relationship

$$K = c\tau$$

From eqn. (4) can be derived the relationship $\Delta\nu/\Delta\lambda = \nu/\lambda$ and from this the half-intensity width $\Delta\lambda$ can be expressed as the frequency difference $\Delta\nu$. During the coherence time the difference $\Delta\nu$ exhibited by the frequencies occurring may amount at most to half an oscillation.

$$|\Delta\nu \cdot \tau| \leqslant \tfrac{1}{2}$$

From this one can write for the coherence length

$$K \leqslant \frac{\lambda^2}{2\Delta\lambda} \tag{5}$$

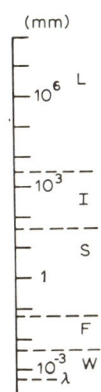

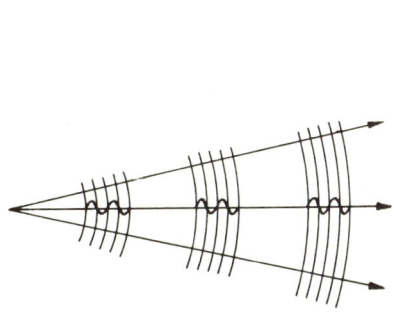

Fig. 5. The emission of a train of waves.

Fig. 6. Coherence length of some light sources: W, white light; F, filtered white light; S, spectral lamp; I, monoisotopic spectral lamp; L, laser.

The emission intervals of the individual atoms are independent of those of other atoms, with the result that light emitted from two different atoms and especially from different light sources is in general not coherent. In laser light sources a fixed phase relationship exists between the individual emitting atoms because of the nature of the excitation conditions, so that even two different lasers can emit coherent radiation. Equation (5) gives the relationship between the coherence length and the spectral width of the light. It follows from this that white light has a very limited coherence length of a few μm. With monochromatic filters the coherence length can be raised by the power of ten, e.g. with a special interference filter with a half-intensity width of 5 nm at wavelength 600 nm it can be raised to approximately 0.07 mm. An important improvement has been made possible by spectral lamps, with which one can obtain a coherence length of more than 1 m. All previous light sources have, however, been excelled by far by the laser (Fig. 6).

To produce coherent light beams the radiation emitted from a single light source must be split up into two (or more) beams. There exist different possibilities; the most important of which are wave-front division, amplitude division and wave-front shearing (Sect. 3.D). The coherence between two light beams has fundamental significance

References pp. 536—546

for interferometry, because only coherent light beams are capable of producing interference effects.

(C) INTERFERENCE

By interference one understands the superposition of two or more light waves with the result that, depending on the position of observation, the intensity of the light is seen either to increase or to decrease. The superposition of two light waves with amplitudes A_1 and A_2 and phases φ_1 and φ_2 gives the resultant state of oscillation

$$\Phi* = A_1 \exp(i\varphi_1) + A_2 \exp(i\varphi_2)$$

Clearly perceptible interference phenomena only occur when the interfering light beams are coherent, i.e. the light beams must have the same frequencies ($\nu_1 = \nu_2 = \nu$, respectively $\lambda_1 = \lambda_2 = \lambda$) and constant phase relations ($\varphi_1 - \varphi_2 = \Delta\varphi = $ const.), and when the difference in the distances travelled is smaller than the length of coherence ($s_1 - s_2 = \Delta s < K$). Under these conditions (Fig. 7)

$$\Phi* = (A_1^2 + A_2^2 + 2A_1 A_2 \cos \Delta\varphi)^{1/2} \exp(i\varphi*) \tag{6}$$

The light intensity is the square of the oscillation amplitude.

$$I = A^2 \tag{7}$$

For complete destructive interference

$$I* = A_1^2 + A_2^2 + 2A_1 A_2 \cos \Delta\varphi = 0$$

and this is possible only for $A_1 = A_2 = A$. Assuming equal amplitude

$$\phi* = A \exp(i\varphi) [1 + \exp(i\Delta\varphi)] = 2A \cos \tfrac{1}{2}\Delta\varphi \exp[i(\varphi + \tfrac{1}{2}\Delta\varphi)] \tag{8}$$

The resultant light vector is displaced in phase by $\tfrac{1}{2}\Delta\varphi$ from the individual vectors and its amplitude is

$$A* = 2A \cos \tfrac{1}{2}\Delta\varphi$$

The intensity resulting from the interference of two waves becomes

$$I* = 4A^2 \cos^2 \tfrac{1}{2}\Delta\varphi \tag{9}$$

It is now obvious that

$$\Delta\varphi = (2g-1)\pi \qquad \text{destructive interference } (I = 0), \text{ and} \tag{10}$$
$$\Delta\varphi = 2g\pi \qquad \text{maximum reinforcement } (I = 4A^2)$$

402

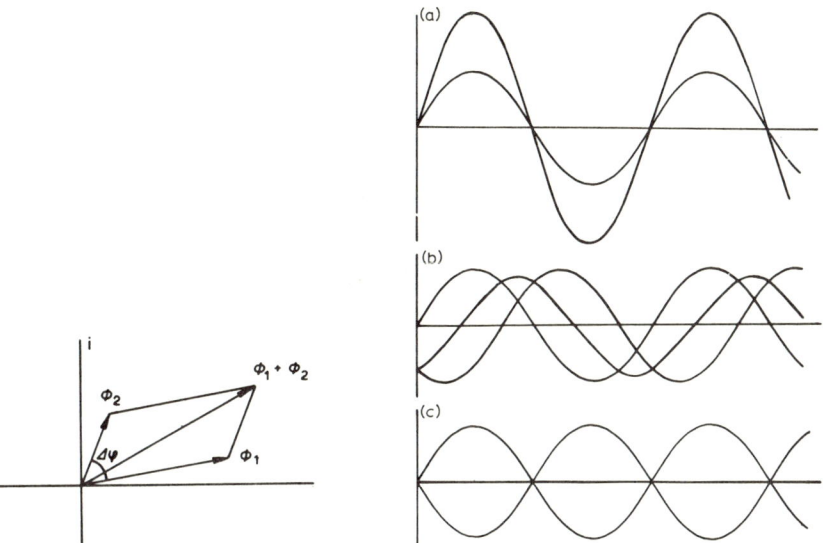

Fig. 7. Addition of two states of phase in the plane of the complex number.

Fig. 8. Superimposition of two waves: (a) $\Delta\varphi = 2g\pi$; (b) $\Delta\varphi$ of no particular value; (c) $\Delta\varphi = (2g-1)\pi$.

where g is an integer (Fig. 8).

These interference phenomena form the basis of all interferometric measurements. It is therefore important to produce clear and interpretable interference patterns. One distinguishes various arrangements which are commonly used in the interference apparatus for analytical uses. The most important of these are described as follows.

(D) INTERFERENCE ARRANGEMENTS

(1) Wave-front division

The wave-front of the radiation emanating from a light source is defined as the plane of points, that vibrate in equal phase without phase difference. According to the form of the wave-front one can speak of, for example, a spherical wave or a plane wave. We shall consider only plane waves, obtained theoretically from an infinitely distant light source, in practice obtained by the use of a collimator.

With wave-front division (aperture division) the interference effect is obtained by inserting a diaphragm in the wave-front, as was done, for example by Young [2] and Rayleigh [13]. The diaphragm contains at least two apertures depending on the number of desired coherent beams (Fig. 9). The amplitude of the light wave is not altered by passing through the apertures. To obtain interference phenomena the coherent beams are brought together again by means of mirrors, prisms or other optical image elements. In order to produce a plane wave-front these arrangements use a collimator with a narrow entrance slit placed before the diaphragm (Fig. 10). The apparatus for wave-front division needs comparatively a slight optical lay-out, and is also rather insensitive to external vibrations, temperature fluctuations and the like. However, the entrance slit must be severely limited in order to obtain sufficient phase coincidence of the wave-fronts leaving from the individual points of the slit [39]. The slit may lie perpendicular to the plane of the illustration (Fig. 11). At a distance b symmetrical about the axis lie two slit points each producing through the objective a plane wave-front inclined to the other at an angle ϵ. The wave-fronts resulting from both slit-points will only appear as one if the phase differences occurring within the entire diaphragm width a_D are not greater than

$$\Delta\varphi = 2\pi(\Delta s/\lambda) = \pi \qquad \text{or} \qquad \Delta s = \tfrac{1}{2}\lambda$$

i.e.

$$\sin\tfrac{1}{2}\epsilon << \frac{\lambda}{2a_D}$$

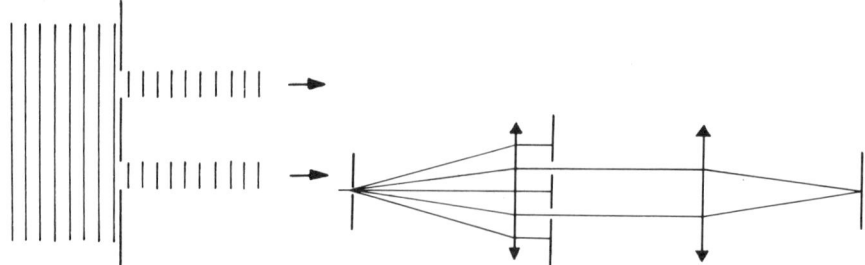

Fig. 9. Use of a diaphragm for dividing a plane wave-front.

Fig. 10. Rayleigh's arrangement for interference.

404

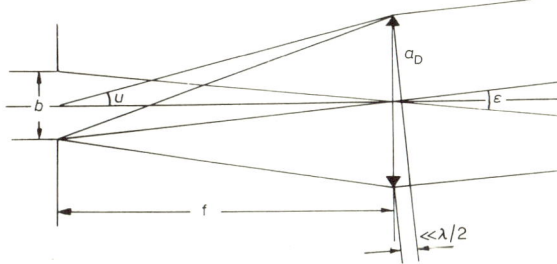

Fig. 11. Conditions for coherence at the slit.

As the slit lies in the focal plane of the objective

$$\sin \epsilon = b/f$$

Defining u as the half-aperture angle, then

$$\sin u = a_D/2f$$

Thus, good contrast in the interference pattern is obtained only when the slit-width fulfils the coherence condition

$$b << \frac{\lambda}{2 \sin u} \tag{11}$$

The splitting of the wave-front can also be accomplished without a diaphragm, for example, by using a Fresnel biprism instead (Fig. 12). In practical applications the object of investigation itself can be used to produce wave-front interference in a similar way to the Fresnel biprism (Gouy's interference method is based on this procedure, Sect. 7.C). It is essential for wave-front division that the previously uniform front is geometrically split up so that at least two separate beams of equal amplitude are formed representing separate fronts.

The intensity distribution in the interference pattern can be found by the following additional inspection. The diaphragm apertures are divided into narrow bands of width ξ. Within a light beam inclined to the axis at an angle α there is a phase difference $\Delta\varphi$, where $\sin \alpha = \Delta s/\xi$ and with eqn. (3) becomes

$$\Delta\varphi = 2\pi \frac{\sin \alpha}{\lambda} \xi$$

This phase difference leads to mutual extinction of the light passing

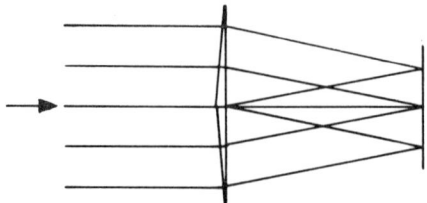

Fig. 12. The production of interference patterns with Fresnel's biprism.

through the bands of width ξ when $\Delta\varphi = 2\pi g$. Integrating over the state of phase $\phi = \exp(i\Delta\varphi)$ of all elementary bands and bearing in mind that the diaphragm consists of two apertures with the width a and distance d apart, so we get

$$\phi_\alpha = \int_0^a \exp(i\Delta\varphi)\,\mathrm{d}\xi + \int_d^{d+a} \exp(i\Delta\varphi)\,\mathrm{d}\xi$$

Abbreviating by the expression

$$2\pi\,\frac{\sin\alpha}{\lambda} = L \tag{12}$$

one gets

$$\phi_\alpha = \frac{1}{iL}\,[\exp(iLa)-1] + \frac{1}{iL}\,[\exp(iL(d+a))-\exp(iLd)]$$

$$= \frac{1}{iL}\,[\exp(iLa)-1]\,[1+\exp(iLd)]$$

$$= \frac{2}{L}\,\sin\tfrac{1}{2}La\,\exp(\tfrac{1}{2}iLa)\,(2\cos^2\tfrac{1}{2}Ld + i\cdot 2\sin\tfrac{1}{2}Ld\,\cos\tfrac{1}{2}Ld)$$

$$= \frac{2}{L}\,\sin\tfrac{1}{2}La \cdot 2\cos\tfrac{1}{2}Ld\,\exp[\tfrac{1}{2}iL(d+a)]$$

The intensity at an angle α to the axis is given by

$$I = 4\,\frac{\sin^2\left(\pi\,\dfrac{\sin\alpha}{\lambda}\right)}{\left(\pi\,\dfrac{\sin\alpha}{\lambda}\right)^2}\cos^2\left(\pi\,\frac{\sin\alpha}{\lambda}\,d\right) \tag{13}$$

406

It can easily be seen that the extinction $I = 0$ occurs for

$\frac{\sin \alpha}{\lambda} a = g\ (\neq 0)$, i.e.

$$\sin \alpha = g\frac{\lambda}{a}\ (g \neq 0) \tag{14a}$$

but also for

$\frac{\sin \alpha}{\lambda} d = \frac{1}{2}(2g-1)$

$$\sin \alpha = (2g-1)\frac{\lambda}{2d} \tag{14b}$$

The intensity distribution for the case $d = 4a$ is shown in Fig. 13. The angular spacing of the interference fringes is given by eqn. (14b). When α_1 and α_2 are the appropriate angles for two successive minimum-points and f is the focal length of the system, then the spacing w of these points is given by

$$w = f(\sin \alpha_2 - \sin \alpha_1) = \frac{f\lambda}{d} \tag{15}$$

w indicates the spacing of the interference fringes in the interference pattern produced by wave-front division.

(2) Amplitude division

In amplitude division the light is partially reflected and partially transmitted at a beam-splitter plate with a partially transmitting

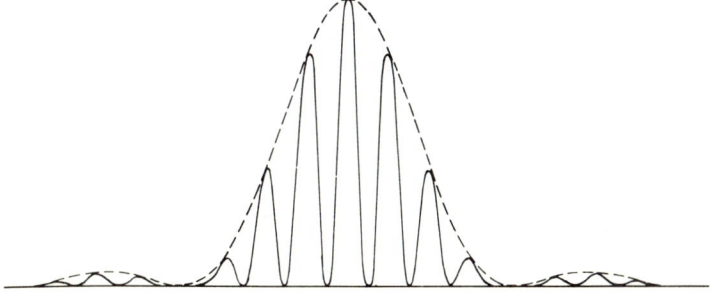

Fig. 13. The distribution of intensity in the Rayleigh interference pattern.

layer. Such layers can be evaporated on to glass plates in a vacuum. Aluminium, silver or chromium is normally used in the applications of interest here.

The waves resulting from amplitude division (Fig. 14) are coherent because of their entirely similar history before their separation. With a plane beam-splitter their wave-fronts are similar to that of the impinging wave and they are therefore geometrically similar to each other. But their intensities R (on reflection) and T (on transmission) are diminished in comparison to that of the incident light I. When one neglects the loss due to absorption, $R + T = I$. In general one chooses $R = T$ and speaks of a half-transmission beam-splitter plate. The same type of plate is also used to recombine the two beams to produce interference.

Let us consider two coherent waves with plane wave-fronts which propagate at an angle ϵ and interfere in a plane E. E lies perpendicular to the bisected angle of the surface normal to the wave-front, and the line of intersection of two wave-fronts lies perpendicular to the plane of the figure (Fig. 15). Lines of equal intensity are produced in E, that lie parallel to the line of intersection of the wave-fronts. Depending on the phase difference $\Delta\varphi$ obeying eqn. (10) these lines are either light or dark. It is enough to consider the intensity of these points from E lying in the plane of the figure. When at a point P_0 the

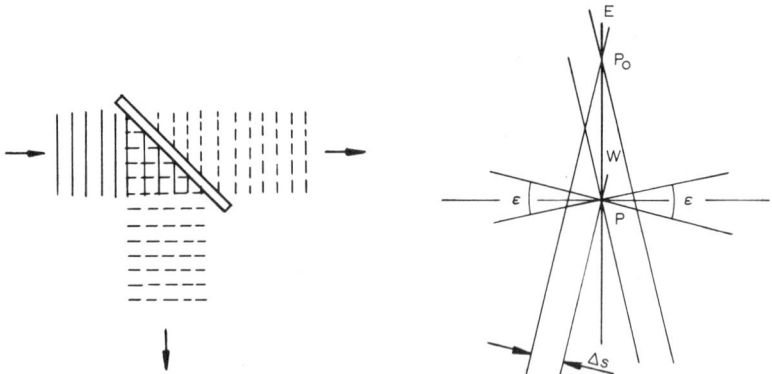

Fig. 14. Amplitude division of a plane wave-front.

Fig. 15. Representation of interference between two plane wave-fronts.

408

phase difference $\Delta\varphi = 0$ and therefore the maximum intensity $I = 4A^2$ prevails, then it follows that at a point P at distance w from P_0 the phase difference is $\Delta\varphi = 2\pi\Delta s/\lambda$ where $\Delta s = w \sin \frac{1}{2}\epsilon$.

From eqn. (9) the intensity at P is given by

$$I = 4A^2 \cos^2\left(\frac{\pi w}{\lambda} \sin \frac{1}{2}\epsilon\right) \qquad (16)$$

It is obvious that for monochromatic light a periodic variation in intensity exists, in which each maximum reaches the intensity $I = 4A^2$. Maximum values exist for $\cos^2[(\pi w/\lambda) \sin \frac{1}{2}\epsilon) = 1$, i.e. for

$$w = g \frac{\lambda}{\sin \frac{1}{2}\epsilon} \qquad (17)$$

For $g = 1$ w gives the fringe spacing

$$w = \frac{\lambda}{\sin \frac{1}{2}\epsilon} \qquad (18)$$

The advantage of amplitude division lies in that there is no critical limitation on the entrance aperture. One can work with large apertures for the light beam so that one has altogether very much greater energy available than with the wave-front division.

The arrangements with amplitude division most important for analytical purposes are those of Jamin, Michelson and Kösters, as well as Mach and Zehnder.

In the two-plate arrangement of Jamin (Fig. 16) the light falls obliquely on a glass plate, the front surface of which is coated with a semi-transparent layer and the rear surface of which is fully reflecting. At the front surface part of the light is reflected, the rest being refracted, reflected by the rear surface and leaving the plate parallel to that reflected from the front surface. Both light beams are reflected again at a second glass plate similar to and parallel to the first. As they leave the second plate, both beams have travelled different paths of equal length and have undergone the same number of reflections and transmissions. The amplitude of each of the two interfering light beams is given as RT^2. From this the ratio of reflection to transmission $R : T$ for optimum light gain can be calculated. Thus when conventionally talking of "half" transmission plates, the ratio $R : T$ need not be unity. In the vaporization procedure it is possible to obtain the desired optimal value of the coating.

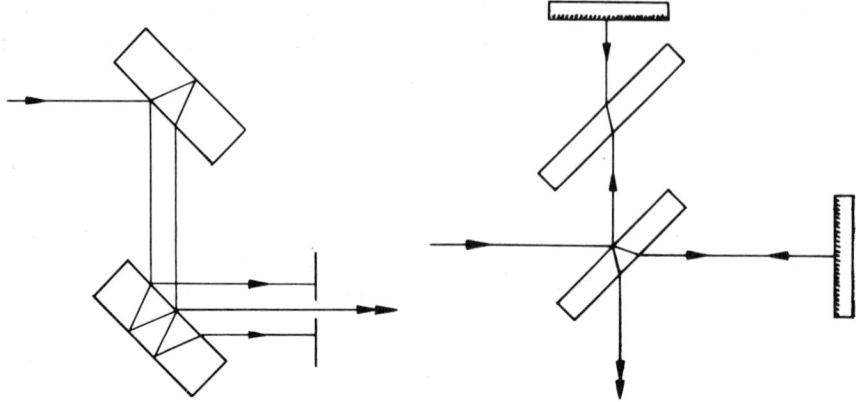

Fig. 16. The interference arrangement of Jamin.

Fig. 17. The interference arrangement of Michelson.

In the apparatuses of Michelson (Fig. 17) and of Kösters (Fig. 18) only one partially transmitting surface serves both for dividing and also for re-combining the interfering light beams. The Michelson type makes possible a wide separation of the paths of both beams and consequently also a large cross-section of the object field. It is suitable above all for homogeneity testing on objects of all three states of matter. In contrast, the light beams in the Kösters type run parallel to each other in a way similar to the Jamin type. This type is used predominantly for dilatometric measurements.

The apparatus derived by Mach and Zehnder (Fig. 19) from the Jamin type likewise produces widely separated paths of the interfering light beams. Its use corresponds roughly to that of the Michelson

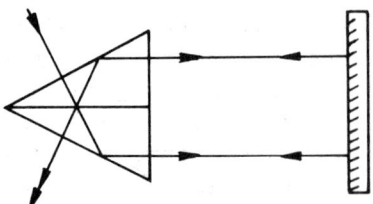

Fig. 18. The interference arrangement of Kösters.

410

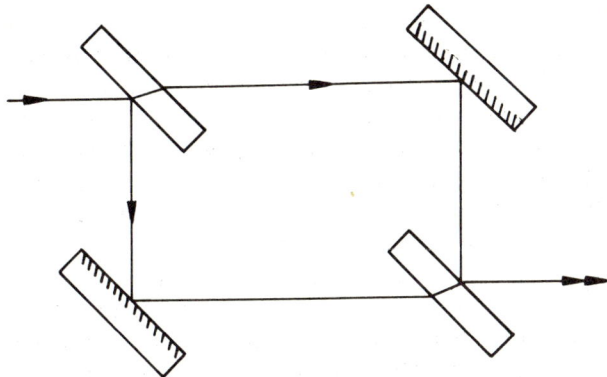

Fig. 19. The interference arrangement of Mach and Zehnder.

type. But as the light passes through the object field only once, it often shows conditions more favourable for interpretation. The expenditure on optics and adjusting work is obviously considerably greater.

The angle ϵ between the interfering wave-fronts can be altered in the Michelson and Mach—Zehnder interferometers by inclining one of the mirrored surfaces. By this the direction and, from eqn. (18), the spacing of the interference fringes can be adjusted at will. This possibility renders the interference method with amplitude division highly adaptable towards the object of investigation and makes it very valuable for quantitative homogeneity testing. Especially with $\epsilon = 0$ one has an infinite fringe spacing, i.e. a uniformly illuminated image plane, and very slight inhomogeneities in the object can be displayed as differences of brightness.

(3) Wave-front shearing

The principle of wave-front shearing can be considered as a special case of amplitude division, see for example Françon [37], Bryngdahl [40] and Yamamoto [41]. Also in shearing the wave-front is divided by suitable optical arrangements into two similar wave-fronts. While complete separation is produced between both fronts in amplitude division, in shearing both part-fronts are set only a small distance apart, so that they run through approximately the same object volume (Prasad [42]). Shearing methods are suitable therefore only for investigating inhomogeneous substances, or testing surfaces.

One distinguishes different types of shearing according to whether the wave-fronts are positioned, relative to one another (a) laterally (lateral shearing), (b) rotated (rotational shearing) or (c) altered in the size ratio (radial shearing). For analytical purposes, the last two procedures mentioned are of interest only in very special cases of gradient fields, so that the discussion will be limited to lateral shearing.

One prefers to employ polarized light when using the shearing method. Light is said to be polarized, or more exactly linearly polarized, when the light vector representing the electromagnetic radiation vibrates in a straight line (Fig. 20). This assumption has already been made in Sect. 3.A for simplifying the discussion. There the general state of vibration, with the oscillation occurring in random directions in a plane at right angles to the direction of propagation, was by convention split up into two linearly polarized vibrations, mutually perpendicular.

Linearly polarized light is produced when natural light is allowed to pass through a polarizer. A polarization film is usually used for this today, made from a dichroic high polymer material, e.g. poly-(vinyl alcohol).

The shearing effect is normally produced by means of a birefringent material. A substance is said to be birefringent when the velocity of light passing through it is dependent on the state of polarization and the direction of the light. Birefringent substances possess a preferred direction, the "optical axis", in which the velocity of light is independent of the state of polarization. In all other directions of

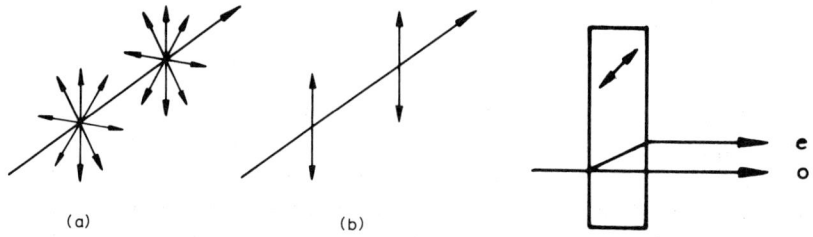

(a)   (b)

Fig. 20. Direction of the vibration of the light-vector (a) for unpolarized light, and (b) for linearly polarized light.

Fig. 21. Birefringence by a quartz plate: o = ordinary ray, e = extraordinary ray.

412

transmission the light is split into two mutually perpendicularly polarized components, one of which, the ordinary ray, has the same velocity as that in the direction of the optical axis, while the other, the extraordinary ray, has its velocity altered according to the direction. This makes it possible to split light into two parts, polarized at right angles to each other (Fig. 21). The most frequently used birefringent substances are calcite, crystalline quartz and ADP crystals. With combinations of prisms or plane-parallel plates of birefringent material very effective elements can be manufactured for splitting light into mutually perpendicularly polarized beams, of which the most widely used is the Wollaston prism, which gives a symmetrical deflection of both beams (Fig. 22) and the Savart-plate composed of two equally thick plane parallel plates which produces a compensation of optical path difference between both beams (Fig. 23). In both figures the arrow gives the direction of the optical axis of the crystal; the circle indicates the direction of the optical axis at right angles to the plane of the drawing; the dotted arrow indicates the optical axis running through the plane of the illustration at 45°, so that the extraordinary beam emerges from the plane of the drawing.

(4) Diffraction

Diffraction is a special case of interference (Françon [37]). This phenomenon refers to those effects in the image plane which result from introducing a diaphragm into the image-forming ray. Since the edges of the imaging lenses act as diaphragms, the formation of any

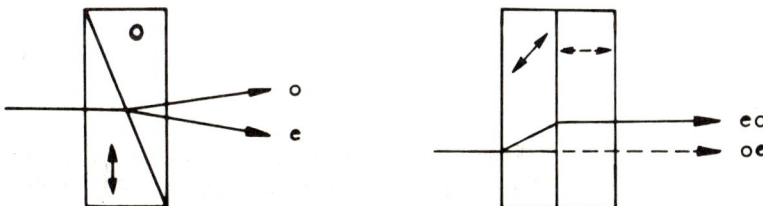

Fig. 22. Birefringence by a Wollaston prism.

Fig. 23. Birefringence by a Savart plate.

optical image is necessarily associated with diffraction phenomena. However, diffraction can only be clearly observed when the limiting aperture is of the order of the wavelength of the light.

Diffraction is self-evident from the Huygens' theory, according to which each point of the wave becomes the point of origin of a new spherical wave, which interferes with the other spherical waves. If a plane wave-front falls on a narrow slit, for example, then the points of the slit will become the centres of new waves capable of interference. An angle α with $\sin \alpha = \lambda/a$, is defined in such a way that for each light beam emerging from the slit there exists a second beam at the same angle possessing a phase difference of $\Delta s = \frac{1}{2}\lambda$ where they meet in the focal plane of the imaging lens (diffraction plane). At this point all beams are cancelled out pair-wise producing darkness (Fig. 24).

Generally the distribution of intensity in the diffraction plane, resulting from a plane wave falling on the slit (width a), can be deduced by analogy with the intensity distribution in the interference pattern according to eqn. (13). In the diffraction plane

$$\phi_\alpha = \int_0^a \exp(i\Delta\varphi)\, d\xi = \frac{1}{iL}[\exp(iLa)-1] \text{ with } L = 2\pi \frac{\sin\alpha}{\lambda} \text{ [eqn. (12)]}$$

This gives

$$\phi_\alpha = \frac{2}{L}\sin\tfrac{1}{2}La\,\exp(\tfrac{1}{2}iLa)$$

$$= \frac{\lambda}{\pi \sin\alpha}\sin\left(\pi \frac{\sin\alpha}{\lambda}a\right)\exp\left(i\pi \frac{\sin\alpha}{\lambda}a\right)$$

thus for the intensity

$$I = \frac{\sin^2\left(\pi \dfrac{\sin\alpha}{\lambda}a\right)}{\left(\pi \dfrac{\sin\alpha}{\lambda}\right)^2} \tag{19}$$

For $(\sin\alpha/\lambda)a = g$ $(\alpha \neq 0)$ there are minima $(I = 0)$, between which lie maxima, with height decreasing as α increases. For a point at distance w from the centre $\sin\alpha = w/f$, so for

$$w = g\frac{f\lambda}{a} \qquad (g \neq 0) \tag{20}$$

414

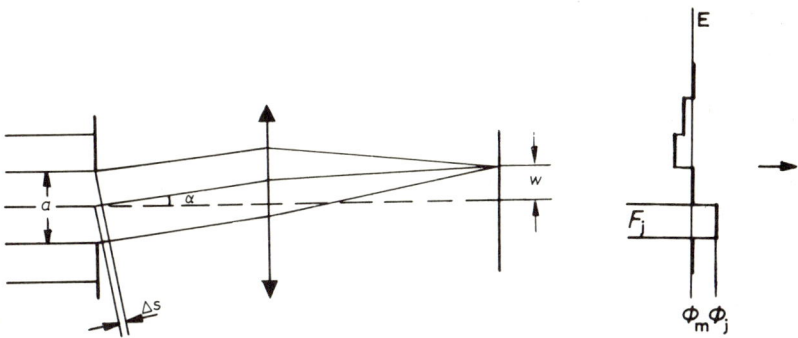

Fig. 24. Diffraction at a slit.

Fig. 25. Wave-front behind a transparent heterogeneous object.

the intensity reaches a minimum. The distribution of intensity corresponds to a curve which is marked in Fig. 13 as a dotted line.

In the case of an aperture which is assumed to be a slit, the diffraction fringes run as straight lines, parallel to the image of the slit.

Diffraction interference has proved useful for analytical purposes only to the extent that the heterogeneity of an object can be made more clearly visible by modifying the diffraction pattern by means of the optical devices just mentioned. These methods are known in microscopy as the dark-field procedure and the phase-contrast procedure (Zernike [24]). But they are also applied in macroscale investigations (Wolter [43]). In the case of these, as with other interference methods for testing homogeneity, it is necessary to transform phase differences in the object (invisible to the eye) into intensity differences in the interference pattern. The changes of intensity in the image plane after conversion into a diffraction interference pattern are based on the following correlation. Whenever a plane wave-front passes through a heterogeneous object phase differences occur, i.e. the states ϕ_j pertaining to the points P_j in a plane E beyond the object differ because of their different phases φ_j. Their amplitude is maintained if the object is free of absorption, so that there occur no differences of brightness in the image and the phase differences do not become visible, A_j = const. = 1. These become visible only after conversion into the diffraction pattern. For the points in the plane E we can specify a mean state of phase ϕ_m (Fig. 25), which we find by vectorial addition of the single vectors, always after multiplication with the relevant scalar area element F_j

$$\phi_m = \frac{1}{F} \oint \exp(i\varphi_j)\, dF = A_m \exp(i\varphi_m)$$

The state of phase ϕ_j differs at individual points in the plane E from the mean state of phase ϕ_m, because of heterogeneity.

$$\phi_j = \phi_m + \Psi_j \tag{21}$$

The difference vector Ψ_j is expressed by

$$\Psi_j = \phi_j - \phi_m = \exp(i\varphi_j) - A_m \exp(i\varphi_m)$$

$$= [1 + A_m^2 - 2A_m \cos(\varphi_j - \varphi_m)]^{1/2} \exp(i\psi_j) \tag{22}$$

The undiffracted light which is focussed at the centre of the diffraction pattern corresponds to the mean vector ϕ_m. If one eliminates the undiffracted light from the image by means of an opaque diaphragm (Gayhart and Prescott [44]), the amplitude A_m of the mean vector is reduced to zero (Fig. 26(a)). For every point in the object there remains in the image only the component characterized by the difference vector Ψ_j

$$\phi_j^* = \phi_m + \Psi_j$$

The intensity I_j^* relating to the image of P_j after inserting the optical devices, according to eqn. (22), amounts to

$$I_j^* = 1 + A_m^2 - 2A_m \cos(\varphi_j - \varphi_m) \tag{23}$$

The intensity is dependent on the phase φ_j. Phase-changing parts of the object are projected by the diffracted light in the focussing plane, as areas of varying brightness. With a homogeneous object field (φ_j = const. = φ_m) the image plane appears dark after the introduction of the diffraction diaphragm since generally $A_m \approx 1$ and $I^* = (1 - A_m)^2 \approx 0$. Therefore one speaks of the dark-field procedure.

The phase-contrast method of Zernike and Ingelstam [45,46] operates on the same principle. In this a translucent layer is introduced instead of the opaque diaphragm. This layer is produced by vaporization on to a glass plate. Its thickness is of such dimensions that it transmits the undiffracted light with a phase difference of $\frac{1}{2}\pi$ compared to the diffracted light. That corresponds to a rotation of

416

the mean vector in Fig. 26(b) of 90°. Thus $\phi_m = A_m \exp(i\varphi_m)$ now becomes $\phi_m^* = A_m \exp[i(\varphi_m + \frac{1}{2}\pi)]$.

The state of phase in the image of P_j is now given by the addition of the additional vector Ψ_j according to eqn. (22), to the rotated mean vector ϕ_m^*

$$\phi_j^* = \phi_m^* + \Psi_j = A_m \exp[i(\varphi_m + \frac{1}{2}\pi)]$$

$$+ (1 + A_m^2 - 2A_m \cos(\varphi_j - \varphi_m)^{1/2} \exp(i\psi_j) = A_j^* \exp(i\varphi_j^*)$$

and with it the appropriate intensity

$$I_j^* = A_j^{*2} = 1 + 2A_m^2 - 2A_m \cos(\varphi_j - \varphi_m) \tag{24}$$

$$+ 2A_m(1 + A_m^2 - 2A_m \cos(\varphi_j - \varphi_m))^{1/2} \cos(\psi_j - \varphi_m - \frac{1}{2}\pi)$$

is independent on the phase φ_j. Phase differences are therefore seen as differences in brightness.

In the field-absorption procedure, described by Erdmann [47], the diffraction centre is not changed, but a partial absorption is effected in the surrounding field. As a result of insertion of a field-absorption plate, ϕ_m remains unaltered and Ψ_j becomes $\Psi_j^* = A \exp(i\psi_j)$ where $A < 1$. Thus

$$\phi_j^* = \phi_m + \Psi_j^* = A_m \exp(i\varphi_m) + A \exp(i\psi_j)$$

$$= (A_m^2 + A^2 + 2A_m A \cos(\varphi_m - \psi_j)^{1/2} \exp(i\varphi_j^*)$$

and the intensity becomes

$$I_j^* = A_m^2 + A^2 + 2A_m A \cos(\varphi_m - \psi_j) \tag{25}$$

This expression is formally similar to eqn. (23).

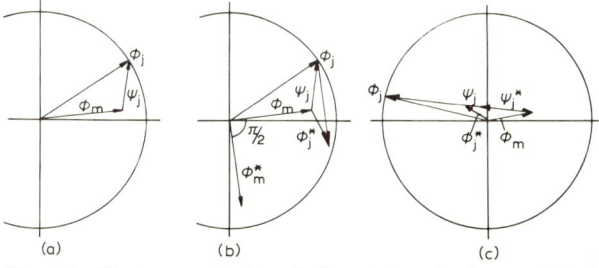

Fig. 26. Vector analysis of the state of phase with diffraction interference: (a) dark-field technique; (b) phase-contrast technique; and (c) field-absorption technique.

In the dark-field procedure very small intensity values are obtained only whenever $\varphi_j - \varphi_m = 2g\pi$ and especially when $\varphi_j = \varphi_m$ and in any case when $A_m \approx 1$. On the basis of these conditions good contrast appears in the interference pattern with points of almost total extinction, above all when only small phase differences $|\Delta\varphi| < \frac{1}{2}\pi$ occur in the object. This is similarly valid for the phase-contrast procedure. On the other hand, in the field-absorption procedure diminishing intensity occurs for $\varphi_m - \psi_j = (2g-1)\pi$ and $A = A_m$. Ψ_j must therefore have an opposite direction to ϕ_m which means that ϕ_j also has an opposite direction to ϕ_m (Fig. 26(c)). Consequently, figures of high contrast only arise when phase differences of at least π occur. This method is suited best to objects which have large phase differences and thereby complements both the other diffraction interference methods.

(E) THE REFRACTIVE INDEX

The refractive index is an important physical constant for a material, the determination of which represents the principal task in analytical interference measurements. Its definition derives from the speed of light. The speed of light in a vacuum c_{vac} is a natural constant, but as a result of the interaction of the light with matter the velocity of light decreases as it passes through any matter, and therefore according to eqn. (4) its wavelength also decreases.

The frequency is determined by the oscillation process giving rise to the photons, and this is independent of the surrounding matter. Therefore, to characterize the spectral region of light it would be more correct to indicate the frequency than the wavelength (dependent on matter). However, the wavelength differences for gases are very small, e.g. the wavelength of the green mercury line is 546.225 nm in a vacuum, on the other hand in "standard air" it is 546.074 nm. (Standard air is dry air of the usual composition with 0.03% carbon dioxide at $15°C$ and 760 Torr, as laid down by the international agreement in spectroscopy.) In analytical interferometry it is normal to quote the wavelength relating to standard air in giving the spectral region of the light.

The refractive index of a material is defined as the ratio of the speed of light in a vacuum to that in the material in question

$$n = \frac{c_{vac}}{c} \tag{26}$$

418

Accordingly we have $n \geqslant 1$. When measured under prescribed standard conditions of pressure, temperature and the wavelength of the light, it is a characteristic constant of the material. These conditions are indicated by corresponding indices. Thus $n_\lambda^{t,p}$ is the refractive index of a material at $t°C$ and p Torr for the wavelength λ nm.

Pressure and temperature affect the refractive index in many various ways depending on the aggregate state of the substance. These influences are dealt with separately in Sect. 5.C.

The variation of the refractive index with wavelength is called dispersion (Born and Wolf [34], Joffe [48]). A material's dispersion is closely related to its absorption characteristics. Whereas the refractive index in a spectral region which is free of absorption gradually increases as the wavelength decreases, it rises sharply in the close proximity of an absorption peak only to fall off again steeply at wavelengths which lie just beyond the absorption peak. According to whether $dn/d\lambda < 0$ or > 0 we speak of normal dispersion or anomalous dispersion.

Colourless materials possess no absorption peak in the visible spectral region. Their refractive indices can be represented very approximately by a dispersion equation

$$n^2 - 1 = k_0 + \frac{k_1}{\lambda^2 - \lambda_1{}^2} + \frac{k_2}{\lambda^2 - \lambda_2{}^2} \tag{27}$$

where λ_1 and λ_2 represent the nearest absorption peaks in the UV and IR region, respectively, while k's are constants for the substance. By far the majority of substances which are examined by interferometric methods exhibit normal dispersion behaviour in the visible region, so that one can frequently use a simplified practical dispersion equation without employing the last term in eqn. (27).

In the special case of gases n is very little different from 1, so that one can use the approximation $n^2 - 1 = 2(n-1)$. Since $n-1$ is a very small number it is general to use the refractive power m instead of the refractive index of a gas,

$$m = (n-1)\, 10^6 \tag{28}$$

From this and eqn. (27) one arrives at Cauchy's dispersion equation

$$m = m_0 \left(1 + \frac{k}{\lambda^2}\right) \tag{29}$$

The constants m_0 and k are given for some gases in Table 1 (λ in nm).

As well as eqn. (27) the Hartmann dispersion equation may also be used for all transparent substances

$$n = n_0 \frac{k'}{(\lambda - \lambda_0)^\kappa} \tag{30}$$

where the constants n_0, k', λ_0 and κ are calculated from refractive index measurements at four different wavelengths. Often κ is put equal to unity for the sake of simplification, so that one need measure only three values of n to determine the constants. For the most important gas, air, different writers and institutions have produced more specific dispersion equations; they relate to standard air at 15°C. Edlén [49,50] has given a very precise equation valid for wavelengths beyond the visible region. After division by 1.000162 in order to relate it to carbon dioxide-free standard air, Edlén's dispersion equation reads

$$m = 83.4078 + \frac{24{,}056.40}{130 - \sigma^2} + \frac{159.94}{38.9 - \sigma^2} \tag{31}$$

In this $\sigma = 1/\lambda_{vac}$ and λ_{vac} is the wavelength in nm in vacuo. Consideration of the IR region of the spectrum led Peck and Reeder [51] to make certain corrections to the relationships.

A general equation which takes into consideration the dependence on pressure, temperature and composition was set out by Owens [52]. In order to arrive at the refractive power of dry normal air, free of carbon dioxide at 0°C and 760 Torr eqn. (31) must be multiplied by $1 + \alpha t = 1.0549$ where $\alpha = 0.003661$ and $t = 15$°C.

Because of the refractive index's dependence on wavelength it is

TABLE 1

Constants for Cauchy's dispersion equation, eqn. (29), for some gases (according to Born and Wolf [34])

Gas	m_0	$k \cdot 10^{-2}$	Gas	m_0	$k \cdot 10^{-2}$
Air	287.9	56.7	Oxygen	266.3	50.7
Argon	279.2	56	Nitrogen	291.9	77
Helium	34.8	23	Methane	426	144.1
Hydrogen	136	77	Ethane	736.5	90.8

important to agree universally on a definite reference wavelength in the central visible spectral region. The green Hg-e-line $\lambda_e = 546.1$ nm has been chosen internationally as such [53] and one denotes the main refractive index by n_e and the main refractive power by m_e. As a crude characterization of the dispersion properties one defines the mean dispersion Δn as the difference of the refractive indices at the hydrogen lines F and C, $\Delta n = n_F - n_C$, or, more recently, at the cadmium lines F′ and C′, $\Delta n = n_{F'} - n_{C'}$. For gases one uses the refractive power difference $\Delta m = \Delta n \cdot 10^6$ instead of Δn. Frequently one calculates the Abbe number

$$\nu = \frac{n_D - 1}{n_F - n_C} \quad \text{or} \quad \nu_e = \frac{n_e - 1}{n_{F'} - n_{C'}}$$

Specific material constants have been sought which are related to the refractive index but are independent of these external influences. A starting point has been offered by Clausius and Masotti's relationship for the electrical polarizability of a molecule in an electric field

$$\gamma = \frac{3}{4\pi L} \frac{\epsilon - 1}{\epsilon + 2}$$

where ϵ is the dielectric constant dependent on the frequency of the field and $L = 2.687 \cdot 10^{19}$ is Loschmidt's number (the number of molecules per cm^3). For molecules with zero dipole moment γ is a constant. On the other hand γ shows a temperature dependency for dipolar molecules, so that the following considerations are not strictly valid for these. According to the Maxwell relationship, one has $\epsilon = n^2$. Between L, the density ρ, the molecular weight M and Avogadro's number $N = 6.023 \cdot 10^{23}$ (the number of molecules in one mole) there exists a relationship

$$LM = \rho N$$

And so from this

$$\gamma = \frac{3M}{4\pi\rho N} \frac{n^2 - 1}{n^2 + 2} \tag{32}$$

By using the constant value $\frac{4}{3}\pi N\gamma/M = r$ or $\frac{4}{3}\pi N\gamma = R$ one obtains the relationship found independently by Lorentz and Lorenz

$$r = \frac{n^2 - 1}{n^2 + 2} \frac{1}{\rho}, \qquad R = Mr = \frac{n^2 - 1}{n^2 + 2} \frac{M}{\rho} \tag{33}$$

r is called the "specific refraction" and R the (molar) refraction. Since γ is a constant (for dipole-free molecules) r and R are also constants and indeed are independent of the state of matter. Accordingly, there arises a simple possibility of finding the approximate refractive index of a gas or vapour. The (molar) refraction can be determined from refractometric and pyknometric measurements in the liquid state. Avogadro's law is valid for gas-like states, if one assumes, as a simplification, that they behave like ideal gases. According to this all ideal gases possess the same molar volume, which for a normal gaseous state is $V_M = 22416 \text{ cm}^3 \text{ mol}^{-1}$. From this the density of the gas is $\rho = M/V_M$, and from eqn. (33) with $n^2 + 2 \approx 3$ it follows that

$$n^2 \approx 1 + 3 \, \frac{R}{V_M}$$

For the refractive power, putting $n^2 - 1 = 2(n-1)$, one gets the simple relationship

$$m \approx \frac{1.5R}{V_M} \cdot 10^6 = 66.92R \tag{34}$$

A further important property of the refraction (and of the specific refraction) is their additivity when materials are mixed. If several materials with refractions R_j are contained in a mixture in the respective concentrations c_j then eqn. (35) is valid for the refraction of the mixture

$$R = \sum_j c_j R_j \tag{35}$$

Whereas a corresponding relationship eqn. (54) also exists in the case of gases for the refractive power (which is easier to measure), the relationship eqn. (35) is of special significance for liquids. Recently Akobjanow [54] gave a relationship which shows an even better constancy than the Lorentz–Lorenz expression: $r = [n - (1/n)]/\rho$. For water from 15 to 100°C, for example, the Abokjanow relationship gives a specific refraction between 0.584 and 0.585, whereas for the Lorentz–Lorenz expression it varies between 0.205 and 0.208. Since the refraction, like the refractive index, depends on the wavelength of the light, the dispersion between two wavelengths, e.g. between C and F is defined by

$$\Delta R = R_F - R_C$$

(F) THE MEASUREMENT OF REFRACTIVE INDEX BY INTERFERENCE

For a light beam travelling through a medium of path length l mm and refractive index n, the product

$$s = ln \tag{36}$$

is defined as the optical path length of the beam on this path. For two coherent light beams which cover the optical path lengths s_1 and s_2 between the points where they are divided and re-combined, the difference in their optical path lengths, the optical path difference

$$\Delta s = s_2 - s_1 \tag{37}$$

determines the phase difference between the interfering beams at the point of re-combination, so according to eqn. (3)

$$\Delta \varphi = 2\pi \frac{\Delta s}{\lambda} \tag{38}$$

It will depend on the value of $2\Delta s/\lambda$ (as indicated by eqns. (10)) whether light or darkness will predominate at this point. Where there is a continuous change in the optical path difference, the bright and dark fringes in the interference pattern are shifted continuously.

The technique of interference measurement is based on the determination of the optical path difference. This results in a comparative measurement against a known or at least a constant optical path length.

An interference arrangement for the measurement of refractive index is set up in such a way that both interfering light beams cover equal geometric path lengths. The optical elements in the two beams have identical properties, in as far as they influence the optical path length.

If one beam passes through a test substance with the optical path length ln and the other through a reference substance with $l_0 n_0$ and if, moreover, they cover identical geometric paths $l = l_0$, then this results in an optical path difference

$$\Delta s = l(n - n_0) \tag{39}$$

The optical path length is measured as a multiple h of the wavelength λ, $\Delta s = h\lambda$. The symbol h denotes the fringe number stating by how many fringe spacings the interference pattern has been displaced.

Thus

$$l(n - n_0) = h\lambda \tag{40}$$

This is an important basic formula for all interferometric measurements. On some instruments the fringe number h is indicated directly by the displacement of the interference pattern against a measuring scale. On other instruments it is determined by restoring the original phase equality of the two interfering beams, using an optical compensator which is introduced into the path of one of the beams. This method, while being more exact, necessitates more expensive apparatus.

With heterogeneous objects the distribution of the refractive index or its gradient in the object can be determined from the form of the interference fringes.

The path difference can generally be measured with a double-beam interferometer to $\frac{1}{50}$ to $\frac{1}{100}$ of a wavelength. From eqn. (40) it follows that there is an error of the magnitude of $\Delta(n - n_0) = \lambda l^{-1} \Delta(h)$ in determining the refractive index difference.

For wavelength $\lambda = 546.1$ nm, path length $l = 1000$ mm of the cuvettes and error $\frac{1}{50}\lambda$ there results $\Delta(n - n_0) = 1 \cdot 10^{-8}$.

When using a triple-beam interferometer or the Moiré method an increase in accuracy of nearly a factor of ten is possible.

4. Apparatus and accessories

The optical apparatus built and available commercially for interferometry will now be briefly reviewed. Besides these there are numerous interference arrangements, specially designed for particular research problems and with no general application. These, together with their respective applications, are described, as far as is considered necessary, in Sects. 6 and 7. Reference should be made to the monographs of Candler [60], Twyman [61], Löwe [62] and Joffe [48].

(A) LABORATORY INTERFEROMETERS FOR THE MEASUREMENT OF HOMOGENEOUS SUBSTANCES

(1) The Rayleigh—Löwe interferometer

The Rayleigh—Löwe interferometer works on the principle of wave-front division (Fig. 27). Generally white light is used but for

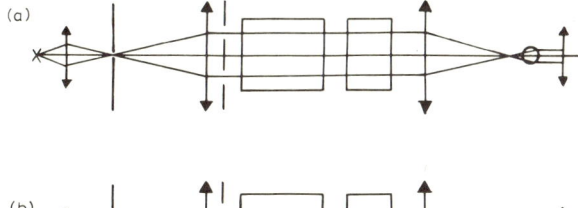

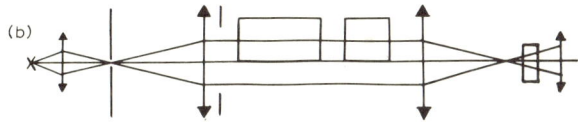

Fig. 27. The Rayleigh—Löwe interferometer: (a) horizontal cross-section; (b) vertical cross-section.

calibration purposes monochromatic light is also used. The beam-splitting diaphragm is situated immediately behind the collimator which has a vertical entrance slit obeying the coherence condition (11). This produces two coherent light beams running parallel, which suffer a change in their phase relationship during the course of their path through the cuvette and compensator. They are re-combined in the focal plane of the telescope objective to give an image of the entrance slit. The image of the slit shows broad diffraction fringes on both sides of its central maximum, and in addition interference fringes can be seen parallel to the entrance slit image. The intensity distribution at right angles to the direction of the slit for mono-chromatic light is given by eqn. (13) and is shown in Fig. 13. The spacing w between interference fringes is calculated from eqn. (15) with a being the distance between the apertures.

$$w = \frac{f\lambda}{a}$$

For example for the laboratory interferometer LI 3 of VEB Carl Zeiss Jena, $a = 12$ mm and $f = 300$ mm and so for green light ($\lambda = 546.1$ nm) a fringe spacing $w = 0.014$ mm is obtained. With a ten-fold magnification the interference fringes are even more clearly discernible. For measurement one magnifies these 100- to 200-fold with the help of a cylindrical lens with its axis vertical, magnifying only in a horizontal direction, while leaving the fringes optically unaltered in the vertical direction. Through this the overall diminution in the light density with magnification is kept within reasonable limits. Moreover, the cylindrical lens brings the advantages that without separate accessories one can produce a stable interference pattern as reference marks and observe layers or other perturbations in the vertical direction in the cuvette.

The interference fringe-spacing is proportional to the wavelength, so that white light gives rise to a coloured pattern. With phase equality between interfering beams the interference pattern shows a bright white fringe (zero-order maximum) in the centre. On either side the first-order minima for all colours lie close together and one can observe black lines in the pattern. With increasing order-number the minima for differing wavelengths become more widely separated, so that the fringes in the interference pattern become more and more colourful. The zero-order maximum is almost colourless compared with the other maxima, and the intense black neighbouring minima make it easy to distinguish and hence to use as a reference mark. A path difference $\Delta s = h\lambda$ between the two interfering beams produces a displacement in the interference pattern of the amount

$$w = f\frac{h\lambda}{a} \tag{41}$$

to that side on which the phase was retarded (Fig. 28). From the displacement w, with equal length of the cuvette for test and reference substances, the refractive index difference can be obtained from eqn. (40), to give

$$n - n_0 = \frac{a}{lf}\,w \tag{42}$$

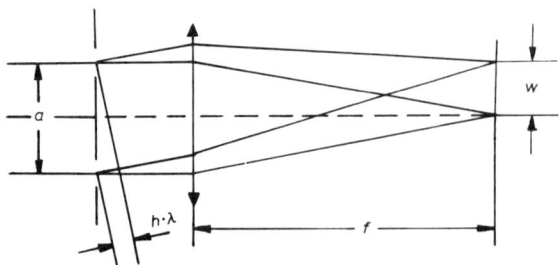

Fig. 28. Representation of the shift of the interference fringes due to a difference of phase.

On account of the difficulty involved in the direct measurement of very small values of w, all known Rayleigh interferometers possess a compensation arrangement by which the phase difference, imposed on both the light beams passing through the cuvettes, is returned to zero. The simplest and most commonly used compensator, that of Löwe, consists of two identical small plane-parallel glass plates which are inclined to the direction of the light and parallel to each other, and are passed through by both the light beams. By rotating one plate the optical path length of one light beam and hence the phase difference between both beams can be altered. The optical path difference adjusted by compensation is given, using the symbols shown in Fig. 29, by

$$\Delta s = d[\sqrt{N^2 - \sin^2(\alpha - \delta)} - \sqrt{N^2 - \sin^2\alpha} - \cos(\alpha - \delta) + \cos\alpha] \quad (43)$$

It can be seen therefore that the correct adjustment of the compensator is to return the zero-order maximum to the original position it had when the phase difference was zero. With this compensation arrangement the measurement of the displacement w of the interference pattern, is referred back to that of the rotation angle δ of the compensator. The rotation is effected by a lever with a micrometer drive, from which the interferometer value i can be read off. The relationship between the interferometer value and the optical path difference is found by calibrating the interferometer (Sect. 5.B). In the compensation procedure it is necessary to mark the original position of the interference pattern for a phase difference of zero. To

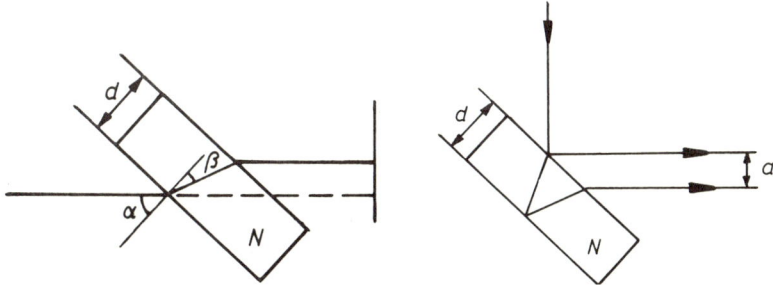

Fig. 29. Representation of the optical path length in the Lowe compensator.

Fig. 30. Spacing between the beams after leaving out the beam-splitter plate in the Jamin interferometer.

References pp. 536—546   427 facilitate this operation with the Rayleigh—Löwe interferometer a second system of interference fringes is produced from two coherent light beams which run parallel to the first pair below the cuvette and compensator with their phase state remaining unaltered by the compensation process. Rayleigh—Löwe interferometers possess, because of the nature of the compensation measurement, a very high sensitivity of measurement, which can be augmented yet again by means of metallized cuvettes [64,65], while they are at the same time very insensitive to external effects. They are firmly established for both gas and liquid analyses.

(2) The Jamin interferometer

The Jamin interferometer works on the principle of amplitude division (Fig. 16). An incandescent lamp serves as a light source, illuminating the first beam-splitter plate through a condenser lens. If d is the thickness and N the refractive index of the plate then the distance between both coherent beams for an incident angle α is given by (Fig. 30)

$$a = \frac{d \sin 2\alpha}{\sqrt{N^2 - \sin^2\alpha}} \tag{44}$$

and specifically for $\alpha = 45°$

$$a = \frac{d}{\sqrt{N^2 - 0.5}} \tag{45}$$

On leaving the second plate the beams selected for interference run parallel. They are examined through a telescope, in whose focal plane they produce the interference pattern. The spacing w of the interference fringes is, as with the Rayleigh double-slit method, dependent on the distance a^* between the beams entering the telescope. Favourable conditions for the interference pattern are obtained when the second plate, opposite the first, is so inclined at a horizontal axis as to include an angle ϵ between its surface normal and the plane of the incident light on the first plate (Schönrock [66]). With the notation of Fig. 31 and with $\sin \alpha = a/y$ and $\tan \epsilon = a^*/y$ it then follows that $a^* = a \tan \epsilon/\sin \alpha$. The fringe spacing w is given by combining eqns.

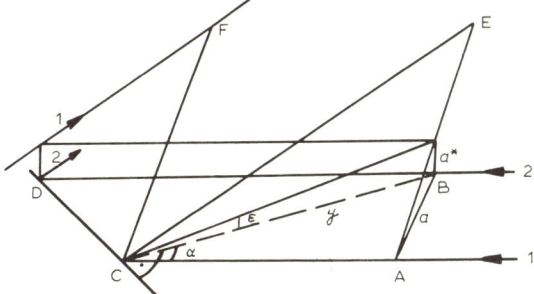

Fig. 31. Spacing between the coherent beams after leaving out the second beam-splitter plate in the Jamin interferometer: AB = a, spacing between the coherent beams leaving the first plate; C, point where beam 1 meets plate 2; D, meeting point of beam 2 with plate 2; E, a point on beam 1 after reflection at C; F, a point on the part of beam 1 interfering with beam 2. AEFC lie in one plane, CD is a straight line on the front surface of the plate 2, which is not itself drawn in the diagram.

(15) and (44)

$$w = \frac{f\lambda}{a^*} = \frac{f\lambda\sqrt{N^2 - \sin^2\alpha}}{2d \tan \epsilon \cos \alpha} \tag{46}$$

Small angles of tilt ϵ give large distances between interference fringes, which can be seen with the unaided eye or with a low-power magnifying lens. The fringe spacing is permanently fixed by the apparatus manufacturer. With the plates parallel one has $\epsilon = 0$ and the fringe spacing is infinite, i.e. the field of vision is uniformly illuminated and is not suitable for measurement. Cuvettes and compensator are subject to the same conditions as with the Rayleigh—Löwe interferometer (Sect. 4.A(1)). Either the Löwe compensator, with an optical path difference given by eqn. (43), or the original Jamin compensator is used. In the Jamin compensator two identical glass plates arranged at equal but opposite angles in the two coherent light beams are rotated simultaneously. The path difference in a Jamin compensator is given by an equation derived from (43) when α is replaced by $(-\alpha - \delta)$ in the second and in the last term on the right-hand side.

A reference fringe system may likewise be used as a zero mark [67]. But usually the fringes are adjusted at such a spacing that the shift of the interference pattern can be read directly off a scale and the compensator dispended with. Rayleigh and Jamin interferometers have been investigated by Kolomiitsov [68] from a common view point.

(3) The Michelson—Kinder interferometer

This type of interferometer [69] also uses the principle of amplitude division, but in a different arrangement — that of Michelson (Fig. 17). Since it is advantageous that both coherent beams run through two cuvettes lying side by side, a mirrored surface is placed immediately behind the beam-splitter. Both the light beams are reflected from mirrors after passing through the cuvettes and the compensator. They fall again on the beam-splitter at which they are recombined. The layout of the optical elements is shown in Fig. 32. In order to obtain equal optical path lengths for both the interfering beams, one of them is reflected at the silvered part of the front surface and the other at the silvered rear surface of a plane parallel glass plate. The refractive index and the thickness of this mirror plate are used to produce the same phase difference as results by passage through the beam-splitter plate.

The compensator in this apparatus consists of two pairs of glass plates — a double Löwe compensator. One plate of each pair, one before the test cuvette, the other before the reference cuvette, is permanently fixed. Both the other plates, with a common axis of rotation, make possible the compensation of the phase difference. The four-plate compensator has the advantage that the fringe displacement is almost linearly dependent on the angle of rotation. The reading scale is therefore graduated directly in fringe numbers h for λ = 546.1 nm.

The interference fringe spacing is adjusted exactly by opposed rotation of a pair of glass wedges in the path of one beam. The optical path difference produced by this is compensated for in the other

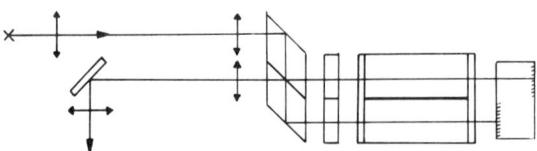

Fig. 32. The Michelson—Kinder interferometer.

TABLE 2

Technical data for some commercial fixed interferometers

Type and manufacturer	Principle	Max. cuvette length	Smallest detectable refractive index difference	Range (refractive index difference)
Laboratory/interferometer LI 3 (Zeiss, Jena, G.D.R.)	Rayleigh— Löwe	3 m eff. 1 m	$6 \cdot 10^{-9}$ $1.8 \cdot 10^{-8}$	0.000018 0.000055
Laboratory/interferometer ITR-1 (U.S.S.R.)	Rayleigh— Löwe	1 m	$1.8 \cdot 10^{-8}$	0.000109
Jamin-interference refractometer (Bellingham and Stanley, London, England)	Jamin	50 cm	$1 \cdot 10^{-7}$	No information
Interferometer ISK-453 (U.S.S.R.)	Jamin	50 cm	$4 \cdot 10^{-8}$	~ 0.0001
Interference refractometer (Zeiss, Oberkochen, G.F.R.)	Michelson— Kinder	50 cm eff.	$1.2 \cdot 10^{-8}$	0.00024

beam by a pair of movable glass wedges. An additional interference pattern which may be used as a reference mark is produced, as with the Rayleigh interferometer, by light beams passing below the cuvettes.

Table 2 gives a summary of the most important data for some commercial interferometers.

(B) PORTABLE INTERFEROMETERS FOR THE MEASUREMENT OF HOMOGENEOUS SUBSTANCES

(1) The Rayleigh—Löwe autocollimating interferometer

In all portable interferometers the light is reflected so that it passes through the cuvettes at least twice. By this means the actual length of the cuvette is diminished but the precision of measurement is not affected. Moreover in Rayleigh—Löwe autocollimating interferometers the optical collimator serves at the same time as the telescope, thereby enabling the length of the instrument to be reduced still further. The light is reflected onto the entrance slit, Fig. 33, which again must obey the coherence condition (11). Interferometers of this type possess a compensator similar to that in the laboratory type Rayleigh—Löwe interferometers, and also need to be calibrated depending on the substance to be examined. They give lower precision than the laboratory apparatuses, since the cuvette length and the focal length of the telescope objective must be relatively short. The advantage of autocollimating interferometers is that they are easily transportable due to their smaller dimensions and lower weight, for which reason they are suitable for special measurement tasks, e.g. on expeditions or as explosion-proof models for use in mines [70].

(2) The Jamin—Doi interferometer

Doi [17] proposed a modification to the autocollimating Jamin interferometer, in which the light is reflected parallel to but displaced from the original light path, making it possible to combine both plates of the Jamin interferometer in one, so that external disturbing influences are strongly reduced (Fig. 34). The cuvette is divided into three parts. The central part takes the gas specimen, while both the outer parts are used to take the reference gas. Since the collimator and the telescope systems are less bulky than in the Rayleigh interferometer the Jamin—Doi instrument is much more compact. The spacing of the interference fringes can be adjusted such that it is possible to read the fringes directly on a scale. The scale is marked, in percentage methane for use in mines or in percentage halothane for anaesthetic measurements. The compensator is dispensed with. The shift of the zero-order fringe indicates the percentage in the binary

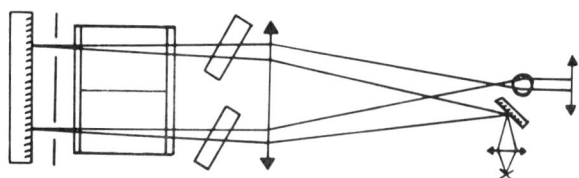

Fig. 33. The Rayleigh—Löwe autocollimating interferometer.

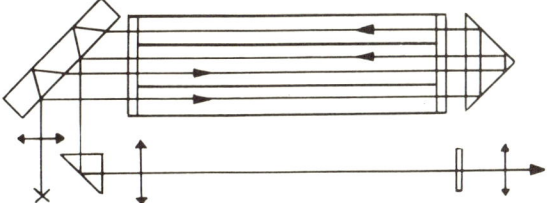

Fig. 34. The Jamin—Doi interferometer.

mixture. The range and precision of measurement of these types of apparatus are determined by the effective cuvette length. In most commercial apparatus this is twice the actual length [71]. In the mine-gas interferometer "Gasi" [72] however, a six-fold gain over actual length is achieved by making use of multiple reflection in the cuvette. The measurement accuracy remains the same with this short cuvette, so the "Gasi" instrument is smaller than other interfero-meters — it is in fact the smallest interferometer manufactured for analytical purposes.

(C) INTERFEROMETERS FOR THE MEASUREMENT OF STRATIFIED SUBSTANCES

Stratified heterogeneous substances are characterized by the refractive index gradient at each point having the same direction or at least a definite, e.g. axially symmetric, directional distribution. The refractive index is a constant, therefore, within any one surface at right angles to this direction. Stratified substances are met with for example, in electrophoresis, diffusion and sedimentation investiga-tions. While in both the former methods the refractive index gradient lies in the vertical direction, in sedimentation investigations it lies parallel to the direction of the effective gravitational force. This dis-tinction has no significance in principle for the optical process but the differing operational techniques lead to a completely different apparatus design. Details of typical electrophoresis apparatus, diffu-sion apparatus and ultracentrifuges have been left out here. The dis-cussion will, therefore, be restricted to describing the interference optical arrangements of these instruments.

Most apparatuses for the investigation of stratified substances are based on the Toepler schlieren arrangement, in which the object is

TABLE 3

Technical data for some commercial portable interferometers

Type and Manufacturer	Principle	Usage	Weight (kg)	Limits of error	Measurement range
Liquid interferometer ITR-2 (U.S.S.R.)	Rayleigh—Löwe	Liquids	15	$2.5 \cdot 10^{-7}$ $-4 \cdot 10^{-6}$	0.00065 -0.01
Interference gas analyser IGA (U.S.S.R.)	Jamin—Doi	CH_4, CO_2 O_2	2.17	0.3% CH_4 0.3% CO_2 0.5% O_2	6% CH_4 6% CO_2 21% O_2
Mine-gas interferometer SchI-5 (U.S.S.R.)	Jamin—Doi	CH_4, CO_2	1.35	0.3% CH_4	6% CH_4
Mine-gas interferometer (Zeiss, Oberkochen, G.F.R.)	Jamin—Doi	CH_4, CO_2	1.15 1.15	0.1% CH_4 1.0% CH_4	10% CH_4 100% CH_4
Riken gas indicator Type 17 (Riken Keiki, Tokyo, Japan)	Jamin—Doi	CH_4 or others	1.0	0.1% CH_4	6% CH_4
Mine-gas interferometer DI-2C (Czechoslovakia)	Jamin—Doi	CH_4, CO_2	1.0	0.1% CH_4	100% CH_4
Mine-gas interferometer "Gasi" (Zeiss, Jena, G.D.R.)	Jamin—Doi	CH_4, CO_2	0.75	0.25% CH_4	6% CH_4
Anaesthetic gas measurement apparatus halanometer (Zeiss, Jena, G.D.R.)	Jamin—Doi	Halothane	0.75	0.2% Hal.	5% Hal.

inserted in a parallel beam of light between a collimator and an image-forming system (Fig. 35). The schlieren optical techniques of Lamm [73], Longsworth [74], Philpot [75] and Svensson [76], dependent on light diffraction, are somewhat similar, incorporating minor variations of this apparatus, as are the interference methods of Rayleigh [13], Svensson [19], Philpot and Cook [77] and of Gouy [9].

The Rayleigh—Svensson interferometer requires monochromatic light, because a great number of parallel interference fringes must be produced, and with good contrast. The entrance aperture of the collimator takes the form of a grille with a large number of vertical slits whose grating constant fulfils the relationship given by Svensson [19] for the superposition of the maxima. A diaphragm with two apertures situated in the path of the parallel beam produces two light beams which pass one through the cuvette containing the stratified material and one through the adjacent cuvette filled with homogeneous material. In the image plane of the grating is produced a broad interference fringe system which alters as soon as the optical path difference of the interfering light beams is altered. To produce interference fringes which can be easily interpreted one uses, in addition, an anamorphotic imaging system (Fig. 36) whose cylindrical lens possesses an axis parallel to the grille slits. An interference pattern with vertical fringes is obtained which reproduces the height of

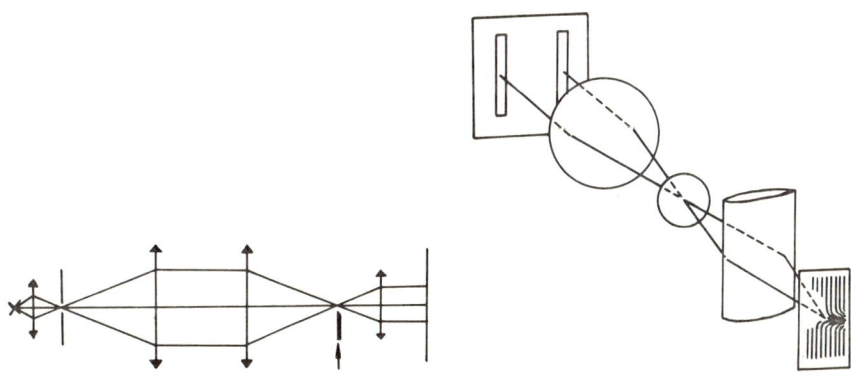

Fig. 35. Toepler's schlieren arrangement.

Fig. 36. Projection arrangement of the Rayleigh—Svensson interferometer.

the cuvette, but laterally an image of the grille slits showing the interference. A continuous change of refractive index with height in the test cell gives rise to interference fringes in the form of continuous curves representing the refractive index n vs. the height x. The $n—x$ curves usefully supplement the gradient curves $dn/dx/x$, recorded with the above mentioned schlieren optical procedures, when it comes to interpreting them and taken together they improve the overall precision. Most commercial instruments of this type permit both the interferometric measurement of refractive index distribution according to Rayleigh and Svensson and also the schlieren optical determination of Philpot and Svensson for the gradient distribution. In the Rayleigh—Svensson arrangement the use of suitable accessories allows both interfering beams to pass through the test cell with a very small difference of height between them. Wiedemann [78] achieved this by using a tilted entrance slit, an oppositely inclined cylindrical lens and two biprisms in the parallel beam. Another possibility is given by insertion of inclined plane parallel glass plates into the path of the beam immediately before and after the cell [79]. With this arrangement the refractive index difference between adjacent layers is recorded directly as a function of the height, i.e. the gradient curve is produced by interferometry. Not only is the interferometric recording about fifty times more sensitive than the schlieren optical method of Philpot and Svensson, but it offers the great advantage that it records the distribution of both the refractive index and the gradient simultaneously. A good example is the Rayleigh—Svensson system. Here four beams are allowed to pass through the diaphragm, and are brought in pairs to interfere. A biprism in the parallel path of the beams behind the cells ensures that the two interference patterns do not superimpose. The course of the light beams in parallel paths is evident from Fig. 37.

The Jamin interference principle (Fig. 16) was developed by Antweiler [80] for the investigation of stratified substances. It is used only in the smaller apparatus, since it would otherwise entail unacceptably high expenditure on large optical components of high precision. The essential advantage over the Rayleigh—Svensson arrangement lies in the greater light intensity. Moreover one can work with white light, which gives a clearly visible zero-order fringe making the distribution of the refractive index easy to follow.

A summary of the electrophoresis and diffusion instruments on the European market is given in Table 4.

436

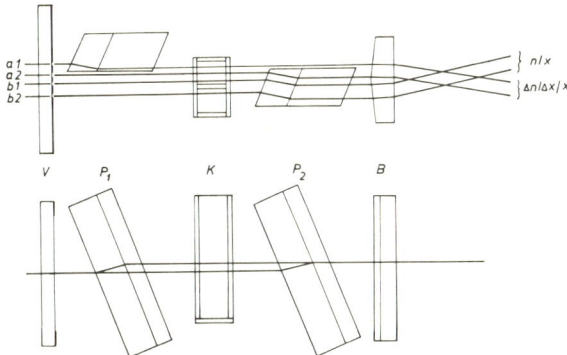

Fig. 37. Simultaneous interference arrangement for the Rayleigh—Svensson interferometer. V, Plate with four slits; P_1, P_2, plane parallel plates; K, cuvette; B, biprism.

(D) INTERFEROMETERS FOR THE MEASUREMENT OF RANDOMLY HETEROGENEOUS SUBSTANCES

In a transparent medium regions with different refractive index from their surroundings are described as optically heterogeneous, or as "schlieren" ("streaks"). For the determination and investigation of schlieren in a material one often uses schlieren instruments (which according to Toepler [23,81] depend on the observation of light deviation in the penumbral field (Fig. 35)). These lead to clear images which, however, are difficult to interpret quantitatively. Whenever the quantitative interpretation of the schlieren pattern is important, therefore, one employs an interferometer. Since, in general, one must illuminate a large object field uniformly, interference arrangements with wave-front division cannot be considered. The most common interferometers for homogeneity testing use amplitude division (Sect. 3.D(2)) in the Michelson or Mach—Zehnder arrangement, in both of which the object is placed in one branch of the interferometer. Monochromatic light is used, though white light may also be used on rare occasions [82]. The problem of the optimum visibility of the interference and its dependence on the entrance aperture shall not be discussed here [63,83—90]. The coherence length of the light source limits the permissible difference in the optical path lengths of the two interfering light beams. As the insertion of the object alters the optical path length in one branch, a corresponding alteration in the

TABLE 4

Some commercial apparatus for electrophoresis or diffusion investigations

Type and manufacturer	Technique		Rayleigh—Svensson		Cuvette height (mm)
	Philpot—Svensson	Other techniques [a]	n/x	$dn/dx/x$	
Electrophoresis apparatus FOKAL B (Strübin and Co., Basle, Switzerland)	+	−	+	−	120
Electrophoresis apparatus (PHYWE, Göttingen, G.F.R.)	+	+	+	−	100
Spinco Model H (LKB, Stockholm, Sweden)	+	+	+	+	86
Schlieren equipment 80 with diffusion apparatus (Zeiss, Jena, G.D.R.)	+	+	+	+	80
Electrophoresis apparatus FOKAL F (Strübin and Co., Basle, Switzerland)	+	−	+	−	60
Schlieren camera (PHYWE, Göttingen, G.F.R.)	+	+	+	−	50
Micro-electrophoresis apparatus (Boskamp, Hersel/Bonn, G.F.R.)	+	−	+	−	44
Electrophoresis apparatus 35 (Zeiss, Jena, G.D.R.)	+	−	+	−	35
Diffusion interferometer (Zeiss, Oberkochen, G.F.R.)	−	−	Jamin	−	20

[a] e.g. the scale method of Lamm, the schlieren-scanning method of Longsworth or the interference techniques of Gouy.

second branch must also take place. This is produced in the Michelson interferometer by moving the reference mirror in the direction of the light beam. By this means the optical path length of the light beam which passed twice through the beam-splitter plate is simultaneously compensated.

In the Mach—Zehnder interferometer the optical paths are symmetrical by virtue of the design principle. It is only necessary to introduce a compensation plate into one branch when the object in the other branch (e.g. the window of a wind-tunnel) alters the optical path length to such an extent that the path difference exceeds the coherence length. Since laser light sources possess a coherence length far exceeding the dimensions of the interference apparatus, their use with the Michelson and the Mach—Zehnder interferometers brings the great advantage that any compensation of optical path length can be dispensed with [91]. The two branches of a Michelson interferometer, for example, can have quite different lengths, and the very difficult adjustment to obtain equal optical path lengths is now superfluous.

Adjusting these apparatuses means producing an interference pattern suitable for the intended investigation. One obtains this in the Michelson interferometer by tilting one of the two final mirrors at a slight angle $\frac{1}{2}\epsilon$ to an axis at right angles to the direction of incidence of the light. The two interfering wave-fronts are thus set at an angle ϵ to each other, the direction and magnitude of ϵ giving the direction and spacing of the interference fringes according to eqn. (18). The fringe pattern can in this way be adapted to suit the object: with marked heterogeneities (large phase differences in the object field) one choses to have fringes close together, and with weak heterogeneities, relatively far apart. A setting of three or four fringes in the image plane is convenient for evaluation and provides at the same time very sensitive conditions for the interpretation. With a homogeneous object field the fringes appear as straight lines. The existence of schlieren in the object field causes a phase delay in the wave-front relative to its surroundings. If, for example, this delay is half a wavelength, $\Delta\varphi = \frac{1}{2}\lambda$ so the interference fringes, which pass through the image of the schlieren, undergo at the position of the schlieren a deflection of $2\Delta\varphi/\lambda$ times one fringe spacing, thus in our example exactly one fringe spacing. The measurement of the interference fringes gives a quantitative expression of the homogeneity of the object field. A special case is with the tilt angle $\frac{1}{2}\epsilon$ equal to zero. This produces an infinite fringe spacing and no interference fringes appear. Consequently, with a homogeneous object the image plane is uniformly illuminated (light or dark). Also this setting is advantageous when a rapid direct test of the wave-front is required. The regions of equal phase retardation of the wave-front appear in the interference pattern as curves of equal density. Alternatively one may use cube corner prisms in order to be independent of any variation in the position of the mirror [92].

In the Michelson interferometer the light is passed twice through the object field. This has on the one hand the advantage of double the sensitivity of measurement for objects with weak schlieren, and on the other hand, however, the disadvantage, that with more strongly interacting objects, the returning wave-front is distorted, which, in the interpretation, renders the exact correspondence of a point in the interference pattern to a point in the object more difficult.

The Mach—Zehnder interferometer is adjusted by rotating and displacing one or more of the four mirrors. Also when white light is to be used, one must use monochromatic light for the adjustment, in order to set initially the position of the interference fringes with approximate compensation for the optical path length. After this one can compensate the optical path difference to zero with white light. The difficult adjustment of the Mach—Zehnder interferometer [93—96] makes this apparatus awkward for rapid investigations with changing objects. The optical arrangement requires, moreover, a very stable construction, since during the investigation the mirrors must not be allowed to move in any way relative to each other. This recommendation must also apply to the Michelson interferometer, though in this case one need not observe it so strictly.

Essentially less sensitive towards external vibrations are those interferometers in which both interfering beams pass through the same optical elements. These are shearing and diffraction interferometers. In the interferometers with lateral shearing [97—100] (Sect. 3.D(3)) the wave-front is divided before or after the object into two fronts side by side and these are re-combined for interference (Fig. 38). A slit or a Wollaston prism serves as an entrance aperture. If one assumes as a simplification that the refractive index in the object in the direction of the transmitted light does not alter, then the phase difference between neighbouring points in the image, i.e. the gradient of the refractive index in the direction of the shearing, is recorded. An additional birefringent element such as a Wollaston prism in the

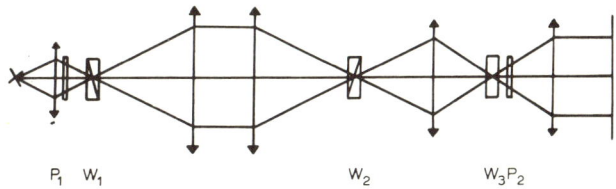

Fig. 38. The lateral-shearing interferometer: P_1 and P_2, polarizers; W_1, W_2 and W_3 Wollaston prisms.

parallel beam or a Savart plate in the convergent beam behind the first image plane produces an inclination of the wave-fronts to one another, so that they form interference fringes at equal distances in the second image plane. Though yielding the gradient distribution they can be interpreted in the same way as the interference patterns produced in the previously mentioned double-beam interferometers. The method is above all very sensitive to small-volume heterogeneities. Arrangements have been described in which the light is passed through the object twice, as well as those through which it passes only once. Shearing interferometers are particularly important for testing lenses and other optical components: the shearing effect is realized either by rotation or by a grating [100,101].

Diffraction interferometers [47] can be constructed on the principle of the Toepler schlieren arrangement (Fig. 35) (Sect. 3.D(4)). Here also either white or monochromatic light can be used for illumination. The entrance aperture must obey the conditions given in eqn. (11). Only one plane wave-front is passed through the object. By inserting a diaphragm in the plane of the diffraction pattern, the additional vector corresponding to any point in the image plane interferes with the mean vector, which is uniform over the whole surface. As a result of this an interference pattern is produced analogous to those with the Michelson or Mach—Zehnder interferometers, when these are set at infinite fringe spacing. The advantage of the diffraction interferometer lies above all in its simple optical construction and its ease of adjustment [102].

(E) ACCESSORIES FOR INTERFEROMETRIC MEASUREMENTS

(1) Light sources

White light from incandescent lamps is used for those interferometric measurements in which it is necessary to indicate the phase

TABLE 5

Some commercial interferometers for the investigation of heterogeneous substances

Type and manufacturer	Interference principle	Object field cross-sectional area
Interferometrical system (Intertech Corp., U.S.A.)	Mach—Zehnder	250 mm ⌀
Interferometer ISK454 (U.S.S.R.)	Mach—Zehnder	225 mm ⌀
Mach—Zehnder interferometer (Novotechnik, G.F.R.)	Mach—Zehnder	200 mm × 140 mm elliptical
Mach—Zehnder interferometer Mod. D 2 (Zeiss, Oberkochen, G.F.R.)	Mach—Zehnder	180 mm × 120 mm elliptical
Differential interferometer D1 (Schrader, Wenden, G.F.R.)	Shearing	100 mm ⌀
Twyman—Green interferometer Model 723 (Perkin—Elmer, U.S.A.)	Michelson	100 mm ⌀
Schlieren apparatus 80 (Zeiss, Jena, G.D.R.)	Diffraction	80 mm ⌀
Interferometer Model 734 (Perkin—Elmer, U.S.A.)	Mach—Zehnder	75 mm ⌀
Mach—Zehnder interferometer NIP 1 (Zeiss, Oberkochen, G.F.R.)	Mach—Zehnder	70 mm ⌀
Interféromètre de Michelson Type 200 (Sopra, Paris, France)	Michelson	40 mm ⌀
Differential interferometer (Hacker and Co., West Caldwell, U.S.A.)	Shearing	

difference of zero. This applies for determinations of concentrations of homogeneous substances and to some extent also for investigations of heterogeneous materials.

Monochromatic light is necessary for setting up instruments as well as for specific measurement problems and photographic recording. A large coherence length is well worth striving for, because this essentially determines the range of measurement of the apparatus. As far as the expense allows, lasers are used with modern interferometers.

The commonest monochromatic light sources are the spectral lamps — commercial discharge lamps filled with a noble gas or a metallic vapour. The place of the formerly widely employed sodium lamp whose doublet at 589.0/589.6 nm leads to periodic disappearances of the interference fringes at high phase differences, is today taken by the mercury lamp. The high-pressure lamp is used when a high intensity is important, and the low-pressure lamp when it is important to obtain a large coherence length with good definition of the lines. The most important mercury line for interferometry is the green e-line at 546.1 nm, as it is narrow, intense and corresponds with the wavelength of maximum visual sensitivity. It is selected from the Hg spectrum by means of a colour filter (a combination of suitable coloured glasses) or by an interference filter (on the principle of interference in thin layers).

For dispersion measurements the Hg-g line and lines from other spectral lamps are also necessary (Table 6).

The formerly widely used high-voltage discharge lamps (Geissler tubes) with their low brightness cannot compete with the modern spectral lamps just mentioned.

TABLE 6

Light sources used in interferometry and their wavelengths

Spectral line	Wavelength (nm)	Light source	Remarks
C	656.3	H_2	Weak line
C′	643.8	Cd	N.P.L. proposed line
	632.8	Laser	
D_1	589.6	Na	} Doublet
D_2	589.0	Na	
d	587.6	He	N.P.L. proposed line
e	546.1	Hg	N.P.L. proposed line
F	486.1	H_2	Weak line
F′	480.0	Cd	N.P.L. proposed line
g	435.8	Hg	N.P.L. proposed line
h	404.7	Hg	N.P.L. proposed line

For the investigation of very rapidly occurring processes (e.g. shock waves) high-intensity arc-light sources have been developed which produce light impulses of the order of μs or even a few ns duration. For recording rapid processes in the time-dilatometry a series of flashes are produced at a high frequency of up to some 10^4 s^{-1}.

(2) Temperature control

The refractive indices of all substances are temperature-dependent. Therefore in all investigations — with the exception of the examination of the thermal conditions and heat discharge — a constant uniform temperature must be maintained throughout the object during the measurement. This applies especially to liquids, since these have very high temperature coefficients (see Sect. 5.C).

It is important in comparative interferometric measurements to maintain both substances at the same temperature. This can be achieved by suspending both cuvettes side by side in one bath. For measurements at room temperature, it is not usually necessary to keep the temperature of the bath constant. With excessive fluctuations in the temperature of the environment, or when very high precision is required the bath should be thermostatically controlled. No great demands are made on the temperature control precision, provided that the temperature coefficients of test and reference liquids are approximately equal. Completely different conditions prevail, however, for electrophoresis and diffusion investigations, since these processes must run for many hours under highly constant conditions. In these cases an exact temperature must be maintained with an efficient thermostat. In order to eliminate the slight periodic temperature fluctuation of the thermostat, the cuvette is suspended in a water bath whose temperature is controlled by a thermostat. If the measurements are to be made with very high precision, the room temperature must be strictly controlled.

(3) Gas mixer for calibration purposes

The calibration of interferometers for concentration determination (Sect. 5.B) requires the preparation of mixtures or solutions of precisely known composition. This is not difficult for liquids, using standard weighing techniques. On the contrary the preparation of definite

444

gas or vapour concentrations is complicated, when suitable instruments are not available.

The following arrangement has proved suitable for this purpose [103,104].

The chief part of the apparatus (Fig. 39) for preparing a known vapour concentration consists of the circulation system P-V-K-4, which can be extended by turning the stop-cocks 1 and 4 to include the adjoining measurement cuvettes of the interferometer at A1 and A2. The circulation system is evacuated with stopcocks 7 and 8 closed. The small evaporation flask V is removed and the test material is weighed in at the liquid state, the flask is reattached, and the stop-cocks 7 and 8 are opened. A temporary opening of stop-cock 6 returns the pressure in the system to normal, by adding the desired carrier gas at A4. Since air is usually used, a drying cartridge is introduced at this point. In the closed system the pump P circulates the mixture long enough to vapourize the liquid sample and homogenize the mixture. The final state can be recognized by a constant reading on the interferometer.

For the preparation of a mixture of two gases the circulation system is again evacuated. The graduated burettes B1 and B2 are filled with one gas component with the help of a mercury reservoir in order that the desired doses can be delivered from these into the circulation system. By introducing the carrier gas at A4

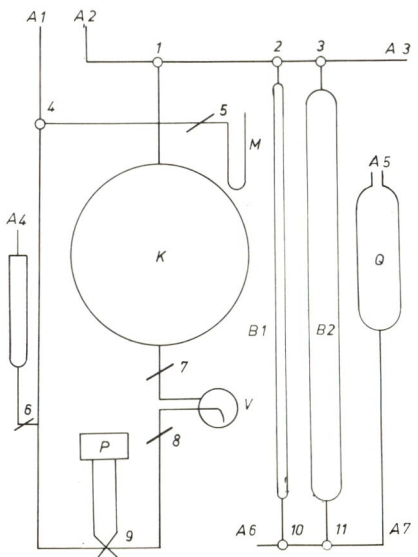

Fig. 39. Diagram of an apparatus for preparing gas mixtures: A_1—A_7 connecting points; B_1 and B_2, gas burettes; K, mixing sphere; M, manometer; P, pump; Q, mercury reservoir; V, vaporizer; 1—11, taps.

a standard pressure can be obtained and checked with the manometer M. The pump P provides for thorough mixing. Very low concentrations may be prepared by using a stop-cock for which the volume of the hole in the cock itself is known.

The apparatus described by Lacy and Woolmington [105] works on the same principle, see also ref. [106].

In the gas mixing apparatus GMA 1 of the Radiometer Company, Copenhagen, both the gaseous components are led through capillaries and then mixed together. The composition of the mixture can be calculated from the dimensions of the capillaries, the gas pressure and the nature of the gas [107—109].

Gases can also be mixed in predetermined concentrations for calibration purposes, with the help of two metering pumps.

Similarly in the gas-mixing device by VEB Junkalor, Dessau, an adjustable difference in the pressure of the two gases is used to control the desired mixing ratio for a calibration gas.

(4) Sorption media

The interferometric analysis of complex gaseous mixtures is usually carried out by the specific separation of single components by means of suitable sorption media [110—113]. For this adsorption (physical binding to the surface) and absorption (chemical binding) are equivalent techniques. On account of the favourable properties of adsorption media, the adsorption technique is preferred, however in any particular case the specific effect is the deciding factor. For gas analysis only one component at a time may be removed from the mixture, whereas all others must pass through unhindered by the sorbant and without chemical side reactions occurring. Also on passing through the sorbant no new component (such as water lost from the sorbant) may appear in the mixture.

Widely used adsorbants are activated charcoal, silica-gel and aluminium oxide, and the zeolites. The zeolites possess a crystal structure with large cavities, in which the gas molecules are retained. The inner surface of 1 g of adsorbant can amount to some 100 m^2. Correspondingly their adsorption capacity is very high and can attain some 0.1 g per 1 g of adsorbant. The adsorption capacity is very strongly dependent on the adsorbing material and the adsorbed gas (adsorbate) as well as upon the temperature and the pressure. Activated charcoal, for example, exhibits a strong adsorption capacity for gases with high

446

boiling points (Brunauer [114]). The adsorption process with the adsorbants mentioned takes only a few seconds under normal conditions, but this time must be allowed for the adsorption process to reach completion. An adsorption tube should be used, whose length is at least four times its diameter. The state of saturation in silica-gel coloured blue with cobalt chloride (blue-gel), is indicated by discoloration. The other adsorbants have generally no externally visible indication of saturation. One must remember to renew the exhausted adsorbant in time. Some adsorbants can be regenerated by heating. Blue-gel regains its blue colour in this way. Zeolite can be regenerated by heating for approximately one half to one hour at $250°-350°C$. Temperatures up to $500°C$ will not destroy the crystal structure so that in practice it can be regenerated as often as desired. It should be noted that the adsorption occurs as an exothermic process. In order to maintain the temperature equality with the reference gas, the sample must, if necessary, be thermostatted.

The special significance of zeolites requires some further remarks [115,116,126]. Synthetic zeolites are alumino silicates. The openings of the cavities in the crystal have a diameter, which is characteristic for the particular type. The zeolites of the types 3A, 4A, 5A, 10X and 13X have openings with diameters of approximately 3, 4, 5, 9 and

TABLE 7

Adsorption of several gases on Zeolites at $20°C$

Gas	Zeolite type				
	3A	4A	5A	10X	13X
Ammonia	+	+	+	+	+
Carbon dioxide	+	+	+	+	+
Sulphur dioxide	+	+	+	+	+
Water	+	+	+	+	+
Methanol	+	+	+	+	+
Ethanol	+	+	+	+	+
Propane	—	—	+	+	+
n-Butane	—	—	+	+	+
i-Butane and isoparaffins	—	—	—	+	+
n-Heptane	—	—	+	+	+
Benzene	—	—	—	+	+
1,3,5-Triethyl benzene	—	—	—	—	+

TABLE 8

Sorption media for gases and vapours

Inorganic species	Ref. *	Sorption media	Sorbed simultaneously *	
Ammonia	1,2,9 9 8	(a) H_2SO_4 1 : 10 diluted (b) $CuSO_4$, anhydrous (c) P_2O_5	HCl, PH_3, H_2S H_2O, amines, halides	(B)
Arsine	9 9 9	(a) $KMnO$, saturated solution (b) $AgNO_3$, 5—10% solution (c) CuO	PH_3, SO_2, H_2S H_2S, hydrocarbons	
Chlorine	9	(a) Soda lime (b) See halogens		(A)
Hydrogen chloride	9 9 9	(a) Soda lime (b) $CuSO_4$, anhydrous on pumice (c) CaO	NH_3, PH_3, H_2S N_2, H_2O	(A)
Hydrogen cyanide	2	H_2SO_4, concentrated		(B)
Hydrogen fluoride	9	NaF		
Iodine in the air	7	K_2CO_3 solution 1 : 7.5 dilution		
Carbon dioxide	2,3,5,6,8,9 3,9 3,5,6	(a) Soda lime (b) NaOH on pumice (c) KOH, solid		(A) (C) (C)
Carbon monoxide	8 8 9 1—5,9	(a) Ag_2O (b) Manganese dioxide—copper (II) hydroxide (c) $K_4Ni_2(CN)_6$ solution (can be regenerated) (d) 125 g CuCl + 265 g NH_4Cl + 750 cm^3 H_2O	H_2, CH_4	
Carbonyl sulphide	1	KOH 30% in 1 : 1 alcohol—water	O_2, hydrocarbons	(C)
Ozone	2,3,5,6,8,9 3,6 3,5,6	(a) Soda lime (b) Manganese dioxide or metal (c) KOH		(A) (C)

Gas	Ref.	Reagent	Removes/Notes	Code
Phosphine	9	(a) CuSO₄ concentrated solution	NH₃, HCl, H₂S	
	9	(b) KMnO₄ saturated solution	AsH₃, H₂S, SO₂	
Oxygen	1—5,9	(a) Alkaline pyrogallol solution	CO₂	
	1,2,9	(b) Alkaline quinhydrone solution		
	1,3,5,9	(c) CrCl₂ solution, 20%	CO, CO₂,	
	1,3,9	(d) Metallic copper in solution of concentrated (NH₄)₂CO₃ + concentrated NH₄OH	C₂H₂, CH₄	
Sulphur dioxide	8	(a) Soda lime		(A)
	1—5,9	(b) KOH solution, 28%		
	9	(c) KMnO₄ solution, 2 N	AsH₃, PH₃, H₂S	(C)
Sulphur trioxide	3	(a) Potassium pyrosulphate		
		(b) H₂SO₄, concentrated		(B)
Carbon disulphide	9	(a) NaOH, 33%		
	3,9	(b) Triethylphosphine, concentrated		(C)
Hydrogen sulphide	8	(a) Soda asbestos		(A)
	9	(b) See AsH₃ (a) and (b)		
	9	(c) NaOH, 33%		(C)
Nitrogen	1,5	(a) Mg powder + CaO + Na	HCl, H₂O	
	1,3,9	(b) Ca + Ca₃N₂	O₂, hydrocarbons	
Nitric oxide	1	(a) Bromate or permanganate solution, acidified	N₂O	
	1	(b) Nitrosulphuric acid	N₂O	
Nitrogen dioxide	8,9	(a) Soda asbestos		(A)
	3,9	(b) Concentrated H₂SO₄		(B)
Nitrous oxide	1—3	(a) CaCl₂, see H₂O (a)		
		(b) See NO (a) and (b)		
Water vapour	2,5,9	(a) CaCl₂	NH₃, N₂O, amines, halide, alcohols	
	8,9	(b) Mg(ClO₄)₂, anhydrous		
	5,8,9	(c) P₂O₅	NH₃, amines, halides	
	5,9	(d) Concentrated H₂SO₄		(B)

TABLE 8 (continued)

Inorganic species	Ref. *	Sorption media	Sorbed simultaneously *
Hydrogen	8	(a) See CO (b)	
	3,5	(b) CuO—asbestos at 300°C	
	3,9	(c) Palladium sponge	
Halogen	6,9	(a) Hg heated	O_3
	9	(b) Alkaline arsenate(III) solution	
	3,9	(c) $FeSO_4$ solution	NO, N_2O
		(d) See halide	
Halide	1–5,9	KOH solution 28%	(C)
Organic species			
Ethane	4	See CH_4 (a)	
Acetylene	2,3	(a) $KI + HgI_2$ in H_2O + KOH	Allylene
	2,3	(b) Cu_2O + KOH	
	8	(c) $CuCl_2$ + KOH	
Diethyl ether	6	Zn, with $Zn_3(PO_4)_2$ in alternate layers	CCl_4, n-C_5H_{12}, n-C_3H_7Cl
Methane	4	(a) Glycerine (pure) mixed with water 1 : 1	C_2H_6
	8	(b) See CO (b)	
Alcohols		(a) $CaCl_2$. See H_2O (a)	
		(b) Alkaline permanganate	
Other hydrocarbons	1–3	(a) Bromine water	
	1,3,4,9	(b) Oleum (H_2SO_4 with 25% SO_3)	(B)
	3	(c) H_2SO_4, concentrated with anhydrous vanadic acid	
	4	(d) $AgNO_3 + H_2SO_4$	(B)

* In column 2 the numbers denote 1, Andress and Wüst [119]; 2, Löwe [120]; 3, Klemenc [110]; 4, Kattwinkel [121]; 5, Mönch [122]; 6, Rausch [123]; 7, Süess [124]; 8, Bush and Loneragan [125]; 9, Lux [111].
In the last column the letters denote:
(A) Parallel adsorptions are possible with these gases on soda-lime or soda-asbestos.
(B) Parallel adsorptions are possible with these gases on H_2SO_4.
(C) Parallel adsorptions can be carried out with these gases on NaOH and KOH.

10 Å, which are of the order of magnitude of molecular diameters. On this favourable circumstance depends the high selectivity of the synthetic zeolites which are therefore called molecular sieves. Molecules smaller than the diameter of the pores are adsorbed whereas large molecules are not taken up. In a mixture containing several gases, which can be adsorbed according to their shape, polar, polarizable, and unsaturated molecules such as H_2O, CO_2, NH_3, CH_3OH and others are preferred. The adsorption of several gases on the zeolites is summarized in Table 7.

With all adsorbants their selectivity must be considered in relation to the particular gas mixture. For really meticulous testing it is necessary to exclude possible parallel adsorptions in each particular case to avoid errors in the measurement.

In addition to the above mentioned adsorbants there are numerous sorption-media for different applications given in the literature. Useful investigations have been made into the effectiveness of soda-lime [117,118]. Some sorption media suitable for interferometry are summarized in Table 8. Here it should also be noted that their specificity is limited. For example, soda-lime or soda-asbestos absorbs from mixtures not only CO_2 but also Cl_2, HCl, O_3, SO_2, H_2S, and NO_2.

Similarly, sulphuric acid is used as a sorption medium for complete or even partial absorption of NH_3, HCN, SO_3, H_2S, nitrogen oxides, H_2O and unsaturated hydrocarbons. The hydroxides NaOH and KOH absorb C_2H_2, CO_2, COS, O_3, SO_2, SO_3, CS_2, H_2S, halogens and halides and other compounds.

The use of these sorption media also necessitates, as with the previously mentioned adsorbants, a critical examination of the specific selectivity with respect to the test gas in question. The information given in Table 8 can therefore only be taken as suggestions and indications for the choice of suitable sorption media.

The partially inadequate specificity of the sorption media usually causes no serious difficulty in their application in practice, since the mixtures to be analysed contain only a few known components of their type, and the parallel-sorptions can be avoided by a suitable choice of media.

5. Methods for interferometric analysis of homogeneous substances

Several comprehensive monographs [48,127—130] deal with methods for the interferometric determination of concentration.

Some of these discuss interferometry as a section of or in conjunction with refractometry of liquids, and therefore also describe the refractometric methods for the determination of refractive index through angular measurements with the aid of prisms. The interferometric techniques are distinguished from these essentially by the principle of the measurement which is about 1,000 times more sensitive and can thus be applied to gases. These interference techniques thus represent precision methods for the determination of refractive index of liquids whereas with gases, which can hardly be analysed at all by angular measurements, they provide a routine method of measurement.

In the following discussion both the theory of the methods and the various possibilities for the interferometric analysis of homogeneous transparent substances are presented. However, the analytical techniques for the investigation of microscopic objects are not discussed. Details of interference microscopy and phase-contrast microscopy are available in the literature [131,132].

(A) PROPERTIES OF SUBSTANCES

By interferometric measurement one can determine, according to eqns. (39) and (40), the optical path difference between two light beams, which on paths of equal geometric length pass through two different materials. In order that a clear interference pattern suitable for measurement be produced, both substances must fulfil the following conditions.

(i) There must be no turbidity due to dust or undissolved particles. The interference contrast is diminished by light scattering.

(ii) Both materials must have the same transmission spectrum. The condition eqn. (10) for extinction in interference depends on the amplitude of the interfering beams being equal at all wavelengths. In order to minimize loss of intensity and at the same time to eliminate troublesome dispersion effects, one uses colourless reference substances and as near colourless as possible test samples. Errors can arise when coloured samples are investigated, since the measurements are made with white light.

(iii) Both materials must be homogeneous. Heterogeneities disturb the wave-fronts and lead to distortion of the interference fringe. They are usually produced by temperature or concentration gradients. The former can be eliminated by improved thermostating

(Sect. 4.E(2)) and the latter by stirring.

(iv) *The reference substance must remain unchanged.* The refractive index of the reference substance must not change during the course of a series of measurements: this can occur for example through evaporation, water absorption, decomposition, polymerization or other processes. The most suitable reference substances for gas analysis are carbon dioxide-free, dry, fresh air or a pure gas. In liquid analysis one prefers a pure liquid with low vapour pressure, preferably not a solution — best of all is distilled water for which very precise refractive index measurements have been made by Tilton and Taylor [133]. The following values have been obtained by interpolation from their detailed tabulated results.

$$n_e^{20°} = 1.3344661 \qquad dn/dt\,(20°) = -8.964 \cdot 10^{-5}$$

$$n_e^{30°} = 1.3334106 \qquad dn/dt\,(30°) = -12.068 \cdot 10^{-5}$$

$$n_{C'}^{20°} = 1.3314594 \qquad n_{F'}^{20°} - n_{C'}^{20°} = 0.0059894$$

$$n_{F'}^{20°} = 1.3374488 \qquad n_{F'}^{30°} - n_{C'}^{30°} = 0.0059583$$

It must further be noted that the phase difference between the interfering light beams, resulting from passage through the cuvettes must lie within the range of measurement of the interferometer. For interferometers with interchangeable cuvettes one can calculate, from eqn. (40), the permitted refractive index difference from the measurement range of the interferometer $\Delta s_{max} = h_{max}\lambda$ and the chosen cuvette length l

$$n - n_0 \leqslant \frac{h_{max}\lambda}{l}$$

For given test and reference substances the maximum permissible cuvette length is then given by

$$l \leqslant \frac{h_{max}\lambda}{n - n_0}$$

Since one can obtain higher measurement sensitivity with greater cuvette length, one should attempt to keep $(n - n_0)$ as small as possible, i.e. to choose a reference liquid with a refractive index almost the same as the test sample. With mixtures it is best in this respect to

454

use the component which is present in the highest concentration. But when the cuvette lengths as demanded by the sensitivity and by the range of measurement are significantly different one can employ the following device: one inserts into the path of the reference beam a glass plate perpendicular to the direction of the light, whose phase difference with respect to air in this same direction $\Delta\varphi_G = h_G\lambda$ has been measured precisely. This glass plate produces a shift of the zero-point and therefore of the range of measurement. In order that the original measurement range 0 to $h_{max}\lambda$ shall be widened to $h_G\lambda$ to $(h_{max} + h_G)\lambda$, without discontinuity, h_G must be chosen so that $h_G \leqslant h_{max}$. If the refractive index of the glass plate $n_G = 1.5$, the thickness d_G can be calculated from $(n_G - 1)d_G = h_G\lambda \leqslant h_{max}\lambda$ so that for $h_{max} = 100$ and $\lambda = 5.46 \cdot 10^{-4}$ mm one obtains $d_G \leqslant 0.109$ mm. Instead of using such a thin glass plate one can use two glass plates whose thicknesses differ by the amount d_G, and by using a range of glass plates of this type, the restriction on the choice of cuvette length can be largely eliminated.

In most cases the measurement range of portable interferometers is fixed from the outset by the permanent built-in cuvettes and therefore the range of application of these instruments is limited.

(B) CALIBRATION PROCEDURES

The calibration provides the relationship between the value i read on the interferometer scale and the composition of the test material.

Calibration is necessary

(i) for interferometers with non-linear compensators, since with these i is not proportional to the phase difference Δs produced by the compensator;

(ii) for liquid mixtures and solutions, since the refractive index does not change linearly with concentration;

(iii) for gas mixtures, in which the refractive indices of all the individual components are not known.

Gas interferometers without compensators and with direct percentage indication require no calibration, only a check (at regular periods) of the zero-point with cuvettes filled with the same substances.

The Δs—i relationship is given by the manufacturer for each interferometer, but it is recommended to check this from time to time. After setting of the zero in white light one compensates with monochromatic light (Hg-lamp $\lambda_e = 546.1$ nm) from one fringe to the next. One obtains the relationship between the phase difference $\Delta s = h\lambda = 5.461 \cdot 10^{-4} h$ of the compensator, for successive integers of h and the interferometer readings i.

With the help of the calibration table accompanying the apparatus, the refractive index difference of substance investigated in cuvettes of length l is given by eqn. (40)

$$n - n_0 = \frac{\lambda}{l} h$$

For gas mixtures the relationship between refractive index and concentration is linear (Sect. 5.D(1)) but this does not hold for liquids. For calibration with liquids one prepares a series of concentrations at close intervals and measures with white light. The concentrations must be carefully checked by a volumetric or gravimetric method, and must cover a sufficiently wide range to cover all the concentrations likely to be met in the test samples.

For gas mixtures this calibration is only required when the refractive index of either component is not known [134]. Gas or vapour mixtures can be prepared by means of suitable mixing apparatus (Sect. 4.E(3)). Because of the dependence of refractive index on temperature and pressure the calibration curves hold only for those conditions of temperature and pressure at which they were established. Examples of calibration curves are shown in Figs. 40—42.

One cannot help noticing the discontinuities in the curves. At regular intervals the curve breaks to a new branch separated from the foregoing by one fringe spacing. This is caused by the different dispersion characteristics of the cuvette contents and the compensator plates. Compensation to zero phase difference occurs only for the mean wavelength λ_e whereas for other wavelengths the zero-order maximum is displaced. Therefore the appearance of the interference fringes is altered in the region of the discontinuities in a characteristic way and can be observed as follows. With the same substance in both interferometer cuvettes (zero condition) one observes the normal interference pattern. The change in the fringe pattern can be clearly seen if an additional component is stirred or dropped slowly into one cuvette, and the shifting interference pattern is continuously compensated. An initially coloured second-order minimum, e.g. to the right, loses its colour until one observes at a particular concentration three almost black minima. As the concentration is increased

456

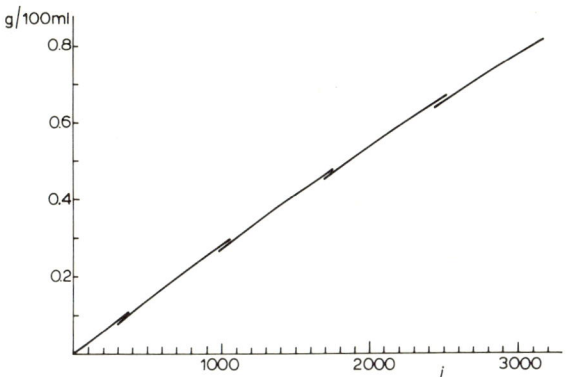

Fig. 40. Calibration curve for caprolactam in water; cell length 40 mm.

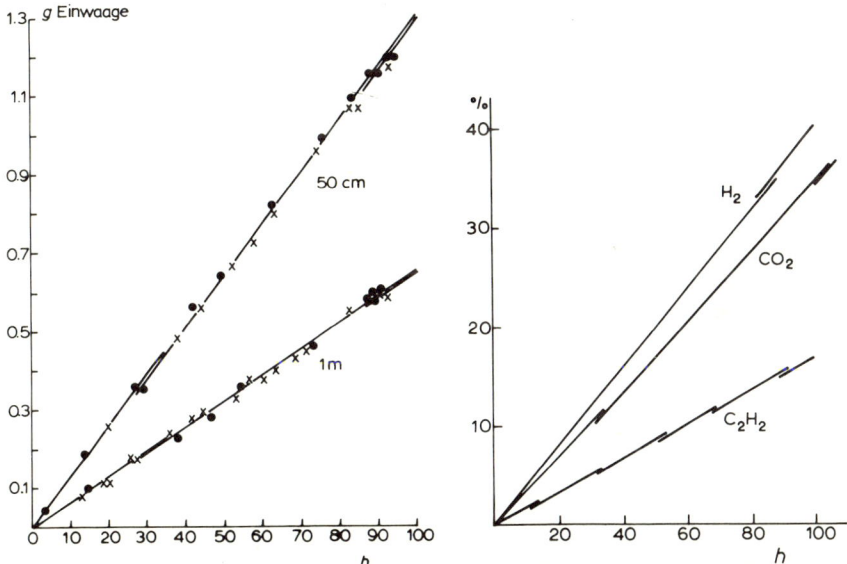

Fig. 41. Calibration curves for halothane in air, cell lengths 50 cm and 1 m.

Fig. 42. Calibration curves for hydrogen, carbon dioxide and acetylene in air; cell length 1 m. For hydrogen the shift of the fringes is in the opposite direction to that for carbon dioxide and acetylene.

further the previously black first-order minimum to the left becomes coloured. The first-order maximum to the right is now flanked by black minima and has the same appearance as the zero-order maximum originally had, and thus may be taken as the centre fringe. This switching of apparent zero-order fringe from one maximum to the next gives rise to the breaks in the calibration curve.

The breaks in the calibration curve can be either forward or backward, i.e. with increasing concentration they can move to the next higher or lower fringe number (Fig. 42).

In each case it must be a whole fringe number. Beyond each break an achromatic position is produced as the concentration increases, at which the fringes regain their original clarity. Breaks and achromatic points appear at approximately equal intervals, with the fringe being used for the zero setting occurring further and further away from the original zero-order maximum. The appearance of break-points was first described by Siertsema [135] (cf. Faust and Marrinan [136]).

The position of the break-points can be calculated from dispersion data [137,138]. Taking $\Delta n = n_{F'} - n_{C'}$ as the average dispersion of a material, one can define the relative effective dispersion of the mixture by

$$D = 100 \frac{\Delta n_2 - \Delta n_1}{n_2 - n_1} \tag{47}$$

The "Compensator dispersion" can be defined by the analogous expression

$$D^* = 100 \frac{\Delta s_{F'} - \Delta s_{C'}}{\Delta s} \tag{48}$$

where Δs gives the appropriately adjusted phase difference of the compensator. D^* can be calculated from the optical data and dimensions for the compensation device. For brevity the quantity

$$\Lambda = 100 \frac{\lambda_{C'} - \lambda_{F'}}{\lambda_e} = 30.00 \tag{49}$$

is introduced. With the achromaticity conditions formulated by Hansen [139]

$$cl(n_{2,C'} - n_{1,C'}) + \Delta s_{C'} + g\lambda_{C'} = 0 \qquad \text{and}$$

$$cl(n_{2,F'} - n_{1,F'}) + \Delta s_{F'} + g\lambda_{F'} = 0$$

458

one can calculate the break-points and the achromatic points from

$$h = q \frac{D + \Lambda}{D - D^*} \tag{50}$$

where $q = g + \frac{1}{2}$ for breaking points and $q = g$ for achromatic points, g = any integer. The sign of q, and hence of g, is positive for a "forward" break and negative for a "backward" break.

It is important to observe these break-points very carefully, during both the calibration and also the analysis in order to avoid making errors in the measurement. In practice the apparent ambiguity of the compensation setting can be reduced to unequivocal concentration values by means of the calibration curve. In many cases the occurrence of break-points can be avoided by a suitable layout of the compensator of the interferometer arrangement [140,141].

(C) TEMPERATURE AND PRESSURE

The refractive indices of substances are dependent on temperature and pressure. The influence of these factors is very different for gases and liquids and the effect on gases will be dealt with first. The refractive power of an ideal gas is directly proportional to the prevailing pressure p (in Torr) and inversely proportional to the temperature T (in K)

$$m^{t,p} = \frac{p}{760} \frac{273}{T} m^{0,760}$$

t is the temperature in °C. We can write

$$m^{0,760} = \frac{760}{p} \left(1 + \frac{t}{273}\right) m^{t,p} \tag{51}$$

For brevity we introduce F, a factor dependent on the external conditions

$$F = \frac{760}{p} \left(1 + \frac{t}{273}\right) \tag{52}$$

F can easily be estimated from a nomogram such as that in Fig. 43.

However, for real gases a correction factor $(1 + \kappa(p - 760))$ is introduced into eqn. (51). Since the constant κ is in most cases for gases smaller than $1 \cdot 10^{-6}$ this correction can in practice be neglected [142]. In the interferometric measurement the fringe number

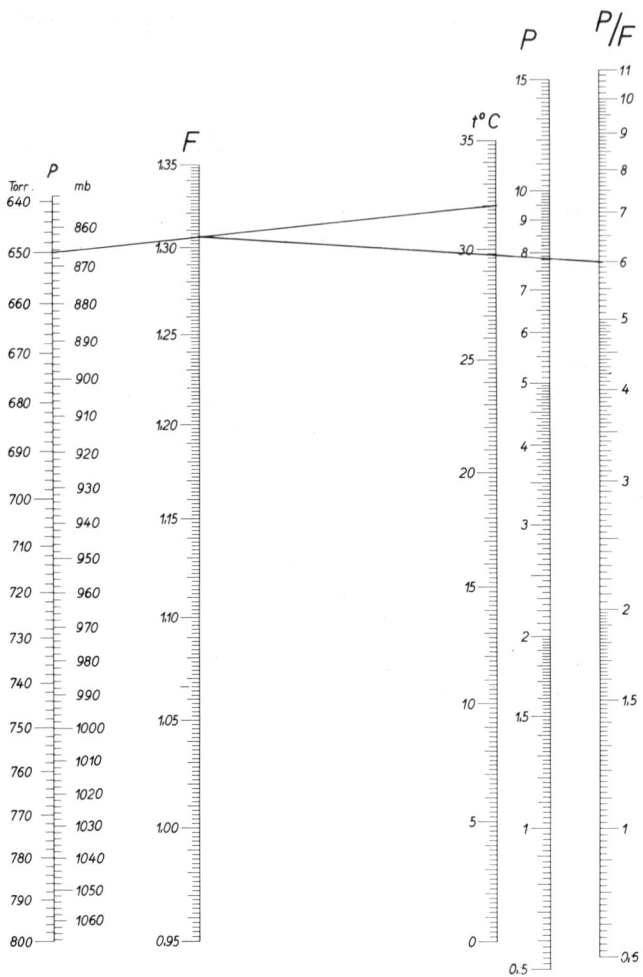

Fig. 43. Nomogram for estimating the factor F and for multiplication of a number by F. As an example, for p = 650 Torr, t = 32° C, P/F = 6.0; one can read off that F = 1.306 and P = 7.84.

h is determined under the prevailing external conditions, and in consequence so also, according to eqn. (40), is the difference in the refractive index. For results of various determinations to be comparable the refractive index difference must be reduced to the standard state for a gas, 0°C, 760 Torr.

460

From eqns. (51) and (52) this is given by

$$n^{0,760} - n_0^{0,760} = \frac{\lambda}{l} Fh \qquad (53)$$

To simplify the way of writing these values, n and m are always written without superscripts for the standard state.

The refractive indices of liquids vary relatively little with pressure, e.g. for water by about $1.4 \cdot 10^{-5}$ per 760 Torr [143], and the normal fluctuations in atmospheric pressure affect only the 7th decimal place. These are therefore only significant for precise measurements when one is not measuring against a reference liquid subjected to the same fluctuations. In contrast one must pay careful attention to the temperature. The temperature coefficient of the refractive index $dn/dt = n'$ lies for most liquids between $1 \cdot 10^{-4}$ and $6 \cdot 10^{-4}$ per degree. With highly sensitive measurements to the seventh decimal place it is of invaluable advantage to make the measurement against a reference liquid with an approximately equal temperature coefficient, and thus involve only very low expenditure on thermostating. One chooses as reference liquid the pure solvent, or, with a mixture, the component of highest concentration (as recommended by the theory in Sect. 5.A), so that the difference between the temperature coefficients of the reference liquid, n_0' and that of the test sample, n', is very small, $\Delta n' = n' - n_0'$. Both liquids must be at the same temperature. If with this proviso a refractive index difference $\Delta n = n - n_0$ is measured at a temperature t, then Δn is altered by a temperature change Δt, by the amount $\Delta n' \cdot \Delta t$. For this value not to exceed the limit of error δ_l attainable with cuvette length l, the temperature difference must remain

$$\Delta t \leqslant \delta_l / \Delta n'$$

These limiting values for Δt are given in Fig. 44 for the individual cuvette lengths. For example in Table 9, referring to aqueous sugar solutions, measured against water in different cuvette lengths for particular measurement ranges, $\Delta n'$ gives the maximum apparent difference in the temperature coefficients for these ranges. It is evident that in general one can work at room temperature without additional thermostating. Obviously one must avoid strong temperature fluctuations in the working environment. If over a long period it is necessary to have a higher constancy of temperature, so one must join a thermostat to the temperature bath. In this fashion a temperature con-

References pp. 536—546

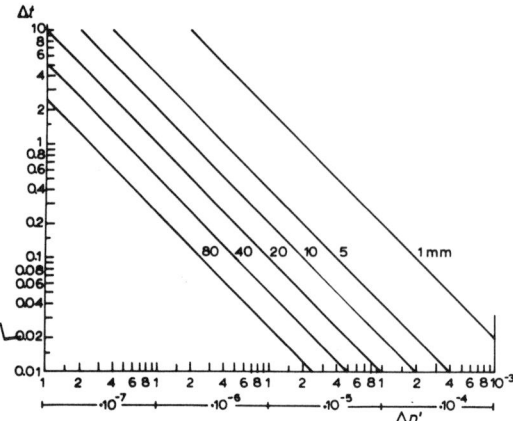

Fig. 44. Permissible temperature variation for liquids.

stancy in the double cuvette is attainable, which allows one to make the most of the measurement sensitivity.

(D) DETERMINATION OF CONCENTRATION

(1) Binary gas mixtures

The determination of concentrations in binary gas mixtures requires no special comments, when one employs a direct-reading interferometer. As previously discussed, the reference gas is filled

TABLE 9

Range of measurement and maximum difference of the temperature coefficients for aqueous sugar solutions

Cuvette length (mm)	Range of measurement $(n - n_0)$	Range of measurement (% sugar)	Max. $\Delta n'$
80	0.00063	0.44	$-5 \cdot 10^{-7}$
40	0.00125	0.87	$-1 \cdot 10^{-6}$
20	0.0025	1.7	$-2 \cdot 10^{-6}$
10	0.005	3.4	$-4 \cdot 10^{-6}$
5	0.010	6.8	$-8 \cdot 10^{-6}$
1	0.05	30.9	$-4.2 \cdot 10^{-5}$

into the reference cuvette. If the test material contains a single additional component, for which the scale is divided into percentages, so one makes observations only at that temperature and pressure for which the scale was calibrated, or corrects readings accordingly. For an instrument calibrated for 25°C and 760 Torr the measurements made at other temperatures will be multiplied by the factor

$$F' = \frac{760}{p} \left(1 + \frac{t-25}{298} \right)$$

(see eqn. (51) as well as Table 10).

Interferometers, not possessing a percentage scale, may be used for any gas mixture but they require individual calibration (Sect. 5.B). The difficult preparation of known gas mixtures can be avoided, when the refractive indices of the individual components are known.

The refractive index n of a gas mixture, according to the law of Biot and Arago, in most cases equals the sum of the refractive indices of the components

$$n = \sum c_j n_j \qquad \text{or} \qquad m = \sum c_j m_j \qquad \text{with} \qquad c_j = 1 \tag{54}$$

where c_j denotes the concentration of the component j.

If one of the components of the binary gas mixture serves as the reference gas, $n_1 = n_0$, then

$$n - n_0 = c_2(n_2 - n_1)$$

TABLE 10

Values of the factor $F' = \dfrac{760}{p} \left(1 + \dfrac{t-25}{298} \right)$

p Torr	$t°$ C								
	20	21	22	23	24	25	26	27	28
740	1.010	1.014	1.017	1.020	1.024	1.027	1.030	1.034	1.037
745	1.003	1.007	1.010	1.013	1.017	1.020	1.023	1.027	1.030
750	0.996	1.000	1.003	1.006	1.010	1.013	1.016	1.020	1.023
755	0.990	0.994	0.997	1.000	1.004	1.007	1.010	1.014	1.017
760	0.983	0.987	0.990	0.993	0.997	1.000	1.003	1.007	1.010
765	0.976	0.980	0.983	0.986	0.990	0.993	0.996	1.000	1.003
770	0.970	0.974	0.977	0.980	0.984	0.987	0.990	0.994	0.997

and according to eqn. (53) the percentage content, $P = 100c_2$, of the second component is given by

$$P = \frac{100\lambda F}{l(n_2 - n_1)} h \tag{55}$$

where n_1 and n_2 are the refractive indices of the pure gases in the standard state. They are given as refractive powers $m = (n-1) \cdot 10^6$ for a series of gases in Table 11.

Eqn. (55) is valid only as long as no break-points appear in the measurement range. Otherwise the position and direction of the breaks in the calibration curve must be calculated from eqn. (50) or determined by direct observation. If a measured fringe number h lies beyond a break-point, h must be reduced by the number g of break-points between 0 and h, with regard to the sign, i.e. h must be replaced by $(h-g)$ in eqn. (55). The general equation for the determination of percentage of a gas in a binary mixture reads as follows

$$P = \frac{100\lambda F}{l(n_2 - n_1)} (h - g) \tag{56}$$

With eqn. (56) a calculated calibration curve or table can be drawn up, which gives the appropriate percentage contents for the measured h values directly.

It is not always possible, as hitherto assumed, to choose as reference gas one of the components of the mixture. It may be that the particular gas is not available in sufficient purity, or that another pure gas is more cheaply and more readily available. Nevertheless the binary mixture can still be analysed. To maintain the range of measurement one must take care that the refractive index of the reference gas is as close as possible to that of the mixture. With eqn. (54) we now have

$$n - n_0 = n_2 - n_0 - c_1(n_2 - n_1)$$

and with eqn. (56)

$$P_1 = 100 \frac{n_2 - n_0}{n_2 - n_1} - \frac{100\lambda F}{l(n_2 - n_1)} (h - g)$$

$$P_2 = 100 \frac{n_0 - n_1}{n_2 - n_1} + \frac{100\lambda F}{l(n_2 - n_1)} (h - g) \tag{57}$$

464

TABLE 11

Refractive powers of some gases and vapours

Chemical symbol	Name	m_e	$\Delta m = m_F - m_C$	Reference
Inorganic gases and vapours				
—	Air, dry and CO_2-free	293.11	3.27	E, O
NH_3	Ammonia	379	8.0	LB
Ar	Argon	283.14	3.5	LB
BF_3	Boron trifluoride	413.4	3.54	WR
BCl_3	Boron trichloride	1404.0		LB
Br_2	Bromine	1184.9		ICT, LB
DBr	Deuterium bromide	621.10	13.65	L
HBr	Hydrogen bromide	622.14	14.17	L
Cl_2	Chlorine	784.0	13.9 [a]	LB, ICT, N [a]
DCl	Deuterium chloride	450.15	7.31	L
HCl	Hydrogen chloride	451.27	7.69	L
$(CN)_2$	Cyanogen	820.9	15.8	WR
HCN	Hydrogen cyanide	430.2	6.6	WR
D_2	Deuterium	137.58	2.16	F
He	Helium	34.95	0.18	ICT, LB
I_2	Iodine	2200		ICT
HI	Hydrogen iodide	925.8	25.72	ICT, LB
CO_2	Carbon dioxide	450.6	5.79 [a]	ICT, LB, N [a]
CO	Carbon monoxide	336.1	5.5	ICT, LB
Kr	Krypton	428.7	5.9	ICT, LB
Ne	Neon	67.25	0.28	ICT, LB
O_3	Ozone	520.0		ICT, LB
O_2	Oxygen	272.27	3.83 [a]	LB, N [a]
D_2S	Deuterium sulphide	647.73	15.79	L
H_2S	Hydrogen sulphide	650.68	15.67	L
SO_2	Sulphur dioxide	664.0	13.8	ICT
SF_6	Sulphur hexafluoride	765.8	5.1	WR
CS_2	Carbon disulphide	1477	70	LB
SiH_4	Silanes, monosilane	833.0	19.9	WR
Si_2H_6	disilane	1657.4	41.1	WR
SiF_4	Silicon tetrafluoride	568.3	4.1	WR
N_2	Nitrogen	299.14	3.33	LB
NO	Nitric oxide	295.5	3.9	ICT, LB
N_2O	Nitrous oxide	510.0	7.2	ICT, LB
NF_3	Nitrogen trifluoride	483.3	5.2	WR
H_2O	Water vapour	252.7	3.8	ICT, LB
H_2	Hydrogen	139.37	1.81	F
Xe	Xenon	705.5	12.7	ICT, LB

TABLE 11 (continued)

Chemical symbol	Name	m_e	$\Delta m = m_F - m_C$	Reference
Organic gases and vapours				
CF_4	Carbon tetrafluoride	494.3		WR
CF_2Cl_2	Freon 12	1132	17	H
CCl_4	Carbon tetrachloride	1781.9	30.6	R
$COCl_2$	Phosgene	1149.0		LB
$CHCl_3$	Chloroform	1412	19.9	N
CH_2Cl_2	Dichloromethane	1156.8	19.8	R
CH_2Br_2	Dibromomethane	1435.0		LB
HCOOH	Formic acid	574 (D)		r
CH_3F	Fluoromethane, methyl fluoride	443.1	5.6	R
CH_3Cl	Chloromethane, methyl chloride	770.8	12.4	R
CH_3Br	Bromomethane, methyl bromide	951.9	18.9	R
CH_3I	Iodomethane, methyl iodide	1291.4	34.8	R
CH_3NH_2	Aminomethane, methylamine	683.4	12.2	R
CH_4	Methane	443.3	6.7	ICT, LB
CH_3OH	Methanol	563.0	8.5	R
C_2Cl_4	Tetrachloroethylene	2009 (D)	40 [a]	LB, N [a]
C_2HCl_3	Trichloroethylene	1705	34.0 [a]	r, N [a]
$CHBrCl \cdot CF_3$	Halothane	1582	23.8	N
C_2H_2	Acetylene	600.7	11.9 [a]	LB, N [a]
$C_2H_2Cl_2$	Dichloroethylene	1473 (D)		LB
C_2H_3N	Acetonitrile	759.2	11.9	R
C_2H_4	Ethylene	719.8	16.3	WR
$(CH_2)_2O$	Ethylene oxide	750.2	11.4	R
CH_3CHO	Acetaldehyde	774 (D)		r
$C_2H_4Cl_2$	Dichloroethane	1415 (D)		ICT
C_2H_5Cl	Ethyl chloride	1095.6	18.0	R
C_2H_5Br	Ethyl bromide	1261.3		LB
$C_2H_5NO_2$	Nitroethane	1169		LB
C_2H_6	Ethane	764.8	11.8	LB
C_2H_5OH	Ethanol	870	10.8	N
$(CH_3)_2O$	Dimethyl ether	888.3	14.4	R
$(CH_2)_2C$	Propadiene	1060.5	28.1	WR
C_3H_4	Methyl acetylene	973.9	21.3	WR
C_3H_5Cl	Chloropropylene	1444 (D)		ICT, LB
C_3H_6	Propylene	1057.1	23.1	WR
$(CH_2)_3$	Cyclopropane	959.8	16.3	R
$(CH_3)_2CO$	Acetone	1086.7	19.3	R
$C_3H_6O_2$	Ethyl formate	1193 (D)		r
$C_3H_6O_2$	Methyl acetate	1193 (D)		ICT
C_3H_8	Propane	1080.8	15.5	WR

TABLE 11 (continued)

Chemical symbol	Name	m_e	$\Delta m = m_F - m_C$	Reference
C_3H_7OH	Propanol	1172 (D)		r
C_4H_8	α-Butylene	1370.4	28.5	WR
$C_4H_8O_2$	Ethyl acetate	1586 (D)		ICT
C_4H_{10}	n-Butane	1390.1	21.9	WR
C_4H_9OH	n-Butanol	1454 (D)		r
$C_4H_{10}O$	Diethyl ether	1554	23.6	N
C_5H_{10}	Amylene	1693 (D)		ICT, LB
$C_5H_{10}O_2$	Propyl acetate	1805 (D)		r
C_5H_{12}	n-Pentane	1779	29.8	Y
$C_5H_{11}OH$	n-Amyl alcohol	1789 (D)		r
C_6H_5Cl	Chlorobenzene	2084 (D)		r
C_6H_6	Benzene	1759	43	N
C_6H_{12}	Cyclohexane	1856 (D)		r
$C_6H_{11}OH$	Cyclohexanol	1959 (D)		r
$C_6H_{12}O_2$	Butyl acetate	2114 (D)		r
C_6H_{14}	Hexane	2032 (D)		LB
C_7H_8	Toluene	2092		r
$C_7H_{12}O$	Methylcyclohexanone	2175 (D)		r
$C_7H_{14}O_2$	Amyl acetate	2423 (D)		r
C_7H_{16}	Heptane	2303 (D)		r
C_8H_{10}	Xylene	2410		r
$C_{10}H_{12}$	"Tetralin"	2934 (D)		r
$C_{10}H_{18}$	"Decalin"	2934 (D)		r

Those values denoted with [a] correspond to the source similarly identified. The source indications signify: E, Edlén [49,50]; F, Frivold and co-workers [144]; H, Horvath [145]; ICT, International Critical Tables [146]; L, Larsén [147]; LB, Landolt and Börnstein [148]; N, Nebe [130,138]; O, Owens [52]; R, Ramaswamy [149]; WR, Watson and Ramaswamy [150]; Y, Yamshikov [151]; r, Calculated from the specific refraction according to eqn. (34). Values followed by (D) indicate index values given as m_D. The values of m_e from refs. F and WR and most of those from L are also given in LB.

In contrast to eqn. (56) the quotients $(n_2 - n_0)/(n_2 - n_1)$ or $(n_0 - n_1)/(n_2 - n_1)$ now appear as constants, which determine the shift of the measured value away from the zero-point. The sensitivity of measurement, which is qualified by the factor $(h - g)$, is the same as before.

An important consideration affecting the attainment of higher precision is the choice of cuvette length, since this, in eqns. (56) and (57), is the only factor open to choice by the experimenter. The precision and range of measurement of an interferometer are usually given in terms of phase differences or refractive index differences per cuvette length. For gas analysis however, it is the concentration differences that are of interest. These can be calculated from the formula given, when one substitutes the corresponding values for h and l. The refractive index differences of any two gases are constants, which one can find from Table 11. A quick determination of the limiting values of concentration is facilitated by Fig. 45. With differing cuvette length as a parameter the precision of measurement according to eqn. (55) is plotted on the left-hand ordinate against the refractive power difference on the abscissa, whereas the right-hand ordinate indicates the measurement range. If a reference gas not contained in the mixture is used, then the range, as read off the diagram, is displaced by the amount $100(n_2 - n_0)/(n_2 - n_1)$, where n_0 and n_1 are taken as being both greater or both less than n_2.

Finally, mention should be made of a dynamic method of inter-

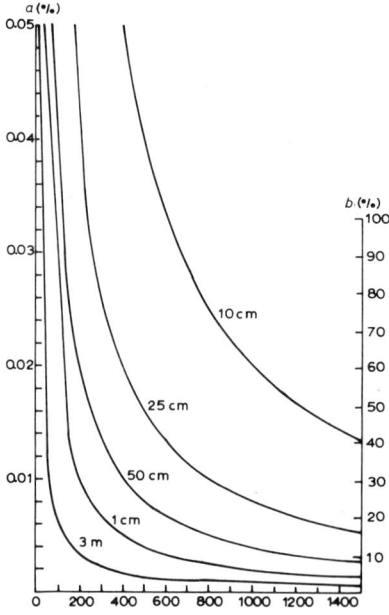

Fig. 45. Diagram for choosing the optimum path length for the cell: a, attainable accuracy of measurement; b, permissible region for measurement; abscissa, refractive power difference $m_2 - m_1$.

ferometric gas analysis described by Bärwinkel and Hartkämper [152]. A known amount of the gas sample, being less than the volume of the cuvette, is allowed to flow into the cell, and the resulting shift of the interference pattern is measured. The concentration can be calculated from the ratio of the sample volume to the cuvette volume, and the interferometer reading. An advantage of this method is that even small volumes of samples, insufficient to fill the cell completely under normal conditions, can still be measured.

(2) Ternary gas mixtures

The relative interferometric measurement of gas mixtures yields directly only one measurement value, the phase difference. With mixtures of several gases one must necessarily undertake further measurements to accomplish a complete analysis.

The most frequently employed purely interferometric technique is the sorption method. In this one component is removed from the mixture by means of a specific sorption medium (sorbent). The specificity of the sorption medium in relation to the components of a mixture is a crucial factor in the success of the analysis; the sorption medium must remove one component (sorbate) of the mixture completely, without altering the others and without giving off another gas through chemical reaction (see Sect. 4.E(4)). In Table 7 information is given about sorption media, whose suitability with regard to their specificity in a particular case must be tested. For an analysis, the cuvette is filled first with the test material as it comes, and then again after the sorption of one component. The reference cuvette is filled with one suitable component of the remaining mixture. In addition one should consider the prerequisites for the reference gas laid down in Sect. 5.A.

Granular sorption media are much to be preferred since in general they do not release any vapour. They are poured into vertical cylindrical or U-shaped glass tubes through which the gas sample is passed. For liquid sorbents one uses wash bottles. The necessary quantity of sorbent depends on its sorption capacity, and also on the temperature and pressure and obviously on the amount of the sorbate. This must of necessity be determined by a preliminary investigation. The sorbent must be renewed before its capacity is exhausted, sometimes indicated by a change of colour.

Cuvettes with three tubes, lying side by side, are available for the analysis of ternary mixtures by the sorption technique. They speed up the measurement procedure as the sample gas flows through two cuvette tubes in succession without the setting of the taps being altered. Figure 46 shows an arrangement in which the reference gas flows through on the left, the test material after sorption of one component flows through the middle and the unaltered test material flows through the right-hand tube. Through displacement of the cuvette one changes the tube in each beam path and measures accordingly the phase difference between the middle and left tubes (Δs_I) and between right and middle tubes (Δs_{II}).

Denoting n_1, n_2, n_3 as the refractive indices of the components, $n_0 = n_1$ that of the reference gas, and $\bar{c}_1$ and $\bar{c}_2$ the concentrations of the components 1 and 2 after sorption of component 3 ($\bar{c}_1 + \bar{c}_2 = 1$ and $\bar{c}_2 = c_2/(1-c_3)$), the following relationships hold

$$\Delta s_I = l\left(\sum_{1,2} c_j n_j - n_0\right) = c_2 l(n_2 - n_1)$$

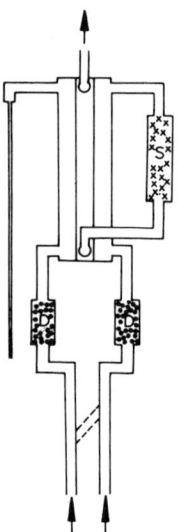

Fig. 46. Arrangement for the analysis of ternary gas mixtures with a triple-compartment cell: S, sorbent; D, drying agent.

$$\Delta s_{II} = l\left(\sum_1^3 c_j n_j - \sum_1^2 \bar{c}_j n_j\right) = l[c_3(n_3 - n_1) + c_2(n_2 - n_1)] - \Delta s_I$$

$$= c_3[l(n_3 - n_1) - \Delta s_I]$$

With $\Delta s_I = \lambda F h_I$, $\Delta s_{II} = \lambda F h_{II}$, taking into account variation in temperature and pressure, it follows from

$$c_2 = (1 - c_3)\bar{c}_2 \qquad \text{with} \qquad \bar{c}_2 = \frac{\lambda F}{l(n_2 - n_1)} h_I$$

that

$$c_3 = \frac{\lambda F h_{II}}{l(n_3 - n_1) - \lambda F h_I} \tag{58}$$

Another method for a purely interferometric analysis of a ternary mixture arises out of the differing dispersion characteristics of the components. As described in Sect. 5.B, the discontinuities in the calibration curve of a binary mixture are dependent on dispersion effects. Their position is altered when a third component appears in the mixture.

Corresponding to the definition (47) the relative effective dispersion of a mixture of the two pure gases i and j is denoted by

$$D_{ij} = 100 \frac{\Delta n_i - \Delta n_j}{n_i - n_j}$$

In particular for a mixture of test material (n) and reference gas (n_0)

$$D_0 = 100 \frac{\Delta n - \Delta n_0}{n - n_0} \tag{59}$$

From this, and with eqn. (53), it follows that

$$\Delta n - \Delta n_0 = \frac{\lambda F}{100 l} D_0 h \tag{60}$$

Without imposing any fundamental limitation on the technique, one can assume that one of the three gases can serve as reference gas, e.g. $n_0 = n_1$, $\Delta n_0 = \Delta n_1$. Since for ternary mixtures $n = \sum_1^3 c_j n_j$, it follows from eqn. (53) that

$$c_2(n_2 - n_1) + c_3(n_3 - n_1) = \frac{\lambda F}{l} h$$

and from this by multiplication by $D_{3,1}/100$ that

$$c_2(n_2 - n_1)\frac{D_{3,1}}{100} + c_3(\Delta n_3 - \Delta n_1) = \frac{\lambda F}{100l} D_{3,1} h$$

From eqn. (60) with $\Delta n = \Sigma_1^3 c_j \Delta n_j$ it follows that

$$c_2(\Delta n_2 - \Delta n_1) + c_3(\Delta n_3 - \Delta n_1) = \frac{\lambda F}{100l} D_0 h$$

and by difference that

$$c_2\left[\Delta n_2 - \Delta n_1 - (n_2 - n_1)\frac{D_{3,1}}{100}\right] = \frac{\lambda F}{100l}(D_0 - D_{3,1})h$$

Moreover introducing $D_{2,1}$ and correcting h by the number of break-point, one obtains

$$c_2(n_2 - n_1)(D_{2,1} - D_{3,1}) = \frac{\lambda F}{l}(D_0 - D_{3,1})(h - g)$$

$$\tag{61}$$

$$c_3(n_3 - n_1)(D_{3,1} - D_{2,1}) = \frac{\lambda F}{l}(D_0 - D_{2,1})(h - g)$$

With these formulae the concentrations of the three gaseous components can be calculated from the measured values $(h - g)$ and D_0. One finds the dispersion D from eqn. (50) to be

$$D_0 = \frac{D^*H + \Lambda}{H - 1} \tag{62}$$

In this $h_b/q = H$ gives the achromatic length in the calibration curve, and $H \gtrless 0$ depending on whether $q \gtrless 0$. The determination of D_0 depends in practice on the determination of the achromatic length and therefore on the observation of the first break-point beyond it. If one assumes that the interferometrically measured value h lies beyond the first break-point h_{b1}, h_{b1} and therefore also H can be determined. It is only necessary, after measuring the shift of the fringes, to slowly displace the sample gas by the reference gas (component 1), for the break-point to be observed, as described in Sect. 5.B.

This technique does not work for small concentrations giving shifts below the first break-point. But precisely for $h < h_{b1}$ the direct determination of the dispersion is available as a second measurement to assist in the analysis. One requires for this technique two mono-

472

chromatic light sources giving a long and a short wavelength of visible light. The cadmium lamp with monochromatic filters for selecting the C' or alternatively the F' lines is ideally suitable. The fringe spacing in the interference pattern is, according to eqn. (15), directly proportional to the wavelength, i.e. $w/w_e = \lambda/\lambda_e$. One can measure w by means of the compensator, in that one displaces the monochromatic interference pattern by one fringe spacing from the zero-point and this displacement is expressed as the interferometer reading i. The difference $\Delta w = w_{C'} - w_{F'}$ is an instrumental constant for the interferometer. For the two cadmium spectral lines

$$\Delta w = w_e \frac{\lambda_{C'} - \lambda_{F'}}{\lambda_e} = 0.300\, w_e$$

The compensator can in general only compensate precisely for an intermediate wavelength, whereas for longer and shorter wavelengths an exact coincidence with the initial position cannot be obtained simultaneously. If $i_{C'}$ and $i_{F'}$ are the interferometer values, for compensation with $\lambda_{C'}$ and $\lambda_{F'}$ respectively, then we denote $\Delta i = i_{C'} - i_{F'}$ as the chromatic difference of the test-gas for the lines C' and F'.

As an illustration Fig. 47 shows the relationship for the case $H > 0$. For $\Delta i = g\Delta w$, where g is any integer, there exists an achromatic point. This always happens for the zero-point $i = 0$, $\Delta i = 0$. With changing i in white light a first-order minimum in the interference pattern becomes more and more coloured. At the first break-point the chromatic difference Δi takes half the value of the difference Δw. In general one can assume $\Delta i = q\Delta w$ where again q is defined as $\pm \frac{1}{2}$ for the first break-point. With concentrations beyond the first break-point as Δi increases still further, one of the neighbouring fringes (for C' or F') shifts closer to the measured value i_e, so that one refers to this for the measurement of the chromatic difference Δi (with monochromatic light the order number of an individual fringe is indistinguishable) and beyond the first break-point a lack of ambiguity cannot be guaranteed. Since in this region the break-points can be observed as described previously, the two techniques complement each other to advantage.

Measurement of the chromatic difference is thus limited to concentrations below the first break-point. Since the measured value remains relatively small, one can to a good approximation assume a linear relationship between i and the fringe number h. This is given by

$$\frac{\Delta i}{i} = \frac{q\Delta w}{i_{b1}}, \qquad q = \pm\frac{1}{2}$$

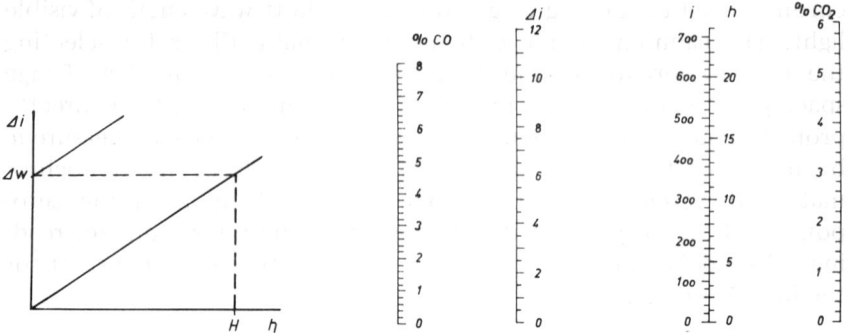

Fig. 47. Representation of the achromatic length.

Fig. 48. Nomogram for the analysis of CO—CO_2—air mixtures.

or if one replaces i by h

$$h_{b1} = \frac{q \Delta w}{\Delta i} h, \qquad q = \pm\frac{1}{2}$$

For the achromatic length $H = h_{b1}/q$ this becomes

$$H = \frac{\Delta w}{\Delta i} h \qquad\qquad (63)$$

By measuring h and Δi, and calculating first H and then D_0 (using eqn. (62)) the concentration of each component can then be determined from eqn. (61).

In order that D_0 can be obtained sufficiently accurately, the i value must be measured with white light and with the two spectral lines immediately one after the other, more than once, if possible. For the mean value the error should not amount to more than $\frac{1}{50}$ to $\frac{1}{100}$ of the fringe spacing so that the uncertainty in the result should not exceed one percent of the measured value. Even if this technique does not give the highest precision, it does represent a simple, quick, and inexpensive method. The evaluation is made easier by tables or graphical aids (Fig. 48).

In addition to the purely interferometric technique, ternary mixtures can also be determined by measuring another physical or chemical property in addition to the refractive index. A prerequisite is that the refractive index of each component be known. If, for

474

example, in addition to n, the gas density ρ of the unknown sample is measured, then the concentrations c_j of the individual components are given by the solution of the simultaneous equations

$$\sum c_j n_j = n \; ; \qquad \sum c_j \rho_j = \rho \; ; \qquad \sum c_j = 1 \tag{64}$$

The mathematical evaluation can be by-passed, for example by the use of a triangular nomogram (Fig. 49).

Since the development of small specifically indicating gas detection tubes has reached an analytically useful state, they have been widely used for determinations in the field of individual components. The complementary interferometric measurement is carried out on the unaltered mixture. In some earlier procedures, one component of the mixture was removed by, for example, a gravimetric or a combustion method and determined by volume contraction. The mixture was thus reduced for the interferometric measurement to a binary system.

(3) Liquids

What has been said about the analysis of gases applies to a large extent to the analysis of liquids, but several distinctions must be observed.

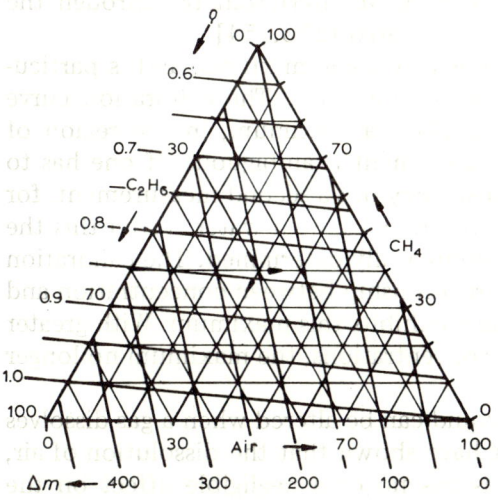

Fig. 49. Triangular diagram for the analysis of CH_4–C_2H_6–air mixtures.

In the measurement of liquids one must pay great attention to the control of temperature (Sect. 5.C). This is important in investigations on water and aqueous solutions and in the more difficult measurements on organic liquids whose temperature coefficients are many times greater. Besides, some solvents have high vapour pressures which lead to evaporation and therefore to the formation of schlieren. The constant-temperature bath required fulfils yet another purpose with those interferometers which work with a fixed interference pattern for reference marks.

For a double cuvette in air quite small errors in the alignment of the cuvette windows can make the measurement impracticable. However, if the cuvette is lowered into a thermostating liquid whose refractive index is approximately similar to that of the measured liquids and the light beams producing the reference interference pattern likewise run through the temperature bath, the error in the angular alignment of the cuvette windows is practically compensated and does not disturb the measurement.

The refractive indices of solutions and liquid mixtures do not vary linearly with concentration, so that one must always refer to a calibration. Literature values are not of much use since almost all previous tables for $n = n(c)$ refer to the Na-D line which, being a doublet, is hardly suitable for interferometry. Equipment is, however, available which extends the possibility of measurement of refractive index on liquids with the accuracy of interferometry through the whole visible spectrum up to the infrared [153,154].

A further peculiarity is shown by some mixtures, and is particularly marked for methanol—water mixtures. The calibration curve $n = n(c)$ for these mixtures displays a maximum, in the region of which the interferometric measurement is ambiguous. If one has to measure in this region one must carry out a second measurement, for example after diluting the mixture 1 : 1 with water. From this the concentration can then be unequivocally determined. The calibration curve $i = i(c)$, which gives the relationship between concentration and the interferometer reading, also exhibits this maximum. With greater cuvette length, i.e. for lower concentrations, the maximum no longer appears (Fig. 50(a) and (b)).

The refractive index of a liquid can be altered when a gas dissolves in it. Investigations on water have shown that the dissolution of air, oxygen, hydrogen or coal gas has no or a negligible effect on the refractive index, but the effect is quite marked with carbon dioxide,

476

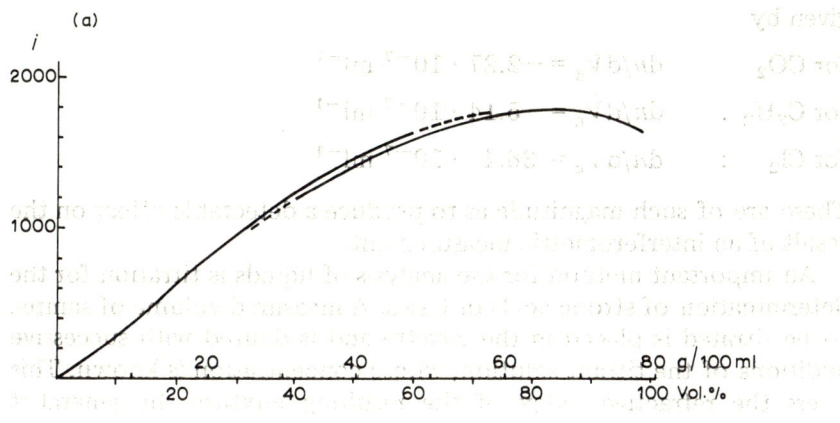

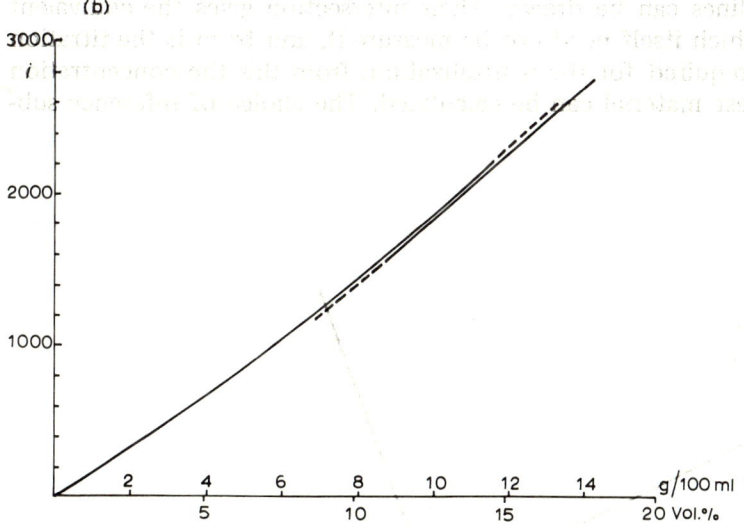

Fig. 50. Calibration curves for ethanol—water mixtures: cell lengths are (a) 1 mm, and (b) 5 mm.

acetylene and chlorine. If V_g is the volume of gas dissolved in 100 ml of water, the variation of refractive index per ml of dissolved gas is given by

for CO_2 : $dn/dV_g = -2.27 \cdot 10^{-7}$ ml^{-1}

for C_2H_2 : $dn/dV_g = 5.14 \cdot 10^{-7}$ ml^{-1}

for Cl_2 : $dn/dV_g = 36.4 \cdot 10^{-7}$ ml^{-1}

These are of such magnitude as to produce a detectable effect on the result of an interferometric measurement.

An important method for the analysis of liquids is titration for the determination of strong acids or bases. A measured volume of sample to be titrated is placed in the cuvette and is diluted with successive additions of the titrant solution, whose concentration is known. This alters the refractive index of the resulting mixture. In general it decreases until after passing the neutralization point when it increases again. The resulting interferometer values are plotted in a diagram (Fig. 51). It is enough to measure a few points from which the straight lines can be drawn. Their intersection gives the equivalent point (which itself need not be measured), and from it the titration volume required for the neutralization; from this the concentration of the test material can be calculated. The choice of reference sub-

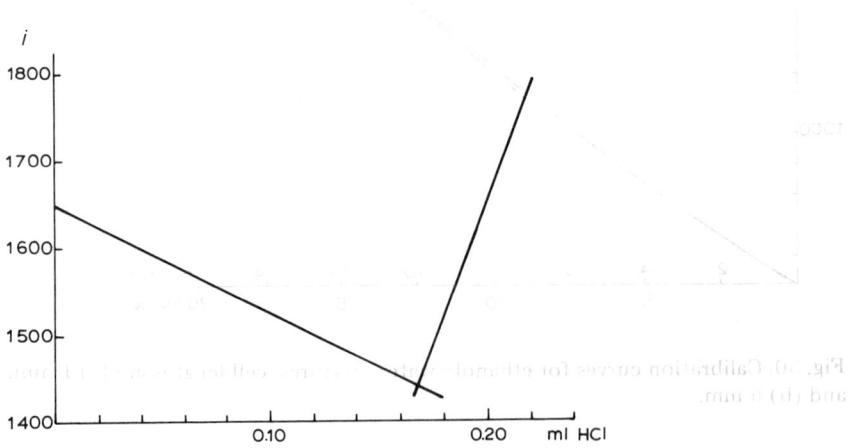

Fig. 51. Change of refractive index during the course of a titration (from Vajgand and Todorovski [241]). 3.20 ml of 0.1186 M sodium hydroxide titrated with 2.294 M hydrochloric acid.

478

stance is not important, as only the shape of the titration curve is of interest and not the absolute values of refractive index. The interferometric titration gives a quick, precise and inexpensive technique, which, with certain restrictions, may be used specifically for the analysis of acids and bases. It is of course not necessary to construct a calibration curve.

Most concentration determinations on liquids are carried out by refractometry in the narrower sense, that is by measurement of the deviation of light at a prism or the critical angle of total reflection. For the methodology of refractometry there exists a wealth of literature [48]. A precision of about $1 \cdot 10^{-5}$ refractive index units can be obtained, so that for higher precision interferometry must be employed. The methods of refractometry in the investigation of binary mixtures, in the determination of the concentration of a single component in mixtures which cannot be measured directly as well as in the analysis of ternary mixtures can largely be transformed to interferometry.

The indirect determination of a concentration is necessary when the sample material is not suitable for direct measurement in the interferometer due to turbidity or light absorption. The constituent to be measured is extracted from the mixture into a specific solvent, and determined as a solution in that solvent. A weighed-out quantity of the sample is added to a measured quantity of solvent. After a certain time, which can be shortened by shaking or stirring, the clear solution is separated off by, e.g. distillation, centrifuging, and measured interferometrically. By means of the calibration curve for this solution, the original concentration can be calculated. An example is the determination of blood alcohol (Sect. 6.G). In the practical application of this technique one must beware of side reactions in each individual case, which could falsify the result.

The methods of analysis of ternary mixtures by refractometry are also familiar in interferometry. Well-known examples are the determination of alcohol and extract contents in beer, wine, brandy and other spirits. Since two quantities, alcohol and extract, are to be determined one requires in addition to the refractive index a second quantity such as the density, to be measured to complete the characterization. Other examples are mixtures of oils or fats. In this a melting point determination is most often taken in addition to the refractive index measurement. Compared with these refractometric measurements the interferometric method offers higher precision, but there are no new aspects not discussed in Sect. 5.D(2) for ternary gas mixtures.

The sorption technique is in general not applicable to liquids, but the purely interferometric techniques of break-point determination or the measurement of the achromatic length can be used. The linearity existing between refractive index and concentration according to eqns. (53) and (55) for gases, however, does not apply here. The basic relationship (40) remains valid and by combination with the definition eqn. (59) gives

$$n - n_0 = \frac{\lambda}{l} h$$

$$\Delta n - \Delta n_0 = \frac{n - n_0}{100} D_0$$

(65)

D_0 can be expressed according to eqn. (62) in terms of the achromatic length H. H can be determined, as described for the observation of the first break-point, by successive dilution of the test material. The simplest method for liquids is the direct measurement of the chromatic difference Δi. The simple ratio (63) is also valid here for not too large measured values. Introducing this into eqn. (65) one obtains the determination equations for n and Δn.

$$n = n_0 + \frac{\lambda}{l} h$$

$$\Delta n = \Delta n_0 + \frac{\lambda}{100} h \frac{D^* \Delta wh + \Lambda \Delta i}{\Delta wh - \Delta i}$$

(66)

The relationship between n, Δn and the concentration of the components must be determined by means of a calibration.

6. Applications of interferometry for the analysis of homogeneous substances

When compared with the majority of chemical procedures, interferometric measurements offer the great advantage of simplicity. The analytical requirements of each individual case are generally fairly obvious and less complicated than might be expected from what has been discussed previously. The rapidity with which the measurements may be carried out, together with their high precision has lead to a very wide range of applications for analytical interferometry. The following discussion can only point to examples and to some areas where interferometric measurements play an important role. As the methods have now become well established in certain fields, references to the older literature will be largely avoided. Mention will be made here simply to the comprehensive works of Bauer and Fajans [160], Williams [161], Löwe [128], Joffe [48], and Nebe [130].

(A) TECHNICAL GASES AND VAPOURS

Both Haber and Löwe are considered to have been the first to introduce analytical interferometry into technology. Haber used the interferometer for the determination of the ammonia content in gases in his plant for the synthesis of ammonia, which was to be so important in the development of agriculture. This is a typical example of the determination of one component in the final product of an industrial gaseous mixture. Similar conditions apply to the production of numerous gases. Of some considerable significance is the analysis of mixtures of the noble gases which have found wide technological application in gas-filled fluorescent lamps.

Naturally, purity control of a product also plays an important role. In such cases, interferometric measurements are particularly simple, as under conditions which have been previously laid down, one need only give a limiting value for the shift of the interference band, which must not be exceeded if the desired degree of purity is to be met. As an example one might cite the measurements by Bell and Krantz [162] made on nitrous oxide intended for use as an anaesthetic.

The final and intermediate products from gas and coke works have been thoroughly investigated with the emphasis lying on the determination of benzene in coal gas [163,164]. Although there exist numerous methods and apparatuses for the analysis of gases, interference measurements are used again and again on account of their speed and precision. Further specialized examples are the measurement of fuel—air mixtures immediately before ignition in the internal combustion engine, the testing for gas permeability of balloon materials or membranes, and problems of gas supply to gas-heated stoves for the home and for industrial buildings. In the last case, the layout of the ventilation system also plays a major role which is also important for many other industrial plant and machines. Section 6.C deals with noxious and toxic gases in more detail.

One should also make mention of exhaust gases which can contain significant proportions of valuable unused gas. Interferometric measurements can, for example, give information as to whether the loss of fuel gas is so high that the construction of a regeneration plant would be justified. Kutz [165] has drawn attention to the difficulties associated with the investigation of chlorine and hydrogen chloride at temperatures up to 100°C.

The relationships involved in the analysis of two-component mixtures [166] are simple. But more complicated systems can also be analysed interferometrically. Thus the concentration of a volatile component in a liquid mixture may be determined when one measures the vapour concentration over the liquid in equilibrium with it, and calculates the concentration in the liquid from the partial pressure in the vapour phase [167]. Roth et al. have dealt with another interesting application in the field of hydrocarbon chemistry [168]. They determined the concentrations of hydrocarbons with three, four or more carbon atoms by using activated charcoals with differing adsorptive powers towards different hydrocarbons. This is in fact a chromatographic procedure with an interferometer as a detector, just as was used by Kögler [169]. Löwe [170] has mentioned a series of further applications.

(B) EXPLOSIVE GASES AND VAPOURS

Organic solvents are used in almost all branches of chemistry and industry. They are present as ingredients of the final product, for example, in paints, lacquers and adhesives, and they may be encountered in the production process anywhere in chemical industry, or may be used for cleaning processes. As most solvent vapours can form explosive gaseous mixtures with air it is advisable to handle them with considerable caution. A mixture with air will be capable of exploding when the gas or vapour concentration exceeds a certain minimum level but on the other hand does not exceed an upper limit which is determined by the amount of oxygen needed for combustion. Between these lower and upper limits the mixture can be exploded by an electrical spark.

The awareness of these sources of danger has lead to safety precautions which above all demand strict control of a danger zone through both objective and subjective tests, among which interferometric methods are to be found. They are used first and foremost for the detection of fire damp in underground mines. Furthermore the testing of emptied tankers, tank wagons, fuel containers and so on for explosive amounts of residual gas is absolutely essential before repairs can be carried out by welding, with adequate safety. A knowledge of the explosion limits is essential for a reliably objective control of the danger area. Values are presented in Table 12 for some gases mixed with air at $20°C$ and 760 Torr, taken from Nabert and Schön [171]. Schön and Steen have collected data of particular relevance for mixtures of gases for anaesthetics [172]. When using these data one should remember that they are subject to various additional conditions such as energy and temperature of the spark, oxygen content of the air, and temperature of the environment, which can result in a change from the tabulated values. These relationships have been dealt with at length [173,174].

In those places where any kind of work has to be carried out, the concentrations of combustible gases must lie below the lower explosive limit, since with concentrations above the upper limit, mixing with air can very rapidly lead to the formation of explosive mixtures. Interferometers are exceptionally suitable for routine control of work places, rooms or containers for the presence of small amounts of combustible gases. One should take great care that in zones of possible explosion danger only explosion-proof interferometers are employed. This applies to the fire-damp interferometer, which is, however, normally only calibrated in percent of methane in air. If other gases or vapours are to be determined, recalibration will be necessary in order to convert the read percent-methane values into corresponding values for the gas in question. This recalibration can be effected by calculating the relevant conversion factor from the refractive index for the gases. Conversion factors for over fifty gases are given in ref. 130.

The largest number of interferometers finds application in dangerous underground mining operations. In coal mines one sometimes encounters releases of methane in considerable quantities. Even in normal mining operations larger or smaller amounts of methane escape from the surrounding rock so that there is always some methane present in the mine air. As methane at concentrations between the limits of 5.0 and 15.0 vol.% in air forms explosive mixtures, a single spark in the presence of such a mixture can set off a violent explosion. In order to be sure to avoid such an occurrence,

TABLE 12

Explosion limits for some gases and vapours when mixed with air at 20° C and 760 Torr [171]

Formula	Name	Explosion limits			
		Vol.%		$g\ m^{-3}$	
		Lower	Upper	Lower	Upper
Inorganic					
NH_3	Ammonia	15	28	105	200
$(CN)_2$	Cyanogen	6.0	32	130	700
HCN	Hydrogen cyanide	5.4	46.6	60	520
CO	Carbon monoxide	12.5	74	145	870
CS_2	Carbon disulphide	1.0	60	30	1900
H_2S	Hydrogen sulphide	4.3	45.5	60	650
H_2	Hydrogen	4.0	75.6	3.3	64
Organic					
CH_2Cl_2	Dichloromethane	13	22	450	780
CH_3Cl	Methyl chloride	7.1	18.5	150	400
CH_3Br	Methyl bromide	8.6	20.0	335	790
CH_4	Methane	5.0	15.0	33	100
CH_3OH	Methanol	5.5	31	73	410
$CHCl{=}CCl_2$	Trichloroethylene	7.9	—	430	—
C_2H_2	Acetylene	1.5	82	16	880
$CH_2{=}CCl_2$	1,1-Dichloroethylene	5.6	13	220	530
$CH_3 \cdot CN$	Acetonitrile	3.0	—	50	—
C_2H_4	Ethylene	2.7	28.5	31	330
C_2H_4O	Ethylene oxide	3.0	100	55	1820
$CH_3 \cdot CHO$	Acetaldehyde	4	57	73	1040
$CH_2Cl \cdot CH_2Cl$	1,2-Dichloroethane	6.2	16	250	660
C_2H_5Cl	Ethyl chloride	3.6	14.8	95	400
C_2H_5Br	Ethyl bromide	6.7	11.3	300	510
C_2H_6	Ethane	3.0	12.5	37	155
C_2H_5OH	Ethanol	3.5	15	67	290
$CH_3 \cdot C{\equiv}CH$	Methylacetylene	1.7	—	28	—
$CH_3 \cdot CCl{=}CH_2$	2-Chloropropene	4.5	16.0	140	510
C_3H_6	Propylene	2.0	11.7	35	210
$(CH_2)_3$	Cyclopropane	2.4	10.4	40	185
$(CH_3)_2CO$	Acetone	2.5	13.0	60	310
$HCO_2 \cdot C_2H_5$	Ethyl formate	2.7	13.5	80	410
$CH_3CO_2 \cdot CH_3$	Methyl acetate	3.1	16	95	500
C_3H_8	Propane	2.1	9.5	39	180

TABLE 12 (continued)

Formula	Name	Explosion limits			
		Vol.%		$g\ m^{-3}$	
		Lower	Upper	Lower	Upper
C_3H_7OH	Propan-1-ol	2.1	13.5	50	340
C_4H_8	But-1-ene	1.6	9.3	35	220
$CH_3CO_2 \cdot C_2H_5$	Ethyl acetate	2.1	11.5	75	420
C_4H_{10}	n-Butane	1.5	8.5	37	210
C_4H_9OH	Butan-1-ol	1.4	10	43	310
$(C_2H_5)_2O$	Diethyl ether	1.7	36	50	1100
C_5H_{10}	Pent-1-ene	1.4	8.7	40	260
$CH_3CO_2 \cdot C_3H_7$	Propyl acetate	1.7	8	70	340
C_5H_{12}	n-Pentane	1.4	7.8	41	240
$C_5H_{11}OH$	Pentan-1-ol	1.3	10.5	47	380
C_6H_5Cl	Chlorobenzene	1.3	7.0	60	330
C_6H_6	Benzene	1.2	8.0	39	270
C_6H_{12}	Cyclohexane	1.2	8.3	40	290
$CH_3CO_2 \cdot C_4H_9$	Butyl acetate	1.2	7.5	58	360
C_6H_{14}	n-Hexane	1.2	6.9	42	250
C_7H_8	Toluene	1.2	7.0	46	270
$CH_3CO_2 \cdot C_5H_{11}$	Amyl acetate	1	10	60	550
C_8H_{10}	o-Xylene	1.0	6.0	44	270
$C_{10}H_{12}$	Tetrahydronaphthalene	0.8	5.0	45	275
$C_{10}H_{18}$	Decalin	0.7	4.9	40	280

one takes care to control thoroughly and regularly at least all those stretches in which work is in progress. In addition, routine measurements of the methane concentrations are made in order to point out in good time dangerous build-ups of concentration. For this purpose both stationary gas analysers set up at specific road junctions underground and also portable gas-measuring apparatuses are used. The two types of apparatus are complementary to each other. The stationary apparatus monitors the mine air at a particular point continuously and transmits the results of the measurements to a central office above ground. The small portable instrument is used to control the mine air on the spot in bore-holes and cracks in the rock before and after blasting, after mine fires, and from time to time also in seldom used galleries. For such purposes fire-damp interferometers are in widespread use.

The air in mines is usually tested for the presence of methane and carbon dioxide. For routine measurements one makes the justified assumption that the nitrogen—oxygen ratio in the mine air does not differ from that in the open air, and is therefore constant. On the other hand the moisture content of the air does vary and would influence the measurements if it were not removed from the air sample by a specific adsorption agent such as silica gel. Carbon dioxide is removed from the mixture by absorption onto soda-lime or soda-asbestos. The fire-damp interferometer is therefore fitted with two absorption cartridges. For the determination of the carbon dioxide content the methane content is determined first in a sample from which the carbon dioxide has been absorbed and then the measurement is repeated on a sample which still contains the carbon dioxide. As the refractive index of carbon dioxide is very close to that of methane, one measures in the second case the sum of the methane and carbon dioxide contents. The difference in the two values gives the carbon dioxide content alone. In some mines other interfering gases occur in addition to methane and carbon dioxide, which may have an effect on the explosion characteristics of the mixture and on the results of the measurements [175]. The most dangerous of these is hydrogen. As the refractive index of hydrogen is less than that of air, the presence of hydrogen results in a lower reading with the interferometer, so that in mines which always or even sometimes contain hydrogen the use of interferometers is not reliable. In such cases the air in the mine must be controlled by other methods.

Somewhat less dangerous is the presence of higher hydrocarbons such as ethane. Indeed, for ethane the lower explosion limit is less than that for methane, but its refractive index is very considerably higher and so the interferometer is capable of indicating amounts below 0.1 vol.%. The indication on the interferometer is not specific, but represents a summation of the effects of all the components, and the presence of higher hydrocarbons such as ethane will cause a larger methane content to be indicated than is in fact present. Should larger amounts of these gases be present one must fall back on accurate laboratory methods of analysis in order to identify the gases and to specify their relative concentrations.

The invaluable assistance of the interferometric method to the problem of the analysis of mine gases, due to its speed, simplicity and reliability, is clear, while on the other hand the limitations of the

486

procedure, curbed by the additional demands that it be portable and explosion-proof, are recognized.

(C) TOXIC GASES AND VAPOURS

A number of gases and vapours are injurious to man when inhaled and under unfavourable conditions may even be lethal. The toxicity of the individual gases varies considerably one from another and in each case is dependent on the concentration. As such gases may be encountered in production processes in the chemical industry and also in the laboratory, they have received particularly close attention [176—178].

Maximum permitted levels for the concentrations of gases and vapours have been laid down in all industrial nations. Risk or even injury to the health of those working in such an environment is to be reckoned with if these levels are exceeded. These Threshold Limit Values (German: Maximale Arbeitsplatz-Konzentrationen = MAK values) have been found empirically, partly from direct experience and partly from experiments with animals, and they are being continually revised. A few aspects of the problem of establishing these values will be mentioned here.

The concentration of an injurious gas which a man can inhale in safety during an eight-hour working period varies from one individual to another, and also depends on the state of health of the person concerned. These subjective factors must necessarily be ignored when establishing a general value. The most important factor relating to the toxic effect is the amount of the noxious gas taken into the body, in other words, the volume of air inhaled. This in turn depends on the nature of the work. When considerable physical effort is to be exerted, lower T-L values must be adopted than would normally be the case. One must also consider that the injurious gases are often not present for a long period of time, but during recurrent shorter intervals. The duration of the exposure to the gas and of the intervals between exposure are important. A T-L value which is given for continuous inhalation over a long period can generally be exceeded without danger when the exposure is only for a short period. For this reason permitted peak concentration values (German: Zulässige Spitzen-Konzentrationen = ZSK values) for specified lengths of exposure are often quoted in addition to the T-L values.

For some gases the T-L values lie below the limit of detection of the interferometric method so that one must resort to other methods of determining these concentrations. The accompanying Table 13 includes only those gases and vapours for which the T-L value is more than ten times the limit of detection of an interferometer with a one-meter cell. For comparison the American Threshold Limit values (1961) of the American Conference of Governmental Industrial Hygienists are quoted as well as the G.D.R. standards TGL 22310 from 1968. While the values adopted in the G.F.R. [179] are largely comparable with the American values, the T-L values decided on in Russia are almost without exception lower [180].

The T-L values apply not only to all kinds of work places, enclosed rooms, factory buildings and spraying chambers, but also in the open air, in sewage works, mines, water works, and so on [181]. In addition, checks on T-L concentrations must also be made in rooms and plants in which the composition of the air must be controlled routinely for the sake of safety from explosions. Table 13 also includes permitted peak concentrations (ZSK values) from the G.D.R. standard applicable to short working periods in the presence of toxic gases.

(D) BIOLOGICAL AND CLINICAL GASES

Interferometric analysis has been applied in the field of biology and agriculture to studies on the respiration of several gases, mainly carbon dioxide, oxygen, nitrogen, ammonia and hydrogen sulphide by plants and animals. The measurement of assimilation by plant-like cultures is carried out in closed rooms and concerns the determination of the carbon dioxide content [182]. Gas samples from store rooms and cold stores for fruit have also been analysed, as described by Griffiths and Davis [183].

The interferometric measurement of the carbon dioxide content of the air in animal stalls has proved itself as a useful rapid method in the field of animal husbandry [184]. It enables one to draw conclusions concerning the effectiveness of the existing fresh-air ventilation in the animals' quarters and also concerning other parameters of interest for animal hygiene. Basically the same problem is encountered when designing ventilation systems for communal living quarters, for example in submarines.

Interferometric analysis has found extensive and varied application in medicine [185]. In this connection must be mentioned the

488

TABLE 13

Threshold limit values for some gases and vapours [179]

Formula	Name	Interferom. detection limit (cm³ m⁻³)	MAK values Threshold limits cm³ m⁻³	mg m⁻³	TGL mg m⁻³	ZSK values mg m⁻³	Maximum exposure (min)
CO_2	Carbon dioxide	120	5000	9000	9000	36000	30
SF_6	Sulphur hexafluoride	38	1000	6000			60
$CHCl_3$	Chloroform	16	50	240	200	500	30
CH_2Cl_2	Dichloromethane	21	500	1750	500	2000	30
$CCl_2=CCl_2$	Tetrachloroethylene	9	100	670	500	1500	30
$CCl_2=CHCl$	Trichloroethylene	12	50	520	250	1000	30
$CHCl=CHCl$	1,2-Dichloroethylene	15	200	790	250	500	30
C_2H_5Cl	Ethyl chloride	23	1000	2600	2000	5000	30
C_2H_5Br	Ethyl bromide	19	200	890	500	1000	30
C_2H_5OH	Ethanol	32	1000	1900	1000	4000	60
$CH_3 \cdot C \equiv CH$	Methylacetylene	17	1000	1650			
$CH_3 \cdot CO \cdot CH_3$	Acetone	23	1000	2400	1000	2000	30
$CH_3CO_2 \cdot CH_3$	Methyl acetate	20	200	610	200	800	30
C_3H_7OH	Propanol	21	400	980			

TABLE 13 (continued)

| Formula | Name | Interferom. detection limit ($cm^3\ m^{-3}$) | MAK values Threshold limits | | TGL $mg\ m^{-3}$ | ZSK values $mg\ m^{-3}$ | Maximum exposure (min) |
			$cm^3\ m^{-3}$	$mg\ m^{-3}$			
$CH_3CO_2 \cdot C_2H_5$	Ethyl acetate	14	400	1400	500	2000	30
C_4H_9OH	Butanol	16	100	300	200	400	30
$(C_2H_5)_2O$	Diethyl ether	15	400	1200	500	2000	30
$CH_3CO_2 \cdot C_3H_7$	Propyl acetate	12	200	840	400	1000	30
C_5H_{12}	Pentane	12	1000	2950			
C_6H_{12}	Cyclohexane	12	300	1050			
$CH_3CO_2 \cdot C_4H_9$	Butyl acetate	10	200	950	400	1000	30
C_6H_{14}	Hexane	11	500	1800			
$C_6H_5CH_3$	Toluene	11	200	750	200	800	30
$C_7H_{12}O$	Methylcyclohexanone	10	100	460			
$CH_3CO_2 \cdot C_5H_{11}$	Amyl acetate	9	100	525	200	1000	30
C_7H_{16}	Heptane	9	500	2000			
$C_6H_4(CH_3)_2$	Xylene	10	200	870	200	800	30
$C_{10}H_{12}$	Tetrahydronaphthalene	7			100	300	30
$C_{10}H_{18}$	Decalin	7			100	400	30

interferometric method for the determination of the conversion rate by measuring the in- and exhaled gases, developed by Wollschitt and co-workers [186]. This particular example will be gone into in more detail as the procedure and the interpretation of the results may be considered typical for three-component analyses, in this case for oxygen, nitrogen and carbon dioxide. The energy conversion in the body is proportional to the quantity of oxygen used up. This in turn is determined from the volume of air inhaled and from the difference between the oxygen contents of the in- and exhaled air. The next step is to determine the respiratory quotient, RQ, which is defined as the ratio of carbon dioxide produced to oxygen consumed in the breath. The basic procedure described in Sect. 5.D(2) is followed, with carbon dioxide being absorbed from the exhaled air and the resulting mixture being compared against inhaled air as reference standard to give the decrease in the oxygen content. In the second step the carbon dioxide-containing exhaled air is measured against the carbon dioxide-free exhaled air. After correction based on a factor determined in the first step, this second step gives the production of carbon dioxide. From the two values one can easily calculate the respiratory quotient. A nomogram constructed by Eckoldt [187] simplifies the interpretation of the data very considerably and renders almost all calculation unnecessary. The procedure has been used not only in clinical research [188—190] but also in veterinary studies [191,192].

A high sensitivity interferometer has been used in conjunction with a spirograph in further development of gas analytical methods for testing the functioning of the lungs. While the spirograph supplies measured volumes of oxygen to the lungs, the use of the interferometer allows the oxygen—nitrogen mixture to be measured continuously. Carbon dioxide is quantitatively absorbed. From the results one can arrive at the residual volume of the lungs and also at other important conclusions concerning their functioning [193—197].

Finally, the use of the interferometer for the measurement of anaesthetic gases should be cited [198—201]. Vaporizers for anaesthetic purposes are calibrated and checked almost exclusively by interferometric methods. Halothane vaporizers need to receive particular attention. Halothane, $CF_3CHBrCl$, has been widely used in recent years as a very powerful anaesthetic, as it offers a number of important advantages compared to other anaesthetic gases and hence considerably simplifies the administration of the anaesthetic. Nevertheless, halothane is toxic in too high a dose, and the concentration administered must be controlled with very high precision. This applies first and foremost to the concentration indicated by the vaporizer [202—207] but also for the complete anaesthetic circulation system [207,208]. A small anaesthetics interferometer (the halanometer) has been introduced alongside the laboratory interferometer for the routine checking and control of vaporizers, and is calibrated directly in vol.% of halothane in oxygen or oxygen—nitrous oxide mixtures [209]. This instrument is also employed in the operating theatre itself for monitoring the halothane concentration in gas mixtures for anaesthesia. Measurements of carbon dioxide levels during narcosis have been made by Mau [210].

(E) SPECIAL INVESTIGATIONS ON GASES IN RESEARCH

On account of the importance attached to the refractive index as a specific physical constant of pure materials, its measurement constitutes an important application of interferometric techniques. The most important results are summarized in Table 11 (Sect. 5.D(1)). As individual gases and vapours become of increasing interest for particular investigations, the data from these measurements are being continually added to. Anaesthetic gases (halothane and penthrane) were mentioned in the preceding section. Horvath carried out measurements on Freon 12 CF_2Cl_2, as part of his investigation of the applicability of the Rayleigh Dispersion Function for light in Freon 12 as compared to air [145]. Green and Robinson used the rapid interferometric method for the analysis of germane—hydrogen mixtures [211]. Interferometric measurements of the refractive index of the surrounding air have arisen as a result of precise measurements of length using etalons [212].

Ashton et al. [213] have utilized the interferometric measurement of refractive index for the determination of virial coefficients. According to the kinetic theory of gases the van der Waals theory can be applied to real gases

$$(p + a/V^2)(V - b) = RT \tag{67}$$

where p, V and T are the pressure, volume and absolute temperature of the gas, respectively, $R = 8.314$ J K^{-1} mol^{-1}, the universal gas constant, and a and b are constants for the particular gas in question. Making permitted approximations for small values of p one can set

492

$V = RT/p$ in the equation and obtain

$$pV = RT + Bp \qquad (68)$$

B is called the second virial coefficient and is a physical constant for the gas in question. A related physical constant is the electrical polarizability, γ. If one substitutes in eqn. (32) for $M/\rho = V$ the above expression $V = B + RT/p$, then

$$B = \frac{4\pi N\gamma}{3[(n^2 - 1)/(n^2 + 2)]} - \frac{RT}{p}$$

The reference cell is evacuated for this measurement, and the interference bands are counted as the pressure in the sample cell is changed. From measurements of refractive index and pressure at various temperatures the second virial coefficient B has been determined for ethylene and neopentane.

Efforts have been made to increase the accuracy of the technique. Mention should be made of the experiments with triple-slit interferometers, which may be based on the Rayleigh type [214—216], on the Jamin type [217,218] or on an arrangement similar to that used in the Mach—Zehnder interferometer. Interference between three beams of light of known relative intensities results, according to a relationship analogous to eqn. (13), in the formation of interference bands of particularly high contrast, which makes possible a higher accuracy than can be obtained with double-beam instruments.

Another method aimed at achieving higher accuracy has been developed by Ingelstam [219]. He utilized diffraction interference based on the principle of phase-contrast. The cell was similar to that used in the triple-beam interferometer, with a centre compartment for the sample flanked on both sides by compartments for the reference substance. By using photomultipliers as detectors, precision of the order of 0.0004 λ was achieved [220]. This meets the requirements for highest precision determinations of refractive indices for gases and liquids [45].

Brief mention will be made here of the many arrangements which have been described for automatic measurement on continuously flowing samples. Various photographic methods have been used to record changes in refractive index on photographic plates [221,222]. More recently, photo-electric methods have been applied [223—226] and interferometers have also been fitted with digital read-outs [227].

An interesting spectro-interferometer for measuring the variation in dispersion with wavelength has been suggested by Roshdestvenski [228]. It is based on the concept of recording the phase-shift of the light caused by changes in refractive index, directly as a function of wavelength. The interference pattern resulting from the illumination of a double-beam interferometer with two-compartment cell, falls on the entrance slit of a spectrograph in such a way that the interference fringes are at right angles to the slit. The spectrogram is then broken up by linear interference fringes which run at right angles to the direction of the spectral lines when there is no difference in the dispersion of the sample and the reference substance. When there is a difference in the dispersions then the interference fringes are curved. In the case of anomalous dispersion, when of course there is absorption of light in that region of the spectrum, there occur strongly marked discontinuities in the interference fringes which are called hooks, and which may be easily interpreted. While Roshdestvenski used a Jamin interferometer for this procedure, Starcev [216] employed a Rayleigh triple-slit instrument for an investigation of the caesium vapour doublet at 455.5 and 459.3 nm. Gorbanj and Sislovski [229] combined an ordinary double-beam Rayleigh interferometer with a spectrograph and recorded directly the dispersion curve of an aqueous iodo-eosine solution at a concentration of $2.5 \cdot 10^{-5}$ g ml^{-1}. Shukhtin [230] used Roshdestvenski's method to determine the gas-density and the free-electron concentration during an electrical discharge in a 1-m tube filled with argon which served at the same time as one of the two cells for the double-beam interferometer. Legay extended this method for the measurement of dispersion in the IR [231]. Tskhai and Mandelstam [232] investigated the variation of dispersion and hence of refractive index within sodium and strontium flames in the region of the absorption lines, with a combination of a Mach—Zehnder interferometer and a spectrograph.

(F) WATER, HEAVY WATER AND AQUEOUS SOLUTIONS

Refractive indices of liquids are usually measured in technical applications with refractometers based on the principle of measuring the critical angle, with which one may obtain an accuracy in the fourth or fifth decimal place. When higher accuracy is demanded the interference method may be used. A popular application is to investigations on water in many fields of technology and research.

494

The Löwe liquid interferometer was soon used to monitor drinking water and effluent, and yielded valuable information about the purity of the water, the effectiveness of the filters and so on, to the water engineers. Fresh water in rivers and also the salt content in the oceans and in the ocean currents can be measured interferometrically. Particular attention has been devoted in balneological investigations to the determination of salts in mineral-spring waters. Interferometric measurements on water have also been used in agricultural chemistry. Only the comprehensive review by Berl and Ranis [233] and the monograph by Löwe [128], which survey the extensive early literature, will be cited here. Zieglgänsberger [234] and also Beitner [235] have reported the determination of residual saccharose concentrations in boiler feed-water.

All of these investigations dealt with very low concentrations of dissolved substances. The ideal reference standard is always pure water, doubly distilled if need be. Water possesses all the necessary properties of a reference standard and makes possible reliable and quick measurements. Of course the interferometric method does not make it possible to identify qualitatively the salts present, and this is a limitation of the method.

The increasing importance of isotopes in research has lead since the 1930's to many interferometric determinations of the refractive index of heavy water and, based on these measurements, to the determination of heavy water in ordinary water. Dychno [236] has reviewed this field. The refractive indices are, for H_2O $n_e^{20} = 1.334\ 466$ and for D_2O $n_e^{20} = 1.329\ 64$. The relatively large difference makes it possible to detect very small amounts of D_2O in water. The techniques for this application have been considerably refined in recent years, for example by Ingelstam et al. [237] who used phase-contrast methods. Blaga et al. [238] and Mercea [239] used double-beam interferometers. Since the introduction of the interferometric technique for liquid samples a very wide range of aqueous solutions has been investigated. Calibration curves or tables of refractive index versus concentration data have been published in standard reference works and in individual publications. The dependence on temperature has also been considered [240]. For the measurement of sparingly soluble substances liquid cells of 500 mm length have been built, with which an accuracy lying in the eighth decimal place of the refractive index, corresponding to 1 mg l^{-1}, may be achieved.

Extensive measurements have been made on mixtures of water

with other liquids. For some of these mixtures the calibration curve of refractive index as a function of concentration exhibits a maximum (see Sect. 5.D(3)). For methanol—water mixtures this maximum corresponds to a concentration of 46.8 g/100 ml, whilst for other mixtures such as ethanol, acetone, acetaldehyde, acetic acid, sulphuric acid, nitric acid and so on, the maximum generally lies above 60 g/100 ml.

Further applications include the measurement of concentrations in aqueous solutions in industrial control laboratories and also in the investigation of chemical reactions, electrolytic processes, osmotic effects, absorption phenomena and also exchange effects. As these processes play an important role in metabolism in nature, interferometric measurements have found application in the fields of biology, biochemistry, agriculture and agricultural chemistry, plant breeding, and foodstuffs research.

(G) OTHER LIQUIDS

While interferometric methods have been widely used for measurements on water and aqueous solutions, their application to non-aqueous solutions and mixtures has been more restricted. Reasons are quite obvious: as reference substance a pure solvent is required; the vapour pressure of the reference liquid should not be too high and the temperature coefficient should be low. Water meets these requirements much better than do organic solvents. Nevertheless organic solutions can be handled by interferometry.

Chemical reactions between organic substances were followed interferometrically at an early date. The application in the field of fatty-acid chemistry to studies on liquid cocoa butter has enabled criteria to be established for discriminating between different types. Both purity control and concentration measurements on organic liquids are important applications [242]. As organic solvents are generally quite volatile, gas-tight cells must be used [243]. Interferometric analysis has been used as detector both in gas and in liquid chromatography [244].

There are a number of possibilities arising from aspects of the technique which apply to liquids but not gases. Thus in liquids titrations with various end-point indicators can be followed interferometrically to advantage. Titrations with interferometric refractive index measurement in aqueous solutions and mixtures were first described by Berl and Ranis [245]. This method of end-point detection has been applied more recently by Marti and Aliod [246] to the titration of very dilute chloride—iodide solutions with silver nitrate as titrant. Vajgand et al. [247—250] have used the method with considerable success in non-aqueous titrations, particularly in glacial acetic acid.

Interferometric measurements are often made on biological and physiological fluids. The refractometric or interferometric test of the serum—protein fraction is often carried out in connection with an electrophoretic serum—protein determination [251,252]. In addition, measurements on specially prepared serum offer the possibility of a test reaction for the presence of cancer cells [253]. The studies by Hirsch and many others on antigen-antibody reactions for diagnostic purposes are not followed up at the present time as they have attracted not only support but also strong criticism. They are only mentioned here as being of interest from the point of view of technique. On the other hand the interferometric determination of blood-alcohol levels has real significance as a precise control method [254—256]. As the procedure is of general interest it will be briefly summarized here according to refs. 257 and 258. The customary clarification of the serum by centrifugation is not used since the refractive index is dependent on a number of factors. Instead the alcohol is separated from the blood and obtained as a clear aqueous solution by adding 1 g of blood to 10 ml of water and distilling from this a sufficiently large amount (4 ml) as to contain with certainty all the alcohol. In the second step of the procedure this clear distillate is measured against distilled water as reference liquid in the interferometer. The percent blood-alcohol content is read off the calibration curve. The agreement with results obtained by Widmark's chemical method is good. When on occasion changes in the distillate, and therefore deviating values, are observed with the interferometer, due to non-alcoholic components of the blood such as acetone, which may also apply to other methods for testing blood, the interferometric measurement still remains a rapid method for checking the values obtained by another method.

The development of polymer chemistry has lead to an increasing use of high-precision interferometry. Refractive index n and its increment, i.e. its derivative with concentration dn/dc have been measured as characteristics of polymer solutions [259—261]. Kalz et al. [262] investigated the effect of the chlorine content on the solution properties of poly(vinyl chloride) and found there to be a definite relationship with the dn/dc data.

Side by side with the practical analytical applications which are so numerous that they cannot all be mentioned here, continue studies of the basic principles of interferometric refractive index measurements which are indispensible for further research. To these belong the measurement of refractive index of new substances, solutions and mixtures [263], and the study of the behaviour of the refractive index under unusual temperatures and pressures. Waxler and Weir [143] determined interferometrically the changes in the refractive index of benzene, carbon tetrachloride and water at pressures up to 1100 bars. Belonogov and Gorbunkov [264] have measured the refractive index of liquid hydrogen in the temperature range 20—31 K and at pressures between 1 and 9 atm, using a Rayleigh interferometer.

The dependence of the refractive index on the wavelength, i.e. the dispersion, has been measured for numerous substances by specially developed interferometric methods. Geffken and Kruis [265] investigated the dispersion of water in the spectral region 200 nm to 1 m. The combination of a Rayleigh interferometer with a spectrograph which they used was later further developed by Roshdestvenski [228] (see Sect. 6.E). The method has been selected by other authors [266—268] for the measurement of anomalous dispersion in organic compounds and in coloured solutions. Römer [269] measured the refractive index of absorbing samples with a modified Jamin arrangement. By using polarization effects he was able to match the intensities of the two coherent light beams and thus to obtain sufficient contrast in the interference pattern to make accurate measurements possible. The method was used in particular for measurements on colloidal gold solutions. Alperovič [270] also used a double-beam interferometer to measure the dispersion of solutions of fluorescent dyes such as Rhodamine, which is of importance in the field of non-linear optical effects.

(H) OPTICAL GLASSES AND THIN LAYERS

The refractive index is a characteristic property not only of liquids and gases but also of transparent solids. Solid substances may be identified by their refractive index, in many cases unambiguously. The refractive index of a transparent solid is generally measured by the method of deflection using a prismatic sample, or by the critical-angle method. However, when a particularly high sensitivity is demanded, interferometric measurements may be useful [271—274]. A method developed by Obreimov [48,275] depending on the principle of Fresnel diffraction enables even small pieces of glass to be measured by immersion. Used in conjunction with a monochromator this method allows the refractive index to be recorded as a function of wavelength.

In addition to the variation of refractive index with wavelength, the variation with temperature and pressure has also been studied. Bååk [276] has measured the temperature coefficients of 34 optical glasses at wavelengths λ_e, λ_g and at the laser wavelength 633 nm, over the temperature range 25 to 75°C, and for some glasses up to 125°C. Samples were prepared in the form of small-angle wedges about 6 mm thick and the Fizeau interference patterns resulting from the wave-fronts reflected from the front and back surfaces were measured. Similar investigations were carried out by Parker and Popov over the range —20 to +80°C [277]. Green [278] has measured the temperature coefficients of glasses over the very wide temperature range —194 to +250°C with a polarizing interferometer. A simple method based on a shearing interferometer with polarizers is described by Andréasson and co-workers [279].

Strain due to pressure or tension acting on one side of a piece of glass causes stress birefringence which is usually investigated by polarizing optical procedures. Schuster and Reitmayer [280] have measured changes in refractive index under such conditions using a Michelson interferometer. The effect on the refractive index of glasses and crystals when a high pressure acts uniformly on all sides, has also been investigated interferometrically [281,282]. Interferometric measurements on crystals are usually carried out under the microscope and will not be discussed here, but Rayleigh and Jamin interferometers have been used for crystal analysis [283—285].

The measurement of the refractive index of a thin layer is always associated with the measurement of the thickness. Both measurements may be made interferometrically. Mönch et al., who discussed this topic at length [129,286] prefers a Kösters arrangement (Fig. 18) but in principle any double-beam interferometer or an interference microscope can be used [132,287]. The method applies to any non-absorbing layer not showing multiple reflection between the boundary surfaces of the layer. If the layer is on a good, even surface, for example on part of one surface of a plane parallel glass plate, then this surface is coated with a reflective coating. At the edge of the layer the coating shows a step with height equal to the thickness of the layer, d. This step is measured in a Kösters interferometer as a fringe shift $h_1\lambda$, which gives a thickness $d = \frac{1}{2}h_1\lambda$ on account of the double light path (Fig. 52). If the glass plate is reversed and replaced in the light beam, then the path difference between the interfering rays is given by $2dn = h_2\lambda$. From this it follows that

$$n = h_2/h_1 \tag{69}$$

The refractive index can also be measured without going through the mirror-coating procedure by bringing the transparent layer into one path of the interferometer while the other remains free [288,289]. When an autocollimating arrangement is used then the relationship $2d(n-1) = h_1\lambda$ applies. The thickness of the layer, however, must be known or be determined separately. With $d = \frac{1}{2}h_1\lambda$ the refractive index is then calculated from

$$n = 1 + h_2/h_1 \tag{70}$$

Other procedures for measuring thicknesses are described by Menzel [290].

If it is not possible to prepare a metallic reflective coating, the refractive index of the layer may be determined by two successive measurements in transmitted light at different angles of incidence to the layer. When the layer is at right angles to the incident light beam, as in the case described above, then $d(n-1) = h_1\lambda$. When the layer is rotated through an angle δ there is a change in the optical path

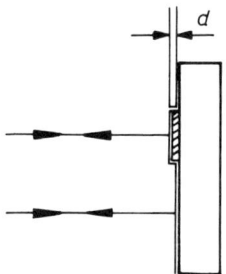

Fig. 52. Representation of a refractive index measurement of a thin layer.

length given by

$$\Delta s = d(\sqrt{n^2 - \sin^2 \delta} - \cos \delta - n + 1) = h\lambda$$

according to eqn. (43), with $\alpha = 0$. The fringe-shift h in this expression is readily obtained by observing and counting the interference fringes which pass the eye-piece as the layer is rotated through the angle δ. The phase difference by passage through the rotated layer relative to passage through air is $(h_1 + h)\lambda = h_2\lambda$. It follows that

$$d(\sqrt{n^2 - \sin^2 \delta} - \cos \delta) = h_2\lambda$$

On rearrangement with

$$d = h_1\lambda/(n-1)$$

this gives

$$n = \frac{h_1^2 - 2h_1 h_2 \cos \delta + h_2{}^2}{h_2{}^2 - h_1{}^2} \tag{71}$$

Measurements of this type have been made on lamellae, thin films, evaporated layers and in particular, dielectric layers [291]. Using this technique Shumate [272] measured thin plates of glass only a few tenths of a millimeter in thickness, and also barium titanate whose refractive index of 2.42 is so high that it cannot be measured by refractometric techniques.

Fleischmann and Schopper [292,293] have suggested a possible approach to making measurements on absorbing layers. The markedly different intensities of the two interfering rays after passage through the layer seriously reduce the contrast of the interference pattern. By inserting and rotating an additional optical polarizing filter the intensities of the two rays can easily be balanced. This arrangement was first developed for use with double-beam interferometers of the Rayleigh type and has since been used to increase the measurement accuracy with triple-beam interferometers. Triple-beam interferometry with a Jamin arrangement has been used by Hariharan and Sen [294].

The special problems associated with thin evaporated layers have been dealt with by Rouard and Bousquet [295], both for transparent and for absorbing layers. One observes a considerable difference in the refractive indices measured for thin layers when compared with those for compact materials. Depending on a number of external parameters, the refractive index measured for a thin layer is subject to wide variation.

7. Interferometric analysis of inhomogeneous substances

Interferometry has found many and varied uses for the investigation of inhomogeneous distribution of refractive index in gaseous, liquid and solid substances. This review will only deal with such procedures and applications as are relevant to chemical problems. The interesting fields of aero- and hydrodynamics, which have provided much of the incentive to further development of interferometric techniques, and many other applications such as the investigation of sonic and ultrasonic effects, shock waves, thermal conductivity and thermal convection in gases and liquids, and so on, must remain undiscussed.

The optical investigation of inhomogeneous materials serves the chemist primarily as an aid to the illustration of structure and to the analysis of reactions. Interferometric procedures offer the particular advantage that they record the distribution of refractive index in the object or, more accurately, portray the deformation of a linear wavefront on passing through the object, directly as an interference pattern without the necessity of touching or even disturbing the object itself. As the interpretation of the interference patterns is of critical significance if reliable results are to be obtained, the problem of evaluation will receive particular attention in the following section.

(A) METHODS OF INTERFEROMETRIC MEASUREMENT ON STRATIFIED SUBSTANCES

The simplest case of inhomogeneous distribution of refractive index is that in which the refractive index gradient in the object possesses the same direction at all points. Thus in any given layer at right angles to the gradient the refractive index is uniform. In the applications of interest here, the refractive index gradient runs either vertically or radially from a rotational axis. These limitations make possible particularly effective and easily interpreted optical observation and recording systems such as were described in Sect. 4. In addition to interferometric procedures, a number of different methods based on the deflection of light have been developed for these purposes and to some extent are employed together with the interference techniques, such as, for example, Lamm's scale method, the schlieren scanning method of Longsworth and the crossed-slit method of Philpot and Svensson. These procedures give directly only the gradient curve, that is, the magnitude of the gradient along a straight line in the direction of the gradient. In the last two cases mentioned, the accuracy is limited due to diffraction effects; the Lamm's scale method involves tedious calculations. This is in contrast with the interferometric method which not only makes possible a much higher accuracy, but also enables the refractive index distribution and the gradient curve to be recorded directly. The simultaneous recording of both of these curves brings considerable advantages for an accurate evaluation, as, for example, the steep parts of the refractive index curve which are difficult to measure appear as peaks in the gradient and are easily measured.

The most important applications to the studies of layered liquids are to be found in investigations on electrophoresis, diffusion and sedimentation, and are dealt with in Sects. 7.B—7.D. The basically very similar optical arrangements for these interferometric analytical methods are described in Sect. 4.C. The evaluation of the curves in each case is also similar and will be discussed here.

The co-ordinate representing the direction of the gradient will be identified by x. The spatial distribution of the refractive index n is then only dependent on x. An example of a refractive index varying in the direction x and the corresponding gradient curve is shown in Fig. 53(a) and (b). The gradient curve dn/dx vs. x can only be obtained approximately from the curve $\Delta n/\Delta x$ vs. x. If the Δx intervals are chosen too large the curves tend to be flattened out (Fig. 53(c)) and the points of detail of the curve are lost. A good approximation to the true curve is realized only when small Δx intervals are used, but this results in small and rather difficult to measure Δn values. The optimum value for Δx must be determined empirically so as to obtain the highest accuracy in the evaluation of the results.

The Gaussian probability distribution plays an important part in the evaluation of the gradient curve since this can be represented by one or the sum of a number of Gaussian curves. The Gaussian distribution is given by

$$W(x) = (k/\sqrt{\pi}) \exp(-k^2 x^2) \tag{72}$$

where k is a constant. Figure 54 shows the curves for the various fac-

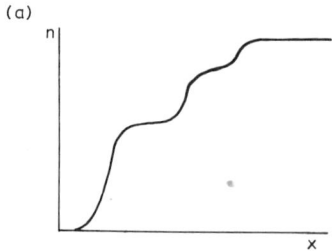

(a)

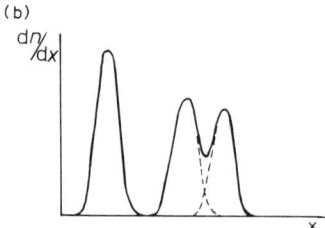

(b)

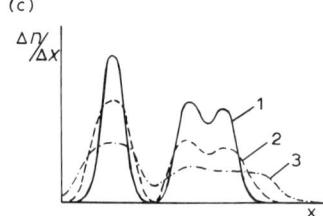

(c)

Fig. 53. The relationship between the curves for stratified media; (a) n vs. x curve; (b) the corresponding dn/dx vs. x curve; (c) the curve of $\Delta n/\Delta x$ against x for too large values of Δx.

tors $k = k_0$, $k = \frac{1}{2}k_0$, $k = \frac{1}{4}k_0$. The area cut by the x-axis from W always has the same magnitude independently of k, namely $\int_{-\infty}^{+\infty} W(x)\,dx = 1$. The measurement of this area is important for the evaluation. It is determined either with a planimeter or by cutting out and weighing. As an approximation, the area under the Gaussian curve may be taken to be that of the triangle defined by the peak and the width at half-height (Fig. 55). The width at half-height $2x_H$ is defined by

$$\tfrac{1}{2}W(0) = (k/\sqrt{\pi})\exp(-k^2 x_H^2)$$

i.e. by

$$2x_H = (2\sqrt{\ln 2})/k \tag{73}$$

504

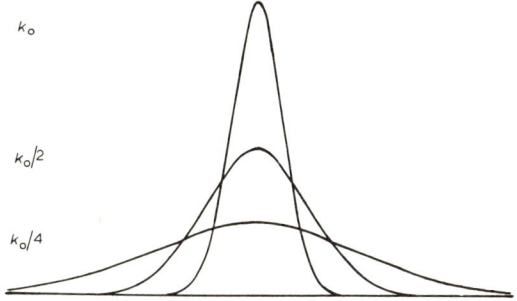

k_0

$k_0/2$

$k_0/4$

Fig. 54. The Gaussian curve for $k = k_0$, $k = \frac{1}{2}k_0$, $k = \frac{1}{4}k_0$.

The area of the triangle obtained in this way has a magnitude

$$F_H = W(0)\, 2x_H = 2\sqrt{(\ln 2)/\pi} = 0.939$$

This factor enables one to calculate the area under the Gaussian curve by multiplying the area of the triangle by 1.064. A more accurate approximation is obtained when one chooses the triangle whose sides are the tangents to the points of inflection on the Gaussian curve. The co-ordinates of these points are

$$x_w = \pm 1/\sqrt{2}\, k$$

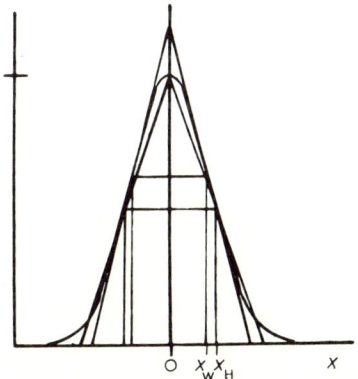

Fig. 55. Triangular approximations to the Gaussian curve.

and

$$W(x_w) = k/\sqrt{\pi e} = W(0)/\sqrt{e}$$
$$= 0.606 \ W(0)$$

From the equations of the tangents to the points of inflection

$$W = \pm\sqrt{2/\pi e} \ k^2(x \pm \sqrt{2}/k)$$

the area of the triangle is given by $F_w = 2\sqrt{2/\pi e} = 0.968$. The correction factor for obtaining the true area under the Gaussian curve is now only 1.033, but constructing the tangents is rather more difficult than measuring the width at half-height. The same factor applies when the area of the rectangle under the height of the inflection points is multiplied by 2

$$F_w = 2 \cdot 2x_w \cdot W(x_w)$$

When the gradient curve contains several overlapping peaks, e.g. the two on the right of Fig. 53(b), it may be resolved into individual Gaussian curves. In so doing one must take care that each Gaussian curve lies symmetrically about its maximum. The sides of the Gaussian curves are practically straight lines between two and nine tenths of their height. The point of intersection of two Gaussian curves lies at half the height of the gradient curve. In Figs. 53(b) and 59 the gradient curves are resolved into shaded Gaussian curves.

Small differences in refractive index give rise to only small shifts in the interference fringes which are difficult to see and to measure. The use of the Moiré technique has proved useful for increasing the sensitivity. Oster et al. [301] and van Oss [302], developing the idea of the Lamm scale, suggested placing a graticule with transparent strips at right angles to the direction of the gradient, in front of the stratified substance, and a second graticule behind the stratified substance at a small angle to the first. The superposition of the grids gives rise to Moiré fringes which exhibit a marked lateral shift when a gradient exists. However, the Moiré effect may also be demonstrated directly from the interference patterns by superposition of a linear fringe system running parallel to the base-line of the interference pattern [303]. This technique is very simple to put into practice at the same time as the measurements are made, by making a double exposure of the gradient curve (fringe spacing w) and the linear interference pattern (fringe spacing w'), or by placing a grid with fringe spacing w' in front of the photographic plate. One can also produce Moiré patterns during subsequent copying or enlarging of the gradient curve by superimposing a linear grid [304]. The Moiré shift Δw_M of the gradient curve is given by the shift of the gradient curve in the direct interference pattern multiplied by $\Delta w_M/\Delta w$. The magnification factor $\Delta w_M/\Delta w$ is arrived at from the smallest common factor $gw = g'w'$ of the fringe spacings, g and g' being whole numbers. Then $w - w' = w'(g' - g)/g$. When the moving fringe system shifts by an amount $w - w'$, the Moiré band moves further by an amount w' (Fig. 56). It follows that the magnification factor $\Delta w_M/\Delta w = w'/(w - w') = g/(g' - g)$. If g is chosen to be equal to $g + 1$, then $\Delta w_M/\Delta w = g$. A convenient practical ratio is $w : w' = 11 : 10$ which gives a magnification factor of 10. A good example of the effect of choosing differing Δx values and varying magnification of the shift by using the Moiré effect, is seen in Fig. 57. The direct interference gradient curve is seen on the edge of the figures. The upper figure, (a), recorded with ratio $w : w' = 9 : 8$, reveals the details of the gradient curve quite clearly, while the lower figure, (b), with a Δx value twice as large and $w : w' = 5 : 4$, records a gradient curve with the details not yet visible.

Measurements of refractive index gradients on stratified liquids have also been used for the determination of thermal diffusion in solutions. A slight separation of the components of the solution has been observed by using a Young—Rayleigh arrangement [305]. Kegeles and Sober [306] have converted a Rayleigh—Svensson arrangement into an automatic interferograph, by making the photographic plate movable in a vertical direction. As the liquid flows through the cell, which has a narrow cross-section, the shift of the bands is continuously recorded. This very sensitive detector has found use in chromatographic investigations.

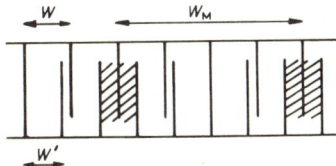

Fig. 56. Representation of the Moiré effect with two rasters.

References pp. 536—546

507

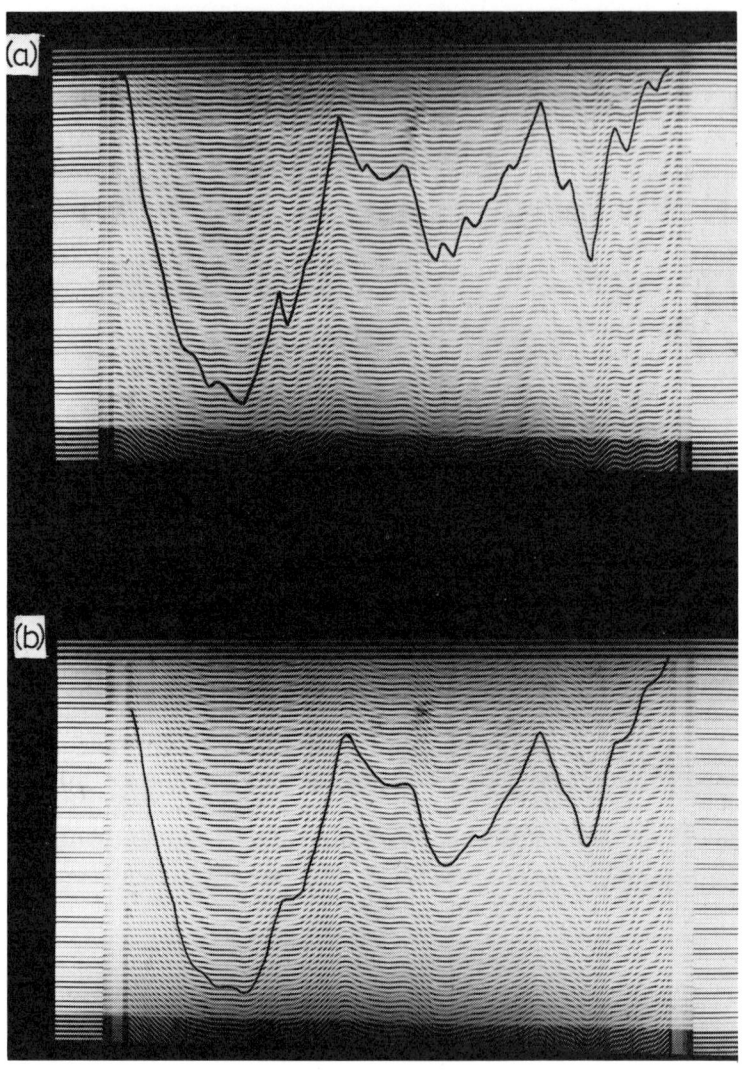

Fig. 57. Moire interference patterns from an object with layered structure, for different values of Δx and different Moiré ratios: (a) Δx = 0.6 mm, 8 : 9, and (b) Δx = 1.3 mm, 4 : 5. (The figure should be viewed from the right-hand side.)

(B) ELECTROPHORESIS

Electrophoresis is the transport of charged high molecular weight particles in an electric field. The velocity v of a particle is proportional to its electric charge q and the electric field E

$$v = kqE$$

Electrophoresis has become particularly useful for the identification of colloidal substances, particularly proteins which differ only in their rate of migration in an electric field. For colloids, however, the charge is not an invariable constant but is strongly dependent on the supporting medium: a colloid particle which might be positively charged in acid solution may change to being negatively charged in basic solution. At some intermediate pH value, the isoelectric point, it shows neither positive nor negative charge ($q = 0$) and does not migrate in an electric field. The determination of the isoelectric point as a specific characteristic of a substance is an important application of electrophoresis.

The migration of a particle with charge q takes place against a mechanical force p_m, which is determined by the velocity v, by the cross-section of the particle and by the viscosity η of the liquid. If we assume, as a very rough approximation, that the particle is spherical with a radius r, then this force is given by the Stokes equation

$$p_m = 6\pi r \eta v \tag{74}$$

The force acting on the particle as a result of the application of the electric field, $p_e = qE$, accelerates the particle until it reaches a velocity v such that the two forces working in opposition, p_m and p_e are equal. External factors remaining constant, the particle reaches a velocity $v = qE/(6\pi r \eta)$. In this equation the parameters q and r are specific for that substance. If in a mixture of colloids two or more substances are present with sufficiently different q/r ratios, they may be separated and identified by electrophoresis. This technique has become an indispensable tool in protein chemistry and for the identification of sera, as it makes it possible to separate the otherwise difficult to identify protein components of the blood serum in an electric field, and to distinguish them optically. Some comprehensive texts on electrophoresis are included in refs. 307—312.

The technique of free-electrophoresis, i.e. electrophoresis of freely movable solution layers, together with the corresponding refractive index and gradient curves, is schematically represented in Fig. 58. The solution under investigation, e.g. a protein solution, is filled into the U-tube until the tube is half full and then a pure buffer solution is poured on top. It is important for the attainment of unambiguous initial conditions that the two layers at the beginning are sharply defined. Of the various types of cells which have been developed, that with a removable upper half has proved most convenient. In this type the upper half of the U-tube is displaced laterally during the filling of both halves, and is only moved over the lower half when electrophoresis is to commence, thus producing a very sharp boundary.

In order to keep the migration velocity constant the electric field along the tube must also be kept constant; this demands the same electrical conductivity in both the solution under investigation and the buffer solution. In order to facilitate the necessary exchange of ions, the two liquids must be dialysed for several hours before electrophoresis. This takes place with a cellophane membrane between the two liquids through which the ions but not the protein particles can diffuse. The time required for dialysis depends on the pH of the buffer solution and on the amount of the substance under investigation, and must be determined experimentally for each application. For the analysis of serum a pH between 8 and 9 is used. The amount of sample required lies between several milliliters (macroelectrophoresis) and 0.1 ml (microelectrophoresis).

For recording the variation in refractive index by comparison with a reference substance provision is made for placing this reference between the two limbs of the U-tube. After the application of the electrical potential the fronts begin to move. Each front should remain as sharp as possible. Interferences due to thermal effects can be largely eliminated by allowing an initial thorough temperature equilibration. The temperature for the measurement should be that

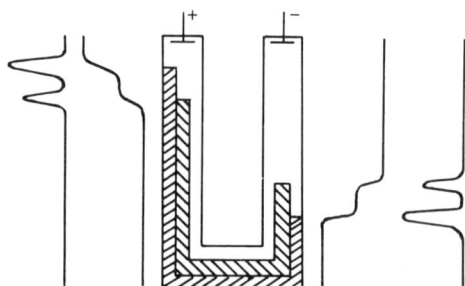

Fig. 58. Schematic separation by electrophoresis and the corresponding refractive index and gradient curves.

corresponding to maximum density of the solvent, i.e. for aqueous solutions +4°C. At this temperature the variation in density between the middle and the edge of the cell, caused by heating during the passage of electric current, is minimized.

As soon as the two halves of the cell are united diffusion starts and the sharpness of the boundary begins to diminish. For this reason, quite apart from saving time, one aims at the shortest possible time for the electrophoretic separation. This in turn demands the use of a relatively high voltage which, however, must at the same time not be too high, otherwise warming of the solution in the tube will give rise to an interference. Optimum separation times depend on the type of cell, the sample and the buffer solutions, and are either given by the manufacturers or must be determined experimentally. These and other possible sources of error together with the steps which should be taken to avoid them, are described in detail in the specialized literature.

When the separation of the individual proteins is sufficiently far advanced, a photographic exposure is taken. The majority of electrophoresis set-ups record the variation of the refractive index n/x interferometrically and at the same time the gradient curve dn/dx vs. x either by the Philpot—Svensson schlieren optical technique, or also interferometrically. In serum protein the following components may be distinguished in the sequence of their mobilities: albumin, α_1-, α_2-, β_1-, β_2-, φ- and γ-globulins. One may assume the same refractive index for each of these components and thus determine the individual concentration of each directly from the two curves. Figure 59 shows the electrophoresis record of a "normal" human serum. The interferometric refractive index curve was taken simultaneously with the gradient curve, obtained by the Philpot—Svensson technique with Wolter phase edges. The line drawn in afterwards is then resolved into Gaussian curves (Sect. 7.A). The area under each Gaussian curve is directly proportional to the concentration of the respective protein, and can be determined by using a planimeter or, approximately, by triangulation. Even simpler is to mark the intersection points of the Gaussian curves and to count the number of interference bands occurring between the intersection points. This number is, to a good approximation, proportional to the area under the Gaussian curve. These figures are then added together and normalized to 100%, from which one can find the relative percentage of each individual protein component.

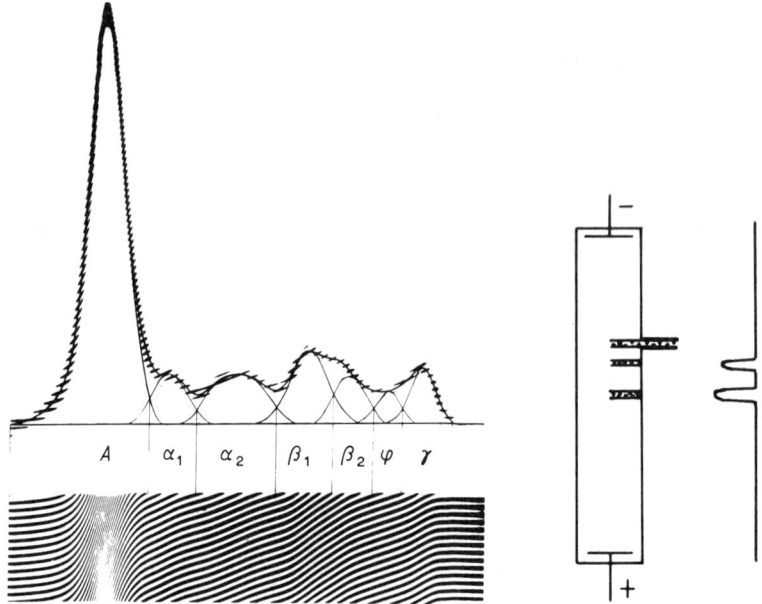

Fig. 59. Electropherogram of normal human serum: upper, Philpot–Svensson gradient curve with interference fringes; lower, interference pattern from the distribution of refractive index. The gradient curve is drawn in afterwards and resolved into the various Gaussian curves. The evaluation gave: A, 25.0 $\equiv$ 59.0%; α_1, 2.7 $\equiv$ 6.4%; α_2, 3.9 $\equiv$ 9.2%; β_1, 4.3 $\equiv$ 10.1%; β_2, 2.6 $\equiv$ 6.1%; φ, 1.4 $\equiv$ 3.3%; γ, 2.5 $\equiv$ 5.9%. Total = 42.4 bands $\equiv$ 100%.

Fig. 60. Diagram of zone electrophoresis.

The commercially available appartuses for electrophoresis make use of different experimental arrangements. The most popular of these is the Rayleigh–Svensson, which in modifications by Wiedemann [313], Svensson [19], and Nebe [79] can be used to record both the gradient curve and the refractive index curve simultaneously (see Fig. 38).

Antweiler [80] used a Jamin interferometer arrangement (Fig. 16), which is suitable for investigations with white light and thus presents the refractive index curve more clearly. Usually, however, the whole of the curve is not recorded at one time, neither is the record of the gradient curve provided. Lotmar [314] has described a Jamin arrangement with only one dividing plate, in which the sensitivity is

512

increased by passing the light through the electrophoresis cell twice (similar to Fig. 34). Wiedemann [308] has given a detailed comparison and evaluation of the individual interferometric arrangements. Curtain [315] has described an apparatus for simplifying the evaluation, with the help of a photoelectric comparator.

A variation of free-electrophoresis is zone electrophoresis, introduced by Bockemüller [316,317]. Here a rectangular cross-section electrophoresis cell is held in a vertical position, and filled with prepared Agar gel. A channel 0.5 mm high in the Agar gel filling receives the sample solution. When the electric field is applied the charged particles move with differing velocity in the Agar gel. Because the bands themselves are very thin the separation is complete in a short time. The variation of the refractive index in the cell is observed visually with the Rayleigh—Svensson method and recorded photographically. The role of the homogeneous reference substance is played by the same Agar gel filling, as the cell is of double width and the channel for the sample under investigation only reaches to the middle of the cell (Fig. 60). The two interfering light beams pass through the homogeneous left half and through the right half of the cell containing the zones. Since the front and rear boundaries of each zone follow one another with very short spaces between them, a variation of refractive index arises which looks very like the gradient curve for free electrophoresis in a U-tube. The methods of evaluation are therefore also similar. One should add here that the Agar filling with the zones cut into narrow bands, can be made the basis for an immuno-electrophoretic method.

(C) DIFFUSION

The tendency of particles in an inhomogeneous liquid or gas to move from regions of higher to those of lower concentration is termed diffusion. In many processes diffusion can cause a serious interference, as for example in the investigation of electrophoresis already mentioned. On the other hand diffusion can be a useful aid in the characterization of macromolecules.

For one-dimensional diffusion in a solution with the concentration c, Fick's first law states that the flow Q per unit time and area is proportional to the concentration gradient

$$Q = -D \frac{dc}{dx} \tag{75}$$

where the factor D is termed the diffusion constant. Combining eqn. (75) with the hydrodynamic continuity equation

$$\frac{dc}{dt} = -\frac{dQ}{dx}$$

one obtains Fick's second law

$$\frac{dc}{dt} = \frac{d}{dx}\left(D\frac{dc}{dx}\right)$$

which can be simplified in cases where one assumes a diffusion constant to remain constant,

$$\frac{dc}{dt} = D\frac{d^2c}{dx^2} \tag{76}$$

This differential equation has the general solution

$$c = k_1 + k_2 \int_0^{x/2\sqrt{Dt}} \exp(-y^2)\,dy \tag{77}$$

the constants k_1 and k_2 are obtained from the following limiting conditions:

(i) The diffusion is assumed to be unrestricted i.e. the diffusion process should not be limited by the walls of the cell, so that we can write, for $x = \infty$ $c = c_1$ and for $x = -\infty$ $c = c_2$ $(c_2 > c_1)$.

(ii) For the initially sharp boundary $x = 0$, i.e. at the time $t = 0$, $c = c_1$ for the region $x > 0$, and $c = c_2$ for the region $x < 0$. Then we can write the particular solution of eqn. (76) as

$$c = \tfrac{1}{2}(c_2 + c_1) - \frac{c_2 - c_1}{\sqrt{\pi}}\int_0^{x/2\sqrt{Dt}} \exp(-y^2)\,dy \tag{78}$$

For low concentrations the change in refractive index is proportional to the concentration, so that

$$n = n_1 + \tfrac{1}{2}(n_2 - n_1)\left(1 - \frac{2}{\sqrt{\pi}}\int_0^{x/2\sqrt{Dt}} \exp(-y^2)\,dy\right) \tag{79}$$

from which the gradient is given by

$$\frac{dn}{dx} = \frac{n_2 - n_1}{\sqrt{\pi}}\frac{1}{2\sqrt{Dt}}\exp(-x^2/4Dt) \tag{80}$$

514

(in this case the sign of the conventional representation of the curve is taken to be positive). Thus eqn. (80) is seen to have the form of the Gaussian curve as in eqn. (72).

The diffusion constant D is almost always determined optically by measuring the refractive index, and therefore in the last two formulae the concentration c was replaced by the refractive index n. Nowadays the more precise interferometric techniques are preferred over the older methods based on refraction, such as Lamm's scale method and the Philpot—Svensson method. The most important of these is the Rayleigh—Svensson method, which combines a number of advantages; high precision interferometric simultaneous recording of the refractive index and gradient curves, short exposure time because of the large aperture, and the possibility of increasing the precision by using Moiré methods. It is necessary to achieve the highest possible sensitivity because the diffusion constant D is dependent on the concentration. In order to determine it unambiguously it is taken as being the extrapolated value for zero concentration, obtained from measurements on very dilute solutions. Low concentrations imply small differences in refractive index which in turn require a high sensitivity for their measurement and evaluation [304]. Fig. 61 shows an example.

The procedure when using the Jamin arrangement for determination of diffusion is the same as that for the investigation of electrophoresis [318]. Even the Mach—Zehnder interferometer has been used to measure diffusion in volatile materials [319].

Several authors have more recently used Gouy's interference method as described by Longsworth [320—322]. A Toepler arrangement with horizontal entry slit (Fig. 35) is used, with the diffusion cell in the parallel light beam. When a concentration gradient exists, one sees in the plain of the slit image, instead of a sharp image of the slit, a number of horizontal interference bands running through a rectangular area (Fig. 62). The bands arise from interference of the light beam passing through the diffusion layer. The steeper the gradient is, the further these bands lie off the axis, and the narrower the diffusion layer is, the greater is the spacing between the bands. The Gouy method offers very high sensitivity using simple apparatus, but the interference pattern is indistinct and can only be evaluated with the help of a computer.

Gradient curves may be recorded by the shearing interference method as developed by Bryngdahl and Ljunggren [40,323,324].

Fig. 61. A Moiré diffusion curve from two interference photographs.

The optical arrangement is very stable, since no separate reference beam is needed (Fig. 63). The diffusion cell takes the place of the object under investigation in the parallel light beam, and its image is reduced in size. The wave-front shearing takes place in the plane of the image in polarized light, with the help of two divider plates similar to those of Savart. By this means high-contrast gradient curves may be recorded. Weinstein [325] succeeded in recording refractive

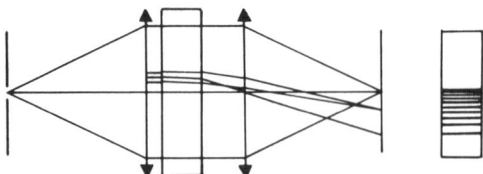

Fig. 62. Diagram of Gouy interference.

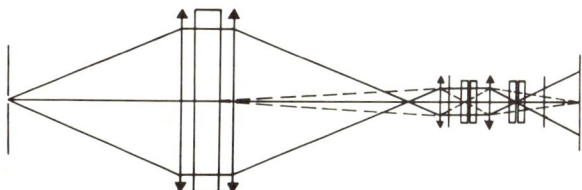

Fig. 63. Diagram of shearing interference.

index and gradient curves interferometrically by using an autocolli-
mated apparatus with a Wollaston prism.

The shearing method using only one Savart plate, as described by
Ingelstam [46], produces a picture with only horizontal linear interference bands in the region of the diffusion layer, which are not very
distinct and which are more difficult to evaluate. Nevertheless this
technique has been used to measure diffusion in dilute solutions of
electrolytes [326,327].

Many special types of cells have been developed with the aim of
facilitating the setting up of a sharp boundary between the two
liquids before the commencement of a diffusion experiment. The
shearing cell with parts which can be slid apart from each other, and
which is usually used for electrophoresis is also suitable for diffusion
measurements. Even better is the sliding-diaphragm cell in which the
denser liquid is first filled in up to the height of the intended boundary, then a thin metal diaphragm is slid across the top, and finally
the less dense liquid is filled in on top of the diaphragm. If the diaphragm is then carefully withdrawn a good sharp boundary is obtained. The best type of cell is considered to be that in which the
boundary may be further sharpened after the two liquids have been
filled in, e.g. by slowly removing the liquid at the height of the
boundary: it is important that liquid is removed from both layers
simultaneously by this technique.

For determining the diffusion constant D photographic exposures
are taken at intervals of the refractive index and/or of the gradient
curve. Equations (79) and (80) form the basis of the evaluation.
Many procedures have been worked out for this evaluation but only
the most important will be summarized here.

From eqn. (80) it follows at the point $x = 0$

$$D = \frac{(n_2 - n_1)^2}{4\pi t (dn/dx)^2} \tag{81}$$

If the gradient is then measured at the point $x = 0$ after a known diffusion time, D can be calculated directly. However it is difficult to determine the exact time for the commencement of diffusion, as it is in practice not possible to produce an infinitely sharp boundary as a starting condition for the diffusion process. The calculated time from the theoretical starting point is always larger than the measured time t^*. The timing error Δt is included in eqn. (81) as $t = t^* + \Delta t$ as an unknown. In order to eliminate this difference, the gradient is measured at various times and is plotted graphically according to

$$1 \Big/ \left(\frac{dn}{dx}\right)^2 = \frac{4\pi D}{(n_2 - n_1)^2} (t^* + \Delta t)$$

The straight line gives Δt by extrapolation to the ordinate 0, and D from the slope.

The values of the gradient dn/dx are found as follows from the refractive index curve (Fig. 64). The coordinates are x', y'; w' is the spacing between the fringes in the interference pattern to be evaluated and Γ is the magnification factor, such that $x' = \Gamma x$ and so on. Corresponding to a small change $\Delta x'$ there will be a change in the ordinate $\Delta y' = \Delta x' \tan \xi = \Delta h w'$ and also a change in refractive index according to eqn. (40) $\Delta n = \Delta h \lambda / d$ (d = the thickness of the diffusion layer in the direction of the light beam). From this it follows with

$$\Delta x = \frac{\Delta x'}{\Gamma} = \frac{\Delta y'}{\Gamma \tan \xi} = \frac{\Delta h w'}{\Gamma \tan \xi}$$

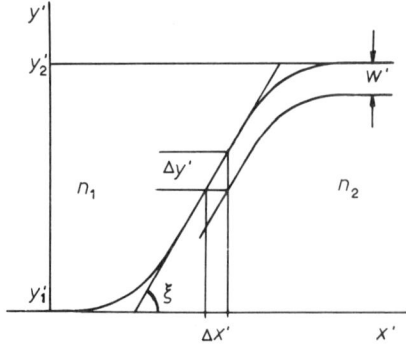

Fig. 64. Representation of the evaluation of the refractive index curve.

and

$$\frac{\Delta n}{\Delta x} = \frac{\lambda \Gamma}{dw'} \tan \xi \qquad \text{or} \qquad \frac{dn}{dx} = k \tan \xi \qquad (82)$$

the factor $k = \lambda \Gamma / dw'$ is a constant for given experimental conditions, so that the gradient can be determined directly from the angle ξ, i.e. from the direction of the tangent to the point of inflection.

Another commonly used method of evaluation starts with the gradient curve. The area F under the gradient curve is measured

$$F = \int_{-\infty}^{+\infty} \frac{dn}{dx} \, dx = n_2 - n_1$$

and in addition the height H, which corresponds to the maximum value of dn/dx at $x = 0$,

$$H = \frac{n_2 - n_1}{\sqrt{\pi}} \frac{1}{2\sqrt{Dt}} \qquad (83)$$

from this it follows that

$$D = \frac{F^2}{4\pi t H^2} \qquad (84)$$

It should be pointed out that the magnification Γ in the x-direction has a direct effect on the area F with the result that the area under the enlarged gradient curve F must be set equal to ΓF. Magnification in the direction of the second co-ordinate is of no consequence, as this effects both F and H equally. In this method as in the previous one, values of $1/H^2$ for several different times are plotted against t in order to overcome an uncertainty in the value for the true start time. D is determined from the slope $\tan \beta$ of the resulting straight line

$$\frac{1}{F^2} = \frac{4\pi \Gamma^2 Dt}{F'^2} \qquad \text{or} \qquad D = \frac{F'^2 \tan \beta}{4\pi \Gamma^2}$$

The area F' remains constant during the diffusion, and is obtained as a mean value from the areas determined with a planimeter, of the individual curves.

A very simple method is to substitute eqn. (83) into eqn. (80),

$$\frac{dn}{dx} = H \exp(-x^2/4Dt)$$

and to solve for D

$$D = x^2 / \left\{ 4t \ln \left(H \middle/ \frac{dn}{dx} \right) \right\}$$

If one measures the width at half height x_H, with ordinate $dn/dx = \frac{1}{2}H$, one obtains directly

$$D = x_H^2/4t \ln 2 = x_H^2/2.77\,t$$

Another very similar method uses the point of inflection of the Gaussian curve instead of the width at half height. Other procedures for evaluating the data are, e.g. the method of moments by Gralén [328], the common point method of Stokes [329] and the partial area method of Tscherkassow and Klenin [330].

All the degradations just discussed assume that diffusion is independent of the concentration and that the system is homogeneous. Only by making these assumptions and by working under constant external experimental conditions one can obtain gradient curves which correspond to ideal Gaussian curves. In fact D is dependent on the concentration [331] becoming greater for higher concentrations. In a diffusion experiment this is apparent in the asymmetry of the gradient curve. If the system is inhomogeneous, e.g. a polydisperse solution, the diffusion of the individual components proceeds with different velocities. This results in a skewed gradient curve which represents the envelope for several overlapping Gaussian curves.

The individual evaluation procedures and details concerning the technical procedure, the cells and other points of technique, are discussed more fully in the literature [332—338].

Mention should be made here of some specialized applications of diffusion experiments. Geddes [332] mentions the following areas: (a) determination of particle size and molecular weight; (b) estimation of the charge on or the ionization of colloidal particles; (c) determination of frictional coefficients; (d) study of solution rates; (e) study of heterogeneous reaction rates; (f) development of the theory of electrolytic solutions; (g) study of the porosity and structure of gels; (h) separation of materials in solution; (i) study of the permeability of membranes. Numerous investigations have dealt with polymeric materials, in particular with organic proteins [320,339—341] and with desoxyribonucleic acid, the molecules of which are the carriers of genetic codes. The determination of structure and particle size of chemical polymerization products also depends on diffusion measurements [342,343]. Interferometric measurements of diffusion in gases have been made at temperatures up to 150°C [344, 345].

(D) SEDIMENTATION

Sedimentation is the movement of particles primarily under the influence of a gravitational field. For an analytical investigation of the principles of sedimentation, gravitational fields are required which are some 1000 times that of the earth. Such fields are achieved with high-speed ultracentrifuges. The determination of sedimentation is an important aid to the characterization of macromolecular substances.

For a theoretical interpretation we shall again take the simplest case of a homogeneous solution. A particle with density ρ_p and of volume V_p in a solvent of density ρ travelling with angular velocity ω at a distance x from the axis of rotation experiences a centrifugal acceleration $\omega^2 x$ and also an effective centrifugal force

$$p_z = V_p(\rho_p - \rho)\, \omega^2 x \qquad (85)$$

This works in opposition to the force of friction p_m according to eqn. (74), with the result that when these two forces are equal a constant rate of sedimentation v is observed, which is proportional to the centrifugal acceleration

$$v = s\omega^2 x \qquad (86)$$

s is the sedimentation constant, and has the dimension of time. When the rate of sedimentation is constant $p_z = p_m$ and for spherical particles with

$$V_p = \tfrac{4}{3}\pi r^3$$

one gets

$$r = \frac{V_p(\rho_p - \rho)}{6\pi\eta s} = \frac{2}{9}\frac{r^3(\rho_p - \rho)}{\eta s}$$

so that the radius will be

$$r = \sqrt{\frac{9\eta s}{2(\rho_p - \rho)}} \qquad (87)$$

If we consider instead of an individual particle, 1 mole of the substance, then in eqn. (85) V_p must be replaced by

$$V_p L = MV$$

where L is the Loschmidt number, M is the molecular weight, and $V = 1/\rho_p$ signifies the partial specific volume in the solution. For this case we can write

$$p_z = M(1 - V\rho)\omega^2 x \tag{88}$$

$$p_m = 6\pi r \eta v L \tag{89}$$

From this it follows, again for spherical molecules, that when the two forces are equal

$$M = \frac{6\pi\eta s L}{1 - V\rho} \sqrt[3]{\frac{3MV}{4\pi L}}$$

The molecular weight M can then be calculated from the sedimentation constant using the relationship

$$M = 9\sqrt{2}\,\pi L \sqrt{V\left(\frac{\eta s}{1 - V\rho}\right)^3} \tag{90}$$

As the idealized assumption that the particles are spherical is in general not true, the coefficient of friction f must be introduced for the force of friction in eqn. (74)

$$p_m = fv = fs\omega^2 x \tag{91}$$

When the two forces given by eqns. (88) and (91) are equal then

$$M = \frac{fs}{1 - V\rho}$$

and since f can be replaced by RT/D, we arrive finally at Svedberg's fundamental equation

$$M = \frac{RTs}{(1 - V\rho)D} \tag{92}$$

From the measured rate of sedimentation one calculates the sedimentation constant s, from which, taken together with the diffusion constant D, the molecular weight may be determined.

In the derivation just given, the gravitational fields have been assumed to be very strong so that the diffusion of the particles does not play any interfering role. However, with relatively weak fields, the effect of diffusion working in opposition to the gravitational field must be considered. The transport of substances by sedimentation in the gravitational field of the ultracentrifuge is directly proportional to the concentration and to the centrifugal force and inversely proportional to the coefficient of friction,

$$Q = \frac{cp_z}{f} \tag{93}$$

Diffusion gives rise to transport in the opposite direction according to eqn. (75). When both factors exactly balance each other, a steady-state sedimentation is observed, which is characterized by the relationship

$$\frac{cp_z}{f} = -D \frac{dc}{dx}$$

From this it follows that

$$\frac{dc}{c} = \frac{p_z}{fD} dx = \frac{p_z dx}{RT}$$

which may be combined with eqn. (88) and integrated to give

$$\ln \frac{c_2}{c_1} = \frac{M}{RT} (1 - V\rho) \tfrac{1}{2} \omega^2 (x_2{}^2 - x_1{}^2)$$

or

$$M = \frac{2RT \ln(c_2/c_1)}{(1 - V\rho)\omega^2(x_2{}^2 - x_1{}^2)} \tag{94}$$

Before steady-state sedimentation can be achieved, relatively long running periods of the ultracentrifuge are necessary, under exactly constant conditions of angular velocity and the temperature. Here again the molecular weight M can be determined from the variation in concentration along a radially mounted cell. For further details of the theoretical interpretation of sedimentation processes, the reader is referred to comprehensive discussions by Nichols and Bailey [346], Hellman [347], Hengstenberg [333], and Meyerhoff [348].

An ultracentrifuge consists basically of a rotor driven by an air-turbine, or sometimes electrically, in which measurement cells are fitted into radially drilled holes. Speeds of over 60,000 r.p.m. can be achieved, and with a radius of 65 mm give rise to accelerations of over 240,000 g (g = the portion of gravity). With smaller radii accelerations of as much as 750,000 g have been reached. The sedimentation cells usually have a cross-section of a sector with an angle of 4°. The rotational axis of the rotator runs vertically, and the cell is illuminated vertically. Different optical arrangements are used based on absorption of light or on the refraction of light. Interferometric techniques were first recommended for the investigation of sedimentation by Philpot and Cook [77] and were later used by several other workers [349—354]. As the measurement of interference is a comparative technique, two adjacent sector cells are required which can be illuminated simultaneously by the optical beam, and as each cell only occupies a very small sector of the circle, relatively long exposures of about 2 min are necessary.

The interferometric techniques employed in the ultracentrifuge are based on the Rayleigh—Svensson or Jamin techniques. These permit the refractive index along the cell to be recorded, and therefore, directly, the concentration. Deflection methods on the other hand (Lamm, Philpot—Svensson) yield the gradient curve for the sedimentation. The two curves are shown in Fig. 65 for three different times during a sedimentation process. It is assumed that both cells are homogeneously filled at the beginning of the experiment. This need not always be the case. Meyerhoff has found that for low molecular weight substances a double-layer cell has proved useful, in which the solution and pure solvent lie one on another such that the separation boundary at the beginning of the sedimentation experiment lies

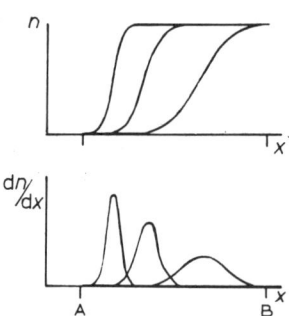

Fig. 65. Distribution of refractive index and its gradient for a sedimentation experiment.

approximately in the middle of the cell. Using either simple cells or double layer ones, one can determine molecular weights of a large variety of materials in an extremely wide range from several hundred to 10^8.

The most important group of substances investigated by this technique is the wide range of proteins. Sedimentation experiments have considerably extended our knowledge of the size and structure of these biologically important substances, and have played an important part in the development of protein chemistry and in medical characterization of sera [355]. In addition, as early as 1949 Nichols and Bailey [346] reported molecular weight determinations on enzymes, hormones, viruses and gelatine. They quoted their measured molecular parameters for a large number of substances both for spherical and for thread-like molecules. Sedimentation experiments on viruses suggested a rod-like shape with a diameter of about 15 nm and a length about 20 times that. The tobacco mosaic virus was subjected to particularly exhaustive investigations, for example ref. 356.

The ultracentrifuge has of course also found application in the analysis of high polymers. Such materials as for example cellulose derivatives and polyvinyl compounds possess thread-like molecules whose structures can only be determined with very powerful ultracentrifuges [333].

With polymolecular substances one can often get a clue as to the distribution of the different types of molecules directly from the sedimentation diagram. In favourable cases a quantitative determination is also possible.

(E) INTERFEROMETRIC METHODS FOR INVESTIGATING THE RANDOM DISTRIBUTION OF REFRACTIVE INDEX IN A SAMPLE

A region in a transparent body of any given state of aggregation, possessing an inhomogeneous distribution of refractive index, which is surrounded by a region of uniform refractive index is termed a schliere (= streak). Of the recognized techniques for the investigation of schlieren in substances, Toepler's knife-edge method (Fig. 35) is noted for its high sensitivity [81]. This technique gives directly a distinct picture of the schlieren-distribution, and is therefore particularly suited, though not exclusively so, for qualitative investigations. For quantitative evaluation, however, interference methods have proved reliable for many decades, and have increased in importance in recent years following the development of laser technology [357—359]. The spacing and the shape of the bands in the interferogram give a direct measure of the deformation of the wave-front as it passes through the object.

Interferometers of the type described in Sect. 4.D are mainly used in this technique (see Wuest [360]). Instrumentation has been designed which allows one to record the Toepler schlieren pattern and the interference pattern of an object simultaneously [361]. Diffraction interferometers with the basic arrangement used for the Toepler technique can be used for either schlieren or interferometer recordings at will [362—367].

While the above mentioned apparatuses yield a representation of the phase of the wave-front after passing through the object, which corresponds to a close approximation to the distribution of refractive index, the shearing interference technique [37,40,368] enables one to determine phase differences within this wave-front, which indicate to a close approximation the distribution of the gradient in the direction of the shearing.

Large-aperture interferometers have been fitted with additional equipment to enable high-speed cinematographic exposures to be recorded [369] for the study of very fast reactions.

The interpretation of interferograms presents an important and very real problem. They may be recorded either visually or photographically. One can measure a curvature of a fringe away from a straight line to about 0.1 of a fringe spacing without any special assistance. A higher sensitivity may be achieved by passing the light several times through the object in the interferometer. Using such an interferometer, Herriott [370] was able to record interference fringes with a spacing corresponding to a path length difference of $\lambda/6$.

When the interference fringes are essentially linear, their intensity may be scanned by the use of slit apertures [371] and may be recorded photo-electrically [372]. A very accurate evaluation is possible when the photographic negative is examined photometrically, for which purpose Dyson [373] has proposed an optical polarization technique for realizing schlieren photography. Dew [374] has described an even simpler arrangement in which he produces a highly magnified picture of the interferogram and, with the help of two photodiodes, focuses on points of equal intensity on the two steep sides of the interference fringe, and thereby achieves an accuracy in the measurement of the band spacing of 0.01 λ.

526

Finally the Moiré interference techniques described in Sect. 7.A may also be used, and have resulted in a number of variations. The Moiré figure from which the absolute deformation of the wave-front can be measured results when a straight-line grid pattern is super-imposed on the interferogram in such a way that the grid lines and the interferogram bands form a small angle with each other. In order to eliminate errors arising from optical imperfections in the inter-ferometer, it is best to use as the grid pattern an interferogram recorded on the same apparatus without the object under study [375,376]. The Moiré figure can be produced by double exposure or by placing the grid pattern in the interferometer. The figure can, however, also be obtained during the photographic copying of the interferogram. Langenbeck [377] described a dual interferometer, a variation of the Michelson apparatus in which two interference pat-terns are produced directly in the instrument and superimposed. Moiré figures quite suitable for evaluation can be quickly focused by adjustment of the mirror in the apparatus.

The methods mentioned so far are, however, of little use for the more common case of irregular variations in the interference fringes. Mention will be made here of two very important approaches to im-proving the accuracy of measurement. Wolter [43,378] described the technique of minimum path length identification which is based on the Toepler arrangement (Fig. 35). When the knife-edge is replaced by a glass plate of which one half causes a phase difference of $180°$ compared with the other, interference at the phase edges results in the appearance of lines of equal shift in the image of the object. The system of lines varies with the position of the phase edge so that the object can be scanned step-wise. The measurement of the very nar-row lines is considerably more accurate than with the schlieren photographs with the original Toepler arrangement.

Finally, Lau and Krug's [379] technique of equidensitometry involving two-dimensional photometry of photographic plates, should be mentioned. One produces from the negative a transparent positive of the same size. When one is laid in exact coincidence over the other, one can observe, because of the non-linear darkening of the photographic plates, lines of equal density which can be distin-guished from the background. They can be copied and measured very accurately [380,381]. Since there is continuous variation of blacken-ing in normal interference pictures, equidensitometry offers some considerable advantages.

The interpretation of interferograms is discussed in detail in the literature in connection with particular problems of measurements. An example will illustrate just how much information can be obtained from an interferometric investigation. Figure 66 was obtained by photographing a glass plate with a Michelson interferometer. The background, which represents a plane wave-front, is adjusted in such a way that the zero-figure appears in the glass plate, i.e. no preferred fringe orientation. One interprets the interferogram as follows.

(i) The area surrounding the glass plate shows a narrow system of parallel interference fringes. The angle between the two wave-fronts, set on the interferometer for the surrounding of the plate is compensated by the glass plate. That means the glass plate causes a deviation of the light beam.

(ii) The direction of the deviation caused by the glass plate lies perpendicular to the direction of the interference fringes of the surroundings.

(iii) The magnitude of the deviation caused by the plate is given by eqn. (18) from the spacing w of the interference fringes and of the wavelength used $\lambda = 546.1$ nm. With a magnification factor $\Gamma = 1$ then $w = 1$ mm, and it follows that

$$\sin \tfrac{1}{2}\epsilon = 5.46 \cdot 10^{-4} \qquad \epsilon = 3.8'$$

When the double passage of the light through the plate is taken into consideration, the value ϵ is approximately the wedge angle of the glass plate.

(iv) In the area of glass plate there is a system of almost circular fringes. The glass plate deforms the plane wave-front into a spherical one: it is acting as a lens.

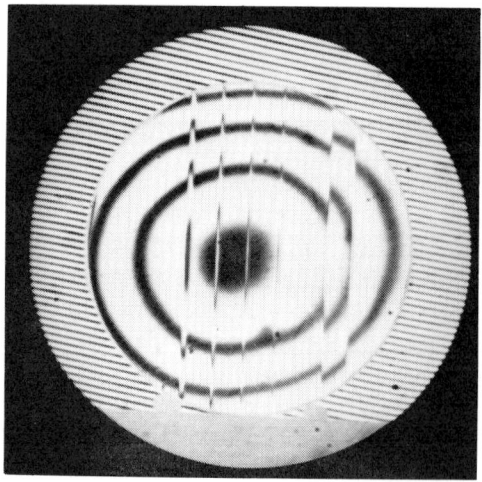

Fig. 66. Interferogram from a glass plate.

(v) The slightly elliptical form of the interference fringe leaves one to conclude that the power of the lens is different in different directions.

(vi) The number of interference rings is a measure of the power of the lens. According to direction, this amounts to between 2.2 and 3.0 rings, averaging 2.6 rings. With a single passage of the light beam this would correspond to an average height $h = 1.3$ rings. With the radius of the plate $r_p = 25$ mm one obtains for the average radius of curvature of the wave-front

$$r_w = \frac{1}{2h\lambda} r_p^2 = 4.4 \cdot 10^5 \text{ mm}$$

The focal length of the lens therefore lies according to direction, between 380 and 520 m.

(vii) One can determine whether the light after passing through the glass plate is converging or diverging, directly on the interferometer, by observing the changes in the fringes when a small adjustment is made on the apparatus. The glass plate has an area on which a layer has been evaporated, and in this area five narrow fringes are again missing. In these areas the wave-front travels faster than the front passing through the coated areas. The distortion of the interference fringes in the area of the missing fringes is outward, the glass plate acts as a concave lens, and the focal length should be preceded by a minus sign.

(viii) The shift in the interference fringes in the area of the missing fringes is equal to $\frac{1}{3}$ of the spacing. As the light beam passes through the glass plate twice, the vaporized layer causes a change in the optical path length compared with air, of $\frac{1}{6}\lambda$.

It is only possible to make an exact determination of the distribution of refractive index in an inhomogeneous substance in particular cases. This applies in the cases discussed in the preceding sections, for layered liquids, possessing a one-dimensional variation of refractive index. Materials in which the refractive index varies in two dimensions can also be measured accurately, as long as it is possible to make the observations by passing the beam of light through the material at right angles to the refractive index gradient. In such a case the refractive index remains constant along any given straight line parallel to the direction of the light beam. As the image is formed of a plane in the test piece at right angles to the direction of the light beam, the interferogram of this plane conveys the full information about the variation of refractive index. If the background or the instrument without test-piece is so adjusted to give an interference pattern free of fringes, then the appearance of a single interference fringe in the image indicates a deformation of the wave-front amounting to one wavelength. In general the difference in optical path length caused by an object is given by eqn. (40)

$$l(n - n_0) = h\lambda$$

where l is the distance travelled in the object, i.e. for a single passage, the thickness of the object, and for a double passage, twice the thickness; n_0 is the refractive index of the surrounding air or liquid or of a homogeneous area in the test piece, when the measurement is made relative to this refractive index; h is the number of interference fringes corresponding to any particular part of the object, and is read off the interferogram. If the background is not free of fringes but is crossed by a number of nearly linear interference fringes, these fringes are curved at points with different refractive index and the magnitude h of the curvature is measured in terms of the spacing between the fringes. When the refractive index n is not itself of primary interest, it can give information about the concentration, density, temperature or pressure in an object.

Thermodynamic investigations often give rise to variation of refractive index about a rotational axis lying not parallel, but at right angles to the direction of the light beam. An evaluation is still possible in such cases. As a simple example, let us assume that within a cylindrical field the refractive index varies with distance from the linear axis such that

$$n = n_0 \qquad\qquad \text{for} \qquad r \geqslant r_f$$

$$n = n_1 - \frac{r}{r_f}(n_1 - n_0) \qquad \text{for} \qquad r \leqslant r_f$$

where n_0 is the refractive index in the surrounding field, r_f is the maximum radius of the field with $n \neq n_0$, and n_1 is the refractive index along the axis of the field (Fig. 67). One can set up a cartesian co-ordinate system in a plane perpendicular to the axis of the field so that the z-axis runs parallel to the direction of the light beam. The optical path difference for a light beam running in a straight line at a distance y from the cylindrical axis when compared with the surroundings, is given by integration along the length of this path.

$$\int_{-r_f}^{+r_f} (n - n_0)\, \mathrm{d}z = h\lambda$$

With

$$z = \sqrt{r^2 - y^2} \qquad \mathrm{d}z = r\, \mathrm{d}r / \sqrt{r^2 - y^2}$$

530

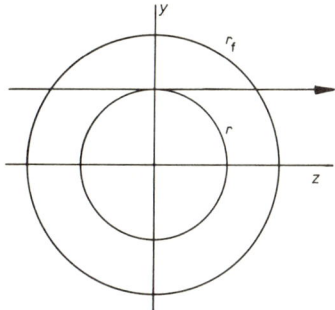

Fig. 67. Representation of the evaluation of a cylindrical distribution of refractive index.

and taking into account that for $r \geqslant r_f$, always $n - n_0 = 0$ it follows that

$$2 \frac{n_1 - n_0}{r_f} \int_y^{r_f} \frac{(r_f - r)\, r\, dr}{\sqrt{r^2 - y^2}} = h\lambda$$

The solution of this integral gives for each distance y the number of fringes corresponding to the optical path difference for a known variation in refractive index, or conversely from the measurement of the number of fringes h and of the maximum field radius r_f, the refractive index n_1 along the axis and hence the overall distribution of refractive index in the cylindrical field.

An evaluation is usually also possible in more complicated cases. In generalized cases of a particular distribution of refractive index one will, as a rule, introduce simplifying assumptions.

(F) SOME APPLICATIONS OF INTERFEROMETRY IN THE INVESTIGATION OF INHOMOGENEOUS DISTRIBUTION OF REFRACTIVE INDEX

Mention has already been made of the numerous applications of interferometry in physics and technology. We must limit this survey to applications for the qualitative or quantitative analysis of materials or of reactions in the field of chemistry. It would also take too much space to discuss the widely varying procedures used for studying different objects and materials, and the often rather difficult interpretations. For an intensive study of any individual field recourse must be had to the appropriate specialized literature.

Since the discovery of schlieren and interference techniques, these have been used for the study of combustion processes, particularly for gas flames. The important factors include the structure of the flame, the variation of temperature within the flame, the combustion velocity and several other parameters. These factors are strongly dependent on the nature of the gases which are burning, their temperature and the proportions in which they are mixed, as well as on other external factors. Gaydon and Wolfhard [382] have written a comprehensive monograph on this subject, which with seventeen pages of literature references is a mine of information for further study. Weinberg's description of optical conditions and methods of investigating flames deserves particular mention [383]. Usually the Mach—Zehnder interferometer is preferred for interferometric investigations of flames and other systems of varying temperatures [384, 385]. Saunders and Smith [386] used the phase-contrast technique described in Sect. 3.D. Shearing methods have also found a wide range of applications [387,388].

In recent years plasma physics has experienced dramatic developments. Interference methods play an important role alongside spectroscopic techniques for studying plasmas. A discharge tube is used which serves at the same time as an interferometer cell, so as to make the evaluation as straight forward as possible. The density of the uncharged gas and the electron concentration in the plasma can both be determined from refractive index measurements [389—396].

Vapour—liquid two-phase processes may be followed optically, particularly in the fields of hydro- and thermodynamics. Both chemical properties and surface exchange effects are important in such cases. Reuter et al. used a Mach—Zehnder interferometer in their investigations of bubbles and mixing zones in liquids [397—399], and it has also been used for the investigation of streaming liquids and stratified solutions [397—400].

Concentration gradients may arise in electrolytes during electrochemical processes. The associated changes in refractive index can be observed and evaluated interferometrically [401,402]. There are some parallels here to the methods used in electrophoresis (Sect.7.B).

Simultaneous measurement of refractive index and light-scattering enable the size of colloid particles to be determined. The disperse solution is compared against a pure solvent in a Rayleigh interferometer [403,404]. This method, which has been used on different

(a)

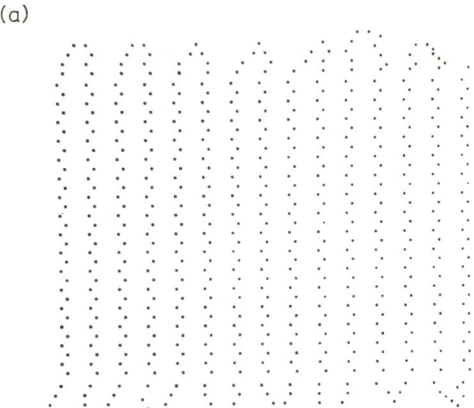

(b)

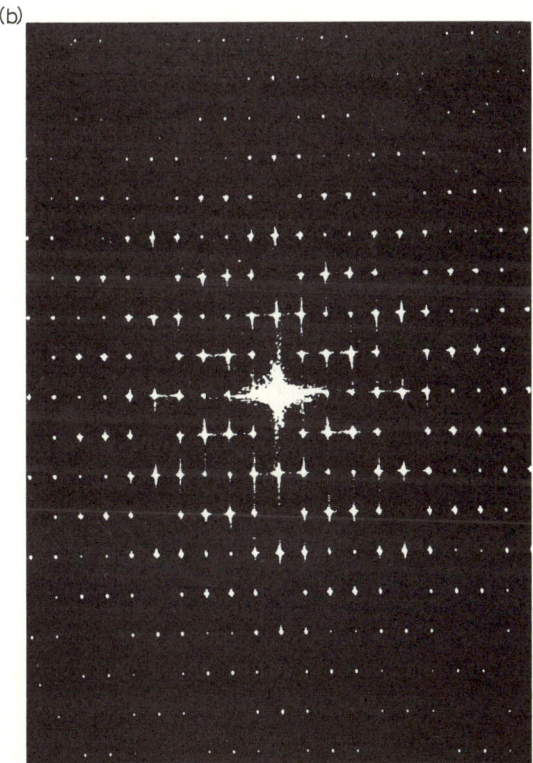

Fig. 68. Diffraction patterns for optical transformation (from Zugenmaier and Bugdahl [422]): (a) diaphragm; (b) diffraction pattern D_1 of the diaphragm; (c) diffraction pattern D_2 of the pattern D_1 (centre blanked out); (d) diffraction pattern D_3 of the pattern D_2 (centre again blanked out).

533

References pp. 536—546

(c)

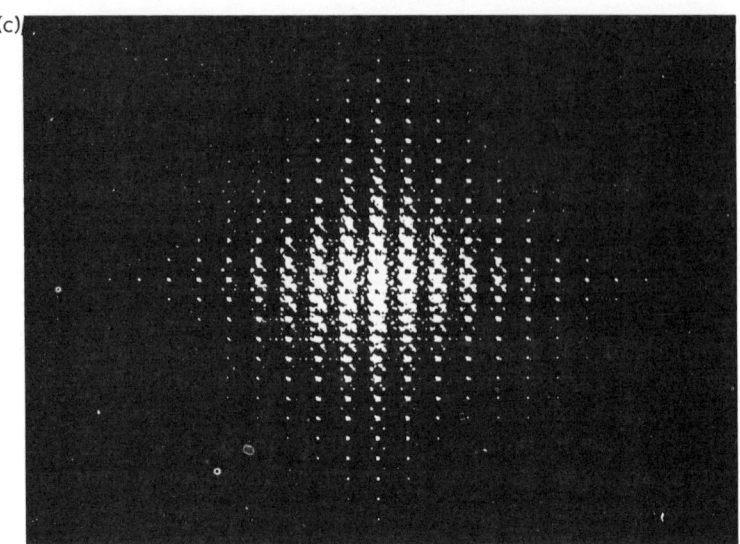

(d)

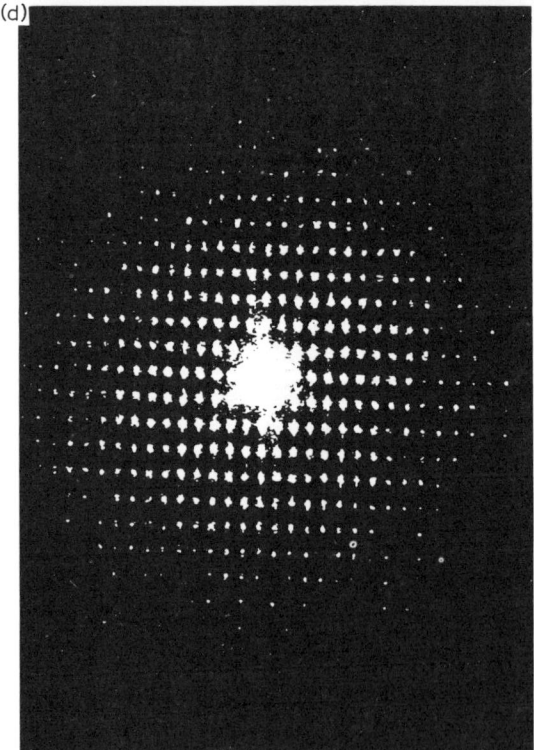

Fig. 68 (contd.)

534

polymeric substances including polystyrene and latex, is an important tool for the characterization of these materials.

Interferometric testing methods are important for the chemistry of optical glasses. The production of glass melts completely free of streaks (schlieren) is a basic requirement of the optical industry. The difficulty of recording and classifying schlieren has prompted numerous suggestions for different measurement techniques using shadow, knife-edge and interference methods [405]. Interference procedures are to be preferred when it is important that the schlieren measurements be accurate and quantitative [406]. The glass test-piece must normally be prepared with two facing, plane surfaces in order that the wave-front should not be disturbed by surface effects [407,408]. Immersion techniques can also be used [409]. Löffler used chemical etching in combination with interference measurements [410]. By using suitable sources and detectors, measurements can also be made at wavelengths outside the visible spectrum [411]. Sims and others [412] and others [413,414] have described the testing of solid glass rods for lasers, for homogeneity. The wide field of testing of optical components such as cell-windows, prisms, lenses, thin coatings, etc., should be considered as purely physical applications, and will not be discussed here [407].

Finally, mention must be made of the technique of optical transformation. X-ray diffraction studies are now an indispensible aid for the elucidation of crystal and also molecular structure, since under certain conditions the X-ray diffraction pattern is uniquely related to the crystal lattice. Bragg suggested a more suitable approach to the elucidation of a molecular structure, in which visible light was used instead of X-rays. In a parallel experiment with visible light, the sample crystal was replaced by a perforated plate and the resulting interference pattern was compared with the X-ray pattern [415]. Bonart and Hosemann used this technique of transformation to longer wavelengths, to aid the interpretation of long-period X-ray diffraction patterns from high polymers [416,417], and Stern [418] determined the degree of crystallinity of polymer fibres in a similar manner. In the meantime comprehensive treatises on optical transformation have appeared [419—421] and the method has now been used for the study of dislocations in crystal lattices [422] (Fig. 68).

References

1 Th. Young, The theory of light and colour, Phil. Trans. Roy. Soc., 92 (1802) 12.
2 Th. Young, A course of lectures on natural philosophy and the mechanical arts, London, 1807.
3 A. Fresnel, Premier mémoire sur la diffraction de la lumière, Letter to the Académie des Sciences, Paris, 1815; Oeuvres complètes d'Augustin Fresnel, Paris, 1866.
4 F. Arago, Compt. Rend., 10 (1840) 813.
5 D. Brewster, Edinburgh Trans., 7 (1817) 435.
6 J. Jamin, Compt. Rend., 42 (1856) 482.
7 J. Jamin, Ann. Chim. Phys., Sér. III, 49 (1857).
8 M. Mascart, Ann. Chim. Phys., 23 (1871) 116.
9 L.G. Gouy, Compt. Rend., 90 (1880) 307.
10 A.A. Michelson, Phil. Mag., 13 (1882) 236.
11 L. Mach, Z. Instrumentenk., 12 (1892) 89.
12 L. Zehnder, Z. Instrumentenk., 11 (1891) 35, 275.
13 Lord Rayleigh, Proc. Roy. Soc., 59 (1896) 198.
14 F. Haber and F. Löwe, Z. Angew. Chem., 23 (1910) 1393.
15 F. Löwe, Physik. Z., 11 (1910) 1047.
16 F. Löwe, Z. Chem. Ind. Kolloide, 11 (1912) 226.
17 U. Doi, Proc. Phys. Math. Soc. Jap., 2 (1920) 38.
18 A. Tiselius, Kolloid Z., 85 (1938) 129.
19 H. Svensson, Acta Chem. Scand., 3 (1949) 1170; 4 (1950) 399, 1329; 5 (1951) 72, 1301.
20 H. Kroepelin, Sitzungsber. Phys. Med. Soz. Erlangen, 58/59 (1926/27) 237.
21 E. Calvet and R. Chevalerias, J. Chim. Phys., 43 (1946) 37.
22 T. Svedberg and K.O. Pedersen, Die Ultrazentrifuge, Steinkopff, Dresden and Leipzig, 1940.
23 A. Toepler, Beobachtungen nach einer neuen optischen Methode, Cohen, Bonn, 1864.
24 F. Zernike, Z. Phys., 36 (1934) 848.
25 J. Dyson, Interferometry as a measuring tool, The Machinery Publishing Co., Brighton, 1970.
26 W. Feussner, Interferenz des Lichtes. Winkelmann's Handbuch der Physik, Vol. 6, Leipzig, 2nd edn., 1906, pp. 878—1031.
27 R.W. Pohl, Phys. Blätter, 17 (1961) 208.
28 E.S. Barr, Appl. Opt., 2 (1963) 639.
31 R.W. Wood, Physical Optics, MacMillan, New York, 3rd edn., 1949.
32 A.N. Sacharevski, Interferometry, Moscow, 1952.
33 M. Françon, Interférences, diffraction et polarisation, in S. Flügge (Ed.), Handbuch der Physik, Vol. 24, Springer, Berlin, 1956.
34 M. Born and E. Wolf, Principles of Optics, Pergamon, Oxford, 3rd edn., 1965.
35 E. Wolf, Progr. Opt., 4 (1965).
36 M. Françon and S. Slansky, Cohérence en Optique, Centre National de la Recherche Scientifique, Paris, 1965.
37 M. Françon, Optical interferometry, Academic Press, New York, 1966; Expériences d'Optique Physique, Gordon and Breach, Paris, 1969.

38 W.H. Steel, Interferometry, Cambridge University Press, London, 1967.
39 B.J. Thompson and E. Wolf, J. Opt. Soc. Amer., 47 (1957) 895.
40 O. Bryngdahl, Progr. Opt., 4 (1965) 39.
41 T. Yamamoto, Opt. Acta, 12 (1965) 230.
42 J. Prasad, Phys. Chem. Glasses, 4 (1963) 112.
43 H. Wolter, Schlieren-, Phasenkontrast- und Lichtschnittverfahren, in S. Flügge (Ed.), Handbuch der Physik, Vol. 24, Springer, Berlin, 1956.
44 E.L. Gayhart and R. Prescott, J. Opt. Soc. Amer., 39 (1949) 546.
45 E. Ingelstam, Transactions of Instruments and Measurements Conference, Stockholm, 1952, pp. 35—39.
46 E. Ingelstam, J. Opt. Soc. Amer., 47 (1957) 536.
47 S. Erdmann, Ein neues, sehr einfaches Interferometer, Dissertation, Aachen, 1951.
48 B.W. Joffe, Refractometric methods of chemistry, Gosudarstvennoe Nauchno-Tekhnicheskoe Izdatel'stvo Khimicheskoi Litaratury, Leningrad, 1960 (in Russian).
49 B. Edlén, J. Opt. Soc. Amer., 43 (1953) 339.
50 B. Edlén, Metrologia, 2 (1966) 71.
51 E.R. Peck and K. Reeder, J. Opt. Soc. Amer., 62 (1972) 958.
52 J.C. Owens, Appl. Opt., 6 (1967) 51.
53 K.J. Habell and A. Jackson, Opt. Acta, 10 (1963) 217.
54 L. Akobjanov, Opt. Acta, 14 (1967) 367.
60 C. Candler, Modern interferometers, Hilger and Watts, London, 1951.
61 F. Twyman, Prism and Lens Making, Hilger and Watts, London, 2nd edn., 1952.
62 F. Löwe, Interferenz-Messgeräte und -Verfahren, Technik, Berlin, 1954.
63 B.S. Thornton, Opt. Acta, 4 (1957) 147.
64 W. Nebe, Jenaer Rundsch., 9 (1964) 36.
65 J. Strassburger, Jenaer Rundsch., 14 (1969) 182.
66 O. Schönrock, Z. Instrumentenk., 62 (1942) 209, 241.
67 P.G. Guest and W.E. Simmons, J. Opt. Soc. Amer., 43 (1953) 319.
68 Y.V. Kolomiitsov, Opt. Spectrosc. USSR, 8 (1960) 370.
69 W. Kinder, Optik, 24 (1966/67) 323.
70 W. Nebe, Jena Nachr., 7 (1956) 215.
71 W. Kinder, Zeiss Werkz., 3 (1955) 78.
72 W. Nebe, Jenaer Rundsch., 10 (1965) 39.
73 O. Lamm, Nova Acta Regiae Soc. Sci. Upsal., 10 (6) (1937) 1.
74 L.G. Longsworth, Ann. N.Y. Acad. Sci., 39 (1939) 105; J. Amer. Chem. Soc., 61 (1939) 529.
75 J.St.L. Philpot, Nature (London), 141 (1938) 283.
76 H. Svensson, Kolloid Z., 87 (1939) 181; 90 (1940) 141.
77 J.St.L. Philpot and G.H. Cook, Research, 1 (1948) 234.
78 E. Wiedemann, Helv. Chim. Acta, 35 (1952) 82, 1895, 2314.
79 W. Nebe, Jenaer Rundsch., 5 (1960) 74.
80 H.J. Antweiler, Chem. Ing. Tech., 24 (1952) 284.
81 H. Schardin, Die Schlierenverfahren und ihre Anwendungen, Ergeb. Exakten Naturwiss., 20 (1942) 303 and references therein.
82 P. Hariharan and R.G. Singh, J. Sci. Instrum., 36 (1959) 323.
83 F.D. Bennet, J. Appl. Phys., 22 (1951) 776.
84 H. Slevogt, Optik, 11 (1954) 366.

85 G. Hansen, Optik, 12 (1955) 5.
86 G. Hansen and W. Kinder, Optik, 15 (1958) 560.
87 G. Schulz and G. Minkwitz, Ann. Phys., 7 (1961) 371.
88 Z. Erdökürti and K. Kántor, Optik, 20 (1963) 243, 304, 325.
89 G. Schulz, Opt. Acta, 11 (1964) 43.
90 M. Françon and S. Mallick, Appl. Opt., 6 (1967) 873.
91 U. Grigull and H. Rottenkolber, J. Opt. Soc. Amer., 57 (1967) 149.
92 M.V.R.K. Murty, J. Opt. Soc. Amer., 50 (1960) 7.
93 W. Kinder, Optik, 1 (1946) 413.
94 R.J. Clark, C.D. Hause and G.S. Bennett, J. Opt. Soc. Amer., 43 (1953) 408.
95 H. Hannes, Optik, 12 (1955) 17.
96 C.S. Chen and J.O. Bird, J. Phys. E, 4 (1971) 157.
97 A. Lohmann and O. Bryngdahl, Appl. Opt., 6 (1967) 1934.
98 D. Malacara and M. Méndez, Opt. Acta, 15 (1968) 59.
99 W.A. Komissaruk and W.I. Janitshkin, Opt. Mekh. Prom., 37 (1970) 29.
100 M.V.R.K. Murty, Appl. Opt., 9 (1970) 1146; M.V.R.K. Murty and E.C. Hagerott, Appl. Opt., 5 (1966) 615.
101 S. Yokozeki and T. Suzuki, Appl. Opt., 10 (1971) 1575.
102 F. Detsch, Jenaer Rundsch., 14 (1969) 193.
103 L. Grupinski, Chem. Tech., 9 (1957) 725.
104 W. Nebe, Chem. Tech., 16 (1964) 673.
105 J. Lacy and K.G. Woolmington, Analyst, 86 (1961) 547.
106 D.N. Hanson and A. Maimoni, Anal. Chem., 31 (1959) 158.
107 D.W. Hill and H.A. Newell, J. Sci. Instrum., 42 (1965) 783.
108 W. Breuer and J. Schreckling, Arch. Tech. Mess. Ind. Messtech., 408 (1970) 5.
109 J. Waclawik, Chem. Anal. (Warsaw), 4 (1959) 329.
110 A. Klemenc, Die Behandlung und Reindarstellung von Gasen, Springer, Berlin, 2nd edn., 1948.
111 H. Lux, Anorganisch-chemische Experimentierkunst, Barth, Leipzig, 1954, pp. 318—348.
112 H. Wollenberg, Erscheinungen an Phasengrenzflächen, Handbuch Lebensmittelchemie, Vol. 2, Springer, Berlin, 1965.
113 H.C. Tsien, Instrum. Contr. Syst., 43 (1970) 118.
114 St. Brunauer, The Adsorption of Gases and Vapors, Princeton Univ. Press, 1943.
115 W. Espe and Č. Hybl, Vak. Tech., 14 (1965) 108.
116 N.N. Awgul and G.G. Muttik, Zeolithes, Nauka, Moscow, 1965. (In Russian)
117 M. Lüder and Ch. Bensow, Anaesthesist, 17 (1968) 257.
118 W. Riedel and K. Steinbreithner, Anaesthesist, 17 (1968) 254.
119 K.R. Andress and K. Wüst, Gasanalyse, in E. Berl and G. Lunge (Eds.), Chemische-technische Untersuchungsmethoden, Springer, Berlin, 8th edn., 1939.
120 F. Löwe, Zeiss Nachr., 5 (1944) 150.
121 R. Kattwinkel, Grubengasanalyse im Kohlenbergbau, Gruyter, Berlin, 1950.
122 C.G. Mönch, Hochvakuumtechnik, Lang, Pössneck, 1950.
123 W. Rausch, Z. Phys. Chem., 201 (1952) 32.
124 H.R. Süess, Beitrag zur Kenntnis der Mischadsorption von Lösungsmitteldämpfen, Dissertation, TH Zürich Nr. 1953, 1953.

125 G.H. Bush and R.J. Loneragan, Analyst, 79 (1954) 371.
126 W. Schirmer, Chem. Tech., 22 (1970) 735; 23 (1971) 23.
127 G. Kortüm and M. Kortüm-Seiler, Refraktometrie und Interferometrie, Hand. physiol.- pathol.-chem. Analyse, Springer, Berlin, 1953, 10th edn., pp. 441—478.
128 F. Löwe, Optische Messungen des Chemikers und des Mediziners, Stein-kopff, Dresden, 1954, 6th edn.
129 C.G. Mönch, Interferenzlängenmessung und Brechzahlbestimming, Teubner, Leipzig, 1966.
130 W. Nebe, Analytische Interferometrie, Geest and Portig, Leipzig, 1970.
131 W. Krug, J. Rienitz and G. Schulz, Beiträge zur Interferenz-Mikroskopie, Akademie-Verlag, Berlin, 1961.
132 H. Beyer, Theorie und Praxis des Phasenkontrastverfahrens, Geest and Portig, Leipzig, 1965.
133 L.W. Tilton and J.K. Taylor, J. Res. Nat. Bur. Stand., 20 (1938) 419.
134 W. Edmondson, Brit. J. Anaesth., 29 (1957) 570.
135 L.H. Siertsema, De Jamin'sche Interferentiaalrefractator, Dissertation, Gro-ningen, 1890.
136 R.C. Faust and H.J. Marrinan, Brit. J. Appl. Phys., 6 (1955) 351.
137 E. Grunwald and B.J. Berkowitz, Anal. Chem., 29 (1957) 124; D. Tinaut, C.R. Acad. Sc. Paris, 258 (1964) 6110.
138 W. Nebe, Jenaer Jahrb., II (1958) 62.
139 G. Hansen, Z. Instrumentenk., 50 (1930) 460.
140 K.A. Chalilulin, Ismeritelnajatech., 8 (1960) 44.
141 R. Torge, Z. Instrumentenk., 75 (1967) 155.
142 D.N. Hanson and A. Maimoni, Anal. Chem., 31 (1959) 77.
143 R.M. Waxler and C.E. Weir, J. Res. Nat. Bur. Stand., 67A (1963) 163.
144 O.E. Frivold, O. Hassel and E. Hetland, Physica, 6 (1939) 972.
145 H. Horvath, Appl. Opt., 6 (1967) 1140.
146 International Critical Tables, Vol. VII, McGraw-Hill, New York/London, 1930.
147 T. Larsén, Z. Phys., 111 (1938) 391.
148 K.-H. Hellwege (Ed.), Landolt-Börnstein Zahlenwerte und Funktionen, Vol. 2, Part 8, Springer, Berlin, 6th edn., 1962.
149 K.L. Ramaswamy, Proc. Indian Acad. Sci., 4 (1937) 675.
150 H.E. Watson and K.L. Ramaswamy, Proc. Roy. Soc. Ser. A, 156 (1936) 144.
151 E.F. Yamshikov, Opt. Spectrosc. USSR, 8 (1960) 29.
152 K. Bärwinkel and A. Hartkämper, Arch. Tech. Mess. Ind. Messtech., 405 (1969) 233.
153 F.J. Schoenes, Z. Angew. Phys., 28 (1970) 362.
154 B. Wilhelmi, Jenaer Rundsch., 14 (1969) 173.
160 N. Bauer and K. Fajans, Refractometry, in A. Weissberger (Ed.), Physical Methods of Organic Chemistry, Interscience, New York/London, 2nd edn., 1949, pp. 1141—1240.
161 W.E. Williams, Applications of Interferometry, Methuen, New York, 4th edn., 1950.
162 F.K. Bell and J.C. Krantz, J. Amer. Pharm. Ass., 29 (1940) 127.
163 F. Müller, F. Freude and F. Kaunert, Gas Wasserfach, 83 (1940) 641.
164 W. Ohme, Erdoel Kohle, 7 (1954) 73.

165 D. Kutz, Beiträge zur Prozessanalyse, Dipl.-Arbeit, Univ. Jena, 1966.
166 P.A. Cole and D.W. Armstrong, J. Opt. Soc. Amer., 31 (1941) 740.
167 E. König, Glueckauf, 15 (1937) 326.
168 F. Roth, W. Ohme and A. Nickisch, Oel Kohle, 38 (1942) 1133.
169 H. Kögler, Chem. Tech., 9 (1957) 400.
170 F. Löwe, Urania, 13 (1950) 204.
171 K. Nabert and G. Schön, Sicherheitstechnische Kennzahlen brennbarer Gase und Dämpfe, Deutscher Eichverlag, Berlin, 2nd edn., 1963.
172 G. Schön and H. Steen, Anaesthesist, 17 (1968) 6.
173 E. von Schwartz, Handbuch der Feuer- und Explosionsgefahr, Jung, München, 5th edn., 1958.
174 J. Calzia, Les substances explosives et leur nuisances, Dunod, Paris, 1969.
175 W. Schuhknecht, W. Stetzer and H. Schinkel, Glueckauf, 93 (1957) 1428.
176 F. Löwe, Chem. Ztg., 68 (1944) 144.
177 J. May, Staub, 19 (1959) 387; 25 (1965) 153.
178 F. Bayer and G. Wagner, Gasanalyse, Enke, Stuttgart, 3rd edn., 1960.
179 Maximale Arbeitsplatz-Konzentrationen gesundheitsschädlicher Stoffe 1961 (MAK-Werte), Staub, 21 (1961) 535 (new edn., 1971); G. Sorbe, Glas Instrum. Tech., 16 (1972) 333.
180 H. Kettner, Staub, 20 (1960) 369.
181 G. Kallenbach and G. Henzschel, WTZ Mitt., 2 (1970) 24.
182 V. Kindt, Jenaer Rundsch., 7 (1962) 114.
183 E.A. Griffiths and R. Davis, Dep. Agr. Sci. Bull., 56 (1926) 78.
184 K. Scholz and H. Pechert, Deut. Landwirt., 7 (1956) 246.
185 A.H. Frucht and E. Otto, Jenaer Nachr., 10 (1965) 4.
186 H. Wollschitt et al., Arch. Exp. Pathol. Pharmakol., 177 (1935) 635.
187 K. Eckoldt, Jenaer Nachr., 10 (1965) 17.
188 K. Wezler and E. Frank, Pflügers Arch. Gesamte Physiol. Menschen Tiere, 250 (1948) 249.
189 F. Bahner and R. Zidek, Z. Gesamte Exp. Med., 47 (1951) 219.
190 O. Skranc and V. Havel, Cesk. Fysiol., 7 (1958) 564.
191 F. Bahner, Pflügers Arch. Gesamte Physiol. Menschen Tiere, 250 (1948) 521.
192 W. Benstz, Arch. Exp. Pathol. Pharmakol., 222 (1954) 575.
193 F. Spengler, Deut. Gesundheitsw., 11 (1956) 686.
194 H. Kleinsorge and L. Jäger, Med. Klin., 56 (1961) 781.
195 D. Weber, Deut. Gesundheitsw., 16 (1961) 2145.
196 L. Jäger, Jenaer Nachr., 10 (1965) 39.
197 Z. Vokáč, Jenaer Rundsch., 12 (1967) 113.
198 S.L. Cowan, H.G. Epstein and S.F. Suffolk, J. Physiol., 101 (1942) P2.
199 K.D. Hall et al., Anaesthesiol., 14 (1953) 38.
200 R.E. Jahn, Analyst, 80 (1955) 700.
201 S. Kobayasi and T. Kinosita, Anaesthesist, 7 (1958) 1.
202 W. Nebe, Jenaer Nachr., 10 (1965) 51.
203 H.J. Brennan, Brit. J. Anaesth., 29 (1957) 332.
204 M. Lüder, Anaesthesist, 13 (1964) 360.
205 M. Lüder and Ch. Bensow, Medizintechnik, 10 (1970) 124.
206 G. Hohmann, Zeiss Mitt., 2 (1960) 89.

207 H.F. Poppelbaum, Uber konzentrationsgeregelte Narkosemittelverdampfer, Habil.-Schr. Deut. Akad. Ärztl. Fortbildg. Berlin, 1969; H.F. Poppelbaum, W. Nebe and W. Bremer, Labortechnik, Engl. Ed., 3 (1970) 18.
208 D. Langrehr et al., Anaesthesist, 19 (1970) 340.
209 G. Lutz, Medizintechnik, 12 (1972) 16.
210 H. Mau, CO_2-Messungen während der Narkose, Dissertation, Berlin, 1969.
211 M. Green and P.H. Robinson, Anal. Chem., 25 (1953) 1913.
212 J. Suska, Exp. Tech. Phys., 17 (1969) 477.
213 H.M. Ashton and E.A. Guggenheim, Proc. Phys. Soc. London Sect. B, 69 (1956) 693; H.M. Ashton and E.S. Halberstadt, Proc. Roy. Soc. Ser. A, 245 (1958) 373.
214 W. Jurjew, Usp. Fiz. Nauk, 47 (1952) 146.
215 R.E. Kinzly, Appl. Opt., 6 (1967) 137.
216 G.P. Starcev, Dokl. Akad. Nauk SSSR, 95 (1954) 1181.
217 P. Hariharan and D. Sen, J. Sci. Instrum., 36 (1959) 70.
218 P. Hariharan, D. Sen and M.S. Bhalla, J. Sci. Instrum., 36 (1959) 72.
219 E. Ingelstam, Ark. Fys., 6 (1953) 287.
220 E. Djurle, Appl. Sci. Res. Sect. B, 4 (1954) 91.
221 Th. Benzinger and G. Kitzinger, AAF Aero Med. Center, Heidelberg, 1946.
222 E. Schulz, Registrierende Interferometer-Anordnung, Dipl.-Arbeit, Halle, 1958.
223 S. Namba, Rev. Sci. Instrum., 30 (1959) 642.
224 T.S. Kolomiitsova and J.V. Novikova, Opt. Spectrosc. USSR, 8 (1960) 189.
225 J. Shamir, R. Fox and S.G. Lipson, Appl. Opt., 8 (1969) 103.
226 J. Neumann, Zeiss Mitt., 5 (1969) 5.
227 W. Kinder et al., Appl. Opt., 7 (1968) 341.
228 D.S. Roshdestvenski, Anomalous Dispersion of Metal Damps, Akad. Nauk SSSR, 1951. (In Russian)
229 J.S. Gorbanj and A.A. Sislovski, Zh. Tekh. Fiz., 25 (1955) 1297.
230 A.M. Shukhtin, Opt. Spectrosc. USSR, 10 (1961) 222.
231 F. Legay, Compt. Rend. Acad. Sci., 240 (1955) 174.
232 N.S. Tskhai and S.L. Mandelstam, Opt. Spectrosc. USSR, 7 (1959) 91.
233 E. Berl and L. Ranis, Fortschr. Chem. Phys., 19 (1928) 1.
234 G. Zieglgänsberger, Zuckererzeugung, 4 (1960) 182.
235 H. Beitner, Uber die Anwendbarkeit des Laborinterferometers zur schnellen Beurteilung von Brau-, Selters- und Kesselwasser, Ing.-Arbeit, Chemie-Ing. Schule, Berlin, 1958.
236 N.M. Dychno, Die Messung der Brechungsdifferenz mit dem Interferometer, in A.L. Schatenstein (Ed.), Isotopenanalyse des Wassers, Vol. 4, Deutsche Verlag Wissenschaft, Berlin, 1960, pp. 233—245.
237 E. Ingelstam, E. Djurle and L. Johansson, J. Opt. Soc. Amer., 44 (1954) 472.
238 L. Blaga et al., Stud. Cercet. Fiz., 15 (1964) 125.
239 J. Mercea, Chem. Ing. Tech., 41 (1969) 508.
240 E. Plško and V. Holba, Chem. Zvesti, 15 (1961) 321.
241 W. Piratzki, Nahrung, 1 (1957) 88.
242 A.M. Zanko and Y.L. Kultevich, Zh. Anal. Khim., 5 (1950) 75; T. Lesiak et al., Chem. Anal. (Warsaw), 16 (1971) 543.

243 J.J. Kipling, Hilger J., 3 (1957) 35.
244 G.L. Stabobinec and S.A. Meckovskij, Zh. Anal. Khim., 16 (1961) 319.
245 E. Berl and L. Ranis, Ber. Deut. Chem. Ges., 61 (1928) 92.
246 J. Marti and C.M. Aliod, Inform. Quim. Anal., 11 (1956) 39.
247 V.J. Vajgand and T.J. Todorovski, Bull. Chem. Soc. Belgrade, 31 (1966) 153.
248 V.J. Vajgand, T.J. Todorovski and F.F. Gaál, Bull. Chem. Soc. Belgrade, 33 (1968) 293.
249 V.J. Vajgand and T.J. Todorovski, Jenaer Rundsch., 15 (1970) 37.
250 T.J. Todorovski, Interferometrijske titracije u različitim rastvaračima, Dissertation, Beograd, 1971.
251 L. Heilmeyer, Interferometrie in E. Bamann (Ed.), Methoden der Fermentforschung, 1940.
252 H.J. Antweiler and G. Klemmer, Z. Anal. Chem., 252 (1970) 256.
253 M.W. Mettenleiter, Nature (London), 145 (1940) 305.
254 S. Wehrli, Mitt. Geb. Lebensmittelunters. Hyg., 45 (1954) 123.
255 H. Elbel, Blutalkohol, Thieme, Stuttgart, 2nd edn., 1956.
256 K. Sellier, Med. Markt, 5 (1957) 414.
257 H. Decker, Deut. Z. Gesamte Gerichtl. Med., 33 (1940) 33.
258 D. Zschocke, Jenaer Rundschr., 6 (1961) 150.
259 E. Elbing and A.G. Parts, Angew. Makromol. Chem., 82 (1965) 270.
260 V. Bugdahl, Kaut. Gummi Kunstst., 22 (1969) 486.
261 I. Baltog, C. Ghita and L. Ghita, Eur. Polym. J., 6 (1970) 1299.
262 G. Kalz, E. Schröder and K. Thinins, Plaste Kaut., 17 (1970) 12.
263 R. Ripan, Z. Székely and G. Kiss-Imreh, Rev. Roum. Chim., 10 (1965) 399.
264 A.V. Belonogov and V.M. Gorbunkov, Opt. Spectrosc. USSR, 14 (1963) 234.
265 W. Geffken and A. Kruis, Z. Phys. Chem., 45 (1940) 411.
266 J.S. Gorbanj and A.A. Sislovskij, Dokl. Akad. Nauk SSSR, 98 (1954) 676.
267 J.S. Gorbanj and A.A. Sislovskij, Dokl. Akad. Nauk SSSR, 108 (1956) 53, 210.
268 I.N. Shkljarevskij, Opt. Spectrosc. USSR, 6 (1959) 508.
269 H. Römer, Z. Angew. Phys., 14 (1962) 631.
270 L.J. Alperovič, Opt. Spectrosc. USSR, 14 (1963) 400.
271 K.J. Hakoila, Ann. Univ. Turku. Ser. A, 12 (1952) 1.
272 M.S. Shumate, Appl. Opt., 5 (1966) 327.
273 A.J. Werner, Appl. Opt., 7 (1968) 837.
274 D.W. Harper and G.B. Boulton, Abstracts ICO 8th Conf., 1969.
275 I.W. Obreimov, Izv. Akad. Nauk SSSR (1945).
276 T. Bååk, J. Opt. Soc. Amer., 59 (1969) 851.
277 C.J. Parker and W.A. Popov, Appl. Opt., 10 (1971) 2137.
278 F. Green, Interferometric Measurement of Temperature Coefficients of Refractive Indices of Optical Glasses, in J.H. Dickson (Ed.), Optical Instruments and Techniques, Oriel Press, Newcastle-upon-Tyne, 1970.
279 S.D.H. Andréasson, S.E. Gustafsson and N.O. Halling, J. Opt. Soc. Amer., 61 (1961) 595.
280 E. Schuster and F. Reitmayer, Glastech. Ber., 34 (1961) 130.

281 R.M. Waxler and C.E. Weir, Nat. Bur. Stand. U.S. Tech. News Bull., (1966) 156.

282 E.D.D. Schmidt and K. Vedam, J. Phys. Chem. Solids, 27 (1966) 1563.

283 P. Wulff, Z. Kristallogr., 77 (1931) 61.

284 E. Menzel, Arch. Tech. Mess. Ind. Messtech., 210 (1953) 157.

285 M.S. Soskin, Opt. Spectrosc. USSR, 11 (1961) 415.

286 G.C. Mönch, O. Böttger and U. Zorll, Längenmessung und Brechzahlbestimmung mit Lichtinterferenzen, Fachbuchverlag, Leipzig, 1954.

287 M.A. Jeppesen and A.M. Taylor, J. Opt. Soc. Amer., 56 (1966) 451.

288 M. Hennig, Wiss. Z. Univ. Halle, 4 (1955) 817.

289 P.A. Flournoy, R.W. McClure and G. Wyntjes, Appl. Opt. 11 (1972) 1907.

290 E. Menzel, Arch. Techn. Mess. Ind. Messtech., 281 (1959) 109; 283/284 (1959) 177.

291 W. Heitmann and G. Koppelmann, Z. Angew. Phys., 23 (1967) 221.

292 R. Fleischmann, Z. Phys., 129 (1951) 275.

293 R. Fleischmann and H. Schopper, Z. Phys., 129 (1951) 285; 130 (1951) 304.

294 P. Hariharan and D. Sen, J. Sci. Instrum., 37 (1960) 417.

295 P. Rouard and P. Bousquet, Progr. Opt., 4 (1965) 145.

301 G. Oster, M. Wassermann and C. Zwerling, J. Opt. Soc. Amer., 54 (1964) 169.

302 C.J. van Oss, J. Sci. Instrum., 41 (1964) 227.

303 W. Nebe, Optik, 21 (1964) 579.

304 V. Bugdahl and B. Rozsondai, Jenaer Rundsch., 12 (1967) 222.

305 J. Chanu, L. Mousselin and F. Parra, Rev. Opt. Theor. Instrum., 41 (1962) 119.

306 G. Kegeles and H.A. Sober, Anal. Chem., 24 (1952) 654.

307 D.H. Moore, Electrophoresis, in A. Weissberger (Ed.), Physical Methods of Organic Chemistry, Interscience, New York/London, 2nd edn., 1949, pp. 1685—1712.

308 E. Wiedemann, Elektrophorese, in Hoppe-Seyler/Thierfelder, Handb. physiol.-pathol.-chem. Analyse, Springer, Berlin, 10th edn., 1953, pp. 54—95.

309 A. Knapp, Chem. Tech., 7 (1955) 708.

310 H.J. Antweiler, Die quantitative Elektrophorese in der Medizin, Springer, Berlin, 2nd edn., 1957.

311 M. Bier, Electrophoresis, Academic Press, New York, 1959.

312 H.-Ch. Gabsch, Elektrophorese, in M. Büchner and H.-Ch. Gabsch (Eds.), Moderne chemische Methoden in der Klinik, Thieme, Leipzig, 2nd edn., 1961, pp. 374—449.

313 E. Wiedemann, Internat. Arch. Allergy Appl. Immunol., 5 (1954) 1.

314 W. Lotmar, Helv. Chim. Acta, 32 (1949) 1847.

315 C.C. Curtain, J. Sci. Instrum., 37 (1960) 190.

316 W. Bockemüller, Z. Naturforsch. B, 13 (1958) 772.

317 W. Bockemüller and A. Oerter, Klin. Wochenschr., 39 (1961) 371.

318 M. Daune, L. Freund and G. Scheibling, J. Chim. Phys., 54 (1957) 924.

319 C.S. Caldwell, J.R. Hall and A.L. Babb, Rev. Sci. Instrum., 28 (1957) 816.

320 S. Shulman, J. Amer. Chem. Soc., 75 (1953) 5846.

321 O.M. Griffith and C.R. McEwen, Anal. Biochem., 18 (1967) 397.

322 J. Strassburger and K.E. Reinert, Jenaer Rundsch., 14 (1969) 188.

323 O. Bryngdahl and St. Ljunggren, J. Phys. Chem., 64 (1960) 1264.

324 O. Bryngdahl, J. Opt. Soc. Amer., 53 (1963) 571.
325 W. Weinstein, Nature (London), 172 (1953) 461.
326 Y. Barrada, Interferometrische Untersuchungen der Diffusionsschicht bei der Elektrolyse, Dissertation, ETH Zürich, 1954.
327 W.J. Thomas and E.K. Nicholl, Appl. Opt., 4 (1965) 823.
328 H. Gralén, Kolloid Z., 95 (1941) 188.
329 R.H. Stokes, Trans. Faraday Soc., 48 (1952) 887.
330 A.N. Tscherkassov and S.T. Klenin, Vysokomol. Soed., 7 (1965) 902.
331 V.N. Zvetkov and S.T. Klenin, Zh. Tekh. Fiz., 29 (1959) 640.
332 A.L. Geddes, Determination of Diffusivity, in A. Weissberger (Ed.), Physical Methods of Organic Chemistry, Interscience, New York, 2nd edn., 1949.
333 J. Hengstenberg, Sedimentation und Diffusion von Makromolekülen, in H.A. Stuart, Physik der Hochpolymeren, Vol. 2, Springer, Berlin, 1953, pp. 411—494.
334 W. Jost, Diffusion, Steinkopff, Darmstadt, 1957.
335 L.O. Sundelöf, Ark. Kemi, 25 (1965) 1.
336 K.F. Tlach, Pharmazie, 22 (1967) 296.
337 V. Bugdahl, Jenaer Rundsch., 12 (1967) 107.
338 V. Bugdahl, Jenaer Nachr., 10 (1970) 257.
339 L.J. Gosting, Advan. Protein Chem., 11 (1956) 429.
340 M. Schönenberger, R. Schmidtberger and H.E. Schultze, Z. Naturforsch. B, 13 (1958) 761.
341 T.E. Thompson and J.L. Oncley, J. Amer. Chem. Soc., 83 (1961) 2425.
342 V.N. Zvetkov and S.I. Klenin, Zh. Tekh. Fiz., 29 (1959) 1393.
343 S. Klenine, H. Benoit and M. Daune, Compt. Rend., 250 (1960) 3174.
344 Ch. Boyd et al., J. Chem. Phys., 19 (1951) 548.
345 R. Varoqui et al., J. Chim. Phys., 2392 (1962) 161.
346 J.B. Nichols and E.D. Bailey, Determinations with the Ultracentrifuge, in A. Weissberger (Ed.), Physical Methods of Organic Chemistry, Interscience, New York, 2nd edn., 1949, pp. 621—730.
347 E. Hellmann, Die Ultrazentrifuge, in Hoppe-Seyler/Thierfelder, Handb. physiol.-pathol.-chem. Analyse, Springer, Berlin, 10th edn., 1953, pp. 95—121.
348 G. Meyerhoff, Angew. Chem., 72 (1960) 699.
349 J.W. Beams, Rev. Sci. Instrum., 34 (1963) 139.
350 H.K. Schachman, Biochemistry, 2 (1963) 887.
351 W.H. Nelson and R.S.T. Nelson, Can. J. Chem., 42 (1964) 731.
352 E. Wiedemann and A. Laforce, Chem. Ing. Tech., 37 (1965) 993.
353 C.H. Chervenka, Anal. Chem., 38 (1966) 356.
354 M. Derechin, Anal. Biochem., 28 (1969) 385.
355 W. Bolt, A. Bolte and N. Lichins, Die Ultrazentrifuge und ihre Bedeutung für die Diagnostik in der Medizin, in M. Büchner and H.-Ch. Gabsch (Eds.), Moderne chemische Methoden in der Klinik, Thieme, Leipzig, 2nd edn., 1961, pp. 575—595.
356 H. Triebel, H. Venner and W. Kayser, Z. Naturforsch. B, 16 (1961) 368.
357 A.K. Oppenheim, P.A. Urtiew and F.J. Weinberg, Proc. Roy. Soc., 291 (1966) 279.
358 D.R. Herriott, Progr. Opt., 4 (1965) 171.
359 H. Schönnagel, Exp. Tech. Phys., 17 (1969) 131.

360 W. Wuest, Arch. Tech. Mess. Ind. Messtech., 144 (1969) 193.
361 W. Horn, Jahrb. Opt. Feinmech., O VI (1961) 143.
362 S.A. Abrukov, Shadow and interference methods for investigation of optical inhomogeneities, Verl. Univ. Kazan, 1962. (In Russian)
363 E.B. Temple, J. Opt. Soc. Amer., 47 (1957) 91.
364 D.W. Holder and R.J. North, Schlieren Methods, HMSO, London, 1963.
365 H. Rottenkolber, Fortschr. Ber. VDI Z., 6 (8) (1965) 1.
366 J.B. Brackenridge and J. Peterka, Appl. Opt., 6 (1967) 731.
367 W. Nebe and M. Rekalić, 10th Int. Conf. Phenomena in Ionized Gases, Donald Parsons, Oxford, 1971, pp. 439—440.
368 O. Bryngdahl, Ark. Fys., 21 (1962) 289.
369 F.D. Bennett, D.D. Shear and H.S. Burden, J. Opt. Soc. Amer., 50 (1960) 212.
370 D.R. Herriott, J. Opt. Soc. Amer., 56 (1966) 719.
371 M. Françon and A. Soulié, Rev. Opt., 42 (1963) 499.
372 G. Nomarski and G. Roblin, Compt. Rend., 261 (1965) 3556.
373 J. Dyson, Appl. Opt., 2 (1963) 487.
374 G.D. Dew, J. Sci. Instrum., 41 (1964) 160.
375 J.W. Gates, Brit. J. Appl. Phys., 5 (1954) 133.
376 R.G. Brooks, J. Maxwell and S.D. Probert, Engineer, 226 (1968) 398.
377 P. Langenbeck, J. Opt. Soc. Amer., 58 (1968) 499.
378 H. Wolter, Ann. Phys., 7 (1950) 341.
379 E. Lau and W. Krug, Die Aquidensitometrie, Akademie-Verlag, Berlin, 1957.
380 H. Haas, Feingerätetechnik, 6 (1957) 312.
381 W. Krug and E. Lau, Phys. Blaetter, 15 (1959) 11.
382 A.G. Gaydon and H.G. Wolfhard, Flames, Chapman and Hall, London, 3rd edn., 1970.
383 F.J. Weinberg, Optics of Flames, Butterworths, London, 1963.
384 G. Schmitz and G. Hottenroth, Z. Phys., 126 (1949) 1.
385 D. Wilkie and S.A. Fisher, Proc. Inst. Mech. Eng., 178 (1963/1964) 461.
386 M.J. Saunders and A.G. Smith, J. Appl. Phys., 27 (1956) 115.
387 J.B. Saunders, J. Res. Nat. Bur. Stand. Sect. C, 69 (1965) 245.
388 O. Bryngdahl, Progr. Opt., 4 (1965) 37.
389 W. Finkelnburg and H. Maecker, Elektrische Bögen und thermisches Plasma, in S. Flügge, Handbuch der Physik, Vol. 22, Springer, Berlin, 1956, pp. 254—444.
390 R.A. Alpher and D.R. White, Phys. Fluids, 2 (1959) 153.
391 W.F. von Jaskowski, Zeiss Werkz., 10 (1962) 107.
392 P.B. Barber, D.A. Swift and B.A. Tozer, Brit. J. Appl. Phys., 14 (1963) 207.
393 S. Takeda and K. Minami, Jap. J. Appl. Phys., 5 (1966) 696.
394 S.S. Medley, J. Appl. Phys., 41 (1970) 142.
395 H.H. Carls and H. Haase, 10th Int. Conf. Phenomena in Ionized Gases, Donald Parsons, Oxford, 1971, p. 62.
396 P. Bogen et al., 10th Int. Conf. Phenomena in Ionized Gases, Donald Parsons, Oxford, 1971, p. 63.
397 H. Reuter, E. Dolling and R. Wessely, Chem. Ing. Tech., 42 (1970) 1109.
398 H. Beer, Chem. Ing. Tech., 43 (1971) 837.
399 F.A. Matekunas and E.R.F. Winter, Chem. Ing. Tech., 43 (1971) 837

400 H. Hannes and R. Weber, Chem. Ing. Tech., 43 (1971) 792.
401 R.N. O'Brien, Rev. Sci. Instrum., 35 (1964) 803.
402 R.V. Bucur and J. Mercea, Rev. Roum. Phys., 13 (1968) 601.
403 D.W. Ovenall and F.W. Peaker, Makromol. Chem., 33 (1959) 237.
404 H. Lange, Kolloid Z., 223 (1968) 24.
405 G. Kaufmann, W. Nebe and H. Viehweger, Jenaer Nachr., 10 (1970) 220.
406 H. Hannes, Glastech. Ber., 29 (1956) 83.
407 J.B. Saunders, J. Res. Nat. Bur. Stand., 53 (1954) 165.
408 A.R. Tynes and D.L. Bizbee, I.E.E.E., J. Quantum Electr., Q3-3 (1967) 459.
409 A. Arnulf and M. Cagnet, Rev. Opt., 34 (1955) 639.
410 J. Löffler, Glastechn. Ber., 37 (1964) 548.
411 F.W. Rosberry, Appl. Opt., 5 (1966) 961.
412 S.D. Sims, A. Stein and C. Roth, Appl. Opt., 5 (1966) 621.
413 E.D. Baird, Appl. Opt., 9 (1970) 465.
414 J.D. Briers, Opt. Laser Technol., (1972) 28.
415 W.L. Bragg, Nature (London), 143 (1939) 678.
416 R. Bonart, Z. Kristallogr., 109 (1957) 309.
417 R. Bonart and R. Hosemann, Makromol. Chem., 39 (1960) 105.
418 P.G. Stern, Kolloid Z., 215 (1967) 140.
419 C.A. Taylor and H. Lipson, Optical Transforms, Their Preparation and Application to X-Ray Problems, Bell, London, 1964.
420 C.A. Taylor, Europ. Polym. J., 2 (1966) 279.
421 G. Bodor and V. Bugdahl, Jenaer Nachr., 10 (1970) 283.
422 P. Zugenmaier and V. Bugdahl, Laser Angew. Strahlentech., 3 (1970) 50.

Index

Abbe number, 421

Absolute stability complex, 272

Absorbance, 217

Absorption spectrum, 217

—, relationship of, to excitation spectrum, 74

Absorptivity, 217

Acetic acid, determination of, by enzyme electrode, 58

Acetol, determination of, by fluorimetry, 180

Acetophenone, decay time of, 144

Acetylacetone, application of, in determination of titanium, by photometric titration, 334, 339

Acetylacetone complexes, phosphorescence of, 157

Acetylcholine, immobilization on ion exchange resin, 65

Acetylcholine reineckate, 59

Acetylene, effect of dissolved, on refractive index of a liquid, 477, 478

—, interferometric calibration curve for, 457

—, sorption of, 450

Acetylthiocholine chloride, application of, in determination of cholinesterase, 64

N-Acetyl-L-tyrosine ethyl ester, phosphorescence, emission and excitation spectra of, 145

Achromatic points, in interferometric calibration curves, 458—459

Acid Alizarin Garnet R, aluminium chelate of, fluorescence excitation spectrum of, 75, 76

Acid—base titrations, comparison of photometric and potentiometric techniques for, 255

—, Higuchi plots of, 260, 262

—, indicated, application of photometric techniques to, 258—270

Acids, determination of, in dark resins, by photometric titration, 265

—, organic, determination of, by fluorimetry, 180

—, weak, resolution of, by photometric titration, 257

Acridinium ion, excited-state acidities of, 136

Acrylamide, application of, in enzyme immobilization, 17

Actinomycin, phosphorescence of, 165

Actinomycin D, fluorescence of, 185

Adenine, determination of, by fluorimetry, 182, 183

—, phosphorescence characteristics of, 163

Adenine nucleoside, fluorescence spectrum of, 183

Adenine nucleotides, fluorescence spectrum of, 183

Adenosine, determination of, by fluorimetry, 182

—, phosphorescence characteristics of, 163

Adrenaline, determination of, by fluorimetry, 180

—, —, by photometric titration, 266

Adsorption, as method of enzyme immobilization, 16

Afterglow, 147

Agarose, application of, in enzyme mobilization, 18

Agricultural chemistry, applications of luminescence in, 83

Air, determination of fluoride in, by photometric precipitation titration, with thorium nitrate, 327

—, determination of organosulphur compounds in, by photofluorimetric titration, 346, 350

—, standard, 418

Air-gap electrode, for ammonia, 46

Air pollution analysis, application of phosphorimetry to, 164

Alcohol, determination of, in blood, by interferometry, 479, 497

—, —, in spirits, by interferometry, 479

Alcohols, sorption of, 450

Alcohol dehydrogenase, phosphorescence of, 166

Alcohol electrodes, 54—55, 62

Alcohol oxidase, application of, 54—55

—, specificity of, 48

Aldehydes, aliphatic, phosphorescence of 4-nitrophenylhydrazones of, 165

—, aromatic, fluorescence of 4-nitrophenylhydrazones of, 165

—, —, phosphorescence of, 130

—, determination of, by photometric titration, with sodium borohydride, 362

—, organic functional group analysis of, by photometric titrations, 352

Aliphatic amines, primary, organic functional group analysis of, by photometric titration, 352

Alizarin Garnet R, aluminium chelate, fluorescence of, 220

—, determination of aluminium by, fluorimetrically, 169, 171

Alkaloidal extracts, determination of acidity of, by photometric titration, 208—209

Alkaloids, opium, determination of, by photometric titration, 266

—, reaction with complex anions, study of, by heterometry, 320

n-Alkyl thiols, determination of, by photofluorimetric titration, with tetra-

mercury acetate fluorescein complexes, 345

Alkyl trimethyl ammonium bromides, determination of critical micelle concentration of, 375—376

Alumina, as adsorbent for enzymes, 16

Aluminium(III), as photometric titrant, for fluoride, 323

—, combining ratios of, with fluoride, 323

—, determination of, by fluorimetry, 169, 170

—, —, by photofluorimetric titration, with EDTA, 348

—, —, by photometric titration with EDTA, 297

Aluminium oxide, application of, in gas analysis by interferometry, 446—452

Amides, determination of, by photometric titration, 266

—, —, —, with calcium hypochlorite, 360

—, organic functional group analysis of, by photometric titrations, 352

Amines, aliphatic, determination of primary, by photometric titration, with 2-ethylhexanal, 356

—, —, organic functional group analysis of primary, by photometric titration, 352

—, aromatic, determination of, by photometric titration, with acetic anhydride, 357

—, —, determination of secondary, by photometric titration, with nitrous acid, 361

—, —, organic functional group analysis of, applications of photometric titrations to, 350, 351, 354, 355

—, primary in presence of tertiary, organic functional group analysis of, by photometric titration, 353

Amino acids, fluorescence characteristics of, 179

—, sulphur-containing, determination of, by photometric titration, 360

Amino acid electrodes, applications of, 53—54

—, stability of, 54

Amino acid oxidase, assay of activity of, 15

D-Amino acid oxidase, specificity of, 48
D-Amino acid oxidase electrode, response of, to D-methionine, 39
—, temperature effects on, 39
Aminoalkylthiophosphorus compounds, determination of, by photometric titration, 361
p-Aminoazobenzene, determination of, by photometric redox titration, with alkaline chromium(II), 340
1-Amino-4-hydroxy anthraquinone, determination of beryllium by, fluorimetrically, 170, 171
Ammonia, determination of, in industrial gases, by interferometry, 481
—, electrodes sensitive to, 7, 10
—, enzyme assay by rate of production of, 14
—, sorption of, 448
Ammonia electrode, air-gap, 46, 52
—, comparison of probes based on, 61
—, preparation of membrane on, by direct polymerization, 27
Ammonium ion, determination of, by fluorimetry, using enzymatic procedure, 176
Ammonium ions, application of electrodes selective to, 13
—, electrode sensitive to, 10
Ammonium ion electrode, determination of L-tyrosine with, 22
Ammonium oxalate, application of, as photometric precipitation titrant, 326
Amperometric EDTA titrations, 284
Amplitude division, 401, 407—411
Amygdalin, response of β-glucosidase membrane electrode to, 36, 37
Amygdalin electrode, applications of, 56—57, 63
—, effect of dialysis membrane thickness on, 41
—, effect of electrode sensor on response of, 41
—, effect of silver ions on, 47
—, pH range of, 38
—, response time of, 34
—, stability of, 31
Amylase, phosphorescence of, 165

Analgesics, determination of, by photometric titration, with nitrous acid, 362
Analytical interferometry, analysis of anaesthetic gases by, 434, 483, 491
—, analysis of fuel gases by, 481
—, analysis of germane—air mixtures by, 492
—, analysis of noble gases by, 481
—, application of, to clinical studies, 491
—, —, to determination of colloidal particle size, 532
—, —, to determination of crystallinity of polymer fibres, 535
—, —, to determination of frictional coefficients, 520
—, —, to determination of lung function, 491
—, —, to determination of ocean salinity by, 495
—, —, to determination of permeability of balloon materials, 481
—, —, to determination of virial coefficients, 492
—, —, to investigation of fruit cold-store atmospheres, 488
—, —, to investigation of water purity by, 495
—, —, to liquid chromatography, 496
—, —, to non-aqueous titrations, 497
—, —, to polymer chemistry, 497
—, —, to study of crystal lattice dislocations, 535
—, —, to study of fats, 479
—, —, to study of flames, 532
—, —, to study of gel porosity, 520
—, —, to study of membrane permeability, 520
—, —, to study of oil mixtures, 479
—, —, to study of optical glasses, 535
—, —, to study of optical transformation, 535
—, —, to titrimetry, 497
—, determination of alcohol by, in blood, 479, 497
—, determination of ammonia by, in industrial gases, 481
—, determination of benzene by, in coal gas, 481

—, determination of carbon dioxide by, in mines, 486

—, determination of distribution of refractive index by, in inhomogeneous substances, 529

—, determination of free-electron concentration by, 494

—, determination of gas density by, in electrical discharges, 494

—, determination of gases by, 462—475

—, determination of halothane by, 491—492

—, determination of heavy water by, in water, 495

—, determination of methane by, 483

—, determination of molecular weight by, 520

—, determination of nitrous oxide purity by, 481

—, determination of organic liquids by, 475—480

—, determination of refractive indices of glasses by, 499

—, determination of refractive indices of sodium flames by, 494

—, determination of saccharose by, in boiler-feed water, 495

—, determination of toxic gases by, 488

—, determination of water by, in methanol—water mixtures, 496

—, historical development of, 391—394

Anamorphatic imaging system, application of, in interferometers, 435

Aniline, determination of, by photometric titration, in presence of 2-naphthylamine, 357

—, fluorescence parameters of, 129

—, photometric titration of, in presence of di-n-butylamine and o-chloroaniline, 256

Anilines, substituted, determination of, by photometric titration, 268

Anthracene, absorption and fluorescence spectra of, 76, 77

—, average lifetime of excited state of, 80

—, determination of, by photometric titration, in presence of fluoranthene, 357

—, fluorescence of, in aereated and non-aerated solutions, 87

—, —, solvent effects on, 132

—, fluorescence parameters of, 128

—, phosphorescence intensity of, external heavy-atom effect on, 134

—, quantum yield of, 79

—, resolution of fluorescence emission of, in presence of quinine, 77

Anthraquinone, delayed fluorescence of, 141

—, detection of, on TLC plates, by phosphorimetry, 165

Antibiotics, application of, in liquid membrane electrodes, 7, 9

Anticoagulants, determination of, by phosphorimetry, 163

Antimony, determination of, by fluorimetry, 168

—, determination of, by photometric titration (stepwise), with bromate, 332, 338

Antimycin, fluorescence of, 185

Antioxidants, determination of, by 2,4,6-tri-t-butylphenoxy radicals, 215

—, determination of oxygen in, by photometric titration, with 2,4,6-tri-t-butylphenoxy free radicals, 352

Antipyrene, determination of, by photometric titration, 266

Anti-Stokes fluorescence, 73

Aperture division, 403—407

Arc wander, effects of, in xenon sources, 154

Arginase, assay of activity of, 15

Aromatic amines, organic functional group analysis of, applications of photometric titrations to, 350, 351, 354, 355

Aromatic compounds, excited-state acidities of, 136

—, phosphorescence quantum efficiencies of, 145

Aromatic hydrocarbons, polycyclic, phosphorimetric examination of, in air, 164

—, polynuclear enhancement of phosphorescence of, 157

Arsenate, determination of, by phosphate ion electrode, 58

Arsenic, determination of, by fluorimetry, 168, 171

550

Arsenic(III), determination of, by photometric redox titration, with $K_3Mo(CN)_8$, 341
—, —, —, with osmium(VIII), 341
—, —, by photometric titration, with alkaline permanganate, 333
—, —, —, with cerium(IV), 332, 338
—, —, by photometric titration (stepwise), with bromate, 332, 338
—, —, by UV photometric titration, with cerium(IV), 213
Arsenic(V), determination of, by photometric titration, 376
Arsenicals, determination of, by photometric titration, 307
Arsine, sorption of, 448
Ascorbic acid, interference of, in determination of glucose, 50
Aspirin, determination of, by phosphorimetry, 163
—, fluorescence of, 185
Atomic absorption inhibition titrations, 369, 372
Atomic fluorescence, 74
Atropine, fluorescence of, 185
Average lifetime, of excited states, table of, 80
Azo dyes, use of, in photometric titrations, 232

Bandpass, equation governing, 105
Barbituric acid, determination of, by photometric titration, 268
Barium, determination of, by photometric titration, with EDTA, 297, 299, 301
—, —, by spectropolarimetric titration, with D(—)CDTA, 370
—, —, —, with PDTA, 368
Barium—calcium mixtures, analysis of, by photometric titration, with EDTA, 282
Barium chloride, application of, as photometric precipitation titrant, 324, 325
Barium perchlorate, application of, as photometric precipitation titrant, for sulphate, 330
Barium(II)—strontium(II) mixtures, analysis of, by photometric precipitation titration, with sodium sulphate, 330

Barium sulphate, light scattering by, dependence on particle size, 317—318
—, suspensions of, scattering by, 224
Barium sulphate particles, charge on, 318, 319
Barium titanate, determination of refractive index of, 501
Bases, organic, determination of, by photometric titration, 266
—, weak, determination of, 267, 268
—, —, photometric titration of, in non-aqueous solvents, 254
Becquerel phosphoroscope, 144, 152
Bentonite, as adsorbent for enzymes, 16
Benzaldehyde, decay time of, 144
—, determination of, in presence of benzoic acid, by phosphorimetry, 160
1,2-Benzanthracene, phosphorescence intensity of, external heavy-atom effect on, 134, 157
Benzene, determination of, in coal gas, by interferometry, 481
—, fluorescence of, in aerated and non-aerated solutions, 87
—, fluorescence parameters of, 128
—, purification of, for fluorimetry, 118
Benzenes, monosubstituted, fluorescence of, 129
1,2-Benzfluorene, enhancement of phosphorescence of, 157
—, phosphorescence intensity of, external heavy-atom effect on, 134
2,3-Benzfluorene, phosphorescence intensity of, external heavy-atom effect on, 134
Benzoic acid, determination of, by photometric titration, 265
—, —, in presence of benzaldehyde, by phosphorimetry, 160
Benzoic acid salts, p-substituted, determination of, 268
Benzoin, application of, to fluorimetric determination of metal ions, 167
—, determination of boron by, fluorimetrically, 170, 171, 174
Benzonitrile, fluorescence parameters of, 129
Benzophenone, decay time of, 144

3,4-Benzpyrene, phosphorescence intensity of, external heavy-atom effect on, 134, 157

Benzyl alcohol, as internal standard, in phosphorimetry, 145

Beryllium, determination of, with 1-amino-4-hydroxy anthraquinone, by fluorimetry, 170

Bicarbonate—carbonate mixtures, analysis of, by photometric titration, 268

Bifunctional reagents, application of, to enzyme immobilization, 16—17

Bilateral slits, 105

Biological pigments, measurement of formation and destruction rate of, 209

Bioluminescence, 90

Biot and Arago, law of, 463

Birefringence, 412—413

Birefrigent substances, optical axis of, 412

Bisdiazobenzidine-2,2′-disulphonic acid, application of, in enzyme immobilization, 17

Bismuth, determination of, by fluorimetry, 168

—, —, by photometric extractive titration, with dithizone, 310

—, —, by photometric titration, with EDTA, 296, 297, 301

—, —, by spectropolarimetric titration, with D(—)PDTA, 371

—, EDTA complex of, UV absorption spectrum of, 277

Bismuth alloys, determination of cerium in, by photometric redox titration, with potassium permanganate, 339

Bismuth—lead mixtures, analysis of, by photometric titration, with EDTA, 276, 277, 278, 296

Bismuth perchlorate, absorption spectrum of, 277

Blaze angle, 100, 101

Blood, determination of alcohol in, by interferometry, 479, 497

—, determination of calcium in, by photometric titration, with EDTA, 297

—, determination of sulphur drugs in, by phosphorimetry, 163

—, determination of urea in, interferences in, 45

Boiler-feed water, determination of saccharose in, by interferometry, 495

Borate, determination of, by phosphate ion electrode, 58

—, —, by photometric redox titration, with alkaline chromium(II), 340

—, —, by photometric titration, 264

Boron, determination of, by fluorimetry, 170, 171

"Break-points" in interferometric calibration curves, 456, 458—459

Bromate, application of, to determination of antimony(III) and arsenic(III), by stepwise photometric titration, 332, 338

Bromobenzene, fluorescence parameters of, 129

p-Bromophenol, photometric titration of, 253

Butadiene, determination of, by photometric titration, with tetracyanoethylene, 357

n-Butanol, purification of, for fluorimetry, 118

Cadmium, determination of, by fluorimetry, 171

—, —, by photometric extractive titration, with dithizone, 310

—, —, —, with PAN, 310

—, —, by photometric precipitation titrant, in alloys, 325

—, —, by photometric titration with EDTA, 298, 301

—, —, by photometric two-phase extractive titration, 302

—, —, by spectropolarimetric titration, with D(—)PDTA, 371

Cadmium ions, electrodes sensitive to, 9

Cadmium selenide detectors, for photometric titrators, 239

Cadmium sulphide detectors, for photometric titrators, 239

Cadmium—zinc mixtures, analysis of, by photometric titration, with EDTA, 275, 296, 297

—, —, —, with EGTA, 299

Calcein, application of, as indicator in spectrofluorimetric titrations, 343, 347
—, determination of calcium by, fluorimetrically, 171
Calcite, birefringence by, 413
Calcium, determination of, by fluorimetric titration, with EDTA, 343, 347
—, —, by photometric precipitation titration, in limestone, 325
—, —, —, in rubber, 326
—, —, by photometric titration with EDTA, 300, 301
—, —, by spectropolarimetric titration, with D(—)CDTA, 370
—, —, —, with PDTA, 368
—, —, in limestone, cement and serum, by photofluorimetric titration, 348
Calcium—barium mixtures, analysis of, by photometric titration, with EDTA, 282, 297
Calcium ions, electrodes sensitive to, 8, 9
Calcium—magnesium mixtures, analysis of, by photometric titration, with EDTA, 275, 296
—, —, —, with EGTA, 299
Calcium—manganese mixtures, analysis of, by photometric titration, with DTPA, 298
Calibration curves, interferometric, discontinuities in, 456, 458
—, phosphorimetric, 156
Caprolactam, interferometric calibration curve for, 457
Carbazole, fluorescence of, in aerated and non-aerated solutions, 87
Carbonate—bicarbonate mixtures, analysis of, by photometric titration, 268
Carbonate ion, determination of, by photometric titration, 253
Carbon dioxide, determination of, by interferometry, in air—CO mixtures, 474
—, —, —, in animal stalls, 488
—, —, —, in mines, 486
—, —, by photometric titration, 267
—, dissolved, determination of, by photometric titration, 253
—, effect of dissolved, on refractive index of a liquid, 476, 477, 478

—, electrodes sensitive to, 7, 10
—, interferometric calibration curve for, 457
—, sorption of, 448
Carbon dioxide gas electrode, application of, 62
Carbon dioxide sensor, application of, in amino acid electrodes, 53
—, —, in urea electrode, 52
Carbon disulphide, determination of, by photofluorimetric titration, in air, 346, 349, 350
—, sorption of, 449
Carbon monoxide, determination of, in air—CO_2 mixtures, by interferometry, 474
—, sorption of, 448
Carbon tetrachloride, Raman bands of, for various mercury lines, 122
Carbonyl compounds, aromatic, intersystem crossing in, 130
—, identification of, in atmospheric dust, by phosphorimetry, 164
Carboxymethyl cellulose, application of, in enzyme immobilization, 18
Carcinogens, phosphorescence of, effect of heavy atoms on, 157
Carminic acid, determination of molybdenum by, in steel, by fluorimetry, 175
—, determination of tungsten by, fluorimetrically, 172
Carriers, for enzyme immobilization, 18
Catalase, immobilized, 4
—, phosphorescence of, 166
Cathodoluminescence, 90
Cauchy's dispersion equation, 419
—, constants for, 420
Cellophane, application of, in enzyme electrode, 25—26
—, effect of, on enzyme electrode stability, 31
Cells fused quartz, 103
—, fused silica, 103
—, native fluorescence of, 103
—, Pyrex, 102
—, spectrophotometric, care of, 115—116
—, spectrophotometric, transmittance of, 114—115

Cellulose, application of, in enzyme immobilization, 18
Cement, determination of calcium, magnesium and iron in, by photofluorimetric titration, 348
Cerium(III), as phototurbidimetric titrant, for fluoride, 322, 323
—, determination of, by fluorimetry, 168, 171
—, —, by photometric redox titration, in bismuth alloys, 339
—, —, —, with potassium permanganate, 339
—, —, by photometric titration, with electrolytically generated Co(III), 338
Cetyl pyridinium chloride, application of, as photometric precipitation titrant, 325
Charcoal, application of, in gas analysis, by interferometry, 446—452
—, as adsorbent for enzymes, 16
Chelates, fluorescence of, analytical applications of, 169, 171
—, fluorescent, quenching of, 169
Chelate exchange titrimetry, 372—375
Chelating agents, optically active, determination of metal ions with, by stereospecific titration, 366
Chelometric titrations, photometric, historical aspects of, 213
—, —, with indicator, theory of, 284—295, 302
—, self-indicating, 272—273, 274—284
—, slope indication, 272, 274—284
—, step indication, 273, 284—302
—, theory of, 270—272
—, use of masking agents in, 276
Chemiluminescence, 90
—, applications of, 342, 347
Chloride, determination of, by photometric titration, with mercury(II) nitrate, 313
—, —, by precipitation titration, 316
Chlorine, effect of dissolved, on refractive index of a liquid, 477, 478
—, sorption of, 448
o-Chloroaniline, determination of, by photometric titration, 264

Chlorobenzene, fluorescence parameters of, 129
Chlorofluorescein, fluorescence of, 130
Chloroform, Raman bands of, for various mercury lines, 122
Chloropromazine, determination of, by phosphorimetry, 163
—, fluorescence of, 185
Chlorophyll, average lifetime of excited state of, 80
—, quantum yield of, 79
Cholesterol, determination of, by fluorimetry, 181
Cholinesterase, assay of, using sulphide electrode, 64
—, determination of, by fluorimetry, 188—189
—, determination of activity of, 12, 14
—, immobilized, applications of, 14
—, substrates for, in fluorimetric assay, 189
Cholinesterase electrode, 59
Chromate, determination of, by fluorimetry, 168
—, —, by photometric titration, 264
Chrome steel, determination of manganese in, by photometric titration, 332, 336
Chromium, determination of, by fluorimetry, 171
—, —, by photometric redox titration, in steel, 338
—, —, by photometric titration, in nichrome alloy, 212, 332, 337
—, —, —, with alkaline permanganete, 334, 339
—, —, by spectropolarimetric titration, with D(—)PDTA, 370
Chromium(II), alkaline, application of, as photometric redox titrant, 340
—, application of, in photometric redox titrimetry, 335
Chromium(VI), determination of, by photometric redox titration, with iron(II), 341
Chromium(III)—thallium(I) mixtures, analysis of, by photometric titration, with cerium(IV), 333

554

Chymotrypsin, immobilized, 4
Citrate, determination of, by photometric titration, 266
Citric acid, determination of, by fluorimetry, 181
Clinical pathology, applications of luminescence in, 83
Cobalt, determination of, by fluorimetry, 171
—, —, by luminol, 179
—, —, by photofluorimetric titration, with oxine-5-sulphonic acid complex, 347
—, —, by photometric extractive titration, with dithizone, 310
—, —, by photometric titration, 212
—, —, —, with EDTA, 298, 299
—, —, by spectropolarimetric titration, with D(—)PDTA, 370
Cobalt(II) perchlorate, application of, as photometric titrant, 313
Cobalt(III), application of, as photometric redox titrant, 338
—, application of oxalate ions of, in photofluorimetric titrations, 344
Cocaine, determination of, by phosphorimetry, 163
Cocoa butter, studies of, by interferometry, 496
Codeine, fluorescence of, 185
Coefficients, side reaction, 272
Coenzymes, determination of, by fluorimetry, 181
—, insolubilized, in electrodes, 60
Co-factor, effects of leaching of, from enzyme electrode, 33
Co-factors, incorporation of, in enzyme electrodes, 60
Coherence length, 400—401
Coherence time, 400
Collisional quenching, 90
Colloids, protective, 221
Colloid particles, determination of size of, by interferometry, 532
Complexes, composition of, by "slope-ratio" method, 270
—, determination of formulae of, by photometric titration, 213
Complexometric titrations, indicator distribution during, 295

—, involving photometric reagents, 302—308
—, photometric, historical development of, 270
—, —, with indicator, titration curves for, 286
—, Ringbom error plots for, 290
—, self-indicating, using complexes of low stability, 284
—, theory of sequential determination of two metals by, 281—284
—, validity of "Higuchi plots" for, 290
—, visual end-point determination in, theory of, 286
—, with dithizone, applications of, 304—305, 307
—, with indicator, precision of, 288
Concentration quenching, 88—89, 143
Conditional stability complex, 272
Conjugation, effect of, on fluorescence intensity, 127—128
Coolants, choice of, for phosphorimetry, 153, 154
Copper, determination of, by fluorescence quenching, 172
—, —, by luminol, 179
—, —, by photofluorimetric titration, with oxine-5-sulphonic acid complex, 347
—, —, by photometric extractive titration, with dithizone, 310
—, —, —, with PAN, 310
—, —, by photometric redox titration, with iodine, 337
—, —, by photometric titration, with EDTA, 274, 296, 297, 298
—, —, —, with triethylene tetramine, 299
—, —, by spectropolarimetric titration, with D(—)PDTA, 370
Copper(II), effect of, on intersystem crossing, 133
Copper—bismuth mixtures, analysis of, by photometric extractive titration, with dithizone, 310
Copper ions, electrodes sensitive to, 9
—, enzyme inhibition by, 48
Copper(II)—iron(II) mixtures, analysis of, by photometric titration with EDTA, 276

Copper(II) perchlorate, application of, as photometric titrant, 313, 362

Copper(II)—zinc mixtures, analysis of, by photometric titration, with dithizone, 305, 308

Coproporphyrin, fluorescence of, 185

Cotton region, 367

Coulometric titrations, application of photometric techniques to, 214

—, photometric detection of end-points in, 333, 338

Coumarins, fluorescence of, 128

Coupling reactions, application of, in enzyme immobilization, 19—20

Critical micelle concentration, determination of, 375—376

Crosslinking, effect of, on response of enzyme, electrodes, 43

Crystal lattice dislocations, study of, by interferometry, 535

Cuvettes, background fluorescence of, 114—115

Cuvette length, choice of, in interferometric analysis, 466—468

Cyanide, determination of, by fluorimetry, with quinones, 167, 170, 171

—, —, by immobilized β-cyano alanine synthase, 13

—, —, by immobilized β-cyanoalanine synthase, 13

—, —, by photometric titration, with mercury(II) nitrate, 313

β-Cyanoalanine synthase, application of immobilized, in cyanide determinations, 13

Cyclohexane, Raman bands of, for various mercury lines, 122

D-(—)-trans-1,2-cyclohexanediamine tetraacetic acid, application of, as spectropolarimetric titrant, 367

Cysteine, determination of, by photometric titration, 360

Dark-field procedure, 415

Deaminase enzymes, assay of, 14

Debye scattering, 224

Decay time, phosphorescence, heavy-atom effect on, 157

Dehydrogenases, assay of, by fluorimetry, 193—195

Delayed fluorescence, 140

2-Deoxy-D-glucose, oxidation of, by glucose oxidase, 2

Destructive interference, 402

Detergents, fluorescence of, 118

Deuterium lamp, 94

Dewar flask, quartz, for phosphorimetry, 149, 152, 153

Dextranase, immobilized, 4

Dextrans, thiolated, determination of, by photometric titration, with salyrganic acid, 361

Diakyl aluminium hydride, determination of, by photometric titration, with isoquinoline, 362

Dialysis membrane, application of, in enzyme electrode storage, 28

—, effect of thickness of, on enzyme electrode response, 40—41

Diamagnetic ions, acetylacetone complexes of, phosphorescence of, 157

—, fluorescent chelates of, 169

Diaminocyclohexanetetraacetic acid, thorium complex of, determination of ethane-1-hydroxy-1,1-diphosphonic acid by, 365

2,3-Diaminonaphthalene, application of, to determination of nitrate, 177

—, determination of selenium by, fluorimetrically, 172

Diazo coupling reactions, application of, in determination of phenols and aromatic amines, by photometric titration, 358

Diazo coupling titrations, two-phase, application of photometric titrimetry to, 215

Diazonium salts, applications of, as photometric titrants, 355

1,2,5,6-Dibenzanthracene, phosphorescence of, heavy-atom effect on, 157

4',5'-Dibromofluorescein, phosphorescence parameters of, 148

Di-n-butylamine, determination of, in presence of N,N-diethylaniline, by photometric titration, 256

Dibutyryl fluorescein, application of, in determination of lipase, 190—191

Dichlorofluorescein, effect of micelles on, 376

Dichroic materials, 412

Dichromate, interference from, in fluorimetry, 86

Dichromate titrations, historical evaluation of photometric titrators for, 209

"Dichrotitrator" (of Ringbom et al.), 242

Dicumarol, determination of, by phosphorimetry, 163

Didymium, determination of, in gadolinite, by spectrophotometric titration, 207—208

Dielectric constant, reduction of, in photometric titrations, 252

Dielectric layers, determination of refractive index of, 501

Dienes, Diels—Alder active, determination of, by photometric titration, with tetracyanoethylene, 357

1,3-Dienes, organic functional group analysis of, application of photometric titrations to, 350

Diethyldithiocarbamate, titration of, with copper(II) nitrate, 373

Diethyldithiocarbamate chelates, photometric titration of, 333, 341

Diethyl ether, for fluorimetry, effects of peroxides in, 117

Diffraction, 413—418

Diffusion, investigation of, by interferometry, 393, 513—521

Diffusion constant, determination of, 515, 517

Diffusion measurements, applications of, 520—521

Diffusion studies, application of interferometry to, 433—438

1,5-Difluoro-2,4-dinitrobenzene, application of, in enzyme immobilization, 17

4,4'-Difluoro-3,3'-dinitrodiphenylsulphone, application of, in enzyme immobilization, 17

Digitalis, fluorescence of, 185

o,o'-Dihydroxyazobenzene, application of, to fluorimetric determination of magnesium, 168

2,2'-Dihydroxyazo dyes, application of, to fluorimetric determination of metal ions, 167

2,2'-Dihydroxymethines, application of, to fluorimetric determination of metal ions, 167

4',5'-Diiodofluorescein, phosphorescence parameters of, 148

β-Diketones, application of, to fluorimetric determination of metal ions, 167

Dilatometric measurements, interference arrangements for, 410

Dimerization, effect of, on fluorescence yield, 143

Dimethylaniline, fluorescence parameters of, 129

Dimethylformamide, purification of, for fluorimetry, 117—118

Dimethylglyoxime, determination of nickel with, by photometric titration, 306

—, photometric extractive titrations with, 310

Di-2-naphthyl-thiocarbazone, determination of silver, zinc and mercury with, by photometric extractive titration, 311

2,4-Dinitrophenyl hydrazine, determination of, by photometric redox titration, with chromium(II), 340

Dioxan, reduction of dielectric constant by, in photometric titrations, 252

Diphenadione, determination of, by phosphorimetry, 163

Diphenylhydantoin, determination of, in presence of phenobarbital, by photometric titration, 256, 265

Dispersion, 419

Dispersion equations, 419—420

Distillation, purification of solvents by, for phosphorimetry, 155

Dithiocarbamates, determination of, by photofluorimetric titration, with tetramercuryacetate fluorescein complexes, 345, 349

Dithizonates, primary, absorption maxima of, 303

Dithizone, application of, as indicator in determination of lead, by photometric titration, 300
—, complexometric titrations with, buffers for, 304
—, determination of zinc in oils by, in non-aqueous solution, 215
—, photometric titration of zinc with, in lubricating oil additives, 308
—, reactions of, with metal ions, 303
—, titration curves of, with metals, 304, 305 .
—, titration of, with silver nitrate, in benzene, 373
—, use of, as indicator in two-phase extractive titrations, 302
—, —, as photometric titrant, theory of, 303—304
—, —, in photometric extractive titrations, 310—311
—, visible-region absorption of, 303
Dopamine, determination of, by fluorimetry, 180
Drugs, determination of, by fluorimetry, 183, 185
Dust, atmospheric, identification of carbonyl compounds in, by phosphorimetry, 164
Dynode coating, 105

Edlen's dispersion equation, 420
EDTA, see "Ethylenediaminetetraacetic acid"
Effective stability complex, 272
Egg albumin, phosphorescence of, 165
Electrical polarizability, 421
Electrode, air-gap, for ammonia, 46, 52
—, D-alanine, 54
—, alcohol oxidase, applications of, 58
—, D-amino acid oxidase, temperature effects on, 39
—, ammonium ion specific, 13
—, amygdalin, 38, 41, 47
—, cadmium ion selective, 9
—, calcium-selective, 8, 9
—, cholesterol, 63
—, cholinesterase, 59
—, creatinine, 63

—, cyanide-selective, 12
—, dialysis membrane, 25
—, glass, 4
—, glucose, effect of temperature on, 39
—, glucose oxidase membrane, response of, 36, 37
—, β-glucosidase membrane, response of, 36
—, glutamine, 54
—, —, pH range of, 38
—, iodide, interference with, 47
—, D-isoleucine, 54
—, lactic acid, applications of, 56
—, —, limitations of, 62
—, D-leucine, 54
—, D-methionine, 54
—, nitrate ion, application of, 58
—, nitrate-ion selective, 8
—, nonactin, determination of urea by, 45
—, D-norleucine, 54
—, oxygen, 10
—, —, operation of, 12
—, penicillin, 57, 62
—, —, range of application of, 48
—, —, response time of, 39
—, D-phenylalanine, 54
—, L-phenylalanine, 63
—, phosphate ion, applications of, 58
—, platinum, for measuring oxidative enzyme systems, 46
—, potassium-ion selective, 8
—, silver-ion selective, 8
—, sulphide selective, 8, 64
—, urea, response curve of, 41—42
—, —, stability of, 28—30
—, urease, 59, 63
—, D-valine, 54
Electrodes, alcohol, applications of, 54—55, 62
—, amino acid, applications of, 53—54
—, —, stability of, 54
—, ammonia sensitive, 7, 10
—, amygdalin, applications of, 56—57, 63
—, carbon dioxide sensitive, 7, 10
—, enzyme, applications of, 49—65
—, —, application of thermistors in, 24
—, —, assay of substrates by, 49—58
—, —, bacterial action on, 28

558

—, —, comparison of, 61—65
—, —, effects of co-factor leaching from, 33
—, —, effects of crosslinking on, 43
—, —, effects of dialysis membrane on, 40—41
—, —, effects of enzyme concentration on, 36—38, 42
—, —, effects of gel membrane composition on, 43
—, —, effects of immobilization process on stability of, 32—33
—, —, effects of membrane thickness on response of, 40
—, —, effects of pH on, 38—39
—, —, effects of stirring rate on performance of, 34, 35
—, —, effects of temperature on, 39
—, —, effects of washing time on response of, 43—44
—, —, enzyme content of reaction layer of, 31
—, —, factors affecting response time of, 34, 36—41
—, —, interferences in electrode sensor, 44—47
—, —, involving co-factors, 60
—, —, platinum-based amperometric, 44
—, —, preparation details of, 24—27
—, —, principle of operation of, 1
—, —, purification of enzyme for, 22
—, —, range of substrate determinable by, 44
—, —, selection criteria for enzyme for, 21—22
—, —, shape of response curve of, 41—43
—, —, shift in calibration curve of, 27
—, —, stability of, 27—33
—, —, storage of, 28
—, —, techniques of assay with, 11
—, gas, 10
—, glucose, applications of, 49—51
—, —, comparison of, 61
—, —, stability of, 28—29
—, inert metal, 11
—, ion-selective, 4—15
—, —, analytical range of, 4
—, —, assay of enzymes by, 11—15

—, —, basic equation for the response of, 6
—, —, categorization of, 7
—, —, interferences to, 6, 9
—, —, list of commercially available, 5
—, —, selective coefficient of, 6
—, substrate, for enzyme assay, 59, 63—65
—, urea, applications of, 51—52
—, —, comparison of, 61
—, —, pH range of, 39
—, —, selectivity coefficients for, 51
—, uric acid, applications of , 55
Electromagnetic radiation, absorption of, by solutions, 216—218
Electrophoresis, 509—510
—, application of interferometry to, 433—438, 444
—, investigation of, by interferometry, 393
Emission spectra, fluorescence, 219
Emission spectrum, determination of, in fluorimetry, 121
—, presence of Rayleigh, Tyndall and Raman scattering in, 77
Enzymatic reactions, electrochemical monitoring of, 11
Enzymes, adsorption of, on insoluble supports, 16
—, anaerobic redox titration of, with photometric end-point detection, 333, 341
—, assay of, by ion-selective electrodes, 11—15
—, chemically bound, preparation of, 23—24
—, dehydrogenases, fluorimetric assay of, 193—195
—, effect of concentration of, on various enzyme electrode responses, 37, 42
—, entrapped in polyacrylamide gels, preparation of, 23
—, hydrolytic, determination of, 188—193
—, immobilization of, 3—4
—, immobilized, 23
—, —, properties of, 20—21
—, —, susceptibility of, to inhibitors, 48
—, incorporation of, in enzyme electrodes, 24—27

—, insolubilization of, 3

—, insolubilized on glass, 4

—, instability of, 3

—, methods of immobilizing, 15—20

—, microencapsulation of, 16

—, oxidative, fluorimetric assay of, 195—196

—, phosphatases, fluorimetric assay of, 191—193

—, preparation of soluble "immobilized", 23

—, purification of, 22

—, purity of, 3

—, specificity of, 2—3

—, transaminases, fluorimetric assay of, 193—195

Enzyme deactivation, prevention of, during immobilization, 16

Enzyme electrode, see: "Electrode, enzyme"

Enzyme inhibitors, 48

Enzyme reactions, interferences in, 47—48

Enzymology, application of fluorimetry to, 186—198

Eosin, enhancement of phosphorescence by, 166

EPA, as solvent for phosphorimetry, 154

Epinephrine, fluorescence of, 185

Equidensitometry, 527

Eriochrome Blue-Black R, determination of metal ions with, by photometric titration, 306

Esters, determination of water in, by photometric titration, 351

Ethane, sorption of, 450

Ethane-1-hydroxy-1,1-diphosphonic acid, organic functional group analysis of, by photometric titration, 354

Ethanol, absolute, purification of, for phosphorimetry, 155

—, determination of, by enzyme electrode, 54—55

—, —, by interferometry, in blood, 479, 497

—, fluorescent impurities in, 117

—, Raman bands of, for various mercury lines, 122

—, reduction of dielectric constant by, in photometric titrations, 252

Ethanol—ethyl iodide, as phosphorimetric solvent, 157

Ethers, aromatic, determination of, by photometric titration, with pyridine bromide perbromide, 360

—, determination of water in, by photometric titration, 351

Ethylene, determination of second virial coefficient of, by interferometry, 493

Ethylenediaminetetraacetic acid, applications of photometric titrations with, 296—301

—, determination of, by spectropolarimetric titration, with zinc nitrate, 370

—, determination of copper with, by self-indicating titration, 275

—, photometric titration of iron with, 275

—, simultaneous photometric titration of bismuth and lead with, 278

Ethylenediaminetetraacetic acid titrations, application of photometric techniques to, 213

—, photometric, using indicators, 284—295

—, self-indicating, 274—284

Ethylene dichloride, purification of, for fluorimetry, 118

Evaporated layers, determination of refractive indices of, 501

Excimer formation, effect of, on fluorescence yield, 143

Excimer quenching, 89

Excitation spectrum, 74—75

—, fluorescence, 219

—, multiplicity of peaks in, analytical advantages of, 77

Excitation wavelengths, determination of, in luminescence analysis, 121

Excited-state acidities, 136

Explosion limits, of certain gases mixed with air, 484—485

Extraordinary ray, 412—413

Fats, analysis of, by interferometry, 479

Fatty acids, determination of, on micro scale, by photometric titration, 267

Ferricyanide, application of, in lactic acid electrode, 56

Fibre optics, application of, to photometric titrators, 243
Fick's laws, 513—514
Field absorption procedure, 417
Filters, activation, 99—100
—, application of, in spectrofluorimeters, 120
—, bandpass, 97—98
—, fluorescence of, errors from, 98
—, fluorimeter, care of, 120
—, for use with mercury-vapour lamps, 98
—, neutral tint, 96
—, polarizing, use of, in fluorimetry, 123
—, primary, 99—100
—, secondary, transmittance characteristics of, 96
—, sharp cut-off, 96—97
—, transmittance spectra of, 97
Fire-damp interferometers, 483, 485
Flames, determination of refracting index of, by interferometry, 494
—, study of, by interferometry, 532
Flavanols, application of, to fluorimetric determination of metal ions, 167, 178
Fluoranthene, determination of, by photometric titration, in presence of anthracene, 357
Fluorescein, as fluorescence standard, 124
—, fluorescence characteristics of, 129, 178
—, quantum yield of, 79
—, tetramercurated, determination of sulphur-containing species by, 345
Fluorescein anion, average lifetime of excited state of, 80
Fluorescein-di-(β-D-galactopyranoside), as substrate for β-D-galactosidase assay, 189
Fluorescence, bathochromic shift in, 127
—, concentration quenching of, 88—89
—, delayed, 90, 140
—, dependence of, on concentration, 81
—, —, on environment, 85—90
—, —, on temperature, 81
—, effect of concentration on, 88—89
—, effect of hydrogen bonding on, 136—140

—, effect of oxygen on, 141—142
—, effect of pH on, 134—136
—, effect of temperature on, 140—141
—, effects of environmental factors on, 131—144
—, effects of structure on, 126—131
—, equation relating intensity of to other parameters, 80
—, impurity quenching of, 90
—, influence of substituents on, 128
—, intensity of, relationship between solute concentration and, 221
—, intensity of sources for producing, 81
—, native, applications of, 178—179
—, observed in aerated and non-aerated solutions, 87
—, quartz cuvettes, 114—115
—, oxygen quenching of, 87—88
—, quenching of, 81
—, —, by metal ions, 142
—, —, by solute concentration, 143
—, sensitivity of, to temperature changes, 119
—, solution, 218—221
—, solvent effects on, 131—133
—, standards for, 124
—, structural conditions for, 220
—, temperature quenching of, 86—87
—, transitions associated with, 126—127
—, types of, 73—74
Fluorescence lifetime, 79—80
—, relationship to fluorescence intensity, 80
Fluorescence spectra, 220
Fluorescence titrator, 345
Fluoride, combining ratios with zirconium and aluminium, 323
—, determination of, by fluorimetry, 171
—, —, by photometric precipitation titration, in various media, 327
—, —, —, with lanthanum and yttrium nitrates, 329
—, —, by phototurbidimetric titration, 322
—, —, by spectrofluorimetric phototitration, with thorium(IV), 342, 347
—, —, titrimetrically, effect of silicate on, 331
—, —, —, with thorium(IV), 322

Fluorimeter, schematic diagram of, 82, 91
Fluorimeters, cells for, 102—103
—, —, care of, 115—116
—, —, cleaning of, 116
—, —, detection of deterioration of, 116
—, cell configuration of, 103
—, detectors for, 105—107
—, filter, comparison of common, 108
—, —, some commercial designs, 109—110
—, —, use of, 109
—, filters for, 96—98, 99—100
—, lamps for, hazards associated with, 95
—, mercury phosphor lamps for, 93
—, mercury-vapour lamps for, 92—94
—, slits for, 103—105
—, standardization of, 124
—, xenon lamps for, 94
Fluorimetric methods, limitations of, 85—90
—, sensitivity of, 84
—, specificity of, 84
Fluorimetric titrator, 318
Fluorimetry, application of, to photosynthesis studies, 179
—, assay of dehydrogenases and transaminases by, 193—195
—, complementary nature of, to phosphorimetry, 161
—, determination of amino acids by, 178
—, determination of anticoagulants by, 164
—, determination of cholinesterases by, 188—189
—, determination of dicumarol by, 164
—, determination of drugs by, 185
—, determination of enzymes by, 186—198
—, determination of metal ions by, 167—178
—, determination of organic acids by, 180—181
—, determination of porphyrins by, 184—186
—, determination of vitamins by, 181
—, determination of warfarin by, 164
—, effect of high temperature coefficients in, 119
—, minimisation of photodecomposition in, 120, 123
562

—, reduction of light scattering in, 122—123
—, removal of oxygen in, 88
—, selection of excitation and emission wavelengths in 120—122
—, sensitivity values in, 170
—, solvents for, 116—118
Fluorobenzene, fluorescence parameters of, 129
Fluorofluorescein, fluorescence of, 130
Fluorometers, see: "Fluorimeters"
Fluorometry, see: "Fluorimetry"
Forced dipolar vibration, 223
Formic acid, determination of, by enzyme electrode, 58
—, —, by photometric titration, with alkaline permanganate, 334, 339
Franck—Condon electronic states, schematic representation of, 131
Franck—Codon principle, 131
Fraunhofer lines, 399
Free-electron concentration, determination of, in electrical discharges by interferometry, 494
Free-electrophoresis, 509—510
Free radicals, aromatic, application of, in photometric titrations, 363
Freon 12, interferometric studies of, 492
Fresnel's biprism, interference patterns from, 406
Frictional coefficients, determination of, by interferometry, 520
Fringe number, 423
"Front-surface" excitation, reduction of concentration effects by, 143
Fuel gases, analysis of, by interferometry, 481
Fumaric acid, determination of, by fluorimetry, 181
—, —, by photometric redox titration, with chromium(II), 341

Gadolinite, determination of didymium in, by spectrophotometric titration, 207—208
β-D-Galactosidase, determination of, by fluorimetry, 189

Gallium(III), complex of, with morin, 344
—, determination of, by fluorimetry, 171, 173, 174
—, —, —, in presence of aluminium, 174
Gallium arsenide, determination of impurities in, by photometric extractive titration, with dithizone, 310
Gallium(III)—indium(III) mixtures, analysis of, by photofluorimetric titration, with EDTA, 344, 348, 349
Gas analysis, application of interferometry to, 481
Gas-density, determination of, in electrical discharges, by interferometry, 494
Gases, effect of dissolved, on refractive index of a liquid, 476—478
—, explosion limits for, 484—485
—, refractive powers of, 465—467
Gas mixtures, analysis of, by interferometry, 446—452
—, binary, analysis of, by interferometry, 463—464
—, preparation of, for calibration of interferometers, 445—446
—, refractive indices of, 463
—, ternary, analysis of, by interferometry, 469—475
Gas-permeable membranes, application of, in ion-selective electrodes, 7
Gel lattices, enzyme immobilization by inclusion in, 17
Gel porosity, study of, by interferometry, 520
Germane—air mixtures, analysis of, by interferometry, 492
Germanium, determination of, by fluorimetry, 171, 174
Germanium dioxide, determination of zinc in, by photometric extractive titration, with zinc, 310
Glass, as adsorbent for enzymes, 16
—, determination of refractive index of, by interferometry, 499
—, enzymes immobilized on, 4
—, optical, application of interferometry to study of, 535
Glass electrode, 4
—, critical evaluation of, as probe for urea, 61

Glass fibre chromatograms, examination of, by phosphorimetry, 165
Glass plate, interferogram from, 528
γ-Globulin phosphorescence of, 165
Gluconic acid, production of, by glucose oxidase, 2
Glucose, determination of, 49—50
—, —, by photometric titration, 208
—, —, in commercial samples, 47, 49
—, —, interferences in, 47
α-D-Glucose, oxidation of, by glucose oxidase, 3
β-D-Glucose, oxidation of, by glucose oxidase, 2
Glucose electrodes, applications of, 49—51
—, effect of temperature on, 39
—, stability of, 28—29
—, variation of stability of, with method of preparation, 32
Glucose oxidase, assay of, by iodide selective electrode, 13
—, immobilized, 4
—, —, applications of, 49—50
—, —, —, in oxygen gas electrode, 62
—, —, pH range of, 38
—, oxidation of α-D-glucose by, 3
—, specificity of, 2, 48
β-Glucosidase, assay of, by cyanide-ion selective electrode, 12
—, immobilized, 4
—, —, application of, 56—57
—, —, pH range of, 38
β-Glucuronidase, determination of, by fluorimetry, 190
Glutamate, determination of, 60
Glutamine electrode, pH range of, 38
Glutaraldehyde, application of, in chemically bound enzymes, 23
—, —, in enzyme immobilization, 17
Gold, determination of, by fluorimetry, 170, 171
—, —, by photometric titration, with potassium iodide, 313
Gouy's interference method, 515—516
Grating, blaze of, 101
—, reflection, overlapping orders from, 101

563

—, resolving power of, 100
Gratings, 98, 100—102
—, bandpass widths of, 102
—, typical efficiencies of, 102
—, use of filters in conjunction with, 102
Grating equation, 100
Grease, fluorescence of, 118
Guanine, determination of, by fluorimetry, 182
Guanosine, determination of, by fluorimetry, 182
Gum arabic, application of, as protective colloid, 316

Haematoporphyrin, fluorescence of, 185
Hafnium, determination of, by fluorimetry, 171
Halanometer, 492
Half-intensity width, 400
Halogens, determination of, by photometric precipitation titrant, 326
—, sorption of, 449
Halothane, determination of, by interferometry, 434, 491—492
—, interferometric calibration curve for, 457
Hantzsch reaction, determination of ammonium ion by, 172, 175—176
Hartmann equation, 420
"Heavy-atom effect", 90, 130, 148
—, on carcinogen phosphorescence, 157
—, on phosphorescence decay time, 157
Heavy water, refractive index of, 495
Heptane, purification of, for phosphorimetry, 155
Heterocyclic compounds, fluorescence of, 128
Heterometer, 247
Heterometric—potentiometric curves, application of, 320
Heterometry, 214, 320
Heteropolyacid anions, heterometric titrations with, 320
Hexafluoroarsenate, determination of, by photometric titration, with tetraphenylarsonium hydroxide, 380
Hexafluorophosphate, determination of, by photometric titration, with tetraphenylarsonium hydroxide, 380

Hexane, purification of, for phosphorimetry, 155
Hexokinase, assay of, by fluorimetry, 193
"Higuchi plots", 214
—, of indicated acid—base titrations, 260, 262
—, validity of, for complexometric titrations, 290
Histidine, phosphorescence of, 166
L-(+)-Histidine, application of, as spectropolarimetric titrant, 366, 369
Homovanillic acid, application of, as substrate for peroxidase assay, 195
Horseradish peroxidase, application of, in amino acid electrodes, 53
Huygens' theory, 414
Hydrazine, determination of, in presence of 1,1-dimethylhydrazine, by photometric titration, 267
Hydrobromic acid, applications of, in fluorimetric determination of metal ions, 169
Hydrobromic acid—boric acid mixtures, analysis of, by photometric titration, 268
Hydrocarbons, aromatic, determination of, by photometric titration, with tetracyanoethylene, 357
—, determination of, by interferometry, 486
—, heterocyclic, determination of, by TLC and phosphorimetry, 165
—, polynuclear, determination of, by photometric titration, 357
—, sorption of, 450
Hydrocarbon solvents, purification of, for phosphorimetry, 155
Hydrochloric acid, applications of, in fluorimetric determination of metal ions, 169
Hydrochloric acid—benzoic acid mixtures, analysis of, by photometric titration, 267
Hydrodynamics, application of interferometry to, 532
Hydrogen, interferometric calibration curve for, 457

—, liquid, determination of refractive index of, by interferometry, 498

—, sorption of, 449

Hydrogen bonding, effect of, on fluorescence, 136—140

Hydrogen chloride, sorption of, 448

Hydrogen cyanide, sorption of, 448

Hydrogen fluoride, sorption of, 448

Hydrogen peroxide, enzyme produced, measurement of, 49

Hydrogen sulphide, determination of, by photofluorimetric titration, with tetramercuryacetate fluorescein, 345, 349

—, sorption of, 449

Hydroxide, titration of, using luminol as chemiluminescent indicator, 342, 347

4-Hydroxyacetophenone, detection of, on TLC plates, by phosphorimetry, 165

Hydroxyanthraquinones, application of, to fluorimetric determination of metal ions, 167

p-Hydroxyazabenzene, determination of, by photometric redox titration, with chromium(II), 340

m-Hydroxybenzoic acid, determination of, by photometric titration, 264

5-Hydroxyindole, shift of fluorescence maximum of, 78

Hydroxynaphthoic acid, application of, to fluorimetric determination of metal ions, 167, 169

2-(2-Hydroxyphenyl)benzoxazole, determination of copper by, fluorimetrically, 172

3-Hydroxypyridine, Stokes shift of, 78

8-Hydroxyquinaldine, determination of, beryllium by, fluorimetrically, 171

8-Hydroxyquinoline, see also: "Oxine"

—, application of, in determination of metal ions, by fluorimetry, 171—172

—, chelates of, photometric titration with dithizone, in benzene, 375

—, determination of magnesium by, fluorimetrically, 174

—, fluorescence yields of, in various solvents, 137, 138

8-Hydroxyquinoline-5-sulphonic acid, determination of silver and manganese by, fluorimetrically, 171—172

5-Hydroxytryptamine, phosphorescence of, 144

Hypochlorite, determination of, by photometric redox titration, with alkaline chromium(II), 340

Imides, determination of, by photometric titration, 266

Indicators, complexometric, classification of, 287

—, for photometric precipitation titrations, 321

—, metallochromic, sensitivity of, 289

Indium, determination of, by fluorimetry, 171

—, —, by spectropolarimetric titration, with D(—)PDTA, 370

Indium(III)—gallium(III) mixtures, analysis of, by photofluorimetric titration, with EDTA, 344, 348, 349

Indole, as fluorescence standard, 124

—, phosphorescence of, 144

—, phosphorescence spectrum of, 166

Indoles, fluorescence of, 128

Indoleacetic acid, fluorescence of, temperature sensitivity of, 119

—, temperature quenching of, 86

Indoxyl, fluorescence characteristics of, 178

Indoxyl acetate, application of, in fluorimetric assay of enzymes, 188

Inhibitors, enzyme, 48

Inner cell effect, 81, 89

Inorganic compounds, direct determination of, by luminescence analysis, 169

Insulin, phosphorecence of, 166

Interference, destructive, 402

—, mathematical treatment of, 402—403

—, measurement of refractive index by, 423—424

Interference arrangements, optical diagrams of, 404, 406, 408, 410, 411

Interference contrast, effect of light scattering on, 453

Interference filters, use of, in photometric titrators, 238

Interference measurements, in UV and IR, 399

—, theoretical background of, 397—424
Interference pattern, condition for the establishment of, 453—455
Interferogram, from glass plate, 528
—, interpretation of, 526, 528
Interferometer, Jamin, 428—430, 431
—, Jamin autocollimating, 432—433
—, Jamin—Doi, 432—433
—, Mach—Zehnder, 437, 439, 440, 442
—, Michelson—Kinder, 430—431
—, mine-gas, 433, 434
—, Rayleigh—Löwe, 424—428, 431
—, Rayleigh—Löwe autocollimating, 431—432
—, Rayleigh—Svensson, 437, 438
Interferometers, application of, to diffusion studies, 433—438
—, —, to electrophoresis, 433, 438, 444
—, —, to study of stratified substances, 433—437
—, application of anamorphatic imaging systems in, 435
—, autocollimating, 432
—, calibration of, 455—459
—, commercial, 442
—, diffraction, 440, 441
—, explosion-proof, applications of, 483
—, gas-mixtures for calibration of, 445—446
—, large-aperture, 526
—, lateral-shearing, 440—441, 442
—, portable, 431—433, 434
—, —, limitations of, 455
—, temperature control in, 444
Interferometric analysis, see: "Analytical interferometry"
Interferometric measurements, light sources for, 441, 443—444
Interferometry, analytical, see: "Analytical interferometry"
—, application of, to electrophoresis, 509—513
—, —, to investigations of stratified substances, 502—508
—, —, to sedimentation studies, 524
—, determination of refractive indices of thin-layers by, 499—500
—, measurement of refractive index of liquid hydrogen by, 498

Intersystem crossing, 72, 73, 220
—, effect of heavy halogen substitution on, 130
—, effect of paramagnetic ions on, 133
Iodate, determination of, by photometric redox titration, with alkaline chromium(II), 340
Iodide, determination of, by fluorimetry, 168
—, —, by photometric rdeox titration, with potassium iodate, 339
—, effect of, on intersystem crossing, 133
—, photometric titration of, with iodate, 211
—, quenching effect of, 85
Iodide electrode, interferences associated with, 47
Iodide membrane sensor, application of, 50
Iodine, sorption of, 448
Iodobenzene, fluorescence parameters of, 129
Iodofluorescein, phophorescence of, 130
Ion exchange, immobilization of acetylcholine by, 65
Ion-exchange resins, as adsorbent for enzymes, 16
Ion-pair partition titrations, 377—380
Iron(II), determination of, by fluorimetry, 168
—, —, by photofluorimetric titration, with oxine-5-sulphonic acid complex, 347
—, —, by photometric redox titration, in titanium sponge, 339
—, —, —, with cobalt(III), 338
—, —, —, with potassium dichromate, 337
—, —, by photometric titration, in nichrome alloy, 212, 332, 337
Iron(II), determination of, by spectropolarimetric titration, with D(—)PDTA, 370
Iron(III), determination of, by photometric titration, with 1,2-dihydroxybenzene-3,5-disulphonate, 214
—, —, —, with EDTA, 274, 297
—, —, in aluminium alloys, by photometric titration with EDTA, 296

—, effect of, on intersystem crossing, 133

Iron(III) citrate, production of iron(II) from, in photometric titrations, by photochemical decomposition, 354

Iron(III)—copper(II) mixtures, analysis of, by photometric titration with EDTA, 276

Iron(II)—thallium(I) mixtures, analysis of, by photometric titration, 333

Iron(III)—thiocyanate complex, historical study of, by photometric titrator, 209

Isocitric acid, determination of, by fluorimetry, 181

Jamin interferometer, 428—430, 431

Jamin—Doi interferometer, 432—433

Karl Fischer titration, application of photometry to, 377

Ketones, aliphatic, phosphorescence of 4-nitrophenyl hydrazones of, 165

—, aromatic, fluorescence of 4-nitrophenylhydrazones of, 165

—, —, phosphorescence of, 130

—, determination of water in, by photometric titration, 351

Lactic acid, determination of, by enzyme electrode, 58, 62

Lactic acid electrode, 56

Lacto globulin, phosphorescence of, 166

Lamps, for fluorimeters, hazards associated with, 95

—, mercury—phosphor, 93

—, mercury-vapour, 92

—, xenon, high-pressure, 94

Lanthanoids, determination of, by spectropolarimetric titration, with D(—)-PDTA, 371

Lanthanum(III), as titrant for fluoride, 323, 329

—, electrogenerated, application of, as photometric precipitation titrant, 327

Lead, complex of, with EDTA, UV absorption spectrum of, 277

—, determination of, by fluorimetry, 168

—, —, by photofluorimetric titration, with EDTA, 348

—, —, by photometric precipitation titrant, 325

—, —, by photometric titration with EDTA, 300, 301

—, —, —, with sodium sulphide, 212

—, —, by photometric two-phase extractive titration, 302

—, —, by spectropolarimetric titration, with D(—)CDTA, 371

—, —, —, with D(—)PDTA, 371

—, determination of silver and copper in, by photometric extractive titration, with dithizone, 310

Lead—bismuth mixtures, analysis of, by photometric titration, with EDTA, 276, 277, 278

Lead(II) ions, electrodes sensitive to, 9

Lead perchlorate, absorption spectrum of, 277

Lenses, testing of, by interferometry, 441

Liebig—Denigés titration, determination of nickel in steel by, 326

Ligands, application of, to fluorimetric determination of metal ions, 167

—, monodentate, application of, as photometric titrants, 314

Light, coherence of, 400—401

—, coherent, production of, 401

—, linearly polarized, 412

—, monochromaticity of, 400—402

—, polarization of, 412

—, wave nature of, 397—400

Light scattering, by precipitates, effect of particle size on, 317

—, in luminescence analysis, 122—123

—, polarization of, 123

Limestone, determination of calcium, magnesium and iron in, by photofluorimetric titration, 348

Lipase, determination of, by fluorimetry, 190—191

—, substrates for, in fluorimetric assay, 191

Liquid chromatography, application of interferometry to, 496

Liquid membrane electrodes, 7

Liquids, analysis of, by interferometry, 475—480

567

—, stratified, application of interferometry to investigation of, 393

Lithium, determination of, by photometric titration, with copper(II) perchlorate, 313

Luciferase, specificity of, 2

Luciferin, oxidation of, by luciferase, 2

Lumiflavin-3-acetic acid, titration of, with sodium dithionite, 341

Luminescence, applications of, 83

—, classification of, 90

—, effect of halogen substitution on, 148

—, effect of hetero atom substitution on, 130

—, history of, 71—72, 82

—, instrumentation for, 91—114

—, theory of, 72—73

Luminescence analysis, application of, to compounds in petroleum, 164

—, calibration of phosphorimeters for, 156

—, correction of spectra in, 125—126

—, determination of peroxide by, 179

—, effect of hydrogen bonding in, 136—140

—, effect of pH on selectivity control in, 135

—, phosphoroscopic resolution in, 144

—, selection of excitation and emission wavelengths in, 120—122

—, solvent effects in, 131—133

—, wavelength calibration in, 125

Luminescence instrumentation, cells for, 102—103

—, cell care in, 115—116

—, cell configuration in, 103

—, commercial fluorimeters, 108—110

—, commercial spectrofluorimeters, 111—114

—, comparison of filter and grating instruments, 119—120

—, control of cell temperature in, 119

—, detectors for, 105—107

—, filters for, 96—98, 99—100

—, gratings for, 100—102

—, lamps for, 92—94

—, monochromators for, 95—102

—, slits for, 103—105

Luminol, applications of chemiluminescence of, 342, 347

—, as chemiluminescent indicator in, photometric titrations, 213

—, determination of, by luminescence analysis, 179

Mach—Zehnder interferometer, 437, 439, 440, 442

Magnesium, determination of, by fluorimetry, 168, 172, 174—175

—, —, by photofluorimetric titration, in limestone, cement and serum, 348

—, —, by photometric titration, with EDTA, 298, 298, 301

—, —, by spectropolarimetric titration, with D(—)CDTA, 370

Magnesium—manganese mixtures, analysis of, by photometric titration, with DTPA, 298

Magnetic stirrers, effect of, on efficiency of photometric titrators, 237

Maleic acid, determination of, by photometric redox titration with chromium-(II), 341

Malic acid, determination of, by fluorimetry, 180, 181

l-Mandelic acid, determination of, by spectropolarimetric titration, with sodium hydroxide, 370

Manganese, determination of, by fluorimetry, 172, 175

—, —, by photofluorimetric titration, with oxine-5-sulphonic acid, complex, 347

—, —, by photometric titration, with bromate, 212

—, —, —, with EDTA, 298

—, —, by spectropolarimetric titration, with D(—)PDTA, 370

Manganese(II) ions, quenching by, 142

Manganese—calcium mixtures, analysis of, by photometric titration with DTPA, 298

Manganese—magnesium mixtures, analysis of, by photometric titration, with DTPA, 298

Masking agents, use of, in chelometric titrations, 276

Membrane permeability, study of, by interferometry, 520

568

Mercaptans, determination of, by photo-fluorimetric titration, with tetramer-curyacetate fluorescein, 345, 349
—, —, by photometric titration, with mercury perchlorate, 361
—, organic functional group analysis of, application of photometric titrations to, 350
Mercaptosuccinic acid, determination of, by photometric titration, 361
Mercury, determination of, by fluorimetry, 171
—, —, by photofluorimetric titration, with EDTA, 348
—, —, —, with oxine-5-sulphonic acid complex, 347
—, —, by photometric extractive titration, with di-2-naphthyl-thiocarbazone, 311
—, —, —, with dithizone, 310
—, —, by photometric precipitation titrant, 325
—, —, by photometric titration, with dithizone, 305
—, —, —, with EDTA, 298
—, —, —, with potassium thiocyanate, 313
—, —, by spectropolarimetric titration, with D(—)PDTA, 371
—, emission lines of, in wavelength calibrations, 125
Mercury(II) ions, enzyme inhibition by, 48
Mercury(II) nitrate, application of, as photometric titrant, 313
Mercury—phosphor lamps, 93
Mercury-vapour lamp, 92—94
—, as interferometric light source, 443—444
—, relative intesity of spectral lines of, 92
Mesaporphyrin, fluorescence of, 185
Metal dithizonates, photometric titration of, with diethyldithiocarbamate, 373, 374
Metal ions, determination of, by fluorimetry, 167—169
—, —, by stereospecific titration, with optically active chelating agents, 366
—, fluorescence quenching by, 142

Metallochromic indicators, sensitivity of, 289
Methane, determination of, by interferometry, 483
—, sorption of, 450
Methanol—water mixtures, analysis of, by interferometry, 496
D-Methionine, response of D-amino acid oxidase electrode to, 39
Methoxychlor, resolution of phosphorescence emission of, 160
6-Methoxy-7-hydroxy-1,2-benzopyrone, application of, as substrate for peroxidase assay, 195
4-Methoxyphenol, excited-state acidities of, 136
2,2'-Methylene dibenzothiazole, application of, in determination of zinc, by fluorimetry, 178
1-Methylguanine, determination of, by fluorimetry, 182
Methylphenobarbitone, complex of, with tetramercuryacetate fluorescein, 345
N-Methylpyrrolidine, determination of, by photmetric titration, 266
4-Methyl umbelliferone, esters of, in lipase assay, 191
Methyl Yellow, determination of, by photometric redox titration, with chromium(II), 340
Michaelis—Menten equation, 34
—, application of, to enzyme electrode response, 44
Michelson—Kinder interferometer, 430—431
Microcuvettes, photodecomposition in, in fluorimetry, 123
Microencapsulation, as method of enzyme immobilization, 16
Microscopy, applications of diffraction techniques in, 415
Mine-gas interferometer, 433
Moiré diffusion curve, 516
Moiré effect, 507
Moiré fringes, 506
Moiré interference patterns, 508
Moiré interference techniques, application of, 527

Moiré method, 424
Moiré methods, determination of diffusion constants by, 515
Molar absorptivity, 217
Molar dispersion, 422
Molecular excitation, by UV or visible radiation, 218
Molecular weight, determination of, by interferometry, 520
—, —, by sedimentation studies, 525
Molecular weight distributions, determination of, in polymers, by light scattering titration, 321
Molybdate, determination of, by phosphate ion electrode, 58
Molybdenum, determination of, by fluorimetry, 172, 175
Molybdenum(VI), determination of, by photometric titration, 306
Monoamine oxidase, determination of, by fluorimetry, 196
Monochromatic light, absorption of, by solutions, 216—218
—, unpolarized, scattering of, 223
Monochromaticity, 400—402
Monochromator, blaze of, 120
—, emission, wavelength calibration of, 125
Monochromators, 95—102
Monovalent ions, electrodes selective to, 9
Mordant Blue 9, determination of aluminium by, fluorimetrically, 170, 171
Morin, application of, as indicator in photofluorimetric titrations, 344
—, determination of aluminium by, fluorimetrically, 169
Morphine, fluorescence of, 185
Myristyldimethylbenzylammonium chloride, determination of critical micelle concentration of, 376

Naphthacene, fluorescence parameters of, 128
—, phosphorescence intensity of, external heavy-atom effect on, 134
Naphthalene, enhancement of phosphorescence of, 157
—, fluorescence of, in aerated and non-aerated solutions, 87

570

—, fluorescence parameters of, 128
—, phosphorescence intensity of, external heavy-atom effect on, 134
—, quantum yield of, 79
Naphthalenes, substituted phosphorescence/fluorescence ratios in, 148
1-Naphthoic acid, excited-state acidities of, 136
2-Naphthol, excited-state acidities of, 136
—, fluorescence of, 134—135
Naphthol AS, derivatives of, in fluorimetric assay of phosphatases, 192
2-Naphtholate (ionic), fluorescence of, 134—135
1-Naphthyl acetate, application of, in fluorimetric assay of enzymes, 188
2-Naphthylamine, determination of, by photometric titration, in presence of aniline, 357
Neodymium nitrate, application of, as photometric precipitation titrant, 324
Neopentane, determination of second virial coefficient of, by interferometry, 493
Nephelometry, 221, 223—224
Neptunium, determination of, by photometric redox titration, 337
Nichrome alloy, determination of iron and chromium in, by photometric titration, 332, 337
Nickel, determination of, by fluorimetry, 172, 177
—, —, by photofluorimetric titration, 347
—, —, by photometric extractive titration, with dimethylglyoxime, 310
—, —, by photometric precipitation titration, in steel, 326, 327
—, —, by photometric titration, in presence of cobalt, 276
—, —, —, with EDTA, 274, 296, 298
—, —, by spectropolarimetric titration, with D(—)PDTA, 370
Nickel—cobalt mixtures, analysis of, by photometric titration, with EDTA, 296
Nickel sulphate, detemination of cobalt in, by photometric titration, 332, 336
Nicotinamide adenine dinucleotide phosphate, reduced form, fluorescence of, 187

Nicotine, determination of, by photometric precipitation titration, with silicotungstic acid, 326
—, —, by photometric titration, with silicotungstate, 212
—, heterometric titration of, with silicotungstate, 320
Nitrate, determination of, by fluorimetry, 172, 176
—, —, by photometric redox titration, 335, 341
Nitrate ions, electrode sensitive to, 8
Nitrate ion electrode, application of, 58
Nitric acid—sulphuric acid mixtures, analysis of, by photometric titration, 266
Nitrite, determination of, by photometric redox titration, with alkaline chromium(II), 340
Nitroaminobenzoic acids, determination of isomers of, by photometric titration, 268
p-Nitrobenzene diazonium chloride, application of, as titrant, in photometric titrations, 358
p-Nitrobenzoic acid, determination of, by photometric redox titration, with chromium(II), 340
4-Nitrobiphenyl, decay time of, 144
2-Nitrofluorene, examination of, on TLC plates, by phosphorimetry, 165
Nitrogen, liquid, as phosphorimetry coolant, 154
—, sorption of, 449
Nitrogen dioxide, sorption of, 449
Nitrogen oxides, quenching effect of, 86
1-Nitronaphthalene, determination of, by photometric redox titration, with chromium(II), 340
5-Nitro-1-naphthylamine, determination of, by photometric titration, 256
Nitroparaffins, primary, determination of, by photometric titration, 364
—, —, organic functional group analysis of, by photometric titration, 354
m-Nitrophenol, comparison of photometric and potentiometric titrimetry of, 255

o-Nitrophenol, determination of, by photometric redox titration, with chromium(II), 340
p-Nitrophenol, photometric titration of, 235
—, titration of, in presence of m-nitrophenol, 252, 255, 264
1-Nitroso-2-naphthol, determination of copper with, by photometric titration, 306
Nitrous acid, application of, as photometric titrant, 361—362
Nitrous oxide, determination of purity of, in anaesthetic gases, by interferometry, 481
—, sorption of, 449
Noble gases, analysis of, by interferometry, 481
Nominal wavelength, 103
Nonactin, as sensor membrane, 9
Nonactin electrode, application of, in determination of urea, 45
Non-aqeous solvents, photometric titrations in, 252, 253—254
Non-aqeous titrations, application of interferometry to, 497
Nucleation, influence of, on turbidimetry, 315
Nylon, application of, in enzyme immobilization, 18

Ocean currents, investigation of, by interferometry, 495
Oestrogens, determination of, by fluorimetry, 185
Oils, analysis of, by interferometry, 479
—, determination of zinc in, by non-aqeous photometric titration, 215
Oil additives, determination of zinc in, by photometric titration with dithizone, 308
Olefins, determination of, by photometric titration, with coulometrically generated bromine, 358
—, organic functional group analysis of, applications of photometric titrations to, 350, 353

Opium alkaloids, determination of, by photometric titration, 266
Optical interference, theory of, 397—400
Optical path length, definition of, 423
Optical rotation titrimetry, 215
Optical rotatory dispersion curves, 367
Optical transformation, 535
Ordinary ray, 412—413
Organic acids, determination of, by fluorimetry, 180
Organic compounds, determination of, by fluorimetry, 178—186
—, determination of fluoride in, by photometric precipitation titration, with thorium nitrate, 327, 328, 329
—, determination of mercury in, by photometric titration, with potassium thiocyanate, 313
—, refractive powers of, 465—467
Organic functional group titrations, application of photometric techniques to, 350—365
Organoaluminium compounds, determination of, by photometric titration, with isoquinoline, 362
Osmium, determination of, by fluorimetry, 168
Oxalate, determination of, by photometric precipitation titration, with electrogenerated lanthanum(III), 327
Oxidases, inhibition of, 48
Oxinates, magnesium and aluminium, determination of, by photometric titration, 268
Oxine, complex of, with tetramercuryacetate fluorescein, 345, 349
—, determination of copper with, by photometric titration, 306
Oxygen, determination of, by photometric titration, in anti-oxidants, 352
—, —, —, in organic compounds, 269
—, —, by 2,4,6-tri-t-butylphenoxy radicals, 215
—, effect of, in phosphorimetry, 142
—, —, on fluorescence yields, 141—142
—, measurement of dissolved, by quenching, 87
—, quenching effect of, in fluorimetry, 87—88

—, —, in phosphorimetry, 158
—, sorption of, 449
Oxygen electrode, 10
—, determination of enzyme activity by, 12
Oxygen gas electrode, with immobilized glucose oxidase, 62
Oxyluciferin, luminescence of, 2
—, production of, by luciferase, 2
Ozone, hazard from, from fluorimeter lamps, 95
—, sorption of, 448

Palladium(II), determination of, by photofluorimetric titration with oxine-5-sulphonic acid complexes, 347
Palladium(II) chloride, application of, in organic functional group analysis, by photometric titration, 353
—, complex with pararosaniline, investigation of, by photometric titration, 314
Papain, immobilized, 4
Paramagnetic ions, metal chelates of, phosphorescence of, 149
—, phosphorescence quenching by, 157
—, sensitivity of phosphorescence to, 147
Pararosaniline, complex with palladium-(II) chloride, investigation of, by photometric titration, 314
Penicillin, fluorescence of, 185
Penicillin electrode, 48
—, application of, 57
—, response-time of, 39
Penicillinase, specificity of, 48
Penicillin β-lactamase, immobilization of, 57
Penicilloic acid, detection of, 57
Pen recorders, errors introduced by, in spectrofluorimetry, 125
Pentacene, fluorescence parameters of, 128
Peri-(1,8,9)-naphthoxanthene, phosphorescence spectrum of, 159
Periodate, determination of, by photometric redox titration, with alkaline chromium(II), 340
Permanganate, determination of, by photometric titration, with peroxide, 337

Permanganate titrations, historical evaluation of photometric titrators for, 209
Peroxidase, determination of, by fluorimetry, 195
—, phosphorescence of, 166
Peroxide, determination of, by luminescence analysis, 179
Peroxides, reduction of luminescence by, 117
Persulphate, determination of, by photometric redox titration, with alkaline chromium(II), 340
Pesticides, phosphorescence characteristics of, 163
—, possible enzyme inhibition by, 48
—, resolution of phosphorescence emission of mixtures of, 160
Petroleum, determination of sulphur- and nitrogen-containing compounds in, by phosphorimetry, 164
pH, effect of, on enzyme electrode response, 38—39
pH electrode, 4
Pharmaceuticals, determination of, by phosphorimetry, 163
Phase-contrast, procedure, 416
Phenanthrene, enhancement of phosphorescence of, 157
1,10-Phenenthroline, application of, in liquid membrane electrodes, 7
—, determination of, by photometric titration, 354
Phenindione, determination of, by phosphorimetry, 163
Phenobarbital, determination of, by phosphorimetry, 163
—, —, by photometric titration, 266
—, —, —, in presence of diphenylhydantoin, 256
Phenol, dependence of fluorescence of, on pH, 134
—, excited-state acidities of, 136
—, fluorescence parameters of, 129
—, phosphorescence spectrum of, 166
—, quantum yield of, 79
Phenols, determination of, by photometric titration, 265
—, organic functional group analysis of, by photometric titration, 353, 355

—, photometric titration of, in n-butylamine, with methanolic sodium hydroxide, 254
—, —, with tetrabutylammonium hydroxide, 256
Phenol-2,4-disulphonyl chloride, application of, in enzyme immobilization, 17
Phenolphthalein, fluorescence of, 129
Phenothiazine derivatives, determination of, by photometric redox titration, with cerium(IV), 341
Phenylalanine, fluorescence characteristics of, 178
—, phosphorescence of, effect of solvent on, 165
Phenyl mercury(II) acetate, enzyme inhibition by, 48
Phosphatases, determination of, by fluorimetry, 191—193
—, substrates for, in fluorimetric assay, 192
Phosphate, determination of, by atomic absorption inhibition titration, 369
—, —, by fluorimetry, using enzymatic procedure, 177
—, —, by photometric titration, 376
Phosphate ion electrode, applications of, 58
Phosphine, sorption of, 449
Phosphorescence, 220
—, afterglow, 147
—, application of, to TLC, 165
—, decay time of, 144
—, external heavy-atom effect on, 134
—, history of, 71
—, linearity of analytical curves for, 156
—, mean lifetime of, 147
—, of fused-quartz cuvettes, 115
—, of proteins, 165—166
—, quenching of, 157
—, —, by oxygen, 158
—, sensitivity of, to heavy atoms, 147
—, —, to paramagnetic ions, 147
—, structural effects on, 148—149
Phosphorescence/fluorescence ratio, effect of heavy atoms on, 133
Phosphorescence signal, heavy-atom effect on, 157

Phosphorescence spectroscopy, *see*: "Phosphorimetry", "Phosphorescence" and "Spectrophosphorimetry"

Phosphorescence spectrum, effect on, by various excitation conditions, 158

Phosphorescent compounds, analysis of mixtures of, 158—160

Phosphorimetry, application of, to analysis of petroleum, 164

—, capillary cells for, 155

—, cell positioning in, 154

—, conduction-cooling devices for, 154

—, conversion of spectrofluorimeters for, 155

—, coolants for, 153

—, determination of anticoagulants by, 163

—, determination of aspirin by, 163

—, determination of chloropromazine by, 163

—, determination of cocaine by, 163

—, determination of dicumarol by, 163

—, determination of diphenadione by, 163

—, determination of ethyl biscoumacetate by, 163

—, determination of multicomponent mixtures by, 158—160

—, determination of pesticides by, 163

—, determination of pharmaceuticals by, 163

—, determination of phenindione by, 163

—, determination of phenobarbital by, 163

—, determination of procaine by, 163

—, determination of sulphonamides by, 163

—, determination of sulphur- and nitrogen-containing compounds by, in petroleum, 164

—, determination of sulphur drugs by, 163

—, determination of warfarin by, 163

—, emission resolution in, 159

—, excitation resolution in, 158—159

—, history of, 144

—, instrumentation for, 149—156

—, —, calibration of, 156

—, modulated light source for, 149

—, phosphoroscopic resolution in, 159

—, precision of, 154

—, —, for clear glasses and snows, 161

—, procedure for, 156

—, purification of ethanol for, 155

—, quartz Dewar flasks for, 149, 152, 153

—, reduction of detection limits by, 161

—, removal of oxygen in, 142

—, rotating quartz sample tubes for, 156

—, rotating sample cells for, 146, 161—162

—, —, construction of, 163

—, sample cells for, 153—154

—, sampling system in, 152

—, solvents for, 145, 146, 150

—, spinner assembly for, 162

—, storage of solvents for, 88

—, time-resolved, 159—160

—, use of aqueous solvents in, 155

Phosphoroscope, Becquerel, 144, 152

—, rotating can, 145, 150

Phosphoroscopes, 149, 150—153

Phosphoroscopic resolution, 144

Photocells, for photometric titrations, 238—139

Photochemical decomposition, production of photometric titrants by, 348, 354

Photodecomposition, diagnosis of, in fluorimetry, 122

—, minimisation of, in luminescence analysis, 120, 123

Photodetectors, limited wavelength-response, 237

Photoemissive surfaces, spectral response curves of, 106

Photofluorimetric titration, analysis of indium(III)—gallium(III) mixtures by, with EDTA, 344, 348, 349

—, determination of aluminium(III) by, with EDTA, 348

—, determination of calcium by, with EDTA or EGTA, in limestone and cement, 348

—, determination of cobalt by, with oxine-5-sulphonic acid complex, 347

—, determination of copper by, with oxine-5-sulphonic acid complex, 347

—, determination of iron by, with EDTA or EGTA, in limestone and cement, 348

—, determination of iron(II) by, with oxine-5-sulphonic acid complex, 347
—, determination of lead by, with EDTA, 348
—, determination of magnesium by, with EDTA or EGTA, in limestone and cement, 348
—, determination of manganese by, with oxine-5-sulphonic acid complex, 347
—, determination of mercury(II) by, with EDTA, 348
—, —, with oxine-5-sulphonic acid complex, 347
—, determination of nickel by, with oxine-5-sulphonic acid complex, 347
—, determination of organo-sulphur compounds by, in air, 346, 350
—, determination of palladium(II) by, with oxine-5-sulphonic acid complexes, 347
—, determination of samarium(III) by, with EDTA, 348
—, determination of sulphur-containing species with, by tetramercuryacetate fluorescein, 345, 349
—, determination of thiourea by, with tetramercuryacetate fluorescein complex, 349
—, determination of transition metal ions by, with quinolinol-5-sulphonic acid, 343—344
—, determination of xanthates by, with tetramercuryacetate fluorescein complex, 349
Photofluorimetric titrations, 342—355
—, application of cobalt(III) oxalate ions in, 344
—, application of gallium(III)—morin complexes in, 344
Photoluminescence, 90
Photometric extractive titrations, applications of, 310—311
Photometric ion-pair partition titrations, 377—380
Photometric precipitation titration, determination of anionic surfactants by, 320—321
—, determination of cadmium by, with sodium bromide, in alloys, 325

—, determination of calcium by, with ammonium oxalate, in rubber, 326
—, —, with sodium oxalate, in limestone, 325
—, determination of fluoride by, 327
—, determination of halogens by, with silver nitrate, 326
—, determination of lead by, with sodium sulphide, 325
—, determination of mercury by, with sodium sulphide, 325
—, determination of nickel by, with dimethylglyoxime, in steel, 327
—, —, with silver nitrate, in steel, 326
—, determination of nicotine by, with silicotungstic acid, 326
—, determination of oxalate by, with electrogenerated lanthanum(III), 327
—, determination of sodium polyvinyl sulphonate by, 325
—, determination of sulphate by, 324, 325
—, —, with barium chloride, 329, 330
—, —, with barium perchlorate, 330
—, determination of surfactants by, 325
—, determination of zinc by, with potassium cyanoferrate(II), in rubber, 326
—, heterogeneous light effects in, 317
—, indicators for, 321
—, self-indicating, applications of, 324—327
—, theory of, 317—318
Photometric redox titrations, theory of, 335
Photometric titration, analysis of bicarbonate—carbonate mixtures by, 268
—, analysis of bismuth—lead mixtures by, with EDTA, 276, 277, 278, 296
—, analysis of cadmium—zinc mixtures by, with EDTA, 275, 296
—, —, with EGTA, 299
—, analysis of calcium—barium mixtures by, with EDTA, 282, 297
—, analysis of calcium—magnesium mixtures by, with EDTA, 275, 296
—, —, with EGTA, 299
—, analysis of copper(II)—iron(III) mixtures by, with EDTA, 276

—, analysis of copper(II)—zinc(II) mixtures by, with dithizone, 308

—, analysis of maganesium—manganese mixtures by, with DTPA, 298

—, analysis of manganese—calcium mixtures by, with DTPA, 298

—, analysis of nickel—cobalt mixtures by, with EDTA, 296

—, analysis of nitric—sulphuric acid mixtures by, 266

—, analysis of thorium—copper mixtures by, with EDTA, 276, 277

—, analysis of zinc—copper(II) mixtures by, with dithizone, 305

—, analysis of zinc(II)—mercury(II) mixtures by, with dithizone, 308

—, determination of acidity of alkaloidal extracts by, 208—209

—, determination of acids in dark resins by, 265

—, determination of adrenalin by, 266

—, determination of aluminium by, with EDTA, 297

—, determination of amides by, 266

—, determination of aminoalkylthiophosphorus compounds by, 361

—, determination of p-aminoazobenzene by, with chromium(II), 340

—, determination of analgesics by, with nitrous acid, 362

—, determination of anthracene by, with tetracyanoethylene and cyclopentadiene, 357

—, determination of antimony(III) by, with bromate/bromide, 332, 338

—, determination of antioxidants by, with 2,4,6-tri-t-butyl phenoxy free radicals, 352

—, determination of antipyrine by, 266

—, determination of arsenic(III) by, with bromate/bromidc, 332, 338

—, —, with cerium(IV), 332, 338

—, —, with osmium(VIII) or $K_3Mo(CN)_8$, 341

—, determination of arsenic(V) by, 376

—, determination of arsenicals by, with di-(p-biphenyl)thiocarbazone, 307

—, determination of barbituric acid derivatives by, 268

—, determination of benzoic acid by, 265

—, determination of borate by, 264, 340

—, determination of bismuth by, with EDTA, 296, 297

—, determination of butadiene by, with tetracyanoethylene, 357

—, determination of cadmium by, with EDTA, 298

—, determination of calcium by, with EDTA, 300

—, —, —, in blood, 297

—, determination of carbon dioxide by, 267

—, determination of carboxyl end groups in polyethylene terephthalate by, 269

—, determination of cerium(III) by, with potassium permanganate, 339

—, determination of m- and o-chloroanaline by, 264

—, determination of chromate by, 264

—, determination of chromium(III) by, with cerium(IV), 333

—, determination of chromium (VI) by, with iron(II), 341

—, determination of chromium and vanadium by, in steel, 338

—, determination of citrate by, 266

—, determination of cobalt by, with EDTA, 298

—, determination of copper by, with EDTA, 274, 296, 297

—, —, with iodine, 337

—, —, with 1-nitroso-2-naphthol, 306

—, —, with oxine, 306

—, —, with triethylene tetramine, 299

—, determination of cyanide by, with mercury(II) nitrate, 313

—, determination of dialkyl aluminium hydride by, 362

—, determination of di-n-butylamine by, in presence of N,N-diethylaniline, 256

—, determination of didymium by, in gadolinite, 207—208

—, determination of diethyldithiocarbamate by, with copper(II) nitrate, 373

—, determination of 2,4-dinitrophenylhydrazine by, with chromium(II), 340

—, determination of dissolved CO_2 by, 253

—, determination of dithizone by, with silver nitrate, 373

—, determination of enzymes by, 333, 341

—, determination of fatty acids by, on micro scale, 267

—, determination of ferrocyanide by, with electrolytically generated cobalt(III), 338

—, determination of fluoranthene by, 357

—, determination of fumaric acid by, with chromium(II), 341

—, determination of glucose by, 208

—, determination of hexafluoroarsenate by, with tetraphenylarsonium hydroxide, 380

—, determination of hydrazine by, in presence of 1,1-dimethylhydrazine, 267

—, determination of hydrochloric acid in presence of benzoic acid by, 267

—, determination of p-hydroxyazobenzene by, with chromium(II), 340

—, determination of m-hydroxybenzoic acid by, 264

—, determination of 8-hydroxyquinoline chelates by, with dithizone, 375

—, determination of hypochlorite by, with chromium(II), 340

—, determination of iodate by, with chromium(II), 340

—, determination of iodide by, with potassium iodate, 339

—, determination of iron(II) by, with cerium(IV), 333

—, —, with electrolytically generated cobalt(III), 338

—, —, with potassium dichromate, 337

—, determination of iron(III) by, with EDTA, 274

—, determination of iron(III) and copper by, in aluminium alloys, with EDTA, 296

—, determination of lead by, with EDTA, 300

—, determination of lithium by, with copper(II) perchlorate, 313

—, determination of magnesium by, with EDTA, 297, 298

—, determination of maleic acid by, with chromium(II), 341

—, determination of manganese by, in chrome steel, 332, 336

—, —, with EDTA, 298

—, determination of mercaptosuccinic acid by, with "p-mercuribenzoate", 361

—, determination of mercury(II) by, with dithizone, 305

—, —, with EDTA, 298

—, determination of metal dithizonates by, with diethyl dithiocarbamate, 373, 374

—, determination of metal ions by, with Eriochrome Blue-Black R, 306

—, determination of N-methylpyrrolidone by, 266

—, determination of methyl yellow by, with chromium(II), 340

—, determination of molybdenum(VI) by, with Eriochrome Blue-Black R, 306

—, determination of neptunium by, 337

—, determination of nickel by, with dimethylglyoxime, 306

—, —, with EDTA, 274, 296, 298

—, determination of nitrite by, with chromium(II), 340

—, determination of nitroaminobenzoic acid isomers by, 268

—, determination of p-nitrobenzoic acid by, with chromium(II), 340

—, determination of 1-nitronaphthalene by, with chromium(II), 340

—, determination of o-nitrophenol by, with chromium(II), 340

—, determination of opium alkaloids by, 266

—, determination of oxinates by, 268

—, determination of oxygen by, in antioxidants, 352

—, —, in organic compounds, 269

—, determination of periodate by, with chromium(II), 340

—, determination of persulphate by, with chromium(II), 340

—, determination of 1,10 phenanthroline by, with iron(II), 365

—, determination of phenobarbital in presence of diphenylhydantoin by, 256, 265
—, determination of phenols by, 265
—, determination of phenothiazine derivatives by, with cerium(IV), 341
—, determination of phosphate by, 376
—, determination of picric acid by, with chromium(II), 340
—, determination of polynuclear aromatic hydrocarbons by, 215, 352
—, determination of porphyrins by, 264
—, determination of primary nitroparaffins by, 364
—, determination of pyridine by, 266
—, determination of quinoline by, 264
—, determination of rare earths by, with EDTA, 297
—, determination of silver by, with EDTA, 297
—, determination of substituted benzoic acid salts by, 268
—, determination of sulphaguanidine by, 266
—, determination of tellurium(IV) by, with permanganate, 341
—, determination of thallium(I) by, with cerium(IV), 333, 341
—, determination of thiocyanate by, with mercury(II) nitrate, 313
—, determination of thiourea by, with potassium iodate, 339
—, determination of thorium by, with EDTA, 299
—, determination of titanium by, using iron(III) and acetylacetone, 334, 339
—, determination of p-toluidine by, 264
—, determination of triphenylguanidine by, 265
—, determination of uranium by, with cerium(IV), 338
—, —, with EDTA, 297
—, —, with Eriochrome Blue-Black R, 306
—, determination of urea by, in glacial acetic acid solution, 261, 262, 265
—, determination of vanadium by, in steel, 332, 336
—, determination of vanadium(IV) by, with iron(II) or bromate, 341

—, determination of very weak bases by, 266
—, determination of water by, in ketones, esters and ethers, 351
—, determination of zinc by, in oil, 215
—, —, with dithizone, 305
—, —, —, in lubricating oil additives, 308
—, —, with EDTA, 298
—, determination of zirconium by, with EDTA, 296, 300
—, "duplication method" of, 234
—, non-aqueous, of phenols, 254
—, of p-bromophenol, 253
—, of diethyldithiocarbamate chelates, 333, 341
—, of mixtures of p- and m-nitrophenols, 255
—, of p-toluidine in 1-butanol, 254
—, of two metals, theory of, 282—283
—, of weak bases, in non-aqueous solvents, 254
—, resolution of weak acids by, 257
Photometric titrations, acid—base, 251—270
—, —, applications of, 264—269
—, —, indicated, 258—270
—, —, self-indicating, 251—257
—, application of, to organic functional group analysis, 350—365
—, —, to study of complexes, 213
—, application of diazo coupling reactions in, 358
—, application of non-aqueous solvents in, 214, 215
—, chelometric, sub-division of, 272—273
—, cobalt(II) perchlorate as titrant for, 313
—, comparison of, with potentiometric titrations, 229—230, 236
—, complexometric, historical development of, 270
—, —, indicated, precision of, 288
—, concentration limits of, 252
—, definition of, 207
—, exploitation of kinetic factors in, 333
—, fluorimetric, see: "Photofluorimetric titrations"

578

—, general theory of, 224—236

—, graphical representation of course of, 235—236

—, heterometry in, 214

—, high-precision, 215

—, Higuchi plots in, 214, 260, 262

—, historical development of equipment for, 208—216

—, instrumentation for, 236—250

—, involving free radicals, 215

—, involving indicators, titration curves for, 227—229

—, involving precipitation, see: "Photometric precipitation titrations"

—, location of equivalence-points in, from poor titration curves, 233

—, monodentate ligand, 313—314

—, of "comparison type", 234

—, of weak acids, 214

—, physico-chemical background to, 216—224

—, precision of, 229—230, 234—235

—, redox, 331—342

—, —, theory of, 335, 342

—, reduction of dielectric constant in, 252

—, self-indicating, titration curves for, 224—227

—, self-indicating chelometric, 272—273, 274—284

—, sensitivity of, 227

—, slope indication chelometric, 272, 274—284

—, statistical treatment of curves for, 230—232

—, titration curves of, dilution-induced curvature in, 236

—, two-phase, 308—313

—, two-phase extractive, using EDTA, 302

—, ultraviolet, 213

—, use of two-colour indicators in, 232

—, using dichromatic light, 215

—, with coulometrically generated titrants, 333, 338

—, with EDTA, 213

—, with oxygen-sensitive titrants, 244

Photometric titrators, application of fibre optics in, 243

—, automatic shut-off, 248—249

—, cells for, 237

—, commercially available, 248, 249, 250

—, design considerations of, 237—238

—, design details of, for use with anaerobic titrants, 245, 248

—, desirable attributes of, 237—238

—, development of, 208—216

—, efficiency of, effect of magnetic stirrers on, 237

—, for anaerobic titrations, 244

—, for turbidimetric and fluorimetric work, 248

—, insensitivity of, to magnetic fields, 247

—, manual, details of, 246—248

—, monochromators for, 218

—, of Field and Baas-Becking, 210

—, of Müller and Partridge, 211

—, operational problems with, near UV cut-off region of cell, 234

—, photocells for, 238—239

—, photovoltaic receptors for, 239

—, recording, 249—250

—, silver/bismuth thermocouples for, 247

—, stability problems with, 234

—, two-photocell, 237, 242

—, variable-length light path, 238, 247, 248

—, with immersed light pipes, 248

—, with vertical illumination, 249

Photometry, application of, to chelate exchange titrimetry, 372—375

Photomultiplier tubes, 105—106

—, characteristics of 118—119

—, for detection of porphyrin fluorescence, 184

—, signal-to-noise ratio of, 107

—, sources of noise in, 107

Photopolymerization, application of, in enzyme electrode preparation, 29

—, effect of conditions of, on enzyme electrode stability, 30

Photoresistors, cadmium sulphide, response of, 239

Photosynthesis, fluorimetric study of mechanism of, 179

Phototurbidimetric titration, determination of fluoride by, 322

Photovoltaic receptors, application of, to photometric titrations, 239

Picolinaldehyde-2-quinolylhydrazone, application of, in determination of zinc, by fluorimetry, 178

Picric acid, determination of, by photometric redox titration, with chromium-(II), 340

"Pilot ion" indicator, 276

"Pilot ion" method, 273—274

Piperidine, determination of, by spectropolarimetric titration, with sodium hydroxide, 370

Plane wave, 403

Plane wave-front, amplitude division of, 408

Plants, determination of fluoride in, by photometric precipitation titration, with thorium nitrate, 327

Plasmas, study of, by interferometry, 532

Platinum electrode, determination of, oxidative enzymes by, 46

Platinum ribbon lamp, use of, in photometric titrator, 209

Platinum sensors, comparison of electrodes based on, 61

Polarization, of light, 412

Polyacrylamide gels, application of, in enzyme immobilization, 17

—, with entrapped enzymes, 23

Polyethylene terephthalate, determination of carboxyl end groups in, by photometric titration, 269

Polymer fibres, determination of crystallinity of, by interferometry, 535

Polymers, determination of molecular weight distributions of, by light scattering titration, 321

Polymethylmethacrylate, as matrix for ion-selective electrodes, 7

Polynuclear aromatic hydrocarbons, determination of, by photometric titration, 215

—, —, —, with tetracyanoethylene, 352

—, enhancement of phosphorescence of, 157

Polystyrene, application of, in enzyme immobilization, 18

Polyvinylchloride, as matrix for ion-selective electrodes, 7, 8

Porous glass, application of, in enzyme immobilization, 18

Porphyrins, determination of, by photometric titration, 264

—, fluorescence of, 184—186

Potassium, reduction of interference by, in determination of urea, 45

Potassium cyanoferrate(II), application of, as photometric precipitation titrant, 326

Potassium iodate, determination of thiourea with, by photometric titration, 339

Potassium iodide, use of, in phometric titration of gold(III), 313

Potassium ions, electrodes sensitive to, 8

Potassium permanganate, application of, as photometric titrant, 332

Potassium thiocyanate, application of, as photometric titrant, 313

Potentiometric titrations, comparison of, with photometric titrations, 229—230, 236

—, —, —, in a limiting situation, 255

Precipitation, analytical applications of, 314

Precipitation reactions, examination of, by heterometry, 320

—, self-indicating, titration curves of, 315

Precipitation titration, determination of chloride by, 316

—, general theory of, 314—319

—, photoelectric, historical aspects of, 212

—, photometric, see: "Photometric precipitation titrations"

—, reduction of precipitate solubility in, 317

Pressure, effect of, on refractive index, 459—462

Procaine, determination of, by phosphorimetry, 163

—, fluorescence of, 185

—, organic functional group analysis of, application of photometric titrations to, 350

580

Propoxycaine, organic functional group analysis of, application of photometric titrations to, 350

l-Propylene diamine, determination of, by spectropolarimetric titration, with nickel perchlorate, 370

D-(−)-1,2-Propylenediaminetetraacetic acid, application of, as spectropolarimetric titrant, 367

Protective colloids, 221

—, application of, 316

Proteins, determination of —SH groups in, by photometric titration, 353

—, phosphorescence of, 165, 166

—, separation of, by electrophoresis, 511

Public health, applications of luminescence in, 83

Purines, determination of, by fluorimetry, 181, 182

—, phosphorescence characteristics of, 163

Pyrene, fluorescence of, in aerated and non-aerated solutions, 87

—, yield of triplet states of, in oxygenated solutions, 142

Pyridine, determination of, by photometric titration, 266

Pyridine bromide perbromide, application of, as photometric titrant, 353, 359

Pyridine-2,6-dicarboxylic acid, determination of vanadium(V) with, by photometric extractive titration, 310

Pyridine nucleotides, determination of, by fluorimetry, 182—183

—, enzyme cycling for, 196—197

Pyridine protoporphyrin, fluorescence of, 185

1-(2-Pyridylazo)-2-naphthol, application of, to photometric extractive titrations, 310

Pyrimidines, phosphorescence characteristics of, 163

Pyrrol, fluorescence of, 128

Pyruvate, determination of, 60

Quantum efficiency, 78

—, determination of, 79

Quantum efficiency ratio, of phosphorescence to fluorescence, 130

Quantum efficiency ratios, phosphorescence/fluorescence, for halogenated aromatic hydrocarbons, 148

Quantum yield, 78

—, fluorescence, 221

—, of various compounds, 79

Quartz, as adsorbent for enzymes, 16

—, birefringence by, 413

Quenching, 80

—, by diamagnetic non-transitional metal ions, 142

—, concentration, 88—89, 143

—, excimer, 89

—, impurity, 90

—, mechanism of, 85

—, oxygen, 87—88

—, —, mechanism of, 141

—, solvent, 132

—, temperature, 86—87

Quenching effects, 221

Quercetin, determination of hafnium by, fluorimetrically, 171, 174

Quinine, average lifetime of excited state of, 80

—, resolution of fluorescence emission of, in presence of anthracene, 77

—, temperature quenching of, 86

Quinine sulphate, absorption spectrum of, 76

—, as fluorescence standard, 124

—, fluorescence spectrum of, 75, 76

—, storage of, 124

Quinoline, determination of, by photometric titration, 264

Quinolinium ion, excited-state acidities of, 136

8-Quinolinol, application of, to fluorimetric determination of metal ions, 167

—, determination of aluminium by, fluorimetrically, 169

Quinolinol-5-sulphonic acid, application of, in photofluorimetric titrations, 343—344

Quinone, application of, in glucose electrode, 50

Radiative transition, forbidden, 220

Radioactive samples, determination of fluoride in, by photometric precipitation titration, with thorium nitrate, 327
Raman effect, 74
Raman peaks, identification of, in excitation spectra, 122
Raman scatter, 122
Raman spectrum, application of, in fluorimeter calibration, 125
Rare earths, determination of, by photometric titration, with EDTA, 297
Rayleigh's equation, 223
—, deviation from, 224
Rayleigh—Löwe interferometer, 424—428, 431
Rayleigh scattering, 74, 223
—, in luminescence analysis, 122
Rayleigh—Svensson interferometer, 437, 438
Redox reactions, study of, by photometric titrators, 211
Refraction, 422
Refractive index, 418—422
—, control of factors affecting, 454
—, determination of distribution of, in inhomogeneous substances, 529
—, change of, during titrations, 478—479
—, effect of pressure on, 459—462
—, evaluation of cylindrical distribution of, 531
—, measurement of, by interferometry, 393, 423—424
—, of a gas mixture, 463
—, of a liquid, effect of dissolved gases on, 476—478
—, of glass, determination of, by interferometry, 499
—, of heavy water, 495
—, of thin layers, determination of, by interferometry, 499—500
—, relationship between concentration and, 456, 480
—, temperature coefficient of, 461
Refractive index gradients, investigation of, by interferometry, 502—503
Refractive power, 419
Refractive powers, of various gases, 465—467

Refractometry, application of, in titrimetric analysis, 478—479
Reserpine, phosphorescence of, 144
Resins, determination of acids in, by photometric titration, 265
Resolving power, 100
Resonance fluorescence, 73
Resorcinol, application of, as titrant, in photometric titrations, 358
—, average lifetime of excited state of, 80
Resorufin, application of, in fluorimetric assay of dehydrogenases, 194—195
—, fluoresence characteristics of, 178
Resorufin butyrate, application of, in fluorimetric assay of enzymes, 188
Response curve, shape of, 41—43
Response time, of enzyme electrode, 33
Response—time curves, of enzyme electrodes, 36—37
Rhodamine B, application of, to fluorimetric determination of metal ions, 167
—, determination of gallium by, fluorimetrically, 171
—, determination of gold by, fluorimetrically, 170, 171
—, quantum yield of, 79
—, temperature quenching of, 87
Rhodanase, determination of activity of, 12
Riboflavin, determination of, by fluorimetry, 181
—, quantum yield of, 79
Ribonuclease, immobilized, 4
—, phosphorescence of, effect of solvent on, 165
Ringbom error plot, 290
Rocks, determination of fluoride in, by photometric precipitation titration, with thorium nitrate, 327
Rotating can phosphoroscope, 145, 150, 151
Rubber, determination of calcium in, by photometric precipitation titration, with ammonium oxalate, 326
—, determination of zinc in, by photometric precipitation titration, with potassium cyanoferrate(II), 326

Saccharose, determination of, in boiler-feed water, by interferometry, 495

Salicylaldehydes, application of, to fluorimetric determination of metal ions, 167

Salicylic acid, application of, to fluorimetric determination of metal ions, 167

Salicylidene-o-aminophenol, determination of aluminium by, fluorimetrically, 169

Salyrganic acid, application of, as photometric titrant, 361

Samarium, determination of, by fluorimetry, 172

Samarium(III), determination of, by photofluorimetric titration, with EDTA, 348

Savart plate, 413, 517

Scandium(III), determination of, by spectropolarimetric titration, with D(—)-PDTA, 370

Scattering, Debye, 224

—, Rayleigh, 223

Scattering coefficient, total, 222

Schlieren, 437, 525

Schlieren-scanning, application of, 438

Sea water, determination of sulphate in, by photometric titration, 269

Sedimentation, 521

Sedimentation studies, application of interferometry to, 433—438

Selectivity coefficient, for ion-selective electrode, 6

—, for urea electrodes, 51

—, for halide and pseudohalide electrodes, 9

Selenium, determination of, by fluorimetry, 168, 172

Self-absorption, distortion of fluorescence spectrum by, 143

—, in phosphorescence measurements, 156

Sensitivity values, fluorimetric, 170

Sepharose, application of, in enzyme immobilization, 18

Serum, analysis of, by electrophoresis, 510

—, determination of calcium and magnesium in, by photometric titration, with EDTA, 297

—, determination of magnesium in, by fluorimetry, 175

Serum albumin, phosphorescence of, 165, 166

Shearing interference, 517

Shock waves, light sources for interferometric study of, 444

Shot noise, 107

Side reaction coefficients, 272

Silastic, application of, in enzyme immobilization, 17

Silica gel, application of, in gas analysis, by interferometry, 446—452

—, as adsorbent for enzymes, 16

Silicate, determination of, by atomic absorption inhibition titration, 369

Silicates, determination of fluoride in, by photometric precipitation titration, with aluminium nitrate, 329

—, —, —, with thorium nitrate, 328

Silicone rubber, as matrix for ion-selective electrodes, 7

Silicotungstate, heterometric titration of nicotine with, 320

Silicotungstic acid, determination of nicotine with, by photometric precipitation titration, 326

Silver, determination of, by photometric extractive titration, with di-2-naphthylthiocarbazone, 311

—, —, —, with dithizone, 310

—, —, by photometric titration, with EDTA, 297

Silver—cobalt mixtures, analysis of, by photometric extractive titration, with dithizone, 310

Silver—copper mixtures, analysis of, by photometric extractive titration, with dithizone, 310

Silver ions, effect of, on amygdalin electrode, 47

—, electrodes sensitive to, 8

—, enzyme inhibition by, 48

Silver—zinc mixtures, analysis of, by photometric extractive titration, with dithizone, 310

Slit, distribution of radiant energy emerging from, 104

583

Slits, types of, 105
Slit width, 103
Soaps, determination of critical micelle concentration of, 375—376
Soap solutions, study of micelle formation by, using photometric procedures, 213
Sodium ampicillin, determination of, 57
Sodium borohydride, application of, as photometric titrant, 362
Sodium bromide, application of, as photometric precipitation titrant, 325
Sodium dithionite, as titrant, photometric titrators for, 244
—, titration of lumiflavin-3-acetic acid with, anaerobically, 333, 341
Sodium dodecyl sulphate, determination of critical micelle concentration of, 375—376
Sodium ions, reduction of interference by, in determination of urea, 45
Sodium penicillin, determination of, 57
Sodium polyvinyl sulphonate, determination of, by photometric precipitation titration, with barium hydroxide, 325
Sodium sulphate, application of, as photometric precipitation titrant, 324
Sodium sulphide, application of, as photometric precipitation titrant, 325
Solid-state electrodes, 7
Solubility, investigation of, by interferometry, 393
Solvents, effect of, in phosphorimetry, 165
—, fluorescence grade, comtamination from packaging, 118
—, fluorescent impurities in, 117—118
—, fluorimetric, purification of, 117—118
—, quenchers in, 117
Solvent effects, in luminescence analysis, 131—133
Solvent glasses, prevention of cracking of, 145, 146
Solvent polarity, effect of, in fluorescence, 133
Solvent quenching, 132
Sorbents, application of, in gas analysis, by interferometry, 469—470

Specific refraction, 422
Spectrofluorimeters, see also: "Fluorimeters"
—, characteristics of, 111
—, commercial types of, 111—114
—, comparison of, 112—113
—, detectors for, 118—119
—, gratings for, 100—102
—, phosphorescence attachments for, 155
—, standardization of, 124
—, temperature control in, 87
—, temperature-controlled cell-holders for, 114
—, use of filters in, 120
—, wavelength calibration of, 125
Spectrofluorimetric phototitration, determination of calcium by, with EDTA, 343, 347
—, determination of fluoride by, with thorium(IV), 342, 347
Spectrofluorimetric phototitrations, 342—355
Spectrofluorimetry, see also: "Fluorimetry"
—, control of cell temperature in, 119
—, correction of spectra in, 125—126
—, effect of oxygen in 142
—, reduction of interference in, by use of pH, 135
Spectro-interferometer, 494
Spectrophosphorimeter, schematic diagram of, 151
Spectrophosphorimeters, 145
Spectrophotometric titrations, see also: "Photometric titrations"
—, definition of, 207
Spectrophotometric titrator, 239—241, 243—244
Spectropolarimetric titration, determination of barium and calcium by, with PDTA, 368
—, determination of l-mandelic acid by, with sodium hydroxide, 370
—, determination of trichloroacetic acid by, with sodium hydroxide, 370
Spectropolarimetric titrations, determination of metal ions by, 370—371
Spectropolarimetry, 365—369

Spherical wave, 403

Spinner assembly, application of, in phosphorimetry, 162

Spirits, determination of alcohol in, by interferometry, 479

Standards, fluorescence, commercially available, 124

Standard air, composition of, 418

Standardization, of spectrofluorimeters, 124

Starch gel, application of, in enzyme immobilization, 17

Starch—iodine reaction, historical study of, by photometric titrator, 209

Starch polysaccharides, organic functional group analysis of, by photometric titration, 353

State of phase, 398

Steel, determination of nickel in, by photometric precipitation titration, 326, 327

Steroids, determination of, by fluorimetry, 181

Stokes' equation, 509

Stokes' fluorescence, 73

Stokes' law, 220

Stokes' shift, 78

Streptomycin, determination of, by fluorimetry, 185

Strontium, determination of, by spectropolarimetric titration, with D(—)-CDTA, 370

Strontium(II)—barium(II) mixtures, analysis of, by photometric precipitation titration, with sodium sulphate, 330

Substoichiometry equation, 309

Succinate dehydrogenase applications of, 58

Succinic acid, determination of, by enzyme electrode, 58

—, determination of, by fluorimetry, 181

Sudan III, as indicator, in photometric titration of urea, 262

Sugar solutions, temperature coefficient of refractive index of, 462

Sulfhydryl groups, determination of, by photometric titration, 350

—, —, —, with salyrganic acid, 361

Sulphaguanidine, determination of, by photometric titration, 266

Sulphanilamide, fluorescence of, 185

Sulphanilamides, organic functional group analysis of, application of photometric titrations to, 350

Sulphate, determination of, by atomic absorption inhibition titration, 369

—, —, by photometric precipitation titration, in vulcanised rubber, 324

—, —, —, with barium chloride, 329, 330

—, —, —, with barium perchlorate, 330

—, —, by photometric titration, in sea water, 269

Sulphate, photometric titration of, with barium(II), 212

Sulphide, determination of, by fluorimetry, 172

Sulphide electrode, 64

Sulphonamides, determination of, by phosphorimetry, 163

Sulphonaphtholazoresorcinol, determination of gallium by, fluorimetrically, 173

Sulphur dioxide, sorption of, 449

Sulphur trioxide, determination of, by photometric precipitation titration, 330

—, sorption of, 449

Surface-active agents, analysis of, by photometric titration, 213

Surface area of a particle, variation of, during precipitation, 316

Surfactants, anionic, determination of, by photometric precipitation titration, 320—321, 325

—, cationic, determination of, by photometric precipitation titration, 325

Suspensions, scattering of radiation by, 221

Svedberg's equation, 522

d-Tartaric acid, determination of, by spectropolarimetric titration, 366

Telluric acid, analysis of, by ammonia, using photometric titration, 253, 264

Tellurium, determination of, by fluorimetry, 168

Tellurium(IV), determination of, by photometric redox titration, with permanganate, 341

Temeperature, effect of, on enzyme electrode response, 39

—, —, on refractive index, 454, 459—462

—, quenching effect of, in fluorimetry, 86—87

—, sensitivity of fluorescence to, 119

Terbium, determination of, by fluorimetry, 172, 177

Tetrabutylammonium hydroxide, photometric titration of phenols with, 256

Tetracaine, organic functional group analysis of, application of photometric titrations to, 350

Tetracyanoethylene, application of, as photometric titrant, 357

Tetracycline, fluorescence of, 185

Tetramercuryacetate fluorescein, complexes of, application in photofluorimetric titrations, 345, 349—350

—, determination of sulphur-containing species with, by photofluorimetric titration, 345, 349

Tetraphenylarsonium hydroxide, application of, as photometric titrant, 380

Thallium, determination of, fluorimetrically, 168, 173

Thallium(III), determination of, by spectropolarimetric titration, with D(—)-PDTA, 370

Thallium(I)—iron(II) mixtures, analysis of, by photometric titration, 333

2-Thenoyltrifluoroacetone, determination of samarium by, fluorimetrically, 172

Thermal conductivity, investigation of, by interferometry, 393

Thermal diffusion, investigation of, by interferometry, 507

Thermal luminescence, 90

Thermistor, application of, in enzyme electrodes, 24

Thiamine, determination of, by fluorimetry, 181

Thiethylperazine, determination of, by photometric redox titration, with cerium(IV), 341

Thin layers, determination of refractive indices of, by interferometry, 499—500

Thin layer chromatography, identification of spots in, by phosphorescence, 165

Thioacids, determination of, by photofluorimetric titration, with tetramercuryacetate fluorescein complexes, 345, 349

Thiocholine, determination of, by sulphide sensitive electrode, 12

Thiocyanate, determination of, by photometric titration, with mercury(II) nitrate, 313

Thiols, determination of, by photofluorimetric titration, with tetramercuryacetate fluorescein complexes, 345, 349

—, organic functional group analysis of, by photometric titration, with palladium(II) chloride, 353

Thiophenols, determination of, by photofluorimetric titration, with tetramercuryacetate fluorescein, 345, 349

Thioridazine, determination of, by photometric redox titration, with cerium(IV), 341

Thiourea, determination of, by photofluorimetric titration, with tetramercuryacetate fluorescein complex, 349

—, —, by photometric redox titration, with potassium iodate, 339

Thorium(IV), as phototurbidimetric titrant, for fluoride, 322, 327

—, determination of, by photometric titration, with EDTA, 299, 300

—, —, by spectropolarimetric titration, with D(—)PDTA, 371

—, determination of fluoride with, by spectrofluorimetric phototitration, 342, 347

Thorium—copper mixtures, analysis of, by photometric titration, with EDTA, 276, 277

Thorium nitrate, application of, as photometric precipitation titrant, 324

Threshold limit values, of gases and vapours, 487, 489—490

Thymol blue, application of, in photometric titrations, 258

Time-resolved phosphorimetry, 159—160
Tin, determination of, by fluorimetry, 172, 177
Titanium(III), application of, as photometric redox titrant, 339
—, determination of, by photometric titration, in presence of acetylacetone, 334, 339
Titanium(IV), determination of, by spectropolarimetric titration with D(—)-PDTA, 370
Titanium sponge, determination of iron in, by photometric redox titration, with titanium(III), 339
Titrants, anaerobic, photometric titrator for use with, 245, 248
Titrations, acid—base, indicated, application of photometric techniques to, 258—270
—, —, —, theory of, 257—258
—, amperometric, with EDTA, 284
—, atomic absorption inhibition, 369, 372
—, change of refractive index during, 478—479
—, complexometric, involving photometric reagents, 302—308
—, —, theory of, 270—272
—, EDTA, with indicator, theory of, 284—295, 302
—, fluorimetric, 220
—, metal—EDTA, with optically active indicators, 369
—, photometric, see: "Photometric titrations"
—, redox under anaerobic conditions, 333
—, self-indicating, theory of, 274—284
—, spectrofluorimetric, 342—355
—, spectrophotometric extractive, with high-absorbance reagents, 308—313
—, two-phase reagent—metal, 308—313
—, two-phase extractive, with EDTA, 302
Titration curves, photometric, 224—228
—, —, location of equivalence point from poor quality, 233
—, —, of weak acid—strong base systems, 259
Titrators, photometric, see: "Photometric titrators"

Titrimetry, application of interferometry to, 497
Tobacco mosaic virus, study of, by sedimentation techniques, 525
Toepler schlieren arrangement, 433, 435, 441
Toluene, fluorescence of, in aerated and non-aereated solutions, 87
—, use of, in phosphorimetric calibrations, 156
Toluene-2-isocyanate-4-isothiocyanate, application of, in enzyme immobilization, 17
p-Toluidine, determination of, by photometric titration, 264
—, photometric titration of, in 1-butanol, 254
p-Tosyl-8-aminoquinoline, application of, in determination of zinc, by fluorimetry, 178
Total scattering coefficient, 222
Transaminases, assay of, by fluorimetry, 193—195
Transmission, of solutions, 217
Triboluminescence, 90
Tri-t-butyl phenol, generation of free radicals from, for photometric titrations, 363
2,4,6-Tri-t-butylphenoxy radicals, application of, to photometric titrations, 215
Trichloroacetic acid, determination of, by spectropolarimetric titration, with sodium hydroxide, 370
Trichloro-s-triazine, application of, in enzyme immobilization, 17
Triethyllead chloride, application of, in determination of organo-sulphur compounds, by photofluorimetric titration, 346
Trifluoroacetic acid, determination of, by spectropolarimetric titration, with sodium hydroxide, 370
Triphenylene, examination of, on TLC plates, by phosphorimetry, 165
—, phosphorescence intensity of, external heavy-atom effect on, 134

Triphenylguanidine, determination of, by photometric titration, 265
Triphenylmethane dyes, use of, in photometric titrations, 232
2,4,6-Tripyridyl-s-triazine, determination of, by photometric titration, 354
Trypsin, immobilized, 4
—, phosphorescence of, effect of solvent on, 165
Tryptophan, determination of, in presence of tyrosine, by phosphorimetry, 160
—, fluorescence characteristics of, 179
—, fluorescence spectrum of, 147
—, phosphorescence of, 144, 145
—, —, effect of solvent on, 165
—, phosphorescence spectrum of, 147
Tungstate, determination of, by phosphate ion electrode, 58
Tungsten, determination of, by fluorimetry, 172
Tungsten lamp, 94
Turbidimetry, 221, 222, 224, 315
—, total scattering coefficient in, 222
Turbidity, 222
Tyndall effect, 223
Tyndall scattering, 122
—, effects of, in spectrofluorimetric phototitrations, 342
L-Tyrosine, determination of, 21, 22
Tyrosine, determination of, in presence of tryptophan, by phosphorimetry, 160
—, fluorescence characteristics of, 178
—, interference of, in determination of glucose, 50
—, phosphorescence of, effect of solvent on, 165
—, phosphorescence spectrum of, 166

Ultracentrifuge, 523—524
—, application of, to analysis of high polymers, 525
—, application of interferometry to, 393
Umbelliferone, fluorescence characteristics of, 178
Umbelliferone phosphate, as substrate, in phosphatase assay, 192
Unilateral slits, 105

Uranium, determination of, by photometric titration, with cerium(IV), 338
—, —, —, with EDTA, 297
Uranium(IV), determination of, by photometric redox titration, with cerium(IV), 339
Uranium(VI), determination of, by photometric titration, 306
Urea, determination of, by photometric titration, 261, 262, 265
—, —, in blood, 46, 52
—, —, —, interferences in, 45
Urea electrode, applications of, 51—52
—, effect of electrode sensor on response of, 41
—, interferences associated with, 45
—, pH range of, 39
—, shape of response curve of, 41—42
—, stability of, 28—30
Urease, assay of activity of, 15
—, immoblized, 4
—, —, effect of activity of, on electrode stability, 30
—, specificity of, 3
Urease electrode, 59, 63
Uric acid, interference of, in determination of glucose, 50
—, oxidation of, 55
Uric acid electrode, applications of, 55
Uricase, application of, 55
—, immobilized, 4
—, specificity of, 48
Urine, determination of urea in, interferences in, 45
Uroporphyrin, fluorescence of, 185

Valinomycin, application of, in potassium ion selective electrode, 8, 10
Vanadium, determination of, by photometric redox titration, in steel, 338
—, —, by photometric titration, in steel, 332, 336
Vanadium(IV) or (V) determination of, by photometric redox titration, with iron(II) or bromate, 341
Vanadium(V), determination of, by photometric extractive titration, with pyridine-2,6-dicarboxylic acid, 310
Vibrational relaxation, 219

Virial coefficients, determination of, utilizing interferometric techniques, 492, 493

Vitamin A, determination of, by fluorimetry, 181

—, fluorescence of, 127

Vitamins, determination of, by fluorimetry, 181

Warfarin, determination of, by phosphorimetry, 163

Water, application of, as phosphorimetry solvent, 155

—, determination of, by interferometry, 496

—, —, by Karl Fischer titration, 377

—, —, by photometric titration, in ketones, esters and ethers, 351

—, determination of fluoride in, by photometric precipitation titration, with thorium nitrate, 327

—, Raman bands of, for various mercury lines, 122

—, refractive index of, 454

Water purity, determination of, by interferometry, 495

Water vapour, sorption of, 449

Wave front, 403

Wave-front division, 403—407

Wave-front shearing, 401, 411—413

Wavelength calibration, 125

Wave-trains, 400—401

Wollaston prism, 413, 440

Xanthates, determination of, by photofluorimetric titration, with tetramercuryacetate fluorescein complex, 349

Xanthine oxidase, phosphorescence of, 166

Xenon lamps, high-pressure, 94

—, stability of, 95

Xenon source, effect of arc wander in, 154

Yohimbine, fluorescence of, 185

Yttrium, determination of, by fluorimetry, 172, 177

Yttritum(III), as titrant for fluoride, 323, 329

Zein, phosphorescence of, 165

Zeolites, adsorption of gases on, 447

—, application of, in gas analysis, by interferometry, 446—452

Zeta potential, on barium sulphate, 318, 319

Zinc, determination of, by fluorimetry, 172, 178

—, —, by photometric extractive titration, with dithizone, 310

—, —, —, with PAN, 310

—, —, by photometric precipitation titration, with potassium cyanoferrate(II), 326

—, —, by photometric titration, in lubricating oil additives, 308

—, —, —, with dithizone, 305

—, —, —, with EDTA, 298, 301

—, —, by photometric two-phase extractive titration, 302

Zinc(II), determination of, by spectropolarimetric titration, with D(—)PDTA, 370

Zinc—copper(II) mixtures, analysis of, by photometric titration, with dithizone, 305

Zinc—magnesium mixtures, analysis of, by photometric titration with EDTA, 299

Zinc(II)—mercury(II) mixtures, analysis of, by photometric titration, with dithizone, 308

Zirconium, determination of, by fluorimetry, 172, 178

—, —, by photometric titration with EDTA, 296, 300

Zirconium(IV), determination of, by spectropolarimetric titration, with D(—)-PDTA, 377

Zirconyl—fluoride reaction, 323

Zone electrophoresis, 512